# Mechanisms and Machine Science

## Volume 59

D1811069

**Series editor**

Marco Ceccarelli
LARM: Laboratory of Robotics and Mechatronics
DICeM: University of Cassino and South Latium
Via Di Biasio 43, 03043 Cassino (Fr), Italy
e-mail: ceccarelli@unicas.it

This book series establishes a well defined forum for monographs, edited Books, and proceedings on mechanical engineering with particular emphasis on MMS (Mechanism and Machine Science). The final goal is the publication of research that shows the development of mechanical engineering and particularly MMS in all technical aspects, even in very recent assessments. Published works share an approach by which technical details and formulation are discussed, and discuss modern formalisms with the aim to circulate research and technical achievements for use in professional, research, academic, and teaching activities. This technical approach is an essential characteristic of the series. By discussing technical details and formulations in terms of modern formalisms, the possibility is created not only to show technical developments but also to explain achievements for technical teaching and research activity today and for the future. The book series is intended to collect technical views on developments of the broad field of MMS in a unique frame that can be seen in its totality as an Encyclopaedia of MMS but with the additional purpose of archiving and teaching MMS achievements. Therefore the book series will be of use not only for researchers and teachers in Mechanical Engineering but also for professionals and students for their formation and future work. Indexed in SCOPUS, Ei Compendex, EBSCO Discovery Service, OCLC, ProQuest Summon, Google Scholar and SpringerLink.

More information about this series at http://www.springer.com/series/8779

Burkhard Corves · Philippe Wenger
Mathias Hüsing
Editors

# EuCoMeS 2018

Proceedings of the 7th European Conference
on Mechanism Science

 Springer

*Editors*
Burkhard Corves
Institute of Mechanism Theory, Machine
  Dynamics and Robotics
RWTH Aachen University
Aachen, Nordrhein-Westfalen
Germany

Mathias Hüsing
Institute of Mechanism Theory, Machine
  Dynamics and Robotics
RWTH Aachen University
Aachen, Germany

Philippe Wenger
LS2N
Ecole Centrale de Nantes
Nantes, France

ISSN 2211-0984          ISSN 2211-0992   (electronic)
Mechanisms and Machine Science
ISBN 978-3-319-98019-5          ISBN 978-3-319-98020-1   (eBook)
https://doi.org/10.1007/978-3-319-98020-1

Library of Congress Control Number: 2018950104

This Springer imprint is published by the registered company Springer Nature Switzerland AG
The registered company address is: Gewerbestrasse 11, 6330 Cham, Switzerland

# Preface

The book on hand presents the most recent research results in the area of mechanism science, which were presented at the seventh European Conference on Mechanism Science (EuCoMeS 2018) in Aachen, Germany. This conference series started in 2006 under the patronage of the International Federation for the Promotion of Mechanism and Machine Science (IFToMM, www.iftomm.net). The aim of the conference is to bring together European researchers, industry professionals, and students from the broad range of disciplines related to mechanism and machine science, in order to stimulate the exchange of new and innovative ideas both on the academic and industrial levels.

The first edition of EuCoMeS took place in February 2006 in Obergurgl (Austria). After venues in Cassino (Italy) in September 2008, Cluj-Napoca (Romania) in September 2010, Santander (Spain) in September 2012, Guimarães (Portugal) in September 2014, and Nantes (France) in September 2016, the 2018 edition was held at RWTH Aachen University, Germany, during September 4–6, 2018.

This book is published under the Mechanisms and Machine Science series of Springer and addresses issues related to: biomedical applications, control issues of mechanical systems, dynamics of multi-body systems, experimental mechanics, haptic systems, history of mechanism science, industrial and non-industrial applications, linkages and cams, mechanical transmissions and gears, mechanics of robots and manipulators, and last but not least theoretical kinematics.

In total, we received a number of 57 papers, which were carefully reviewed by three reviewers per paper. Finally, 49 papers were accepted for presentation during the conference and for publication in this book. Thus, we want to express our gratitude to the reviewers who contributed to this process with their experience and scientific background and to all the authors for their contributions to EuCoMeS 2018.

We are grateful to IFToMM, which supported the conference by Young Delegates grants, and the German Research Foundation (DFG) for their support to the conference EuCoMeS 2018. We also thank everyone at IGMR who helped in the organization before, during, and after the conference and the staff at Springer for their support through all the stages of preparing this book.

We very much hope that this book contributes to the scientific progress, opens up discussions, and also inspires those who could not attend the conference to contribute and attend one of the next issues of EuCoMeS.

June 2018                                                                    Burkhard Corves
                                                                            Philippe Wenger

# Organization

## Conference Chair

Burkhard Corves        RWTH Aachen University, Germany

## Conference Co-chair

Philippe Wenger        CNRS–LS2N, France

## Local Chair

Mathias Hüsing        RWTH Aachen University, Germany

## EuCoMeS Organizing Committee

| | |
|---|---|
| Burkhard Corves | RWTH Aachen University, Germany |
| Mathias Hüsing | RWTH Aachen University, Germany |
| Claudia Cornely | RWTH Aachen University, Germany |
| Jascha Paris | RWTH Aachen University, Germany |

## International Scientific Committee

| | |
|---|---|
| Marco Ceccarelli | University of Cassino and South Latium, Italy |
| Paulo Flores | University of Minho, Portugal |
| Manfred Husty | University of Innsbruck, Austria |
| Doina Pisla | Technical University of Cluj-Napoca, Romania |
| Fernando Viadero | University of Cantabria, Spain |

Philippe Wenger                    CNRS–LS2N, France
Teresa Zielinska                   Warsaw University of Technology, Poland

## Under the Patronage of IFToMM

Teresa Zielinska                   Secretary General of IFToMM
Victor Petuya                      Chair of Technical Committee for Linkages
                                     and Mechanical Controls
Federico Thomas                    Chair of Technical Committee for Computational
                                     Kinematics
Paulo Flores                       Chair of Technical Committee for Multibody
                                     Dynamics
Daizhong Su                        Chair of Technical Committee for Gearing
                                     and Transmissions
Lena Zentner                       Chair of Technical Committee for
                                     Micromachines

## With the Support of

Andres Kecskemethy                 Chair of IFToMM Member Organization,
                                     Germany
DFG Germany                        German Research Foundation

# Contents

**Theoretical Kinematics**

# Biomedical Applications

# Contact Characteristics of Medical Forceps Indentation to Soft Tissue

Anastasia Yakovenko[1], Irina Goryacheva[2], and Marat Dosaev[3(✉)]

[1] Institute of Physics and Technology (State University), Dolgoprudny, Russia
[2] Ishlinsky Institute for Problems in Mechanics of RAS, Moscow, Russia
[3] Institute of Mechanics of LMSU, Moscow, Russia
dosayev@imec.msu.ru

**Abstract.** The plane-parallel indentation of the tool with projections into the homogeneous and inhomogeneous biological tissue is modelled. The linear elastic and two-layer half-spaces are used for tissue modelling. The time-dependence of a force applied to the instrument and dependence of the force on the indentation depth are obtained. The shape of projections and distance between them affect these dependences. The influence of speed on the load applied to the stamp is studied. It is shown that increasing the speed of penetration leads to increasing the value of the total applied force. The shape of projections affects the nature of this dependence too.

**Keywords:** Soft tissue · Indentation · Viscoelastic half-space

## 1 Introduction

A progress of laparoscopic surgery leads to new challenges for modern robotics and mechanics. A development of methods and approaches to a transmission of tactile sensations from the working elements of medical instruments to the fingers of the surgeon is the important issue in a design of medical tools. The need to eliminate the lack of tactile sensations in minimally invasive surgery has been repeatedly noted in the literature [1].

A number of works are devoted to the development of mechatronic devices that implement a transfer of information about the contact load between the executive element of the medical robot and the biological tissue by feedback control [2]. For such problems it is necessary to have information about contact stresses between soft biological tissues and working instrument having complicated shape.

When performing laparoscopic operations, it is often necessary to transfer organs or separate tissues of one organ from others. It is important not to damage these tissues, for example, do not squeeze excessively a blood vessel or a nerve. In work [3], stress magnitudes and durations that can be safely applied with a grasper to different tissues are identified. Also in [4], the relationship between tissue trauma and compression stress magnitude and duration during tissue clamping operation was evaluated. Stresses arising in the tissue depend on various parameters, in particular, on the shape of the surface of forceps projections (teeth). In order to control the stresses arising in soft

© Springer Nature Switzerland AG 2019
B. Corves et al. (Eds.): EuCoMeS 2018, MMS 59, pp. 3–10, 2019.
https://doi.org/10.1007/978-3-319-98020-1_1

tissues, it is necessary to give the estimation of the effect on them of various shapes of the surface of the forceps.

An estimate for the indention of an axisymmetric elastic body into a multilayered biological tissue was given in work [5]. In work [6] the contact of a stamp with projections in the form of narrow parallelepipeds with an elastic half-space was considered. The similar problem of the contact a rigid indenter with soft tissue was considered, for example in [7]. In work [8], an axisymmetric contact problem is considered for an elastic layer subjected to normal indentation of a rigid stamp. Also, in [9], the contact problem for the impression of spherical indenter into a non-homogeneous elastic half-space is considered. Paper [10] shows a mathematically describing the mechanical behavior of soft tissues under large deformations using a nonlinear hyperelastic law. However, these works are quite difficult to apply for multiple contacts.

In this paper, a contact of a rigid stamp having several projections (teeth) with a soft tissue is considered. The tissue is modeled by an elastic half-space and by a two-layer half-space. In the second case the outer layer is elastic, the internal layer is viscoelastic one. Analytical dependencies of the force applied to the instrument on the compression rate of the test tissue and on the rate of penetration are obtained. It is shown that the shape of the stamp teeth significantly influences these dependencies.

## 2    The Problem Statement

The indentation of a rigid indenter into soft biological tissue is considered. The tissue is modeled by elastic and viscoelastic bodies (Fig. 1a, b).

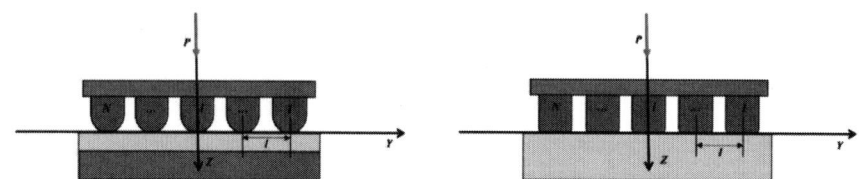

**Fig. 1.** Scheme for indentation of the stamp with projections: (a) the stamp with spherical teeth contacts with two-layer tissue; (b) the stamp with elongated parallelepipeds contacts with a homogeneous half-space

The indenter has on its surface a system of projections of the same shape that are located at the same distance $l$ from each other. The indentation occurs at a given constant speed $v$. The force $P$ is applied upon the indenter that is postured parallel to the tissue surface. Two forms of tool projections are considered: spherical teeth and parallelepipeds having rounded corners elongated along the axis $Ox$ (Fig. 2a, b).

In the case of a rigid spherical stamp, the shape of the contacting surface depends only on the radius and, for small contact areas, can be approximated by the expression: $f(r) = r^2/2R$, where $R$ is the radius of the indenter.

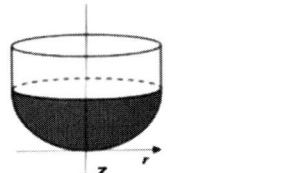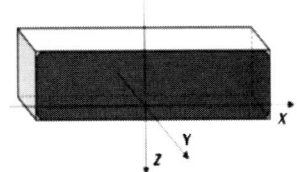

**Fig. 2.** Two types of projection shape: (a) spherical, (b) elongated parallelepipeds having rounded corners

## 3 Tissue Models

To describe the contact condition when loading biological tissue, different models are used depending on the speed of indentation of the stamp. At sufficiently high penetration rates, when it is necessary to take into account the rheological properties of the tissue, it is modeled by a two-layered base (elastic layer on viscoelastic).

A human skin including subcutis can be modeled by the elastic layer. To describe its deformation properties, the Winkler model is used, the relationship between contact pressures $p(x, y)$ and surface displacements $u_{z1}(x, y)$ along the normal to the boundary for which has the following form [11]:

$$p(x, y) = \frac{E_1}{h_1} u_{z_1}(x, y),$$

(1)

where $h_1$ is the thickness of the layer, and $E_1$ is its Young's modulus.

To describe the muscular layer having the thickness $h_2$, the viscoelastic Kelvin-Voigt model is used, for which the relationship between the pressures $p(x, y, t)$ acting on it and displacements $u_{z2}(x, y, t)$ has the following form [11]:

$$p(x, y, t) + T_\sigma \frac{dp(x, y, t)}{dt} = \frac{E_2}{(1 - v^2)h_2} \left( u_{z_2}(x, y, t) + T_\varepsilon \frac{du_{z_2}(x, y, t)}{dt} \right),$$

(2)

$T_\varepsilon, T_\sigma, E_2, v$, are the relaxation time, creep time, Young's modulus and Poisson's ratio of viscoelastic layer correspondingly.

In the case of low rates of deformation and small indentations of the projections in comparison with the thickness of the biological tissue, a model of a homogeneous elastic half-space is used to estimate its stress-strain state, for which the relationship between the normal stresses applied to the boundary $p(x, y)$ and the displacements $u_z(x, y)$ along the normal to the boundary has the form [12]:

$$u_z(x, y) = \frac{1 - v^2}{\pi E} \sum_{i=1}^{N} \iint\limits_{\omega_i} \frac{p_i(x', y')dx'dy'}{\sqrt{(x' - x)^2 + (y' - y)^2}},$$

(3)

$E, v$ are Young's modulus and Poisson's ratio of the tissue correspondingly.

## 4 Case of Elongated Parallelepipeds Indented to the Elastic Half-Space

For projections in the form of parallelepipeds elongated along the Ox axis, Galin's solution [12] is used, according to which the expression for the contact pressures for each projection has the form:

$$p(x,y) = p(x)/(\pi\sqrt{b^2 - y^2}), \tag{4}$$

where $p(x)$ is the distribution of the force acting along the axis $Ox$ normally to the boundary of the half-space, $a$ a $b$ and $(b < < a)$ are the half-length and half-width of each projection, respectively. For elastic displacements, we have the relation: $u(x,0) = p(x)/k$, where the modulus of subgrade reaction is calculated by the formula: $k = \pi E/(2(1 - v^2)\lg(a/b))$.

It is assumed that the projections do not have sharp corners, that is, they have rounded edges. The projections are located at the same distance $l$ and have the same dimensions ($2b$ is the width, $2a$ is the length). We denote indices of the projections: let the rightmost be the $1^{st}$, and the leftmost be $N^{th}$.

In [6], using the relations (3) and (4) and the approach developed in [13], the problem of an indentation of a stamp with a set of projections having the shape of narrow parallelepipeds into an elastic half-space (Fig. 1b) is solved, taking into account their mutual influence. It is obtained that the contact condition for the load distribution and the indentation depth $\delta$, has the form:

$$\delta = \frac{1-v^2}{a\pi E}\left(2\lg(a/b)P_i + \sum_{j=1,j\neq i}^{N} P_j\left(\ln\frac{\sqrt{l^2(i-j)^2 + 4a^2} + 2a}{\sqrt{l^2(i-j)^2 + 4a^2} - 2a} - \frac{1}{a}\left(\sqrt{l^2(i-j)^2 + 4a^2} - l(i-j)\right)\right)\right), \tag{5}$$

The formula (5) provides the distribution of loads on the projections, if you know the value of the introduction of the tool in the base. For example, for three protrusions, the expression for the load applied to the tool has the form

$$P = P_0 + P_1 + P_2 = \frac{a\pi E\delta}{(1 - v^2)} \cdot \frac{4C_1 - C_2 - 6\lg\left(\frac{a}{b}\right)}{2C_1^2 - 2\lg\left(\frac{a}{b}\right)C_2 - 4\lg^2\left(\frac{a}{b}\right)}$$

$$C_1 = \ln\frac{\sqrt{l^2 + 4a^2} + 2a}{\sqrt{l^2 + 4a^2} - 2a} - \frac{\sqrt{l^2 + 4a^2} - l}{a}, \quad C_2 = \ln\frac{\sqrt{l^2 + a^2} + a}{\sqrt{l^2 + a^2} - a} - \frac{2\left(\sqrt{l^2 + a^2} - l\right)}{a}$$

In the dimensionless form, this expression takes the form:

$$\tilde{P} = \frac{\pi}{4}\tilde{\delta} \cdot \frac{4\tilde{C}_1 - \tilde{C}_2 - 6\lg\left(\frac{a}{b}\right)}{\tilde{C}_1^2 - \lg\left(\frac{a}{b}\right)\tilde{C}_2 - 2\lg^2\left(\frac{a}{b}\right)}, \tilde{l} = \frac{l}{2a}, \tilde{\delta} = \frac{\delta}{2a}, \tilde{P}_i = \frac{P_i(1 - v^2)}{4Ea^2},$$

$$\tilde{C}_1 = \ln\frac{\sqrt{\tilde{l}^2 + 1} + 1}{\sqrt{\tilde{l}^2 + 1} - 1} - 2\left(\sqrt{\tilde{l}^2 + 1} - \tilde{l}\right), \tilde{C}_2 = \ln\frac{\sqrt{4\tilde{l}^2 + 1} + 1}{\sqrt{4\tilde{l}^2 + 1} - 1} - 2\left(\sqrt{4\tilde{l}^2 + 1} - 2\tilde{l}\right)$$

The distribution of loads $\tilde{P}$ on teeth projections depending on the indentation depth $\tilde{\delta}$ is shown in Fig. 3 for following parameters values: $\lg(a/b) = 1, \tilde{l} = 0.75$ and $N = 3$. Number of line corresponds to number of the projection.

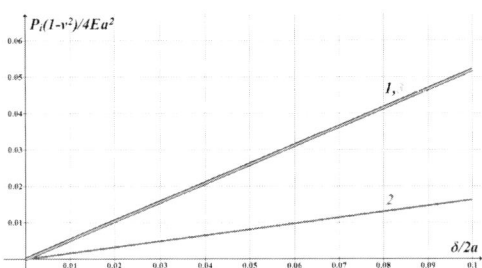

**Fig. 3.** Dependences of loads on the indentation depth.

Thus, the distribution of loads on the projections is symmetrical, and the lowest pressure is observed on the middle one. In the chosen case of three projections, the ratio of the forces acting on the peripheral and middle projection is given by $\tilde{P}_0/\tilde{P}_1 \approx 3.2$. The loads depend on the distance between the projections, and on the ratio of their length to width.

Figure 4 represents the dependence of the total load upon the instrument on the indentation depths for different distances between the projections. The dependences are linear, and as the distance between the projections increases, the value of the applied force also increases for a given value of the indentation depth.

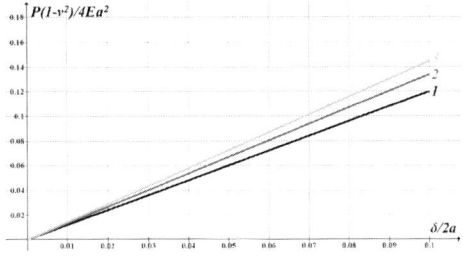

**Fig. 4.** Dependences of total load on the indentation depth for different distance between the projections. Line *1* corresponds to $\tilde{l} = 0.75$, line 2 corresponds to $\tilde{l} = 1$, line *3* corresponds to $\tilde{l} = 1.25$

## 5  Analysis of Indentation to the Two-Layer Tissue

Consider the stamp with rigid spherical projections performing indentation at a constant speed $V$ to the two-layer inhomogeneous half-space (Fig. 1a). All projections are located at the same distance from each other. The condition of contact of each indenter with a two-layered base has the form

$$u_z(x, y, t) = u_{z_1} + u_{z_2} = \delta(t) - f(x, y), \tag{6}$$

where $\delta(t) = Vt$ is the indentation depth, and $f(x, y)$ is the function describing the shape of the surface of the individual projection.

In [6], the problem of indentation of the spherical stamp into two-layered half-space described by relations (1) and (2) with a constant velocity was considered. Using relations obtained in [6], the following dependence of the total load on time in the dimensionless form can be obtained:

$$\tilde{P}(\tilde{t}) = \frac{N\pi\tilde{V}^2}{\frac{H}{E} + (1 - \vartheta^2)} \left( \tilde{t}^2 + 2\left(\frac{1}{B} - 1\right)\left(\frac{1}{B}\left(1 - e^{-B\tilde{t}}\right) - \tilde{t}\right)\right),$$

where $B = \frac{H + (1-\vartheta^2)E}{H + (1-\vartheta^2)TE}$, $H = \frac{h_1}{h_2}$, $E = \frac{E_2}{E_3}$, $T = \frac{T_\sigma}{T_\varepsilon}$, $\tilde{t} = \frac{t}{T_\varepsilon}$, $\tilde{V} = \frac{VT_\varepsilon}{h_2}$, $\tilde{P}(\tilde{t}) = \frac{P(t)}{E_2 R h_2}$, $R$ is the radius of projection, $h_1$, $h_2$ are thicknesses of the layers.

Follow [3, 4] we take the following parameters values $R = 0.015\,\text{m}$, $l = 2.5R = 0.0375\,\text{m}$, $E_1 = 10000\,\text{Pa}$, $h_1 = 0.01\,\text{m}$, $h_2 = 0.03\,\text{m}$, $E_2 = 100000\,\text{Pa}$, $T_\sigma = 0.0000001\,\text{s}$, $T_\varepsilon = 0.00001\,\text{s}$, $V_1 = 0.001\,\text{m/s}$, $V_2 = 0.005\,\text{m/s}$, $V_3 = 0.01\,\text{m/s}$, $B = 1.2224$.

The time-dependence of the total load acting on the tool is shown in Fig. 5a taking into account the restriction on the indentation depth: $\delta \leq h_1$ or $t \leq h_1/V$.

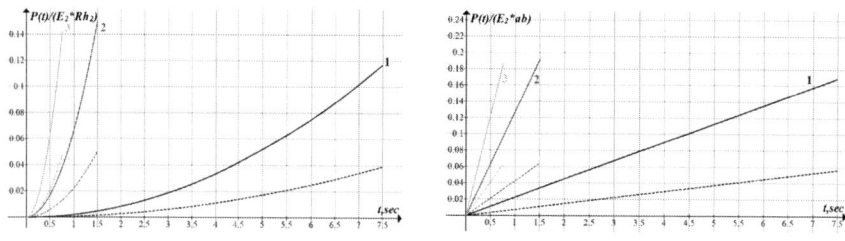

**Fig. 5.** The time-dependence of the total load $\tilde{P}(\tilde{t})$: (a) spherical projections; (b) parallelepiped projections. Curve 1 corresponds to $V_1$, curve 2 corresponds to $V_2$, curve 3 corresponds to $V_3$. The dashed curves show the single projection load.

One can note that increasing of the indentation speed leads to an increase in the applied force. The load applied to the tool depends on the rate of penetration quadratically.

Similarly, we study the indentation of stamp with elongated parallelepiped projections into the two-layered half-space. In this case we must put $f(x, y) = 0$ in (6). The contact area does not depend on time. Then for single projection pressure we obtain:

$$p_i(t) = \frac{K_2 V}{1 + K_1 K_2}\left(t + \left(e^{-\frac{1 + K_1 K_2}{T_\delta + K_1 K_2 T_\varepsilon}t} - 1\right)\left(\frac{T_\delta + K_1 K_2 T_\varepsilon}{1 + K_1 K_2} - T_\varepsilon\right)\right), \qquad K_2 = E_2/h_2(1 - v^2),$$

$K_1 = h_1/E_1$. Then we have the following expression for total dimensionless load:

$$\tilde{P}(\tilde{t}) = \frac{\tilde{V}}{\frac{H}{E} + (1 - \vartheta^2)}\left(\tilde{t} + \left(e^{-B\tilde{t}} - 1\right)\left(\frac{1}{B} - 1\right)\right).$$

Figure 5b shows the time-dependences of the load acting on the tool and on single parallelepiped projections. Increasing in the rate of penetration also leads to an increase of the applied force. In contrast to the spherical projections case, the time-dependence of the total force on the rate of penetration in this case is linear. Therefore, an increase in the indentation speed leads to a multiple increase in the applied load.

# 6 Conclusion

The models of the plane-parallel indentation of the tool with projections into the homogeneous and inhomogeneous biological tissue are developed. The tissue is modeled by the linear elastic and two-layer (elastic and viscoelastic) half-spaces. The dependences of the force applied to the instrument on the indentation depth and time are obtained that allows studying the influence of shape and density of projections on these dependences. One can also determine the maximum pressure value on the most loaded projection. It is established that the lowest pressure acts upon the middle projections, while with increasing distance between the projections the load distribution tends to be uniform. In this case, the total force required to achieve a given amount of penetration increases with increasing distance between the projections.

For the case of high velocities, it is necessary to take into account the relaxation properties of biological tissue. The influence of speed on the load applied to the stamp is studied. It is shown that the speed of penetration increases, the value of the total applied force increases. The shape of projections affects the nature of this dependence. For spherical projections, the dependence is quadratic, and for elongated parallelepipeds it is linear. The obtained results can be used for the purposeful design of the surface of the medical instruments with the purpose of reducing the pressure acting on the tissue from the instrument while providing the necessary degree of the pinch power. The results also can be used to develop the experimental study of the soft tissue properties under various indenter geometries.

**Acknowledgments.** This work was partially supported by the Russian Foundation for Basic Research, project N 16-58-52033 and the Program of the Presidium of RAS I.16.

# References

1. Tholey, G., Desai, J.P., Castellanos, A.E.: Force feedback plays a significant role in minimally invasive surgery: results and analysis. Ann. Surg. **241**(1), 102–109 (2005)
2. Dosaev, M., Selyutskiy, Yu., Yeh, C.-H., Su, F.-C.: Modeling of tactile feedback realized by a piezoelectric drive. Mechatron. Autom. Control **7**, 480–485 (2018). (in Russian)
3. De, S., Rosen, J., Dagan, A., Swanson, P., Sinanan, M., Hannaford, B.: Assessment of tissue damage due to mechanical stresses. Int. J. Rob. Res. **26**(11), 1159–1171 (2007)
4. Wang, J., Qing-Yuan, Y., Li, W., Wang, B.-R., Zhou, Z.-R.: Influence of clamping stress and duration on the trauma of liver tissue during surgery operation. Clin. Biomech. **43**, 58–66 (2017)
5. Yakovenko, A., Goryacheva, I., Dosaev, M.: Estimating characteristics of a contact between sensing element of medical robot and soft tissue, pp. 561–569. Springer (2017)
6. Yakovenko, A.: Modeling a contact interaction between medical instrument and biological tissue. Russ. J. Biomech. (2017)
7. Ting, T.C.T.: The contact stresses between a rigid indenter and a viscoelastic half-space. J. Appl. Mech. **33**(4), 845–854 (2011)
8. Kubenko, V.D., Osharovich, G., Ayzenberg-Stepanenko, M.V.: Impact indentation of a rigid body into an elastic layer. Analytical and numerical approaches. J. Math. Sci. **176**(5), 670–687 (2011)
9. Aizikovich, S.M., Alexandrov, V.M., Kalker, J.J., Krenev, L.I., Trubchik, I.S.: Analytical solution of the spherical indentation problem for a half-space with gradients with the depth elastic properties. Int. J. Solids Struct. **39**(10), 2745–2772 (2002)
10. Liu, Y., Kerdok, A.E., Howe, R.D.: A nonlinear finite element model of soft tissue indentation. Lecture Notes in Computer Science, vol. 3078, pp. 67–76 (2004)
11. Johnson, K.L.: Contact Mechanics. Cambridge University Press, Cambridge (1985). 452 p.
12. Galin, L.A.: Contact Problems. Springer, Dordrecht (2008). 316 p.
13. Goryacheva, I.G.: Contact Mechanics in Tribology. Kluwer, Dordrecht (1998). 344 p.

# Design of a Finger Exoskeleton for Motion Guidance

Eike-Cristian Gerding[1(✉)], Giuseppe Carbone[2], Daniele Cafolla[2],
Matteo Russo[2], Marco Ceccarelli[2], Sven Rink[1],
and Burkhard Corves[1]

[1] Institut für Getriebetechnik, Maschinendynamik und Robotik,
RWTH Aachen University, Aachen, Germany
eike.gerding@rwth-aachen.de,
{rink,corves}@igmr.rwth-aachen.de
[2] LARM: Laboratory of Robotics and Mechatronics,
University of Cassino and Southern Latium, Cassino, Italy
{carbone,daniele.cafolla,matteo.russo,
marco.ceccarelli}@unicas.it

**Abstract.** In this paper, a novel exoskeleton for finger rehabilitation is presented. The exoskeleton is designed as a serial, 2-degrees-of-freedom wearable mechanism that is able to follow human finger motion. Motion tracking is used to characterize the movement of the human finger. Thereby the design requirements are outlined. The mechanism is synthesized, and its simulated motion is compared with experimental data to validate the proposed design.

**Keywords:** Mechanism design · Exoskeleton · Finger motion
Simulation

## 1 Introduction

As the average age of modern population increases, the number of immobility of fingers caused by strokes also increases [1]. The number of strokes in Europe is expected to rise from 1.1 million in 2000 to 1.5 million per year in 2025 [2]. Since the movement of the finger is fundamental to activities of daily living, an exoskeleton for a finger is to be developed for rehabilitation and exercise following an injury or a stroke. Several studies have shown that the rehabilitation after a stroke is faster when using a robotic system compared to conventional rehabilitation methods [3, 4].

The main problem with existing exoskeletons is that they are not adjustable for the fingers of different patients as in [5], bulky such as [3] and with overall equipment not easily transportable such as [6]. Commercial robots, as [6], are considered too expensive for rehabilitation use. [7] has no defined trajectory to move all joints of the finger in a defined way. Existing exoskeletons often use 70% of the motion of a healthy finger [8].

As previous works at the LARM, an anthropomorphic robotic hand has been developed [9]. The LARM Hand IV consists of three fingers with one Degree of

© Springer Nature Switzerland AG 2019
B. Corves et al. (Eds.): EuCoMeS 2018, MMS 59, pp. 11–18, 2019.
https://doi.org/10.1007/978-3-319-98020-1_2

Freedom (DOF) each and can grasp objects of different size. Based on the LARM Hand IV, an exoskeleton with one DOF has been developed [10].

## 2  Design Requirements

The human hand consists of digits, metacarpus and carpus. The fingers consist of three phalanxes, except for the thumb, which consists of two phalanxes. The metacarpus is connected to the proximal phalanx. On the fingers, the second link is the medial phalanx. The third link is the distal phalanx [11]. The joints on the digits between the intermediate and distal phalanx are called distal interphalangeal joints (DIP), and the joints between the proximal and intermediate phalanxes are called proximal interphalangeal joints (PIP). The metacarpophalangeal joints (MCP) connect the proximal and metacarpal phalanxes and the carpometacarpal joints (CMC) connect metacarpal phalanxes and the carpal bones [12]. Flexion reduces the angle between bones or parts of the body, whereas extension increases the angle between the bones of the limb at a joint in the sagittal plane [13]. The planes on the human hand are given in Fig. 1.

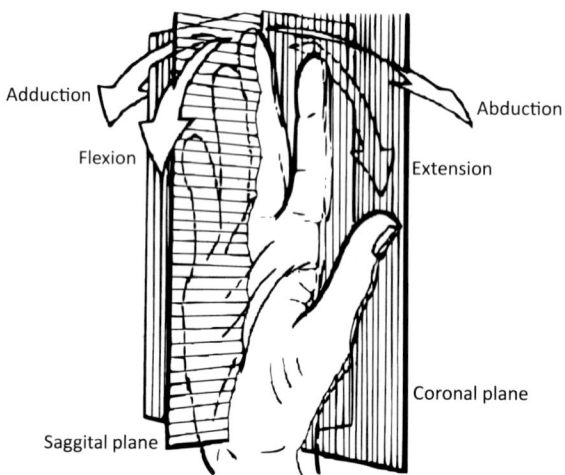

**Fig. 1.**  Relative movements in the human musculoskeletal hand system [16]

The MCP joints have two DOF. The interphalangeal joints PIP and DIP have one DOF each. However, the axis of rotation of these joints is not constant during flexion and extension [14]. Furthermore, the ligaments restrict the movements of the joints. In addition to that, DIP and PIP joint cannot be moved independently from each other [11]. For rehabilitation, it is important to exercise the flexion and extension movements of the finger.

The finger exoskeleton will support rehabilitation exercise of the muscles after an injury or a stroke. For that, the trajectory of the finger will be guided by the exoskeleton with a suitable limited error. To avoid injuries caused by the exoskeleton, unnatural movements of the finger must be avoided. To make this device usable for many people, it should fit to a finger of average length, drivable for individual finger motions, tolerant to misalignment, easy to adjust and simple to use. Technically, it will have few motors, a low weight, low cost and be easy to manufacture. In addition, it will be easy to wear and to transport in different environments.

In order to determine the trajectory of the index finger during experimental tests, a video of the movement has been analyzed. The joints of the index finger have been marked and the lengths between the joints have been measured. After that, the video has been recorded with 30 frames per second. The camera was aligned perpendicular with the side of the finger. For every second frame of the video, the position of the joints have been analyzed accordingly. The distances between the joints were calculated by using the acquired position of the joints. Further, with the measurement of the distance of the joints before recording the video, the measurements of the video could be converted from the units of pixels to millimeters. Figure 2 shows the experimental setup for the finger motion tracking.

From set-up measurements the lengths of the phalanxes of the test subject were identified as 25 mm for the distal phalanx, 28 mm for the medial phalanx and 43 mm for the proximal phalanx. In total, 22 frames have been evaluated for a suitable finger motion characterization. For further calculations, the MCP joint is set as the reference point for the movement. Figure 2 shows the computed trajectory of the finger movement from the experimental test.

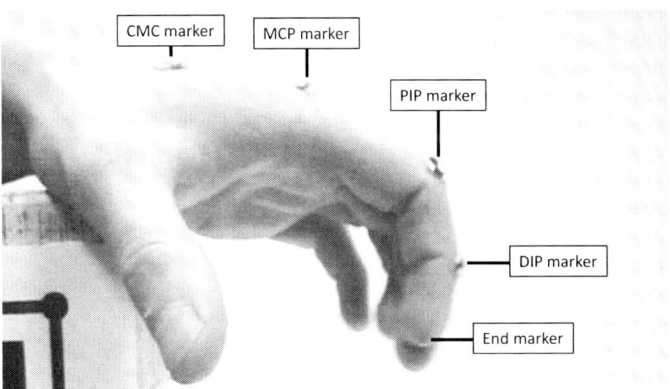

**Fig. 2.** Experimental setup for finger motion tracking

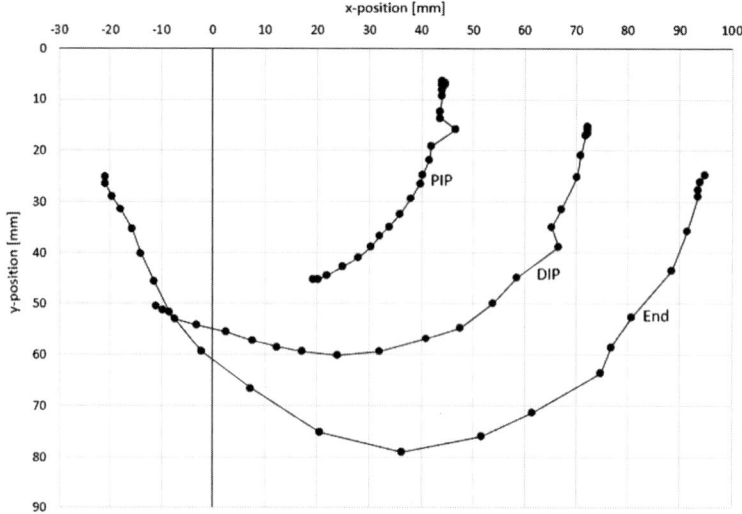

**Fig. 3.** Acquired trajectory of the finger movement by using the markers in Fig. 1

## 3   Exoskeleton Design

Since the exoskeleton is considered efficient with a small number of DOFs as well as compact dimensions, two DOFs are found convenient for a practical solution. As the finger has one independent movement in the MCP joint, and PIP and DIP joints have one independent movement together, the PIP and DIP joint movements can be grouped within a single DOF to design a 2-DOF exoskeleton with proper motion features. Linkages driven by electric motors will be used on the exoskeleton. Figure 4 shows the proposed kinematic scheme of the exoskeleton.

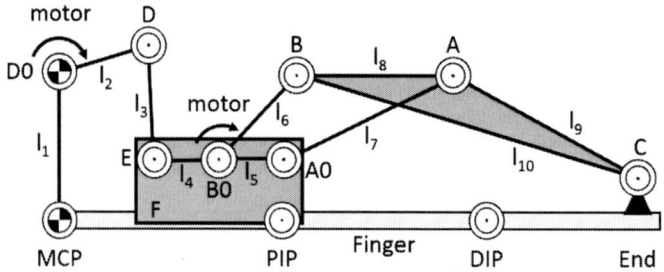

**Fig. 4.** A kinematic scheme of the proposed design with its main parameters

The first linkage D0-D-E-MCP has 4 links and 4 joints with one DOF each. Therefore, this part of the mechanism has one DOF in total. The second linkage (B0-B-A-A0-C) has also one DOF in total, as there are also 4 links and 4 joints with one DOF each. With this linkage, joint C is guided on the trajectory of the fingertip. The exoskeleton can be fixed on the finger at the proximal and distal phalanx as well as the back of the hand. The motors will be installed at the D0 and B0 joint.

For the first kinematic chain, the link lengths have to be chosen in a way that the exoskeleton can be worn on top of the MCP joint and the final position can be reached. Joints E and B0 share the same axes of rotation.

As joint C has a constant distance from the distal digit, its position can be computed. The offset e is on a line from FT to DIP. It chosen to be 10 mm in the direction to PIP and 5 mm perpendicular to that line in upper direction (f). By that, the relative movement between joint C and joint A0 can be calculated. Analog to that, A0 is assumed to have a constant offset a of 5 mm in the direction from PIP to MCP. The offset b is perpendicular to that, being 4 mm in upper direction. With the relative motion, the length of the second kinematic chain can be determined using a synthesis for 3 given positions of the link ABC. The method by Hagedorn was used for the synthesis [15]. The first position has been chosen with 9° of flexion at the MCP joint, 12° of flexion at the PIP joint and 8° of flexion at DIP. The second position with 19° of flexion on MCP, 30° of flexion at PIP and 23° of flexion at DIP. The third position with 34° of flexion at the MCP joint, 54° of flexion at the PIP joint and 51° of flexion at the DIP joint. The positions were taken from the test of grasping movement. As a result of the synthesis, the kinematic parameters are shown in Fig. 4 and given in Table 1.

**Table 1.** Link lengths of the exoskeleton mechanism in Fig. 4

| Link | $l_1$ | $l_2$ | $l_3$ | $l_4$ | $l_5$ |
|---|---|---|---|---|---|
| Length [mm] | 15.0 | 45.0 | 20.0 | 0.0 | 10.2 |
| Link | $l_6$ | $l_7$ | $l_8$ | $l_9$ | $l_{10}$ |
| Length [mm] | 5.1 | 14.0 | 15.8 | 45.0 | 58.1 |

With the link lengths shown in Table 1, a model is created in the software WorkingModel in order to prove the feasibility of the proposed design. All joints are modelled as revolute joints, with the joints MCP and D0 being fixed to the hand. Body F simulates the movement of the PIP joint and the connection to the finger. It is connected through a prismatic joint to MCP and to a path that follows the acquired trajectory of the movement of the PIP joint (path of P1, P2, and P3). Further, the motor in B0 is connected to F. To compare the simulated movement of C with respect to the measured trajectory, the path of the point C (C1, C2, and C3) is also considered in the simulation, but without connection to the exoskeleton. The points C1, C2, and C3 were used for the synthesis of the mechanism (Fig. 5).

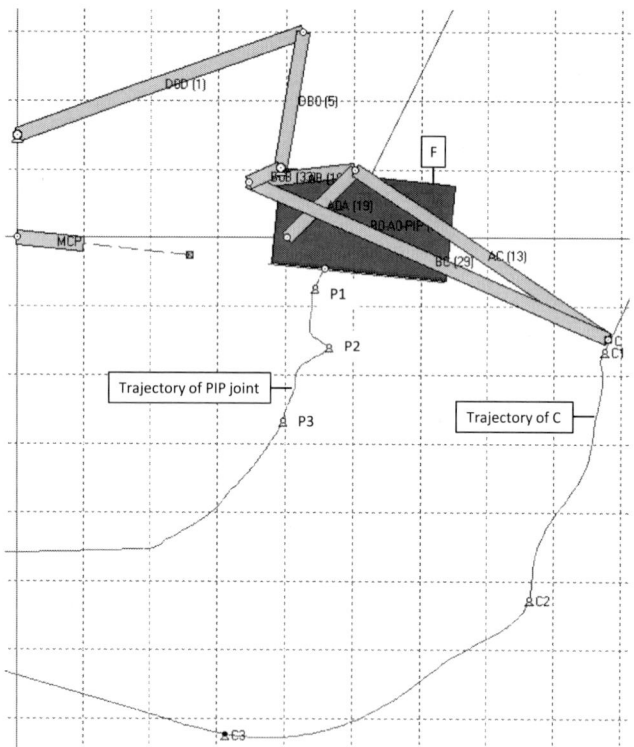

**Fig. 5.** A simulation of the finger movement with exoskeleton design in Fig. 4

## 4   Performance Characteristics

When the angular velocity at D0 is set to a constant value, the required angular velocity at B0 can be calculated by the required position of the joint C. The exoskeleton is put in the given positions from the synthesis and the angles on links D0D and B0B are measured. From the constant angular velocity in D0, the time to reach the positions for D0D can be computed. As B0B has to reach its positions at the same time, the angular velocity at B0B can be calculated. These values can be set as an input for the simulation of the movement. As the rehabilitation movement does not depend on the velocity but on the trajectory, a small angular velocity for the drive in D0 is chosen.

The position of the point C in the simulation is computed. The acquired path in an experimental test is compared to the trajectory of the finger movement (point C'), previously shown in Fig. 3. Figure 6 shows the simulated position of C (continuous line) as compared to the position from the finger movement C' (dots).

From Fig. 6 it can be noted that the position C' on the finger is fulfilled with a small error. The root-mean-square error of 2.66 mm over an overall motion path of 103 mm. This can be considered a very good value for the deviation between measured and computed trajectory. Therefore, the exoskeleton fulfills its demanded motion and can be built as a prototype. The links can be 3D printed, being both lightweight and low-cost.

For the CAD design, some constraints have to be taken in to account. It has to be made sure that the joints of the exoskeleton don not interfere with the finger joints during motion. Further, the mechanism has to allow for a sufficient thickness of the links.

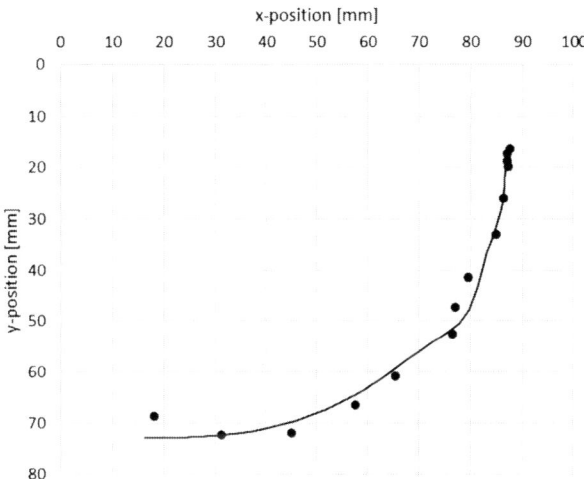

**Fig. 6.** Trajectory during simulation (continuous line) vs measured trajectory (dots)

## 5   Conclusions

In this paper, an exoskeleton mechanism for finger motion guidance was presented. In order to design the mechanism, an exemplary human finger motion was analyzed through motion tracking. Then, a 2-DOF mechanism was synthesized to follow this motion. The design was validated by comparing the motion of the proposed exoskeleton with respect to the experimental data, showing that the trajectory of finger motion is followed with a small error. The resulting design is achieved with compact and wearable features and can be manufactured in an easy and low-cost way. Therefore, it shows great promise for fulfilling the initially set goal of increasing the availability of robotic finger rehabilitation support.

**Acknowledgments.** The first author gratefully acknowledges the Erasmus+ program for the period of study he spent in 2017–2018 at the University of Cassino and South Latium under the supervision of Prof. M. Ceccarelli.

# References

1. Tjahyono, A.P., Aw, K.C., Devaraj, H., Surendra, W., Haemmerle, E., Travas-Sejdic, J.: A five-fingered hand exoskeleton driven by pneumatic artificial muscles with novel polypyrrole sensors. Ind. Robot: Int. J. **40**(3), 251–260 (2013)
2. Kaplan, W., Wirtz, V., Mantel, A., Béatrice, P.S.U.: Priority medicines for Europe and the world update 2013 report. Methodology **2**(7), 99–102 (2013)
3. Agarwal, P., Fox, J., Yun, Y., O'Malley, M.K., Deshpande, A.D.: An index finger exoskeleton with series elastic actuation for rehabilitation: design, control and performance characterization. Int. J. Robot. Res. **34**(14), 1747–1772 (2015)
4. Sale, P., Lombardi, V., Franceschini, M.: Hand robotics rehabilitation: feasibility and preliminary results of a robotic treatment in patients with hemiparesis. Stroke Res. Treat. **2012**, 2–5 (2012)
5. Bataller, A., Cabrera, J.A., Clavijo, M., Castillo, J.J.: Evolutionary synthesis of mechanisms applied to the design of an exoskeleton for finger rehabilitation. Mech. Mach. Theory **105**, 31–43 (2016)
6. Tyromotion GmbH: AMADEO. http://tyromotion.com/en/products/amadeo. Accessed 29 Jan 2018
7. Ates, S., Haarman, C.J., Stienen, A.H.: SCRIPT passive orthosis: design of interactive hand and wrist exoskeleton for rehabilitation at home after stroke. Auton. Robot. **41**(3), 711–723 (2017)
8. Liu, K., Hasegawa, Y., Saotome, K., Sainkai, Y.: Design of an wearable MRI-compatible hand exoskeleton robot. In: International Conference on Intelligent Robotics and Applications, pp. 242–250. Springer, Cham, August 2017
9. Carbone G., Ceccarelli M.: Design of LARM hand: problems and solutions. In: 2008 IEEE-TTTC International Conference on Automation, Quality and Testing, Robotics, AQTR 2008, Cluj-Napoca, pp. 298–303 (2008). (Best Paper Award) J. Control Eng. Appl. Inform. **10**(2), 39–46 (2008)
10. Cafolla, D., Carbone, G.: A study of feasibility of a human finger exoskeleton. In: Service Orientation in Holonic and Multi-Agent Manufacturing and Robotics, pp. 355–364. Springer, Cham (2014)
11. Levangie, P.K., Norkin, C.C.: Joint Structure and Function: a Comprehensive Analysis. F.A. Davis, Philadelphia (2005)
12. Cobos, S., Ferre, M., Uran, M.S., Ortego, J., Pena, C.: Efficient human hand kinematics for manipulation tasks. In: IEEE/RSJ International Conference on Intelligent Robots and Systems, IROS 2008, pp. 2246–2251. IEEE, September 2008
13. Pons, J.L.: Wearable Robots: Biomechatronic Exoskeletons. Wiley, London (2008)
14. Heo, P., Gu, G.M., Lee, S.J., Rhee, K., Kim, J.: Current hand exoskeleton technologies for rehabilitation and assistive engineering. Int. J. Precis. Eng. Manuf. **13**(5), 807–824 (2012)
15. Hagedorn, L., Thonfeld, W., Rankers, A.: Konstruktive Getriebelehre. Springer, Berlin (2009)
16. Mnyusiwalla, H., Vulliez, P., Gazeau, J.P., Zeghloul, S.: A new dexterous hand based on bio-inspired finger design for inside-hand manipulation. IEEE Trans. Syst. Man Cybern.: Syst. **46**(6), 809–817 (2016)

# Development of a Knee Joint Assistive-Mechanism Adapted for Bilateral Roll-Back Motion

Hidetsugu Terada[✉], Koji Makino, Kazuyoshi Ishida,
and Teppei Ogura

University of Yamanashi, Kofu, Japan
{terada,kohjim,isawa,t14jm010}@yamanashi.ac.jp

**Abstract.** To reduce the deployment cost of walking rehabilitation after surgery, a structure of knee joint assistive-mechanism that is adapted for bilateral roll-back motion has been proposed. This has two grooved cams which are not coincident with the locus of the knee joint's imaginary rotation center. However, these generate the self-locking in case of the same motion directions of each follower roller. A design condition has been shown for avoiding these self-locking. X-ray fluoroscopy of a prototype mechanism is used to evaluate the misalignment between the rotation centers of the knee and knee joint mechanism. Then, a load test prototype was used to test the roll-back motion under rated-torque, and some differences in the characteristics of the right and left knee-assistive motions were investigated. Based on roll-back motion data obtained from the Japanese people with an average height, a maximum displacement of 15 mm and an angle of 55° is used for the prototype mechanisms' imaginary rotation centers, and their cam shapes are generated using asymmetrical modified trapezoid motion curves. The results show that the proposed mechanism can be rotated smoothly to generate the exact knee roll-back motion and that its characteristics for right and left knee-assistive motions are almost identical.

**Keywords:** Assistive-mechanism · Roll-back · Knee joint · Bilateral motion
Grooved cam

## 1 Introduction

To improve patients' ability to walk after Total Knee Arthroplasty surgery [1, 2], a knee-assistive robot has been developed to aid rehabilitation [8]. The robot assists with flexion and extension of the knee joint with roll-back motion [6], and it cures abnormal gaits caused by osteoarthritis in the knee. In recent years, two types of that robot which have mirror image shape for the right leg and the left leg have been used in Japanese hospitals. However, to reduce the deployment cost, these hospitals have requested a robot that can be used for both the legs. On the other hand, the conventional assistive robot uses the mechanism which uses grooved cam at an imaginary rotation center based on a knee joint motion [9]. When that conventional mechanism is used to satisfy the request, the shape of grooved cam has a mirror image of a cam plate center line.

© Springer Nature Switzerland AG 2019
B. Corves et al. (Eds.): EuCoMeS 2018, MMS 59, pp. 19–26, 2019.
https://doi.org/10.1007/978-3-319-98020-1_3

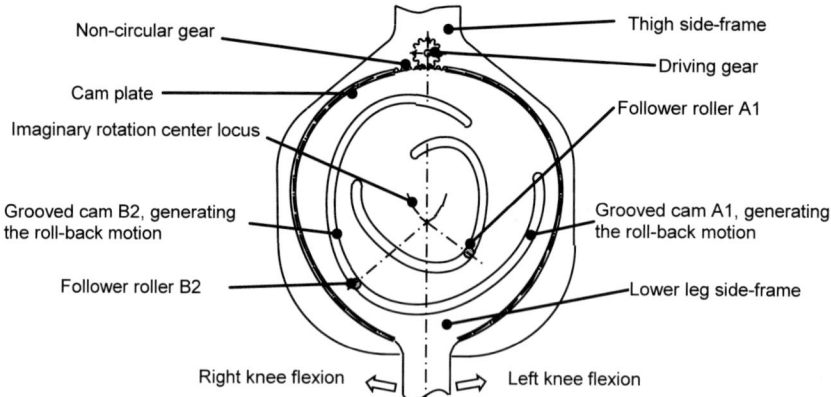

**Fig. 1.** Fundamental structure of assistive-mechanism adapted for bilateral roll-back motion

However, the undercutting occurs on the center section of that cam, and the unstable motion of follower roller will be generated. In other words, unless the follower roller diameter is zero, the conventional mechanism cannot apply to the bilateral roll-back motion.

In this paper, to achieve bilateral roll-back motion, a new structure of assistive-mechanism is designed using grooved cams in which center locus is not coincident with the locus of the knee joint's imaginary rotation center. Especially, the design condition to eliminate the self-locking motion is investigated. Then, the knee flexion misalignment between the imaginary rotation center and this mechanism's rotation center is then evaluated using X-ray fluoroscopy. In addition, the relationship between the knee flexion angle and the torque ratio is evaluated to confirm that the assistive motions for the right and left knees have same characteristics.

## 2  Knee Joint Assistive-Mechanism Design

To assist bilateral roll-back motion, a knee joint assistive-mechanism consisting of a non-circular gear and two grooved cams, as shown in Fig. 1, was designed [10]. This mechanism's structure is fundamentally similar to that of the conventional assistive-mechanisms [8], i.e., the non-circular gear used to generate the roll-back motion, driven by a gear on the center of the thigh side-brace, is symmetrical shape. However, the center locus of grooved cam is not coincident with the locus of the knee joint's imaginary rotation center. The driving gear meshed with the non-circular gear drives the lower leg side plate, and that plate guided by two grooved cams rotates with a roll-back motion.

In this mechanism, based on the knee motion of Japanese people with an average height as $170 \pm 5$ cm [3, 7, 9, 11], the cam profiles and the locus of the non-circular gear's center are both generated by an asymmetrical modified trapezoid acceleration curve (ASMT) [5]. Considering the previous research results, the ratios of acceleration time to deceleration time are determined as 2:3 in the horizontal direction and 3:2 in the vertical direction, which are divided at the sagittal plane of lower leg coordinate system. The non-dimensional displacements on the sagittal plane of lower leg coordinate system $S_x$ and $S_y$ are calculated as functions of the ASMT. A non-dimensional time $T$ is calculated from the knee flexion angle $\theta$ using Eq. (1). The relationships between the non-dimensional time and displacements $S_x$ and $S_y$ in each direction are shown in Fig. 2.

$$T = \frac{\theta - \theta_s}{\theta_e - \theta_s}, (\theta_s \leq \theta \leq \theta_e, \theta_s < \theta_e) \tag{1}$$

Therefore, the horizontal and vertical displacements the roll-back motion on the sagittal plane of lower leg coordinate system, $h_x$ and $h_y$, are defined as mentioned in Eq. (2).

$$h_x = h_{xmax} \cdot S_x, h_y = h_{ymax} \cdot S_y \tag{2}$$

The terminologies used here is listed in Table 1.

**Table 1.** Terminology used for the assistive-mechanism

| Symbol | Terminology |
|---|---|
| $h_{xmax}$ | Maximum horizontal displacement on the sagittal plane of lower leg coordinate system |
| $h_{ymax}$ | Maximum vertical displacement on the sagittal plane of lower leg coordinate system |
| $r_{a1}$ | Initial radial position of follower roller A1 |
| $\theta_{a1}$ | Initial angle of follower roller A1 |
| $r_{b2}$ | Initial radial position of follower roller B2 |
| $\theta_{b2}$ | Initial angle of follower roller B2 |
| $r_{dg}$ | Initial radial position of the driving gear |
| $\theta_{dg}$ | Initial angle of the driving gear |
| $\theta_s$ | Starting roll-back motion angle |
| $\theta_e$ | Ending roll-back motion angle |
| $r_f$ | Radius of the follower roller meshed with the grooved cams |

The locus of the knee joint imaginary rotation center's motion on the sagittal plane of lower leg coordinate system, $\mathbf{P}_{imc}$, is defined as

$$\mathbf{P}_{imc} = r_k e^{j\theta_k}. \tag{3}$$

Here, the length, $r_k$, and direction, $\theta_k$, of this vector are calculated as follows:

$$r_k = \sqrt{h_x^2 + h_y^2}, \theta_k = \mathrm{atan2}(h_x, h_y), (\theta_e \leq \theta \leq \theta_{max}). \tag{4}$$

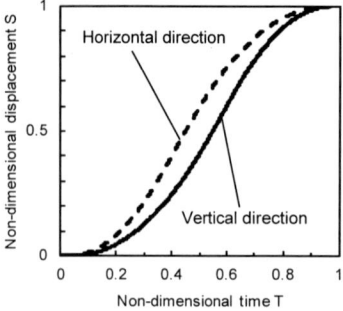
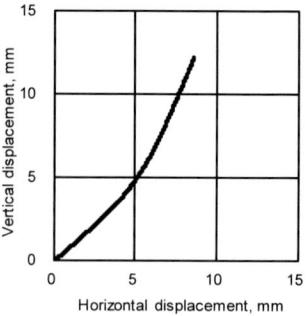

**Fig. 2.** Relationship between the non-dimensional time and displacement

**Fig. 3.** Locus of an imaginary rotation center on the right lower leg on the sagittal plane of lower leg coordinate system

Based on the average knee motion of Japanese people with an average height, Fig. 3 shows the standard imaginary rotation center's locus for the right lower leg. This was calculated based on the maximum roll-back displacement and angle as 15 mm and 55° for the right knee at the maximum right knee flexion angle as 125°, and the left knee has a mirror image motion.

In general, the roll-back motion occurs from 0° to 125°, but the knee joint can be rotated approximately from 125 to nearly 165°, so the maximum knee rotation-angle is defined as $\theta_{\max}$ in this paper. The imaginary rotation locus in the opposite direction has a mirror image shape, and the sign of the maximum x-axis displacement becomes negative, as do the starting and ending roll-back motion and maximum knee rotation-angles. The center loci $\mathbf{P_{a1}}$ and $\mathbf{P_{b2}}$ of the follower rollers A1 and B2 on the leg-side plate are defined as follows:

$$\mathbf{P_{a1}} = r_{a1}e^{j(\theta + \theta_{a1})} + r_k e^{j\theta_k}, \tag{5}$$

$$\mathbf{P_{b2}} = r_{b2}e^{j(\theta + \theta_{b2})} + r_k e^{j\theta_k}. \tag{6}$$

The profiles of the grooved cams are calculated from the envelopes of these center loci [5], and they accord with the locus of the envelope curve that offsets the calculated loci. The tangent vectors of the center loci $\mathbf{P_{a1}}$ and $\mathbf{P_{b2}}$ are defined as in Eq. (7) using the polar complex vectors, in which $\phi_{a1}$ and $\phi_{b2}$ are shown the tangential directions and the $d_{s1}$ and $d_{s2}$ are shown as the each length. Therefore, profiles $\mathbf{P_{cam1}}$ and $\mathbf{P_{cam2}}$ are defined by Eq. (8).

$$\dot{\mathbf{P}}_{\mathbf{a1}} = d_{s1}e^{j\phi_{a1}}, \dot{\mathbf{P}}_{\mathbf{b2}} = d_{s2}e^{j\phi_{b2}} \tag{7}$$

$$\mathbf{P_{cam1}} = \mathbf{P_{a1}} + r_f e^{j\left(\phi_{a1} \pm \frac{\pi}{2}\right)}, \mathbf{P_{cam2}} = \mathbf{P_{b2}} + r_f e^{j\left(\phi_{b2} \pm \frac{\pi}{2}\right)} \tag{8}$$

The plus/minus sign in Eq. (8) represents the direction of contact between the follower roller and the groove.

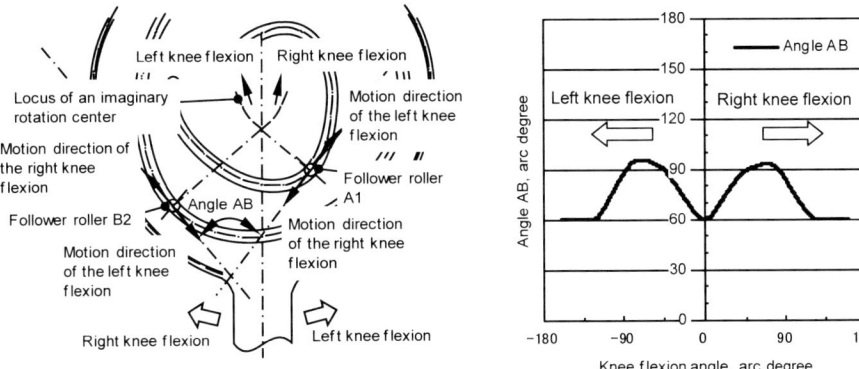

**Fig. 4.** Geometry of the angles between the motion direction follower rollers A1 and B2

**Fig. 5.** Relationship between the knee flexion angle and the angle AB

Based on previous research [4], the non-circular gear's profile is generated by defining the locus of the gear's mesh center $\mathbf{P_{dg}}$ as

$$\mathbf{P_{dg}} = r_{dg}e^{j(\theta + \theta_{dg})} + r_{k}e^{j\theta_{k}}. \tag{9}$$

## 3  Grooved Cam Design

In general, two sets of grooved cam for the planar motion mechanisms are necessary to generate the required motion. However, when the motion directions of the follower rollers coincide with each other, the motion generated by this cam is incorrect, after that, the self-locking occurs. These behaviors have been avoided using following design.

When the imaginary rotation center moves along the bilateral roll-back motion, the motion directions of follower rollers A1 and B2 are defined as shown in Fig. 4, which are calculated from Eq. (7). Then, the motion direction is reversed due to the flexion direction changes caused by the difference between the left and right of the knee. Furthermore, the angle between the two rollers' motion is defined as the angle AB. The follower rollers can be located in every position to satisfy the self-locking avoidance condition. It is clear that self-locking occurs when the motion direction of each roller coincides. Therefore, the condition has to be avoided that defined angle AB is near 0° or 180°. In addition, the allowable angles, which is considered the clearance by manufacturing-errors between rollers and grooves, the elastic deformation, and the pressure angle of cam design [5], are determined as 30° to 150°. However, there is no restriction condition about the radius of the imaginary center. Figure 5 shows the relationship between the knee flexion angle and the angle AB for the arrangement shown in Fig. 4. This satisfies the design requirement by avoiding self-locking at all

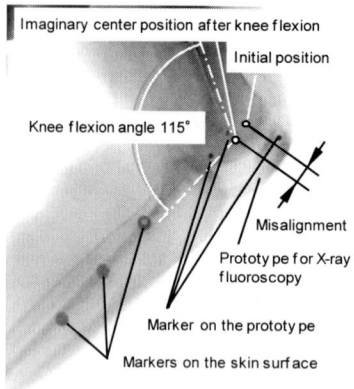

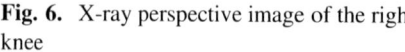

Imaginary center position after knee flexion

Initial position

Knee flexion angle 115°

Misalignment

Prototype for X-ray fluoroscopy

Marker on the prototype

Markers on the skin surface

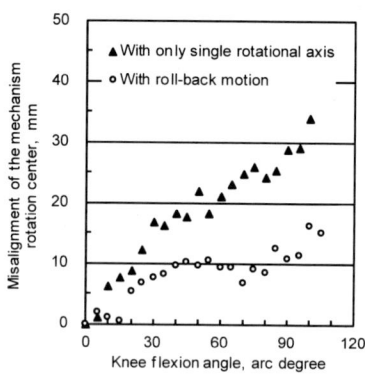

**Fig. 6.** X-ray perspective image of the right knee

**Fig. 7.** Relationship between the misalignment of the mechanism's rotation center for right knee flexion and the knee flexion angle

knee flexion angles, and there is a little difference between the right and left knee flexion behavior which is less than 5°.

## 4  Evaluation by X-Ray Fluoroscopy

To verify the utility of the proposed design, using X-ray fluoroscopy, the misalignment between the initial position and the assistive-mechanism's rotation center was evaluated for knee flexion of a healthy person. For comparison with conventional wearable assistive-mechanisms, using acrylic resin, two types of prototype were made for this test: one, where the cam mechanism was used for bilateral roll-back motion, and the other, with only single rotational axis. Several large markers were arranged on the skin of the thigh and lower leg, and small markers were attached to the prototypes to measure the rotation and position of the lower leg, as shown in Fig. 6. The initial position was calculated from the curvature of the bone contact point, with the knee joint initially in a fully-extended state.

As shown in Fig. 7, the misalignment between the initial position and the mechanism's rotation center for knee flexion was smaller than for a conventional single rotational axis joint. The misalignments were compared using Student's $t$-test, which is commonly-used in medical fields, and it was found that the probability values ($p$-values) were $4.19 \times 10^{-8}$ for the right knee and $1.50 \times 10^{-4}$ for the left knee. These are both well below the significance level of 0.05, so the differences were significant from a medical point of view.

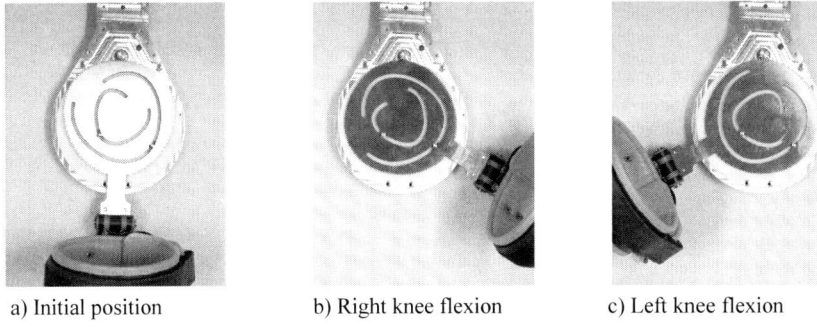

a) Initial position          b) Right knee flexion          c) Left knee flexion

**Fig. 8.** Load test prototype (body side view)

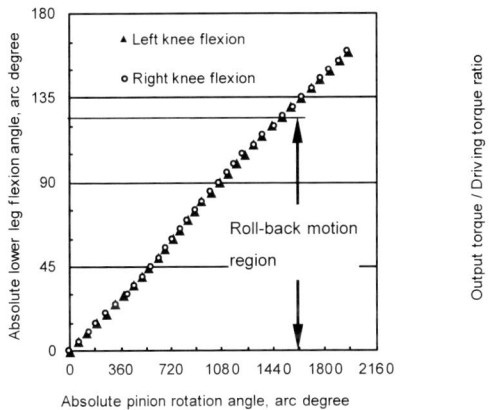

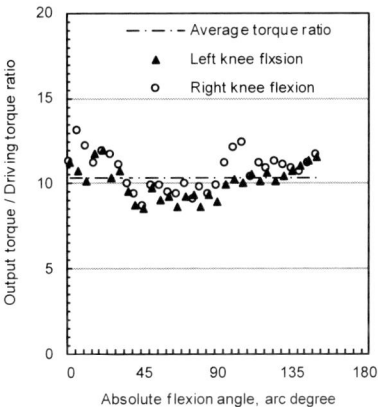

**Fig. 9.** Relationship between the pinion rotation and lower leg flexion angles

**Fig. 10.** Relationship between the lower leg flexion angle and torque ratio

## 5  Evaluation of the Knee Flexion Angle and Torque Ratio

To evaluate the differences between the characteristics of the right and left knee-assistive motions under the rated torque, a load test prototype was made from pre-hardened steel, as shown in Fig. 8(a). This could undergo roll-back motion, as shown in Fig. 8(b) and (c).

The rotation angle of the pinion gear at each lower leg flexion angle was measured using a protractor. The relationship between the pinion rotation and lower leg flexion angle is shown in Fig. 9. The roll-back motion caused a slight distortion in the rotation angle curve in the $0°$ to $125°$ range, and the angle difference between right and left knee is small which is less than 5% of motion range on a lower leg. It was considered that the proposed mechanism has sufficient performance. Then, the generated torque ratio that was the driving torque and the output torque was measured by load cell attached on the loading arm. Figure 10 shows the relationship between the lower leg flexion angle and torque ratio. The average torque ratio was 10.4. Especially, this torque ratio increased at the beginning and end periods of the roll-back motion. These periods were consistent with the required maximum torque during flexion and extension of the knee, and a small difference of torque ratio between right and left knee flexion remained. At this section, the fluctuation of torque ratio caused by the non-circular gear backlash was $\pm 2$. It could be considered that the backlash causes these fluctuations and had to be reduced for clinical trials by patients. Therefore, the addendum modification coefficient of each gear tooth should be adjusted.

## 6   Conclusions

In this paper, to adapt for bilateral roll-back motion, new structure of a knee joint assistive-mechanism and the design condition of a self-locking avoidance were proposed. It was found that this new structure could be rotated to undergo bilateral roll-back motion, and a small difference between right and left knee flexion torque ratio remained due to the non-circular gear backlash. It is necessary to reduce the difference for clinical trials by patients. Therefore, the addendum-modification coefficient of each non-circular gear tooth will be optimized in future works.

## References

1. Cushnaghan, J., Bennett, J., Reading, I.: Long-term outcome following total knee arthroplasty: a controlled longitudinal study. Ann. Rheum. Dis. **68**(5), 642–647 (2009)
2. Hawker, G., Wright, J., Coyte, P.: Health related quality of life after knee replacement. J. Bone Joint Surg. Am. **80**(2), 163–173 (1998)
3. The Japanese Orthopaedic Association: Check point of the prosthetist and orthotist. Igaku-Shoin, p. 262 (1978)
4. Katori, H., et al.: A simplified synthetic design method of pitch curves based on motion specifications for noncircular gears. Trans. Jpn. Soc. Mech. Eng. Part C **60**(570), 668–674 (1994)
5. Makino, H.: Automatic assembly machine kinematics. Nikkan-Kogyo-Shimbun, pp. 29–33, 160-161, 184-186 (1976)
6. Kitta, M. et al.: Artificial knee joint. Japanese patent, JP2010275553A (2010)
7. National Institute of Advanced Industrial Science and Technology: Human body properties database (1998). https://www.dh.aist.go.jp/database/97-98/index.html
8. Terada, H., et al.: Development of a wearable assist robot for walk rehabilitation after knee arthroplasty surgery. In: Applications Advances in Mechanisms Design, Proceedings of TMM 2012, pp. 65–71 (2012)
9. Terada, H., et al.: Developments of a knee motion assist mechanism for wearable robot with a non-circular gear and grooved cams. In: Mechanisms, Transmissions and Applications, pp. 69–76 (2012)
10. Terada, H., et al.: Walking assistive apparatus. Japanese patent, JP2017248375 (2017)
11. Zhu, Y., et al.: Study of wearable knee assistive instruments for walk rehabilitation. J. Adv. Mech. Des. Syst. Manuf. **6**(2), 260–273 (2012)

# A Kinematic Characterization of a Parallel Robotic System for Lower Limb Rehabilitation

Bogdan Gherman[1], Iosif Birlescu[1], Ferenc Puskas[3], Adrian Pisla[1],
Giuseppe Carbone[1,2], Paul Tucan[1], Alexandru Banica[1],
and Doina Pisla[1(✉)]

[1] Technical University of Cluj-Napoca, Cluj-Napoca, Romania
{bogdan.gherman,iosif.birlescu,
paul.tucan,doina.pisla}@mep.utcluj.ro,
carbone@unicas.it
[2] University of Cassino, Cassino, Italy
[3] Electronic April S.R.L., Cluj-Napoca, Romania
april_private@yahoo.com

**Abstract.** The paper presents an innovative architecture of a parallel robot designed for lower limb rehabilitation of post-stroke patients. The work presented in this paper addresses the challenges presented in numerous reports, like in the Multi-Annual Roadmap, ICT 24-28 concerning the need for rehabilitation and assistive devices with the ageing of population, especially in Europe. The robot has a simple design and it is intended to be used especially for bed-ridden patients, even since the acute phase of the stroke, to accelerate the rehabilitation process. A complete kinematic analysis has been performed using the Study parameters of SE(3), and numerical simulations have been presented.

**Keywords:** Lower limb rehabilitation · Parallel robot · Kinematics
Simulation

## 1 Introduction

The population ageing concern has sparkled the interest in the development of new technologies to replace the present human assistance for the elderly (as mentioned in the Robotics 2020 Multi-Annual Roadmap - ICT28). As people age, physical and cognitive issues become more persistent and stroke risk increases. Robotic systems have been developed for lower limb rehabilitation and for assistive purposes. Most of these robotic systems have a serial architecture, in the form of exoskeletons to match exactly the lower limb anatomy. For example, in [5] Kolakowsky et al. have assessed the Ekso™ bionic exoskeleton in the assistance of patients with spinal cord injury. Although the results were positive, the device is to be used by patients in chronic stages of illness. Similar results have been obtained from using other exoskeleton systems like the ReWalk [10], Indego [2], Rex [4], mostly used for assistive purposes. Parallel robots for lower limb rehabilitation have been designed and developed, but most of them target only specific parts of the limb. For example, in [9] the authors have developed a robotic system 3-RUS/RRR used only for ankle rehabilitation, mainly

© Springer Nature Switzerland AG 2019
B. Corves et al. (Eds.): EuCoMeS 2018, MMS 59, pp. 27–34, 2019.
https://doi.org/10.1007/978-3-319-98020-1_4

from standing position. Mohanta et al. in [6] have developed an experimental model for a sitting-type rehabilitation robot. It has three degrees of freedom, all translations, but again it can be used only for ankle rehabilitation.

The study of kinematics in robotics (i.e. the mechanisms motion study) can be achieved with multiple mathematical models such as vector methods, and algebraic methods. The method used in this paper to derive the kinematics (both forward and inverse) is an algebraic method based on the Study parameters of SE(3), which proved very efficient in solving kinematics of parallel complex mechanism in the past [3, 7], and medical robots [8].

The paper is divided as following: Sect. 2 presents the two proposed rehabilitation modules (knee/hip and ankle); Sect. 3 describes the kinematics of the proposed modules with numerical examples; Sect. 4 presents the conclusions and the proposed future work.

## 2    The Proposed Lower Limb Rehabilitation Robot

The authors intend to design a parallel robot used for the gait training of bed-ridden post-stroke survivors. The targeted basic motions intended to be rehabilitated are: the flexion/extension of the hip and knee and the plantar flexion/extension and the forefoot adduction/abduction. The proposed rehabilitation robot has parallel hybrid architecture with two parallel modules chained up together. Each module has two active degrees of freedom, both prismatic joints: the first is designed to perform the flexion of the hip and knee from a supine position; the second is designed to perform the plantar flexion/extension and the forefoot abduction/adduction. For the knee/hip module a fixed coordinate frame XYZ is defined at the base of the platform, and the moving frame $X'Y'Z'$ is located at the anchor point for the ankle. Three kinematic chains are defined: kinematic chain 1 (PRR type) starting from the origin of XYZ and actuated by $q_1$; kinematic chain 2 (PRR type) starting from the origin of XYZ and actuated by $q_2$; kinematic chain 3 (RR type) starting at $t_z$ distance on Z axis having no active joints. The 3 kinematic chains intersect in the moving frame $X'Y'Z'$ producing planar motion. The ankle module fixed frame is in fact the knee module moving frame to allow a global correlation between the ankle moving frame (located at the foot anchor point) and the knee/hip rehabilitation module. For the ankle module 3 kinematic chains are defined: kinematic chain $1'$ (PRR type) starting at the origin of $X'Y'Z'$ and actuated by $q_3$; kinematic chain $2'$ (RPR) without active joints; kinematic chain 3 (PRPRR) actuated by $q_4$. The 3 kinematic chains intersect in the moving frame $X''Y''Z''$ and produce spherical motion. Figure 1a and b illustrate the kinematic schemes of the knee/hip module and ankle module respectively. The parameters within the parentheses are the rotation parameters obtained by substituting trigonometric functions with the tangent of the half angle relations (described in Sect. 3).

Figure 2 presents a CAD model of the robot, showing the patient position and working mode of the robot. The patient lies on the bed in the supine position, with the leg fixed in the leg holder. The ankle module, fixed on the leg holder, together with the

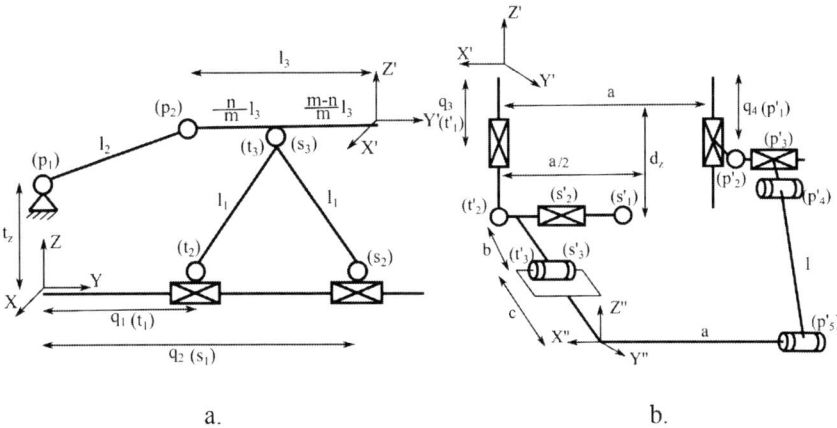

**Fig. 1.** Kinematic schemes of the: a. knee/hip module and b. the ankle module of the lower limb rehabilitation robot

leg itself, wraps the ankle performing the two kinds of motions: flexion/extension and adduction/abduction. The leg holder is guided by the knee/hip module of the robot, but a free translation along a linear guide has been added to allow the motion of the patient's leg.

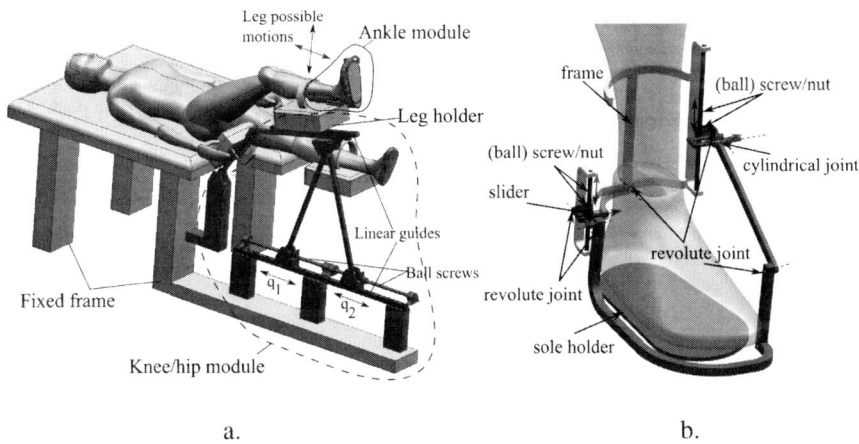

**Fig. 2.** The CAD model of the lower limb rehabilitation robot: a. the knee/hip module; b. the ankle module detail

## 3 Kinematics

The forward kinematic problem was derived for both robotic modules, using the Study parameters of SE(3). Each module is composed of 3 kinematic chains (as stated previously) subject to the constraint equations (Eqs. 1–2) in $4 \times 4$ matrix form:

$$
knee : \begin{cases} K_1 = T_y(t_1) \cdot R_x(t_2) \cdot T_y(l_1) \cdot R_x(t_3) \cdot T_y(\frac{m-n}{m} l_3) \\ K_2 = T_y(s_1) \cdot R_x(s_2) \cdot T_y(l_1) \cdot R_x(s_3) \cdot T_y(\frac{m-n}{m} l_3) \\ K_3 = T_z(t_z) \cdot R_x(p_2) \cdot T_y(l_2) \cdot R_x(p_3) \cdot T_y(l_3) \end{cases} \tag{1}
$$

$$
ankle : \begin{cases} K_1' = T_z(-t_1') \cdot R_y(t_2') \cdot T_y(b) \cdot R_x(t_3') \cdot T_y(c) \\ K_2' = T_z(-d_z) \cdot T_x(-a/2) \cdot R_y(s_1') \cdot T_x(s_2') \cdot T_y(b) \cdot R_x(s_3') \cdot T_y(c) \\ K_3' = T_x(-a) \cdot T_z(-p_1') \cdot R_y(p_2') \cdot T_x(p_3') \cdot R_x(p_4') \cdot T_y(l) \cdot R_x(p_5') \cdot T_x(a) \end{cases}
$$

$$\tag{2}$$

The active joint parameters are $t_1$, $s_1$ for the knee module and $t_1'$, $p_1'$ for the ankle module (which are all translation parameters) while all other parameters represent free motion parameters. To extract the Study parameters from Eqs. 1–2 first the trigonometric functions are rewritten in a rational form using the tangent of half angle formulae $(sin(\alpha_i) = 2t_i/(1+t_i^2), cos(\alpha_i) = (1 - t_i^2)/(1+t_i^2)$, where $\alpha$ is substituted with the rotation parameters t, s, p, $t'$, $s'$, $p'$ corresponding to the constraints defined by Eqs. 1–2), and second the Study parameters are extracted using the ratios described in [3]. The Study parameters for each kinematic chain are then reduced to polynomial form by factorizing the denominator (which is the same for all Study parameters) and then, dividing with the greatest common divisor yielding:

$$
K_1 : <x_0 : x_1 : x_2 : x_3 : y_0 : y_1 : y_2 : y_3> \ = \ <2m(t_2t_3 - 1) : -2(t_2+t_3)m : 0 : 0 : 0 :
$$
$$
(m(l_1 - l_3 - t_1) + l_3n)t_2t_3 + m((l_1+l_3)+t_1) - nl_3 : m(t_2-t_3)(l_1-t_1) + l_3m(t_2+t_3) - l_3nt_3 >
$$
$$\tag{3}$$

$$
K_2 : <x_0 : x_1 : x_2 : x_3 : y_0 : y_1 : y_2 : y_3> \ = \ <2m(s_2s_3 - 1) : -2(s_2+s_3)m : 0 : 0 : 0 :
$$
$$
(m(l_1 - l_3 - s_1) + l_3n)s_2s_3 + m((l_1+l_3)+s_1) - nl_3 : m(s_2-s_3)(l_1-t_1) +
$$
$$
l_3m(s_2+s_3) - l_3ns_3 >
$$
$$\tag{4}$$

$$
K_3 : <x_0 : x_1 : x_2 : x_3 : y_0 : y_1 : y_2 : y_3> \ = \ <2(p_1p_2 - 1) : -2(p_2 - p_1) : 0 : 0 : 0 :
$$
$$
(l_2 - l_3 - dz)p_1p_2 + l_2 + l_3 + dz : l_2(p_1 - p_2) + l_3(p_1+p_2) - tz(p_1+p_2) >
$$
$$\tag{5}$$

$$
K_1' : <x_0' : x_1' : x_2' : x_3' : y_0' : y_1' : y_2' : y_3'> \ = \ <2 : 2t_3' : 2t_2' : -2t_2't_3' : t_2'(t_1't_3' + b + c) :
$$
$$
t_2'(bt_3' - ct_3' - t_1') : t_1't_3' - b - c : bt_3' - ct_3' + t_1' >
$$
$$\tag{6}$$

$$K_2 : <x'_0 : x'_1 : x'_2 : x'_3 : y'_0 : y'_1 : y'_2 : y'_3 > \; = \; <4 : 4s'_3 : 4s'_1 : -4s_1s_3 :$$
$$2(dzs'_1s'_3 + bs'_1 + cs'_1 + s'_2s'_3) - as'_3 : 2(bs'_1s'_3 - cs'_1s'_3 - dzs'_1 - s'_2) + a : \qquad (7)$$
$$2(s'_1s'_2s'_3 - dzs'_3 - b - c) + as'_1s'_3 : 2(bs'_3 - cs'_3 + s'_1s'_2 + dz) + as'_1 >$$

$$K'_3 : <x'_0 : x'_1 : x'_2 : x'_3 : y'_0 : y'_1 : y'_2 : y'_3 > \; = \; <2p'_4p'_5 - 2 : -2p'_4 - 2p'_5 : 2(p'_4p'_5 - 1)p'_2 :$$
$$2(p'_4 + p'_5)p'_2 : (-lp'_5 - p'_1)p'_2p'_4 - p'_1p'_2p'_5 - lp'_2 - p'_3p'_4 - p'_3p'_5 : (-p_1p_5 + l)p_2p_4$$
$$- lp_2p_5 - p_3p_4p_5 + p_1p_2 + p_3 : -2ap'_2(p'_4 - p'_5) + lp'_4p'_5 - p'_2p'_3p'_4 - p'_2p'_3p'_5 -$$
$$p'_1p'_4 - p'_1p'_5 + l : p'_2p'_4p'_5(2a + p'_3) + p'_1p'_4p'_5 - 2ap'_2 + l(p'_4 - p'_5) - p'_2p'_3 - p'_1 >$$

$$(8)$$

Equations (3)–(8) generates ideals of varieties over the constraints of each kinematic chain. The equations within the ideals are homogenized (by multiplying each equation with a factor h), and using elimination theory [1] the free motion parameters are eliminated by computing lexdeg Groebner bases with the following elimination ordering: $h > t_2 > t_3$ for $K_1$, $h > s_2 > s_3$ for $K_2$, $h > p_1 > p_2$ for $K_3$, $h > t'_2 > t'_3$ for $K'_1$, $h > s'_1 > s'_2 > s'_3$ for $K'_2$, $h > p'_2 > p'_3 > p'_4 > p'_5$ for $K'_3$. This computation is also the implicitization for the parametric representation from Eqs. (3)–(8). The computed Groebner bases are presented below ($G_1$ representing the base for $K_1$, $G_2$ for $K_2$, $G'_1$ for $K'_1$ and so on). The bases for $K'_2$ (with 10 generators) and $K'_3$ (with 7 generators) are long and are not presented in detail in this paper.

$$G_1 = <y_0, y_1, x_2, x_3, (l_1^2 m^2 - l_3^2 m^2 + 2l_3^2 mn - l_3^2 n^2 - 2l_3 m^2 t_1 + 2l_3 mnt_1 - m^2 t_1^2)x_0^2 -$$
$$(4l_3 m^2 - 4l_3 nm + 4m^2 t_1)x_0 y_2 + (l_1^2 m^2 - l_3^2 m^2 + 2l_3^2 mn - l_3^2 n^2 + 2l_3 m^2 t_1 -$$
$$2l_3 mnt_1 - m^2 t_1^2)x_1^2 - (4l_3 m^2 - 4l_3 nm - 4m^2 t_1)x_1 y_3 - 4m^2 y_2^2 - 4y_3^2 m^2 >$$

$$(9)$$

$$G_2 = <y_0, y_1, x_2, x_3, (l_1^2 m^2 - l_3^2 m^2 + 2l_3^2 mn - l_3^2 n^2 - 2l_3 m^2 s_1 + 2l_3 mns_1 - m^2 s_1^2)x_0^2 -$$
$$(4l_3 m^2 - 4l_3 nm + 4m^2 s_1)x_0 y_2 + (l_1^2 m^2 - l_3^2 m^2 + 2l_3^2 mn - l_3^2 n^2 + 2l_3 m^2 s_1 -$$
$$2l_3 mns_1 - m^2 s_1^2)x_1^2 - (4l_3 m^2 - 4l_3 nm - 4m^2 s_1)x_1 y_3 - 4m^2 y_2^2 - 4y_3^2 m^2 >$$

$$(10)$$

$$G_3 = <y_0, y_1, x_2, x_3, (l_2^2 - l_3^2 - 2l_3 tz - tz^2)x_0^2 - (4l_3 + 4tz)x_0 y_2 + (l_2^2 - l_3^2 + 2l_3 tz)x_1^2 -$$
$$(4l_3 + 4tz)x_1 y_3 - 4y_2^2 - 4y_3^2 >$$

$$(11)$$

$$G'_1 = <(b^2 - c^2 + t_1'^2)x'_3 + 2t_1 y'_0 + (2b + 2c)y'_1, (b^2 - c^2 + t_1'^2)x'_2 - (2b - 2c)y'_0 + 2y'_1 t'_1,$$
$$(b^2 - c^2 + t_1'^2)x'_1 - 2t'_1 y'_2 - (2b + 2c)y'_3, (b^2 - c^2 + t_1'^2)x'_0 + (2b - 2c)y'_2 - 2t'_1 y'_3,$$
$$(2bt'_1 - 2ct'_1)y'_0 y'_2 \mid (b^2 - c^2 + t_1'^2)y'_0 y'_3 + (b^2 - c^2 + t_1'^2)y'_1 y'_2 - (2bt'_1 + 2ct'_1)y'_1 y'_3 >$$

$$(12)$$

$$G'_2 = <g_1 \ldots g_{10}> ; \quad G'_3 = <g_1 \ldots g_7> \tag{13}$$

To solve the forward kinematic problem using Study parameters for the two modules (knee and ankle) two pure lexicographic Groebner bases ($G^*$ over $G_1, G_2, G_3$, and $G'^*$ over $G'_1, G'_2, G'_3$) are computed with the additional normalizing conditions $(x_0^2 + x_1^2 + x_2^3 + x_3^2 = 1, x_0'^2 + x_1'^2 + x_2'^3 + x_3'^2 = 1)$ and having the monomial ordering $y_0 > y_1 > y_2 > y_3 > x_0 > x_1 > x_2 > x_3$. The final two bases represent the constraint equations for the two rehabilitation modules, and are polynomial functions with Study parameters as variables (both bases contain a univariate polynomial in $x_3$). Due to their length the bases are not presented in this paper. Both bases have computed Hilbert Dimensions 0, which shows that the kinematic solutions are points (no self-motion in the mechanism).

## 3.1 Forward Kinematics

To solve the forward kinematics for both rehabilitation modules one has to evaluate the bases $G^*$ and $G'^*$ with numerical values for the active joint parameters and architectural parameters (e.g. link lengths). Since both bases have univariate polynomials solving them is trivial (find all the solution of the univariate polynomial and substitute them further on) and the computation yields the numerical values for the Study parameters. Reversing the algorithm used to obtain the Study parameters from the DH matrices, gives the displacement and rotation in SE(3) in matrix form. Examples for both modules are presented for the numerical values for the robots links {n = 1, m = 2, $l_2$ = 500, $l_1$ = 600, $t_z$ = 300, $l_3$ = 500}, {a = 130, 1 = 220, c = 150, b = 30, $d_z$ = 150}, and for the active joints {$q_1$ = 500, $q_2$ = 700}, {$q_3$ = 155, $q_4$ = −130}. Table 1 shows the Study parameters numerical solutions for the knee module (4 real solutions, the other 4 being complex and not of interest) and Table 2 shows them for the ankle module (8 real solutions) respectively. The solutions repeat themselves (e.g. sol 1 and 2, 3 and 4 and so on) as a consequence of homogenizing the ideals before computing the Groebner bases in Eqs. (9)–(13).

**Table 1.** Study parameters solutions for the knee module

| Sol no | x0 | x1 | x2 | x3 | y0 | y1 | y2 | y3 |
|--------|--------|---------|----|----|----|----|---------|----------|
| 1 | −0.993 | 0.117 | 0 | 0 | 0 | 0 | 387.17 | 314.37 |
| 2 | 0.993 | −0.117 | 0 | 0 | 0 | 0 | −387.17 | −314.37 |
| 3 | 0.814 | 0.54 | 0 | 0 | 0 | 0 | −517.44 | −154.43 |
| 4 | −0.814 | −0.54 | 0 | 0 | 0 | 0 | 517.443 | 154.438 |

**Table 2.** Study parameters solutions for the ankle module

| Sol no | x0 | x1 | x2 | x3 | y0 | y1 | y2 | y3 |
|--------|-----|-----|-----|-----|-----|-----|-----|-----|
| 1/−2 | −0.927 | −0.370 | −0.035 | 0.014 | −4.311 | 3.616 | 54.765 | −49.657 |
| 3/−4 | −0.267 | −0.962 | −0.010 | 0.036 | −3.790 | 3.015 | −50.510 | 37.010 |
| 5/−6 | −0.035 | 0.014 | 0.927 | 0.370 | 54.765 | −49.657 | 4.310 | −3.616 |
| 7/−8 | −0.010 | 0.036 | 0.267 | 0.962 | −50.511 | 37.010 | 3.791 | −3.014 |

## 3.2 Inverse Kinematics

To solve the inverse kinematics for the rehabilitation modules, the bases $G^*$ and $G'^*$ may be evaluated with (valid) numerical values for the Study parameters, architectural parameters, and solved for the active joint parameters. Figure 3a illustrates a series of trajectories, where the curve A represents the trajectory of the knee and the curve C represents the trajectory of the plantar, with constant orientation of the leg in a hip-knee flexion motion. Figure 3b represents the active joints time diagram for the same motion. Trajectories $B_1, B_2 \ldots B_n$ (Fig. 3a) represent the flexion motions of the leg for various fixed knee positions (different points on the curve A).

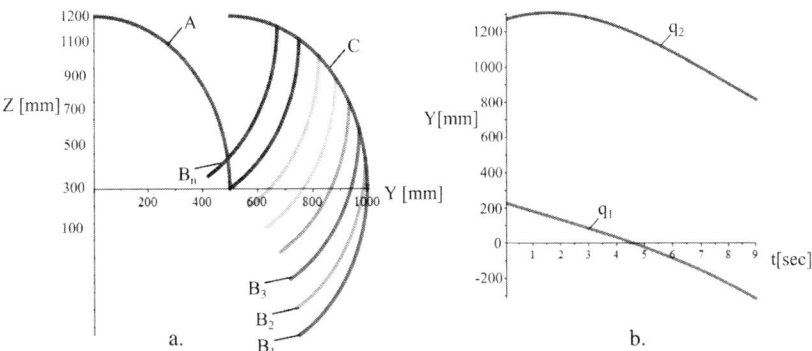

**Fig. 3.** a. Trajectories of the knee/hip module; b. time history diagram for the active joints

## 4 Conclusions

A new robotic architecture designed for the rehabilitation of the bed-ridden patients has been designed and presented in the paper. It has two chained parallel modules, one for the knee/hip and the other for the ankle rehabilitation. The kinematic model has been developed using Study parameters of SE(3). Future work includes a singularity analysis, a walking algorithm and simulation to prove the capacity of the system.

**Acknowledgments.** The paper presents results from the research activities of the project ID 37_215, MySMIS code 103415 "Innovative approaches regarding the rehabilitation and assistive robotics for healthy ageing" cofinanced by the European Regional Development Fund through the Competitiveness Operational Programme 2014–2020, Priority Axis 1, Action 1.1.4, through the financing contract 20/01.09.2016, between the Technical University of Cluj-Napoca and ANCSI as Intermediary Organism in the name and for the Ministry of European Funds.

# References

1. Cox, D., et al.: Ideals, Varieties, and Algorithms: An Introduction to Computational Algebraic Geometry and Commutative Algebra, pp. 276–277. Springer, New York (2007). 2
2. Farris, R.J., et al.: Preliminary evaluation of a powered lower limb orthosis to aid walking in paraplegic individuals. IEEE Trans. Neural Syst. Rehabil. Eng. **19**(6), 652–659 (2011)
3. Husty, M., et al.: Algebraic methods in mechanism analysis and synthesis. Robotica **25**(6), 661–675 (2007)
4. Kilicarslan, A., et al.: High accuracy decoding of user intentions using EEG to control a lower body exoskeleton. In: 2013 35th Annual International Conference of the IEEE Engineering in Medicine and Biology Society (EMBC), Osaka, Japan, pp. 5606–5609 (2013)
5. Kolakowsky-Hayner, S.A., et al.: Safety and feasibility of using the Ekso™ bionic exoskeleton to aid ambulation after spinal cord injury. J. Spine **4**, 003 (2013)
6. Mohanta, J.K., et al.: Development and control of a new sitting-type lower limb rehabilitation robot. Comput. Electr. Eng. **67**, 330–347 (2017)
7. Schadlbauer, J., Walter, D.R., Husty, M.: The 3-RPS parallel manipulator from an algebraic viewpoint. Mech. Mach. Theor. **75**, 161–176 (2014)
8. Vaida, C., et al.: Kinematic analysis of an innovative medical parallel robot using study parameters. In: New Trends in Medical and Service Robots. Mechanisms and Machine Science, vol. 39, pp. 85–99. Springer, Cham (2016)
9. Wang, C., Fang, Y., Guo, S., Chen, Y.: Design and kinematical performance analysis of a 3-RUS/RRR redundantly actuated parallel mechanism for ankle rehabilitation. J. Mech. Robot. **5**(4), 041003 (2013)
10. Zeilig, G.: Safety and tolerance of the ReWalk™ exoskeleton suit for ambulation by Rupal et al. 25 people with complete spinal cord injury: a pilot study. J. Spinal Cord Med. **35**(2), 96–101 (2012)

# Passive Walking Biped Model with Dissipative Contact and Friction Forces

Eduardo Corral[1]($\boxtimes$), Filipe Marques[2], María Jesús Gómez García[1],
Paulo Flores[2], and Juan Carlos García-Prada[1]

[1] MaqLab Research Group, Universidad Carlos III de Madrid, Leganes, Spain
{ecorral,mjggarci,jcgprada}@ing.uc3m.es
[2] MIT-Portugal Program, CMEMS-UMinho,
Department of Mechanical Engineering, University of Minho, Braga, Portugal
{fmarques,pflores}@dem.uminho.pt

**Abstract.** The main purpose of this paper is to present a new planar dynamic model of the biped robot-walking with the supporting foot slippage and contact-impact forces. In contrast to McGeer's passive dynamic models, normal forces and frictional forces acting on the feet and ground have been taken into account in the proposed model. The equations of motion and of the passive dynamic biped are obtained by using the Standard Lagrange multiplier method. The dynamics equations are obtained by forward dynamics. The normal forces acting on the feet of the passive biped are described based on the viscoelastic contact model. In turn, the frictional forces and the slippage is solve by the equations of Bengisu law for dry friction. Finally, results are validated with experimental results of the literature.

**Keywords:** Contact forces · Viscoelastic · Friction force · Passive
Biped · Multibody dynamics

## 1 Introduction

The passive dynamic walking is expected to considerably improve the energy efficiency of the bipedal locomotion. McGeer [1] first introduced the concept of planar passive dynamic walking models and built a class of prototypes. Until now, the types of passive dynamic models varied from simple to complex, and the research varied from simulations to experiments. An uncontrolled 3D passive walking model was first made by Coleman [2]. The Robot Ranger of Cornell University [3] can be highlighted, with three joints in each of its long legs. The prototype developed in the Nagoya Institute of Technology [4], was able to walk about 4000 steps (35 min) without an engine. Moreover, some actuated prototypes have been constructed based on passive dynamics [5]. The biped "PASIBOT" of UC3M is able to walk in a similar way to human with only one actuator/drive [6, 7]. It seems that the mechanical parameters of these walkers work better than the complex control systems of the conventional robots in generating natural looking gaits.

However, most of the studies mentioned above were based on models based on McGeer's assumptions: the impact between the foot of swing leg and the floor is

© Springer Nature Switzerland AG 2019
B. Corves et al. (Eds.): EuCoMeS 2018, MMS 59, pp. 35–42, 2019.
https://doi.org/10.1007/978-3-319-98020-1_5

modeled as fully inelastic and instantaneous and there is no slip between the foot and the floor. Different from the previous studies, Qi et al. [8, 9] discussed the contact stiffness, contact damping, and coefficients of friction between the feet and the ground, and verified their study both through experiments and numerical simulations. Thus, in this paper, a comprehensive multibody system dynamics model for the whole passive walking of a biped with viscoelastic dissipative contact and slippage is presented.

## 2  Methodology

Improving a passive walking biped model requires paying special attention to the feet and floor contact. The shape of the feet and the collision of the feet are important in passive walking. To our knowledge, a viscoelastic contact model for the whole walking process has never been used to analyze passive walking. The collisions in most of the models simulated above are rigid plastic collisions (no-slip and no-bounce: non-smooth).

The main aim of our research is to incorporate a more detailed contact into passive walking to work out the viscoelastic contact influence [10]. For this purpose, a general methodology to handle the contact detection between the feet and ground is developed and implemented. Within the spirit of this methodology, special attention is paid to the contact detection itself, both in terms of computational accuracy and efficiency [11].

The viscoelastic contact between the floor and the feet was examined using the Ristow [12] and Shäfer et al. [13] constitutive law (a modified Hertz contact law) [10]. This viscoelastic model was used because Hertz contact model is restricted to frictionless surfaces and perfectly elastic solids. The friction forces are also taken into account. Both feet can slip at any point of the walking process. The Bengisu equation of friction force model [14] is used. Bengisu and Akay proposed a model capable of capturing the Stribeck effect [15]. Then, the dynamic simulations of multibody models used in the context of this work are carried out using general multibody Matlab code named MUBODYNA [16]. This code is able to perform forward dynamic simulations for spatial multibody systems, using several different multibody formulations [17].

## 3  Dynamic Modeling of the Biped

### 3.1  Geometric Modeling of the Passive Walker

Thus, this work deals with the study of a biped based on the multibody systems formulation on relative coordinates. In a simple way, the planar multibody model considered here includes two legs with round feet that can act with the ground, assuming it to be rigid, flat and smooth. The two legs are linked by a spherical joint. Figure 1, shows the planar model passive dynamic biped on an inclined floor. For simplicity, the model is set as symmetrical, and each leg has a mass of $m$, a length of $l$, and a moment of inertia of $J$ about their mass center $C$. The round feet have a radius of $r$ and the centers of the radius are $O$. The distance between the mass center of each leg $C$ and the spherical joint $S$ is $d$.

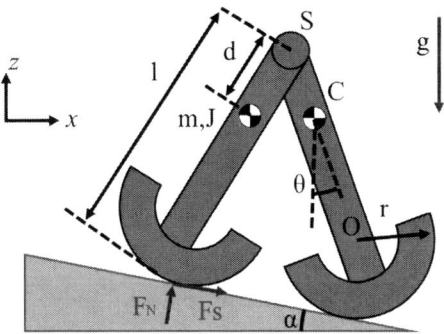

**Fig. 1.** The planar model of the passive dynamic biped

In the present work, the global coordinate system is denoted by *xyz* in which the *z*-axis is parallel to the direction of gravity. Every leg has their own fixed local coordinate system of axis. The rotational angles around y-axis between the leg-fixed or local coordinate system and the global system are denoted by $\theta_1$ and $\theta_2$, respectively. Note that in this case, the y-axis in global frame is parallel to the y-axis in local frame and the Euler parameters can be expressed by:

$$p = \{\cos \theta/2, 0, \sin \theta/2, 0\}^{\mathrm{T}}$$

The model consists of a pair of rigid legs, which are named as leg 1 and leg 2, respectively.

## 3.2   Normal Contact Law

During the walking of the biped, the feet contact with the floor and small penetrations appear, as shown in Fig. 2. The most popular normal contact force model for representing collision is based on the work by Hertz. It should be noted that the Hertz contact theory is restricted to frictionless surfaces and perfectly elastic solids and the energy dissipation during the contact process is not taken into account.

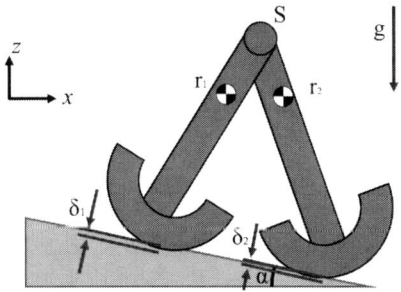

**Fig. 2.** The contact model of the biped feet with small penetrations

In order to describe the possible energy dissipation, many normal dissipative contact force models are developed. The viscoelastic contact between the floor and the feet was examined using the Ristow [12] and Shäfer et al. [13] constitutive law (a modified Hunt and Crossley Contact Model) with their empirical hysteresis damping factor [10]. $F_N = K\delta^n + c\delta\dot{\delta}$, where $F_N$ represents the normal contact force of the contact foot, $\delta$ is the relative penetration or deformation depth, and $\dot{\delta}$ denotes the relative normal contact velocity between foot and the floor. $K$ and $c$ are the generalized stiffness parameter and the hysteresis factor respectively, which are dependent on the radius of the feet and the material properties of the feet and floor. In this model the energy dissipation during the contact process has been taken into account.

By using geometrical parameters and generalized coordinates of the biped, the relative penetration depth between foot of one leg and the floor can be obtained.

When the stance leg is in contact with the floor, the swing leg may scratch the floor as it swings forward, affecting the normal walking of the passive dynamic walker. In experiments, a special floor in order to prevent the swing leg from scratching the floor is built. In the numerical simulations of this paper, it has been assumed that a swing leg is not subjected to the contact forces to avoid scratching the floor. To prevent the scratching of the swing feet and the ground, the contact will occur only when: $\delta > 0$ and $\omega > 0$.

### 3.3 Friction Model

It is known that friction forces of complex nature can be found in all actual mechanical systems that have contacting surfaces with relative motion. The major of models uses Coulomb's dry friction law; however, this model does not take into account the Stribeck effect.

Bengisu and Akay [14] proposed a model capable of capturing the Stribeck effect, as represented in Fig. 3. The model is constituted by two equations (one for the slope and another to describe the Stribeck effect), which are given by

$$F = \begin{cases} \frac{Fs}{v_0^2}\left((\| V_T \| - v_0)^2 + Fs\right)\text{sgn}(V_T) & \text{if } \| V_T \| < v_0 \\ (Fc + (Fs - Fc)e^{-\xi(\|V_T\|-v_0)})\text{sgn}(V_T) & \text{if } \| V_T \| < v_0 \end{cases}$$

in which $\xi$ should be a positive parameter which represents the negative slope of the sliding state.

These models have the particularities that for low velocities the small time step increment is needed, which slows down the simulation, but it has the advantage that for velocities close to zero, the friction force will always be low independently of the displacement. It can also describe the characteristics of dry friction and illustrate the stick-slip phenomenon between contacting bodies, and its coefficients can be gained by doing experiments or consulting the best literature on friction. For our model, the coefficients have the values of $\xi = 1000$ and $v_0 = 0.0001$ m/s. So the Bengisu and Akay dry friction model is adopted in this paper as the friction contact law of the feet and the floor.

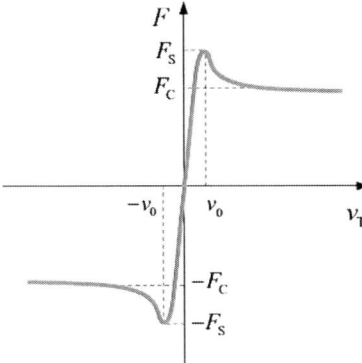

**Fig. 3.** Representation of the Bengisu and Akay model

### 3.4 Dynamic Equations

A multibody dynamic system can be formulated and solved in many ways. In this paper, the equations are implemented in MUBODYNA, a Matlab program.

For this algorithm, the relative coordinates are used. Every body has their body-Fixed or local coordinate system, and to describe their rotation the Euler parameters are used. The translational motion is described in terms of Cartesian coordinates, while rotational motion is specified using the technique of Euler parameters. The kinematic constrains are formulated using generalized coordinates. The constraints equations are associated with kinematic pairs. The program uses Newton-Euler equations of motion, which are augmented with the constraint equations that lead to a system of differential algebraic equations for the motion. It also uses the Standard Lagrange multiplier method to obtain the dynamics.

The algorithm integrator utilized in this work in the resolution of the dynamic equations of motion is the Euler method, which is the most popular and uses the numerical integration method, along with the Runge-Kutta and Adams predictor-corrector. The Euler method allows the use of variable time steps during the integration process. The initial conditions are corrected in the program with standard methodology to minimize and avoid the violation constraints.

### 3.5 Initial Conditions

In order to verify the numerical results of the dynamic model of this paper, the parameters and initial conditions are set from the literature [8]. The parameters of the biped: $m = 1.0$ kg, $J = 0.0096$ kg $\cdot$ m$^2$, $l = 0.40$ m, $d = 0.10$ m, $r = 0.08$ m. The gravitational acceleration: $g = 9.8$ m/s$^2$. The exponent of the modified Hertz contact law is set as: $n = 1.5$. Contact parameters: $K = 1 \cdot 106$ N/m$^{1.5}$; $c = 1 \cdot 107$ N $\cdot$ s/m$^2$; $\mu_0 = 0.50$; $\mu = 0.40$. Slop angle: $\alpha = 0.02$ rad. The initial conditions of the passive biped: $\theta_1 = -0.2479$ rad; $\omega_1 = -0.0052$ rad/s; $\theta_2 = 0.1655$ rad; $\omega_2 = -1.2565$ rad/s.

The linear velocities of the CoM of the leg1: $v_1 = (0.37404, 0, 0.037832)$ m/s and the linear velocities of the CoM of the leg2: $v_2 = (0.49847, 0, 0.058404)$ m/s.

## 4  Numerical Results

In this section, the most interesting numerical results of the model are shown:

Figure 4 shows a set pictures obtained with the animation of the Matlab program on the motion produced by the passive biped-walking robot.

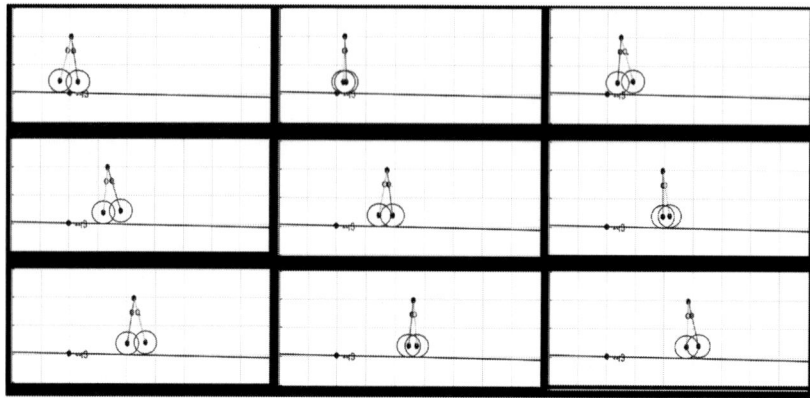

**Fig. 4.** Motion of the passive biped

Figure 5(a) illustrates the phase portraits of one leg relative to a dynamic simulation. It must be said that the response for both legs is the same due the symmetry of the model. In a broad sense, the dynamic behavior of the passive biped-walking robot is consistent with the literature. In particular, Fig. 5(a) indicates that the passive biped gets into stable walking. The end of the lower part of the graph are the hit and the lifting of the foot. It can be noted that during the hitting of the floor, the swing leg changes into a stance one and its angular velocity changes rapidly. It must be highlighted that the dynamic response of the robot is quite sensitive to the initial conditions and to the values of the contact-impact force parameters.

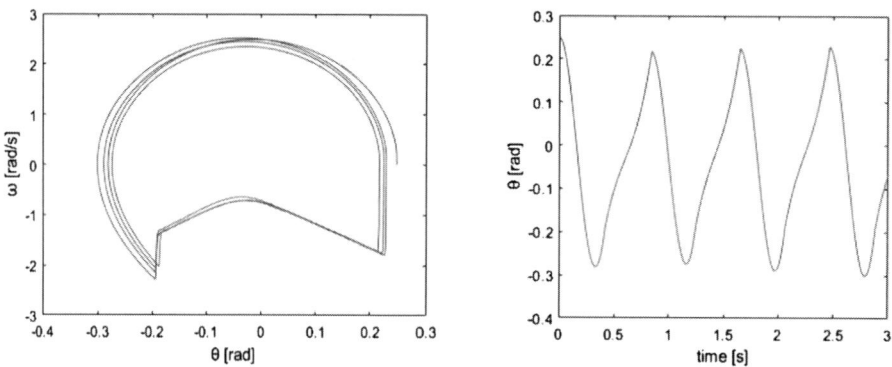

**Fig. 5.** (a) Leg phase portraits (b) Time histories of one leg angle

Figure 5(b) illustrates the time histories of the leg angle. The simulation results show that the motions of the two legs have the same stride and period.

Figure 6(a) shows the time histories of the penetration of the foot and the floor, and Fig. 6(b) depicts the time histories of the contact forces of this foot acting on the floor. These calculation results show that foot is subjected to a large impact both in normal and tangential directions when it hits the floor.

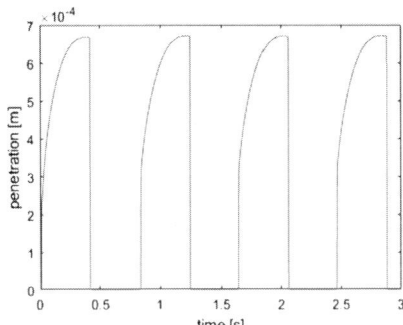 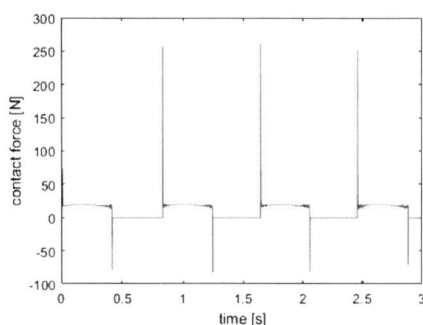

**Fig. 6.** Time histories of: (a) penetration, (b) contact forces acting on foot.

In these graphs, it can be observed how the contact force increases as the penetration increases, until the point where the other foot impacts, producing a phase of double support. In this phase of double support, an impulse is transmitted from one foot to another, greatly increases contact forces, and producing a rebound that switches the stance leg into a swing leg almost instantaneously after the impact.

The contact forces acting on the other leg are identical to that other one, except for a certain phase difference. These results are consistent with the literature [8].

## 5   Concluding Remarks

A planar dynamic model for the whole passive walking of a biped robot-walking with the supporting foot slippage and contact-impact forces has been presented. The results have shown that the contact force and friction force models work properly for this biped model. For these reasons, it can be concluded that the contact forces Ristow and Shäfer type with friction forces Begisu type work correctly in the simulation of the contact/impact between the passive foot and the floor. The model is implemented in a parametric program, which makes the option to do future sensitive analysis possible.

## References

1. McGeer, T.: Passive dynamic walking. Int. J. Robot. Res. **9**(2), 62–82 (1990)
2. Coleman, M.J., Ruina, A.: An uncontrolled walking toy that cannot stand still. Phys. Rev. Lett. **80**(16), 36–58 (1998)

3. Collins, S.H., Ruina, A.: A bipedal walking robot with efficient and human-like gait. In: IEEE International Conference on Robotics and Automation (ICRA), pp. 1983–1988 (2005)
4. Ikemata, Y., Yasuhara, K., Sano, A., Fujimoto, H.: A study of the leg-swing motion of passive walking. In: IEEE International Conference on Robotics and Automation (ICRA), pp. 1588–1593 (2008)
5. Collins, S., Ruina, A., Tedrake, R., Wisse, M.: Efficient bipedal robots based on passive-dynamic walkers. Science **307**, 1082–1085 (2005)
6. Meneses, J., Castejón, C., Corral, E., Rubio, H., García-Prada, J.: Kinematics and dynamics of the quasi-passive biped "PASIBOT". J. Mech. Eng. **57**, 879–887 (2011)
7. Corral, E., Meneses, J., Castejón, C., García-Prada, J.C.: Forward and inverse dynamics of the biped PASIBOT. Int. J. Adv. Robot. Syst. **11**(7), 109 (2014)
8. Qi, F., Wang, T., Li, J.: The elastic contact influences on passive walking gaits. Robotica **29** (5), 787–796 (2011)
9. Qi, F., Bi, L.Y., Wang, T.S., et al.: The experimental study on the contact process of passive walking. Acta. Mech. Sin. **28**(4), 1163–1168 (2012)
10. Machado, M., Moreira, P., Flores, P., et al.: Compliant contact force models in multibody dynamics: evolution of the Hertz contact theory. Mech. Mach. Theor. **53**(7), 99–121 (2012)
11. Flores, P., Ambrósio, J.: On the contact detection for contact-impact analysis in multibody systems. Multibody Syst. Dyn. **24**(1), 103–122 (2010)
12. Ristow, G.H.: Simulating granular flow with molecular dynamics. J. Phys. I Fr. **2**, 649–662 (1992)
13. Shäfer, J., Dippel, S., Wolf, E.D.: Force schemes in simulations of granular materials. J. Phys. I Fr. **6**, 5–20 (1996)
14. Bengisu, M.T., Akay, A.: Stability of friction-induced vibrations in multi-degree-of-freedom systems. J. Sound Vibr. **171**, 557–570 (1994)
15. Marques, F., Flores, P., Claro, J.C.P., et al.: A survey and comparison of several friction force models for dynamic analysis of multibody mechanical systems. Nonlinear Dyn. **86**, 1407–1443 (2016)
16. Flores, P.: MUBODYNA - A MATLAB program for dynamic analysis of spatial multibody systems. University of Minho, Guimarães, Portugal
17. Marques, F., Souto, A.P., Flores, P.: On the constraints violation in forward dynamics of multibody systems. Multibody Syst. Dyn. **39**(4), 385–419 (2017)

# Control Issues of Mechanical Systems

# Mechatronic Model Based Jerk Optimization in Servodrives with Compliant Load

Igor Ansoategui and Francisco J. Campa$^{(\boxtimes)}$

University of the Basque Country UPV/EHU, Bilbao, Spain
{igor.ansoategui, fran.campa}@ehu.eus

**Abstract.** The aim of this work is to develop a mechatronic model for servo-drives that move a compliant load, which is the case of large heavy machine tools. The model integrates the dynamics of the motor, transmission chain and compliant load as well as the dynamics of the control and is used to calculate the maximum overshoot in the load when it reaches the commanded final position as a function of the feed speed and the programmed jerk in the velocity profile. This overshoot must be, in applications as machining, always minimized. The result is a graph that indicates the region of safe values of jerk, where a given overshoot is never surpassed, but also there are several regions where much higher jerk values can be programmed with minimal overshoot. The location of these areas depending on the feed speed and the load natural frequency has been shown for a square sine profile. These optimal values of the jerk have been experimentally validated.

**Keywords:** Mechatronics · Servodrives · Machine tools · Jerk optimization
Path planning

## 1 Introduction

One of the main problems that the designers of drives for positioning heavy loads is the compliance of the load itself, due to the lack of stiffness. The result is the onset of vibrations in the load due to the inertial forces, and also the problem of finding the optimal control gains and path planning in order to avoid them. The result is usually having to employ very low control gains and conservative path planning, with low acceleration and jerk. An example is the industry of heavy machine tools, where having to move heavy rams and columns up to 5 m length and a natural frequency of 10–15 Hz is not uncommon. The use of a mechatronic model integrating the dynamics of the drive, transmission and load together with the control can help improving the design of such systems but also determining the optimal jerk and acceleration in the path planning.

Elemental mechatronic models for servodrives consider just inertial models of the system [4]. These models, provide and initial estimation of the behaviour and frequently results precise enough. However, if the load inertia is very high, the flexibility of the transmission must be taken into account, being the most common approach the use of a 2 degrees of freedom (dof) model of the drive-transmission-load system as the ones proposed by Dequidt [1], Altintas [2] or Caracciolo [3]. However, it may well

© Springer Nature Switzerland AG 2019
B. Corves et al. (Eds.): EuCoMeS 2018, MMS 59, pp. 45–52, 2019.
https://doi.org/10.1007/978-3-319-98020-1_6

happen that the load is also compliant, as in the case of heavy machine tools, so, to have a precise model of the mechanical system, a 3 dof model must be used, as in [4–6]. In these cases, the third dof is the position at the end of the load, which is the part or the machine element that must be moved and is compliant, see Fig. 1. This position is never measured nor considered in the position control due to accessibility problems. That means that the NC of the machine is just capable of controlling the position of the second dof and the relative displacement that happens between the second and third dofs is not avoidable. The result is the onset of vibration in the acceleration/decceleration and specially the appearance of an overshoot in the final position that can compromise the geometry of the part machined.

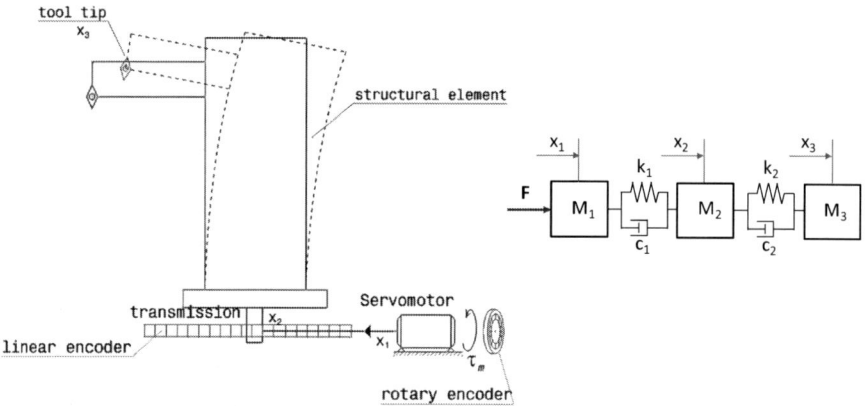

**Fig. 1.** Influence of the column flexibility on the tool error and degrees of freedom: $x_1$, drive encoder position, $x_2$, linear encoder position, $x_3$, tool position. Corresponding 3 dof model.

Here, a 3 dof mechatronic model of a drive with a compliant load is presented and experimentally validated in a test bench. The model is then used to predict the optimal jerk and acceleration values to program in the path planning minimizing vibrations due to inertial forces and overshoot. Hence, the minimization of the vibration is reached without using external devices or modifying the control algorithm of the NC, as the use of open controls is not common in machine tool industry.

## 2   Mechatronic Model of a 3 Dof System

In Fig. 1 the 3 dof lumped parameters model of the mechanical system drive-transmission-load, where $x_1$ is the position of the drive measured at the encoder, $x_2$ is the position measured at the linear encoder at the end of the transmission, and $x_3$ is the position at the end of the load. Masses $m_1$, $m_2$, $m_3$ represents the inertia of the drive, transmission and load respectively, $k_1$ and $k_2$ are the stiffness of the drive-transmission system and the load, and finally, $c_1$ and $c_2$ are the corresponding viscous damping. F is the force equivalent to the motor torque. The equations of motion are:

$$
\begin{bmatrix} m_1 & 0 & 0 \\ 0 & m_2 & 0 \\ 0 & 0 & m_3 \end{bmatrix} \begin{Bmatrix} \ddot{x}_1 \\ \ddot{x}_2 \\ \ddot{x}_3 \end{Bmatrix} + \begin{bmatrix} c_1 & -c_1 & 0 \\ -c_1 & c_1+c_2 & -c_2 \\ 0 & -c_2 & c_2 \end{bmatrix} \begin{Bmatrix} \dot{x}_1 \\ \dot{x}_2 \\ \dot{x}_3 \end{Bmatrix} + \begin{bmatrix} k_1 & -k_1 & 0 \\ -k_1 & k_1+k_2 & -k_2 \\ 0 & -k_2 & k_2 \end{bmatrix} \begin{Bmatrix} x_1 \\ x_2 \\ x_3 \end{Bmatrix} = \begin{Bmatrix} F \\ 0 \\ 0 \end{Bmatrix}
$$

$$(1)$$

The equations of motion in time domain are converted to the Laplace domain, where the three degrees of freedom are related with the force by means of three transfer functions (TF): $TF_1$ relates $x_1$ with F, $TF_2$ relates $x_2$ with $x_1$ and $TF_3$ relates $x_3$ with $x_2$. These transfer functions are represented in a general form in Eq. 2, whose coefficients are shown in Table 1.

$$
TF_i = \frac{n_4 s^4 + n_3 s^3 + n_2 s^2 + n_1 s + n_0}{d_6 s^6 + d_5 s^5 + d_4 s^4 + d_3 s^3 + d_2 s^2 + d_1 s + d_0}
$$

$$(2)$$

**Table 1.** Numerator and denominator coefficients in the transfer functions.

| | $FT_1$ | $FT_2$ | $FT_3$ |
|---|---|---|---|
| $n_0$ | $k_1 k_2$ | $k_1 k_2$ | $k_2$ |
| $n_1$ | $c_1 k_2 + c_2 k_1$ | $c_1 k_2 + c_2 k_1$ | $c_2$ |
| $n_2$ | $m_3 k_1 + (m_2 - m_3) k_2 + c_1 c_2$ | $m_3 k_1 + c_1 c_2$ | $0$ |
| $n_3$ | $m_3 c_1 + (m_2 - m_3) c_2$ | $m_3 c_1$ | $0$ |
| $n_4$ | $m_2 m_3$ | $0$ | $0$ |
| $d_0$ | $0$ | $k_1 k_2$ | $k_2$ |
| $d_1$ | $0$ | $c_1 k_2 + c_2 k_1$ | $c_2$ |
| $d_2$ | $(m_1 + m_2 - m_3) k_1 k_2$ | $m_3 k_1 + (m_2 - m_3) k_2 + c_1 c_2$ | $m_3$ |
| $d_3$ | $(m_1 + m_2 - m_3) \cdot (c_2 k_1 + c_1 k_2)$ | $m_3 c_1 + (m_2 - m_3) c_2$ | $0$ |
| $d_4$ | $(m_1 + m_2) m_3 k_1 + (m_2 - m_3) m_1 k_2$ $+ (m_1 + m_2 - m_3) c_1 c_2$ | $m_2 m_3$ | $0$ |
| $d_5$ | $(m_1 + m_2) m_3 c_1 + (m_2 - m_3) m_1 c_2$ | $0$ | $0$ |
| $d_6$ | $m_1 m_2 m_3$ | $0$ | $0$ |

These transfer functions are integrated then in a mechatronic model programmed in Simulink, where the velocity and position feedback loops are represented, see Fig. 2. Note that $TF_3$ is after the position feedback loop, as there is no direct measurement of $x_3$. The model has been developed thinking in a rotary drive whose rotation is then converted to a translation with a transmission factor $i$ (m/rad). Hence, $TF_1$ must be derived to obtain the motor velocity $\omega_m$ to close the velocity feedback loop: $\omega_m = \dot{x}_1 / i$. F is related to the motor torque by $\tau_m = F \cdot i$. $TF_2$ is also adapted to relate $\omega_m$ and $x_2$. Finally, a simple viscous-Coulomb friction model has been introduced and then identified experimentally.

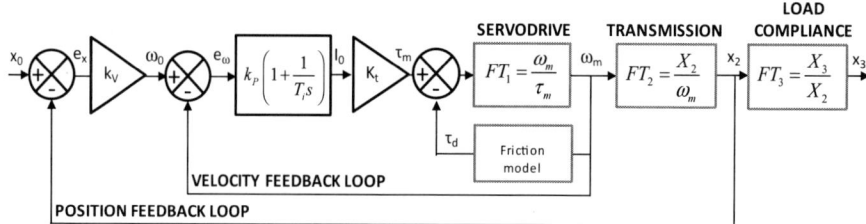

**Fig. 2.** 3 dof mechatronic model of a servodrive with compliant load.

Regarding the control, it is the usual cascade control used in NC machine tool with three cascaded feedback loops of position, velocity and current. The position control is Proportional, the velocity control is Proportional-Integral, and the current control is also Proportional-Integral, although it is not modelled since the loop closing period of this loop is much lower than the dynamics of the velocity and position loops, that is, the electrodynamics will be much faster than the dynamics of the mechanical parts.

## 3  Test Bench Description

A test bench has been set up, see Fig. 3, with a Fagor 42.30A FKM servodrive with rotary encoder and a nominal torque of 6.3 Nm, a ball screw Korta KBS-3210 with a diameter of 32 mm and a lead of 10 mm, a linear encoder Heidenhain Ls 186 MI640 with a resolution of 0.5 µm. The NC is a Fagor 8035. On the table, the load is a mass of 30 kg over two steel plates with a thickness of 1.5 mm to introduce compliance. The position of the mass is externally measured using a linear interferometer HP 5529A.

**Fig. 3.** Test bench: compliant load over a ball-screw drive.

After modelling all the elements of the system, the modal analysis reveals a first mode of 15.1 Hz due to the thin plates. This amplification is clearly seen in $TF_3$ in Fig. 4a. Figure 4b compares the Bode plot of the position closed loop transfer function ($TF_{CPL}$), which relates $x_2$ with the position command $x_0$, with the product $TF_{CPL} \cdot TF3$, which relates the position command with $x_3$. It can be seen how, as expected, although the closed loop is tuned with no resonance and a bandwidth of 8.9 Hz, the addition of a compliant load results in a resonance near its natural frequency that will amplify harmonics of the motion nearby. The influence of the compliant mass mode also appears in $TF_1$ and $TF_2$.

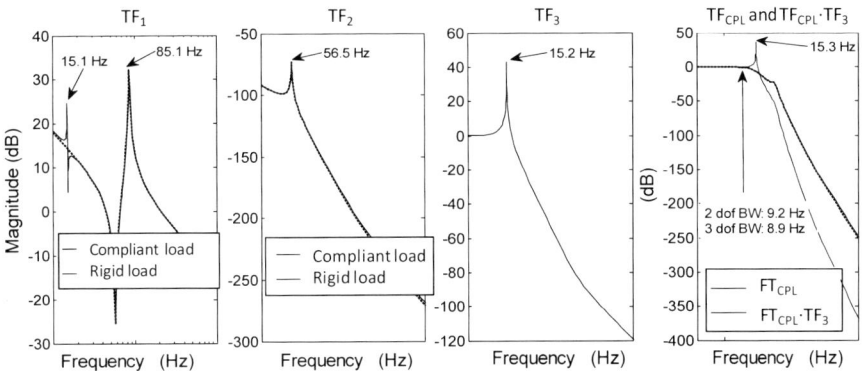

**Fig. 4.** $TF_1$, $TF_2$, $TF_3$ and $TF_{CPL}$ bode diagrams.

## 4   Experimental Validation

Several tests have been made ranging displacements from 20 to 400 mm and feed speeds from 7 to 30 m/min with a square sine velocity profile. For a feed speed of 7 m/min and a displacement from the zero position to 100 mm and back to the zero, Fig. 5 compares the error in the mass position $x_3$, modelled and predicted, and the table position $x_2$, modelled and predicted also, whose difference is better seen in the zoom. Although there are deviations probably due to the damping estimation, the model matches the reality satisfactorily taken into account the simplifications done. It can be

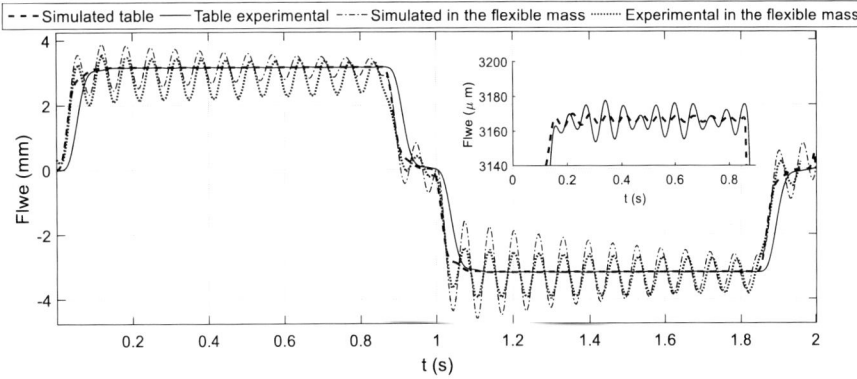

**Fig. 5.** Following error in the table and flexible mass. Vf = 7000 mm/min and $\Delta x$ = 100 mm.

seen how there is a clear vibration that begins in the initial accelerations and also how when the final position is reached, near the 2 s, the position error in the mass reaches values up to 1 mm. All the tests whose motion profile present harmonics near the load natural frequency present this problem. Also, the vibration affects the position of the ball screw table which oscillates up to 20 μm as it can be seen in the zoom between 0 s and 0.9 s.

Several values of the jerk of the motion profile have been tested also. The jerk in a square sine profile conditions the shape of the motion and the appearance of harmonics near the load natural frequency. Figure 6 shows the position overshoot in the mass for several values of the programmed jerk and a feed speed of 15 m/min, experimentally measured and modelled. Above 150 m/s³, the overshoot tends to increase until it stabilizes in 0.8 mm. Below 150 m/s³, the curve has several "valleys", that is, there are values of the jerk that minimize the overshoot and could be considered as optimal, for example at 150 m/s³. As a reference, machine tool manufacturers tend to be conservative with the jerk, and rarely increase it above 30 m/s³ to avoid position overshoots.

On the other hand, to account for the influence of the control, in Fig. 6, the discontinuous line represents the position overshoot calculated with just a mechanical model of the system in open loop. It predicts higher overshoots than measured, probably because there is not a position control trying to compensate the error in the table position and thus in the load position.

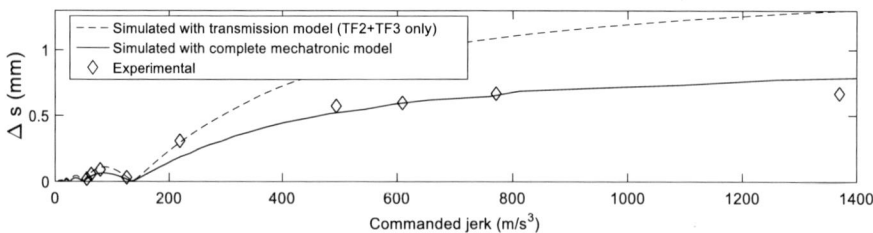

**Fig. 6.** Load overshoot Δs as a function of the programmed jerk, feed speed of 15 m/min.

## 5   Optimal Jerk Path Planning

The use of the graph in Fig. 6 is to enter with a predefined maximum overshoot and obtain the highest jerk, in order to have the most dynamic performance. What happens, is that at low enough jerk values, the predefined overshoot is never surpassed up to a limit, but then after that limit, there will appear several "valleys" in the graph, where with even higher values of the jerk and no overshoot. These isolated areas can be considered for an optimal dynamic performance.

A detailed analysis in frequency domain of the square sine velocity profile reveals that these values happen when the acceleration time of the profile matches the relation with the natural frequency $f_n$ of Eq. 3. Given the math of this profile, the optimal jerk as well as the corresponding acceleration can also be calculated as a function of the feed speed to program and the natural frequency as follows:

$$t_{acc} = \frac{2n+1}{2}\frac{1}{f_n} \quad j = 2V_f\left(\frac{\pi f_n}{2n+1}\right)^2 \quad a = V_f\frac{\pi f_n}{2n+1} \quad n = 1,2,3,\dots \quad (3)$$

In Fig. 7, the measured position of the load in three experimental tests at 30 m/min and a displacement of 400 mm is shown to prove this. E1 is a test at a jerk of 846 m/s$^3$, higher than the first valley and the overshoot reaches 1.8 mm, in E2 the optimal jerk of 256 m/s$^3$ for n = 1 has been used and the overshoot falls to 80 μm, and in E3 a lower jerk of 171 m/s$^3$ outside the valley has been tested and the overshoot rises again up to 220 μm.

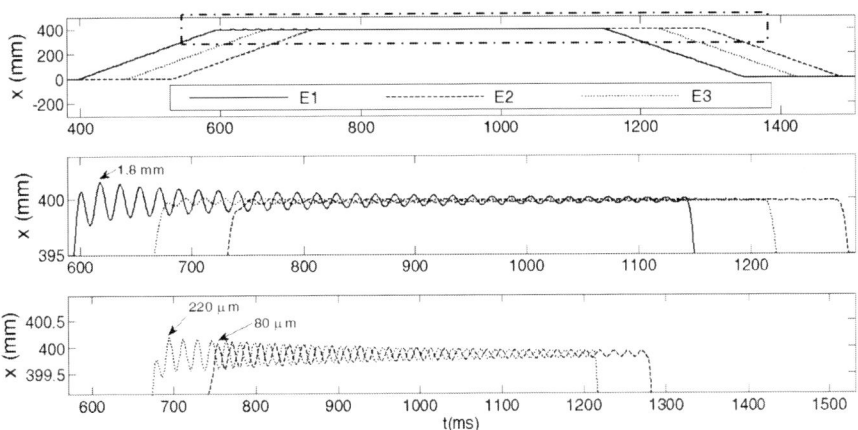

**Fig. 7.** Measured position in tests E1 (j = 846 m/s$^3$), E2 (j = 256 m/s$^3$) and E3 (j = 171 m/s$^3$).

## 6   Conclusions

In this work it has been proved how the use of a mechatronic model that integrates the dynamics of the drive, transmission, compliant load and the control allows estimating the overshoot and vibration that will be present in the load. What is more, that model can be used to calculate the best values of the jerk and acceleration for a motion profile minimizing the overshoot at the end of the displacement. This approach has been tested successfully in a test bench if a compliant load with one predominant mode but can be applied to cases where the load has several modes or the modes of the load and the transmission chain are more similar.

**Acknowledgments.** Authors acknowledge the financial support from the Spanish Government through the MINECO DPI2015-64450-R (MINECO/FEDER, UE) and the support through the project IT949-16, given by the Basque Government.

# References

1. Dequidt, A., Castelain, J.-M., Valdès, E.: Mechanical pre-design of high performance motion servomechanisms. Mech. Mach. Theor. **35**(8), 1047–1063 (2000)
2. Altintas, Y., Verl, A., Brecher, C., Uriarte, L., Pritschow, G.: Machine tool feed drives. CIRP Ann. Manuf. Technol. **60**(2), 779–796 (2011)
3. Caracciolo, R., Richiedei, D.: Optimal design of ball-screw driven servomechanisms through an integrated mechatronic approach. Mechatronics **24**(7), 819–832 (2014)
4. Neugebauer, R., Denkena, B., Wegener, K.: Mechatronic systems for machine tools. CIRP Ann. Manuf. Technol. **56**(2), 657–686 (2007)
5. Wu, S.-T., Lian, S.-H., Chen, S.-H.: Vibration control of a flexible beam driven by a ball-screw stage with adaptive notch filters and a line enhancer. J. Sound Vib. **348**, 71–87 (2015)
6. Fortunato, A., Ascari, A.: The virtual design of machining centers for HSM: towards new integrated tools. Mechatronics **23**(3), 264–278 (2013)

# Fuzzy Logic Controller and PID Controller Design for Aircraft Pitch Control

Erdi Sayar[1(✉)] and Hüseyin Metin Ertunç[2]

[1] RWTH Aachen University, Aachen, Germany
erdi.sayar@rwth-aachen.de
[2] Kocaeli University, İzmit, Turkey
hmertunc@kocaeli.edu.tr

**Abstract.** The goal of this paper is to compare a PID controller and a Fuzzy Logic controller in terms of pitch control of an aircraft. Firstly, derivation of the mathematical model is introduced to define the dynamics of an aircraft. To inspect the performance of the controllers, the PID (Proportional-Integral-Derivative) and FLC (Fuzzy Logic Controller) is proposed for controlling the pitch angle of an aircraft. Simulation results are illustrated in the time domain. In the end, the performances of pitch control are looked into and examined with respect to evaluation criteria of step response so as to determine which control method exhibits better performance depending on pitch angle and pitch rate. It is determined thanks to simulation that FLC shows better performance than PID controller.

**Keywords:** Aircraft · Flight control · Autopilot
Longitudinal dynamic · Fuzzy logic · PID

## 1 Introduction

Classic controllers like PID and PI are used widely in the industrial area due to the fact that they are a rudimentary and robust performance.

In any case, it is challenging to decide ideal parameters of classic controllers. In tuning process for the classic controller, P, I and D parameters must be chosen properly so that step response of closed loop fulfil evaluation criteria such as a minimal rise and settling time and minimal overshoot.

On the other hand, FLC has a well-liked method as it has linguistic rules. FLC consists of some parameters such as linguistic control rules and MFs (Member Functions) have to be tuned for a given system [5].

### 1.1 Background Theory

A control system is basically categorised by its dynamic behaviour. The step response of the controlled system is utilized to demystify the dynamic behaviour.

© Springer Nature Switzerland AG 2019
B. Corves et al. (Eds.): EuCoMeS 2018, MMS 59, pp. 53–60, 2019.
https://doi.org/10.1007/978-3-319-98020-1_7

The step response indicates how the controlled variable behaves. Evaluation criteria such as rise time, settling time, peak time, steady state error and percentage of overshoot can be used to assess the controller's feat [1].

## 1.2   PID Controller

The objective of PID controller is to keep the error as minute as possible. Mathematically a PID controller can be written as:

$$\mu(n) = K_p e(t) + K_i \int e(t)dt + K_d \frac{d}{dt}e(t) \tag{1}$$

where $K_p$, $K_i$ and $K_d$ are proportional, integral and derivative gain parameters, respectively.

## 1.3   Fuzzy Logic Controller

Unlike classic controller, Fuzzy logic is tolerant to imprecision, uncertainty and half-truth. This paves the way for implementing fuzzy logic to nonlinear systems. Fuzzy logic is adaptable since additional functions can be added without changing the system entirely.

# 2   Pitch Control Modeling

By using the mathematical model of the aircraft which is linearized at different conditions, flight control has been designated [3]. This study offers to control the pitch angle of an aircraft so as to stabilize system when the aircraft is nosed up and nosed down. The pitch control system handled in this study is illustrated in Fig. 1 where $X_b$, $Y_b$ and $Z_b$ shows the aerodynamics force components. $\theta$, $\phi$ and $\delta e$ indicate the pitch angle orientation and roll angle orientation of the aircraft in the earth-axis coordinate as well as elevator deflection angle.

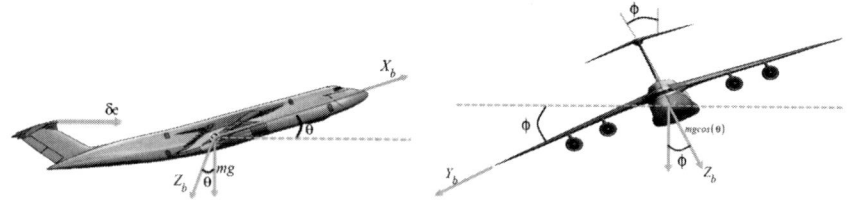

**Fig. 1.** Description of pitch control system

Figure 2 demonstrates the moment, force and velocity components in the body-fixed coordinate of the aircraft. The roll, pitch and yaw moment components are shown as $L$, $M$ and $N$. The term $p$, $q$ $r$ and $u$, $v$ $w$ are the angular rates and velocity components about the roll, pitch and yaw axis respectively. The term $\alpha$ and $\beta$ are the angle of attack and sideslip. Data used for system analysis and modeling is from General Aviation Airplane. The parameters include in dimensional derivatives are considered below [2].

$$Q = 36.8 \text{ lb/ft}^2$$
$$QS = 6771 \text{ lb}$$
$$QS\bar{c} = 38596 \text{ ft} \cdot \text{lb}$$
$$\frac{\bar{c}}{2\mu_0} = 0.016 \text{ s}$$

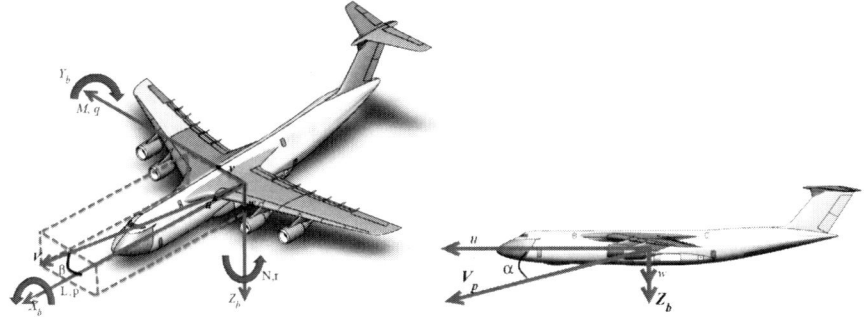

**Fig. 2.** Force, moment and velocity in body fixed coordinate

**Table 1.** Summary of longitudinal derivatives

| Longitudinal derivatives | $F_x$ parameters | $F_z$ parameters | Pitching moment |
|---|---|---|---|
| Rolling velocities | $X_u = -0.045$ | $Z_u = -0.369$ | $M_u = 0$ |
| Yawing velocities | $X_w = 0.036$ | $Z_w = -2.02$ | $M_w = -0.05$ |
| | $X_{\dot{w}} = 0$ | $Z_{\dot{w}} = 0$ | $M_{\dot{w}} = -0.051$ |
| Angle of attack | $X_\alpha = 0$ | $Z_\alpha = -355.42$ | $M_\alpha = -8.8$ |
| | $X_{\dot{\alpha}} = 0$ | $Z_{\dot{\alpha}} = 0$ | $M_{\dot{\alpha}} = -0.8976$ |
| Pitch rate | $X_q = 0$ | $Z_q = 0$ | $M_q = -2.05$ |
| Elevator deflection | $X_{\delta_e} = 0$ | $Z_{\delta_e} = -28.15$ | $M_{\delta_e} = -11.874$ |

Some assumptions are necessary to be regarded. Firstly, the aircraft is in the steady state condition with constant altitude and velocity, hence the lift and weight balance out each other and the thrust and draw eliminate each other. Secondly, changing at the pitch angle does not impact on the speed of the aircraft [4]. In this part, the equation of pitch control system is described. Force Equations

$$X - mgS_\theta = m(\dot{u} + qv - rv) \tag{2}$$

$$Z + mgC_\theta C_\Phi = m(\dot{w} + pv - qu) \tag{3}$$

Moment Equation

$$M = I_y \dot{q} + rq(I_x - I_z) + I_{xz}(p^2 - r^2) \tag{4}$$

Body angular velocities in terms of Euler angles and Euler rates

$$\text{Rolling Rate } p = \dot{\Phi} - \dot{\Psi} S_\theta$$
$$\text{Yawing Rate } q = \dot{\theta} C_\Phi + \dot{\Psi} C_\theta S_\Phi$$
$$\text{Pitching Rate } r = \dot{\psi} C_\theta C_\psi - \dot{\theta} S_\Phi$$
$$\text{Pitch Angle } \dot{\theta} = qC_\Phi - rS_\Phi$$
$$\text{Roll Angle } \dot{\Phi} = p + qS_\Phi T_\theta + rC_\Phi T_\theta$$
$$\text{Yaw Angle } \dot{\Psi} = (qS_\Phi + rC_\Phi)\sec(\theta)$$

Equations 2, 3 and 4 can be linearized by using small disturbance theory. All the variables in the equations of motion are replaced by a reference value plus a perturbation or disturbance.

$$u = u_0 + \Delta u \quad v = v_0 + \Delta v \quad w = w_0 + \Delta w$$
$$p = p_0 + \Delta p \quad q = q_0 + \Delta q \quad r = r_0 + \Delta r$$
$$X = X_0 + \Delta X \quad M = M_0 + \Delta M \quad Z = Z_0 + \Delta Z \quad \delta = \delta_0 + \Delta\delta$$

The reference flight condition is presumed as symmetric and propulsive forces are considered remaining constant. This states that $v_0 = p_0 = q_0 = r_0 = \Phi_0 = \Psi_0 = 0$ Furthermore, if the $x$ axis is initially aligned so that it is along the direction of airplane's velocity vector, then $w_0 = 0$.

Equations can be simplified by linearizing.

$$\left(\frac{d}{dt} - X_u\right)\Delta u - X_w \Delta w + (g\cos(\theta_0))\Delta\theta = X_{\delta_e}\Delta\delta_e \tag{5}$$

$$-Z_u\Delta u + \left[(1 - Z_w)\frac{d}{dt} - Z_w\right]\Delta w - \left[(\mu_0 + Z_q)\frac{d}{dt} - g\sin(\theta_0)\right]\Delta\theta = Z_{\delta_e}\Delta\delta_e \tag{6}$$

$$-M_u\Delta u - \left(M_w\frac{d}{dt} + M_w\right)\Delta w + \left(\frac{d^2}{dt^2} - M_q\frac{d}{dt}\right)\Delta\theta = M_{\delta_e}\Delta\delta_e \tag{7}$$

With Eqs. 5, 6, 7 and parameters of longitudinal stability derivatives, the transfer function changing in the pitch rate with respect to the elevator deflection angle is written as;

$$\frac{\Delta q(s)}{\Delta\delta_e(s)} = \frac{-\left(M_{\delta_e} + \frac{M_{\dot{\alpha}}Z_{\delta_e}}{\mu_0}\right)s - \left(\frac{M_\alpha Z_{\delta_e} - M_{\delta_e}Z_\alpha}{\mu_0}\right)}{s^2 - \left(M_q + M_{\dot{\alpha}} + \frac{Z_\alpha}{\mu_0}\right)s + \left(\frac{Z_\alpha M_q}{\mu_0} - M_\alpha\right)} \tag{8}$$

The transfer function $\Delta\theta/\Delta\delta_e(s)$ can be obtained from Eq. 8 next.

$$\Delta q = \Delta\dot{\theta} \tag{9}$$

$$\Delta q(s) = s\Delta\theta(s) \tag{10}$$

$$\frac{\Delta\theta(s)}{\Delta\delta_e(s)} = \frac{1}{s}\frac{\Delta q(s)}{\Delta\delta_e(s)} \tag{11}$$

thus, the transfer function for pitch control can be written in the following way.

$$\frac{\Delta\theta(s)}{\Delta\delta_e(s)} = \frac{1}{s}\left(\frac{-\left(M_{\delta_e} + \frac{M_{\dot\alpha}Z_{\delta_e}}{\mu_0}\right)s - \left(\frac{M_\alpha Z_{\delta_e} - M_{\delta_e}Z_\alpha}{\mu_0}\right)}{s^2 - \left(M_q + M_{\dot\alpha} + \frac{Z_\alpha}{\mu_0}\right)s + \left(\frac{Z_\alpha M_q}{\mu_0} - M_\alpha\right)}\right) \tag{12}$$

By substituting parameters from Table 1 into transfer function in Eq. 12, following equation can be obtained.

$$\frac{\Delta\theta(s)}{\Delta\delta_e(s)} = \frac{11.7304s + 22.578}{s^3 + 4.9676s^2 + 12.641s} \tag{13}$$

The transfer function in the Laplace domain can be converted into state-space representation below.

$$\begin{bmatrix}\Delta\dot\alpha \\ \Delta\dot q \\ \Delta\dot\theta\end{bmatrix} = \begin{bmatrix}-2.02 & 1 & 0 \\ -6.9868 & -2.9476 & 0 \\ 0 & 1 & 0\end{bmatrix}\begin{bmatrix}\Delta\alpha \\ \Delta q \\ \Delta\theta\end{bmatrix} + \begin{bmatrix}0.16 \\ 11.7304 \\ 0\end{bmatrix}\begin{bmatrix}\Delta\delta_e\end{bmatrix} \tag{14}$$

$$y = \begin{bmatrix}0 & 0 & 1\end{bmatrix}\begin{bmatrix}\Delta\alpha \\ \Delta q \\ \Delta\theta\end{bmatrix} + \begin{bmatrix}0\end{bmatrix} \tag{15}$$

## 3   Classic Controller

The error, which is the difference between the input and the output of closed-loop control, is minimized by PID controller blocks. Acquiring PID parameters can be obtained by automatically tuning or manual tuning. After some trails the values of $K_p$, $K_i$ and $K_d$ are decided to be 28.7421, 11.755 and 10.3955 respectively.

## 4   Fuzzy Logic Controller

The FLC has three basic sub-groups such as fuzzification, fuzzy inference engine (decision logic) and defuzzification stages. The block diagram of the FLC is demonstrated in Fig. 3.

The first block shown in Fig. 3 is fuzzification. It converts input data into fuzzy membership. The rule base and the inference engine perform as human decision-making (Fuzzy Expert), which depends on fuzzy implication and rules from inference. The membership functions (MFs) as well as fuzzy control rules affect control performance substantially. The third block is defuzzification where results from the fuzzy set is defuzzified or converted into a control signal. In

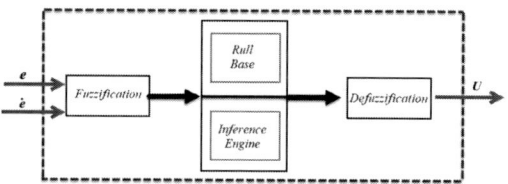

**Fig. 3.** Structure of Fuzzy Logic Controller

the literature, five defuzzification methods exist; centroid, bisector, middle of maximum, largest of maximum, and smallest of maximum [5]. Input signals of the FLC are $e$ (deg) and derivation of output signal, $\dot{y}$ (deg/s), aircraft pitch angle. The values of the $e$, $\dot{y}$ and output are arranged in interval of $[-5\ \ 5]$, $[-7\ \ 7]$ and $[-30\ \ 30]$. These are transformed to the linguistic variables $e$ error and $\dot{y}$ with help of fuzzification. Inputs of the FLC consist of seven linguistic terms PB (Positive Big), PS (Positive Small), PZ (Positive Near Zero), Z (Zero), NB (Negative Big), NS (Negative Small) and NZ (Negative Near Zero) [5]. In other words, the FLC output -the pitch angle- is divided into seven fuzzy set. MFs of pitch angle and derivation of pitch angle are illustrated in Fig. 4 respectively. FLC output and fuzzy control surface are shown in Fig. 5. The triangle membership is used for both input and output and defined in Eq. 16.

$$
f(x, a, b, c) = \begin{cases} 0 & x \leq a \\ \frac{x-a}{b-a} & a \leq x \leq b \\ \frac{c-x}{c-b} & b \leq x \leq c \\ 0 & c \leq x \end{cases} \tag{16}
$$

where a and c locate the "feet" and the parameter b locates the peak of the triangle membership function. Fuzzy logic rules are given in Tables 2 and 3.

The obtained fuzzy set has to be transformed to the signal in such a way that it is sent to the control input. The bisector of the area is used as defuzzification method. The Simulink model of the FLC for the aircraft is shown in Figs. 6 and 7.

**(a)**                                        **(b)**

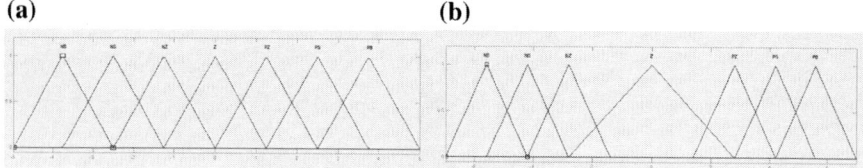

**Fig. 4.** (a) Pitch angle error $e(t)$ and (b) derivative of pitch angle error $\frac{d}{dt}e(t)$

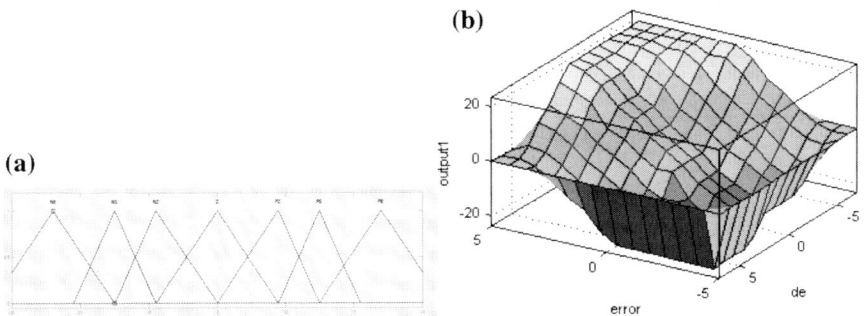

**Fig. 5.** (a) Output signal and (b) fuzzy logic control structure

**Table 2.** Fuzzy logic rule base

| $\dot{y}/e$ | NB | NS | NZ | Z | PZ | PS | PB |
|------|----|----|----|----|----|----|----|
| NB | Z | PZ | PS | PB | PB | PB | PB |
| NS | NZ | Z | PZ | PS | PB | PB | PB |
| NZ | NS | NZ | Z | PZ | PS | PB | PB |
| Z | NB | NS | NZ | Z | PZ | PS | PB |
| PZ | NB | NB | NS | NZ | Z | PZ | PS |
| PS | NB | NB | NB | NS | NZ | Z | PZ |
| PB | NB | NB | NB | NB | NS | NZ | Z |

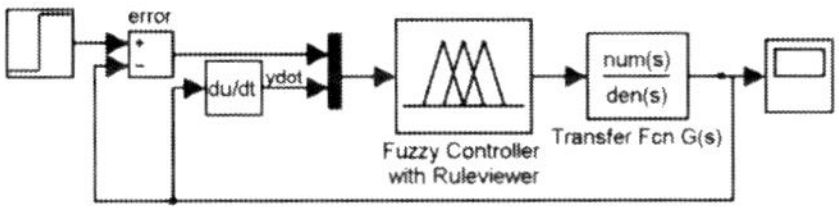

**Fig. 6.** Simulink model of the FLC for pitch control of the aircraft

**Table 3.** Comparision of control performance parameters for pitch angle

| Response characteristics | PID | Fuzzy logic |
|------|------|------|
| Rising time | 0.673 s | 0.579 s |
| Settling time | 3.228 s | 1.187 s |
| Peak time | 1.55 s | 0.73 s |
| Peak value | 1.13 s | 1.1 s |
| Percent overshoot | 12.73 | 7.06 |

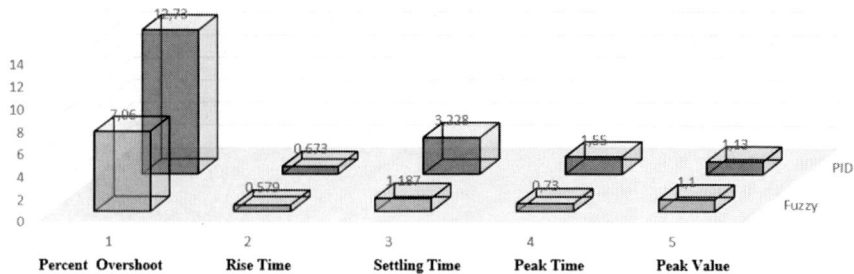

**Fig. 7.** Comparison of control performance parameters for PID and fuzzy logic control

## 5   Conclusion

Modeling which is implemented on an aircraft by using PID and FLC controller to control the given pitch displacement is developed successfully. The proposed control modelings, PID and FLC, are performed in simulation environment in Matlab Simulink. The obtained results from the simulink have been evaluated in terms of time domain specification. Based on the results, the system responses using FLC indicate better performance compared to PID controller. Moreover, implementation of the FLC is much easier than PID methods since it does not require complex mathematical equations.

## References

1. John, S., Rasheed, A.I., Reddy, V.K.: ASIC implementation of fuzzy-PID controller for aircraft roll control. In: 2013 International Conference on Circuits, Controls and Communications (CCUBE), pp. 1–6. IEEE (2013)
2. Nelson, R.C.: Flight Stability and Automatic Control, vol. 2. WCB/McGraw Hill, New York (1998)
3. Promtun, E., Seshagiri, S.: Siding mode control of pitch-rate of an f-16 aircraft. Citeseer (2009)
4. Wahid, N., Hassan, N.: Self-tuning fuzzy PID controller design for aircraft pitch control. In: 2012 Third International Conference on Intelligent Systems, Modelling and Simulation (ISMS), pp. 19–24. IEEE (2012)
5. Zaeri, R., Ghanbarzadeh, A., Attaran, B., Zaeri, Z.: Fuzzy logic controller based pitch control of aircraft tuned with bees algorithm. In: 2011 2nd International Conference on Control, Instrumentation and Automation (ICCIA), pp. 705–710. IEEE (2011)

# Flatness-Based Feedforward Control
# of a Crane Manipulator
# with Four Load Chains

Michael Stoltmann[1], Pascal Froitzheim[2], Normen Fuchs[2],
and Christoph Woernle[1(✉)]

[1] University of Rostock, Rostock, Germany
{michael.stoltmann,woernle}@uni-rostock.de
[2] Fraunhofer Research Institution for Large Structures in Production
Engineering IGP, Rostock, Germany
{pascal.froitzheim,normen.fuchs}@igp.fraunhofer.de

**Abstract.** For a crane manipulator suspending a flexible plate by four chains feedforward control is derived that moves the payload along desired spatial trajectories. Due to the statically indeterminate suspension, the stiffness of the payload has to be taken into account. The actuator coordinates can be algebraically calculated from the desired trajectory of the plate at the position, velocity and acceleration levels exploiting the flatness property of the system. As the system is kinematically redundant, technically meaningful solutions like minimal inclination angles of the chains are determined by optimisation.

**Keywords:** Crane manipulator · Underactuated system
Flatness-based feedforward control · Flexible crane payload

## 1 Introduction

Crane manipulators enable handling large and heavy objects in space by suspending the payload with several cables or chains under computer control. The application standing behind the present investigation is three-dimensional cold forming of steel or aluminium plates by means of a so-called ship building press. For positioning under the press table the plate is suspended by four chain hoists mounted on trolleys of an overhead or gantry crane with two bridges each with two trolleys as schematically shown in Fig. 1. Bridges, trolleys and chain hoists are independently controllable to achieve the desired workpiece position. The four chain attachment points at the plate are flexible in order to isolate the crane gear from process forces.

From a point of view of robotics, a crane manipulator is kinematically indeterminate or underactuated as the position of the payload is not kinematically defined by the position of the actuators, here the bridge and trolley positions and the chain lengths. The objective is therefore to control the actuators in such

© Springer Nature Switzerland AG 2019
B. Corves et al. (Eds.): EuCoMeS 2018, MMS 59, pp. 61–68, 2019.
https://doi.org/10.1007/978-3-319-98020-1_8

a way that the workpiece is moved into desired static equilibrium positions without undesired sway motions. For this purpose the dynamics of the system has to be taken into account. Compared to control of the three-cable crane described in [3] the four-chain suspension of the plate is statically indeterminate [5]. Thus, the stiffness of the chain attachments as well as flexural stiffness of the plate has to be considered. The flatness property of the system is exploited to derive a feedforward control model. A flat system has the property that the state variables and control inputs can be algebraically expressed in terms of the so-called flat outputs and their time derivatives [2].

The paper is organised as follows. In Sect. 2 a dynamic model of the crane manipulator is derived. It is used in Sect. 3 to derive the feedforward control model that incorporates the sway and torsion dynamics of the suspended plate. The feedforward control yields positions of the actuator coordinates in such a way that the platform is moved along desired trajectories in the ideal case of vanishing disturbances. A simulation example is shown in Sect. 4.

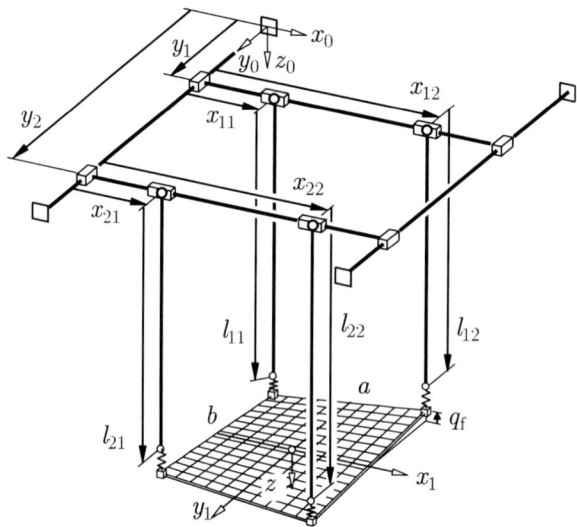

**Fig. 1.** Cable manipulator with four load chains suspending a flexible rectangular plate.

## 2    Dynamics of the Crane Manipulator

The dynamic model of the crane manipulator used for feedforward controller design is formulated in terms of the absolute position coordinates of the flexible plate. The masses of the chains are neglected. The bridge and trolley displacements and the chain lengths are treated as kinematical inputs of the dynamic model.

## 2.1  Coordinates

According to Fig. 1 the independently controllable actuator coordinates are, with $i = 1, 2$ and $j = 1, 2$, the positions of the two bridges $y_i$, the four trolley positions $x_{ij}$ and the lengths of the four chains $l_{ij}$ including the lengths of the unloaded attachment springs. The actuator coordinates are summarised in vector

$$\mathbf{u} = \begin{bmatrix} y_1 & y_2 & x_{11} & x_{12} & x_{21} & x_{22} & l_{11} & l_{12} & l_{21} & l_{22} \end{bmatrix}^{\mathrm{T}}. \tag{1}$$

The spatial position of the flexible plate relative to $K_0$ is described by the six position coordinates of the body frame $K_1$, comprising the three Cartesian coordinates $\mathbf{r}_1 = \begin{bmatrix} r_{1x} & r_{1y} & r_{1z} \end{bmatrix}^{\mathrm{T}}$ of $O_1$, three CARDAN angles $\boldsymbol{\varphi} = \begin{bmatrix} \varphi_1 & \varphi_2 & \varphi_3 \end{bmatrix}^{\mathrm{T}}$ defined in the rotation sequence around the follower axes $z_1$–$y_1$–$x_1$, and a coordinate to describe the elastic deformation of the plate $q_\mathrm{f}$ defined in the following subsection. The position coordinates are summarised in the 7-vector

$$\hat{\mathbf{r}} = \begin{bmatrix} \mathbf{r}_1^{\mathrm{T}} & \boldsymbol{\varphi}^{\mathrm{T}} & q_\mathrm{f} \end{bmatrix}^{\mathrm{T}}. \tag{2}$$

To describe the spatial velocity of the flexible plate relative to $K_0$, the velocity $\mathbf{v}_1 = \dot{\mathbf{r}}_1$ of $O_1$, the angular velocity vector $\boldsymbol{\omega}$ of $K_1$, and the time derivative of $q_\mathrm{f}$ are used, summarised in the 7-vector

$$\hat{\mathbf{v}} = \begin{bmatrix} \mathbf{v}_1^{\mathrm{T}} & \boldsymbol{\omega}^{\mathrm{T}} & \dot{q}_\mathrm{f} \end{bmatrix}^{\mathrm{T}}. \tag{3}$$

The relation between the angular velocity $\boldsymbol{\omega}$ and the time derivatives of the CARDAN angles $\boldsymbol{\varphi}$ is given by a kinematical differential equation of the form [6]

$$\dot{\boldsymbol{\varphi}} = \mathbf{H}_\varphi(\boldsymbol{\varphi})\,\boldsymbol{\omega}. \tag{4}$$

## 2.2  Model of the Flexible Plate

For controller design a simple model of the flexible plate with one elastic degree of freedom describing only the torsional deflection of the plate is used. Compared to other occuring deflection modes, the torsional mode is predominant for the suspension chain force distribution.

### 2.2.1  Kinematics of the Flexible Plate

According to Fig. 2 a rectangular KIRCHHOFF plate (edge lengths $a$, $b$, thickness $h$ and density $\rho$ constant over area $A$) with the origin $O_1$ of the body coordinate system $K_1$ located in the mass center is considered. Assuming small deflections, the torsional deflection of the plate is approximately described by the displacement field

$$w(x, y, t) = \phi(x, y)\, q_\mathrm{f}(t) \quad \text{with} \quad \phi(x, y) = 4\,\frac{x}{a}\frac{y}{b} \tag{5}$$

with the shape function $\phi(x, y)$ describing the displacement field in $z_1$-direction for torsional deflection [1] and the elastic coordinate $q_\mathrm{f}$ corresponding to the deflections of the chain attachment points relative to $K_1$,

$$q_\mathrm{f}(t) = w(\tfrac{a}{2}, \tfrac{b}{2}, t) \equiv w(-\tfrac{a}{2}, -\tfrac{b}{2}, t) \equiv w(-\tfrac{a}{2}, \tfrac{b}{2}, t) \equiv w(\tfrac{a}{2}, -\tfrac{b}{2}, t). \tag{6}$$

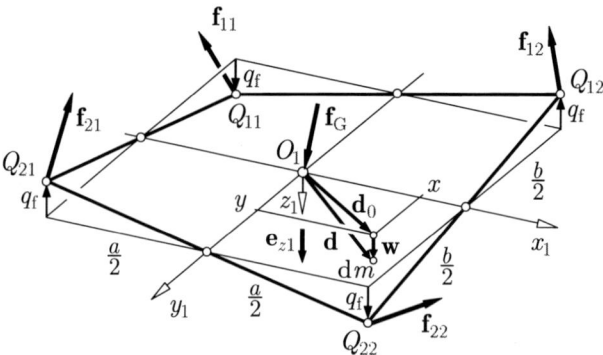

**Fig. 2.** Model of the flexible plate.

According to the floating frame of reference formulation in multibody dynamics [4], the spatial position $\mathbf{r}$ of a mass element $dm \equiv \rho\, h\, dA$ (density $\rho$, area $A$) with coordinates $x$, $y$ in $K_1$ is described by the sum of the position vector $\mathbf{r}_1$ of $O_1$, the vector $\mathbf{d}_0$ from $O_1$ to $dm$ in the undeformed position and the deflection vector

$$\mathbf{w}(x,y,t) = w(x,y,t)\,\mathbf{e}_{z1} = \phi(x,y)\,q_{\mathrm{f}}(t)\,\mathbf{e}_{z1}, \tag{7}$$

with the unit vector $\mathbf{e}_{z1}$ of the $z_1$-axis, thus

$$\mathbf{r} = \mathbf{r}_1 + \mathbf{d} \quad \text{with} \quad \mathbf{d} = \mathbf{d}_0 + \phi\, q_{\mathrm{f}}\,\mathbf{e}_{z1}. \tag{8}$$

Using the tensor representation of a vector product $\mathbf{a} \times \mathbf{b} \equiv \tilde{\mathbf{a}}\,\mathbf{b}$ with the skew-symmetric tensor $\tilde{\mathbf{a}} = -\tilde{\mathbf{a}}^{\mathrm{T}}$, the time derivatives of vectors $\mathbf{d}_0$ and $\mathbf{e}_{z1}$ expressed by the angular velocity $\omega$ are $\dot{\mathbf{d}}_0 = \tilde{\omega}\,\mathbf{d}_0 \equiv -\tilde{\mathbf{d}}_0\,\omega$ and $\dot{\mathbf{e}}_{z1} = \tilde{\omega}\,\mathbf{e}_{z1} \equiv -\tilde{\mathbf{e}}_{z1}\,\omega$. The velocity of $dm$ with respect to $K_0$ then is

$$\dot{\mathbf{r}} = \dot{\mathbf{r}}_1 + \dot{\mathbf{d}}_0 + \phi\, q_{\mathrm{f}}\,\dot{\mathbf{e}}_{z1} + \phi\,\dot{q}_{\mathrm{f}}\,\mathbf{e}_{z1} \quad \text{or} \quad \dot{\mathbf{r}} = \underbrace{\left[\, \mathbf{E} \quad \tilde{\mathbf{d}}^{\mathrm{T}} \quad \phi\,\mathbf{e}_{z1} \,\right]}_{\mathbf{J}(\hat{\mathbf{r}})}\underbrace{\begin{bmatrix} \mathbf{v} \\ \omega \\ \dot{q}_{\mathrm{f}} \end{bmatrix}}_{\hat{\mathbf{v}}}, \tag{9}$$

where $\mathbf{E}$ is the identity matrix and $\mathbf{J}(\hat{\mathbf{r}})$ is the JACOBI matrix of point $O_1$. The acceleration of $dm$ with respect to $K_0$ is

$$\ddot{\mathbf{r}} = \mathbf{J}(\hat{\mathbf{r}})\,\dot{\hat{\mathbf{v}}} + \bar{\mathbf{a}}(\hat{\mathbf{r}},\hat{\mathbf{v}}) \quad \text{with} \quad \bar{\mathbf{a}} = \dot{\mathbf{J}}\,\hat{\mathbf{v}}. \tag{10}$$

### 2.2.2  Equations of Motion of the Flexible Plate

The equations of motion of the plate can be written in the form

$$\mathbf{M}\,\dot{\hat{\mathbf{v}}} = \mathbf{k}^{\mathrm{c}} + \mathbf{k}^{\mathrm{a}} + \mathbf{k}^{\mathrm{P}} \tag{11}$$

with the mass matrix $\mathbf{M}$, the generalised centrifugal and CORIOLIS forces $\mathbf{k}^{\mathrm{c}}$, the generalised applied forces $\mathbf{k}^{\mathrm{a}}$ and the generalised potential forces $\mathbf{k}^{\mathrm{P}}$.

The mass matrix is calculated by

$$\mathbf{M} = \rho\,h \int_A \mathbf{J}(\hat{\mathbf{r}})^{\mathrm{T}}\,\mathbf{J}(\hat{\mathbf{r}})\,\mathrm{d}A = \rho\,h \int_A \begin{bmatrix} \mathbf{E} & \tilde{\mathbf{d}}^{\mathrm{T}} & \phi\,\mathbf{e}_{z1} \\ & \tilde{\mathbf{d}}\,\tilde{\mathbf{d}}^{\mathrm{T}} & \phi\,\tilde{\mathbf{d}}\,\mathbf{e}_{z1} \\ \mathrm{sym.} & & \phi^2 \end{bmatrix} \mathrm{d}A. \qquad (12)$$

Evaluation of the area integrals in (12) under consideration of the shape function $\phi(x,y)$ in (5) yields

$$\mathbf{M} = \begin{bmatrix} m\,\mathbf{E} & \mathbf{0} & \mathbf{0} \\ \mathbf{0} & \boldsymbol{\Theta}_1 & \mathbf{0} \\ \mathbf{0} & \mathbf{0} & m_{\mathrm{ff}} \end{bmatrix} \qquad (13)$$

with the overall mass of the plate $m$, the inertia tensor $\boldsymbol{\Theta}_1$ with respect to $O_1$, the generalised mass of the flexible coordinate

$$m_{\mathrm{ff}} = \rho\,h \int_{-\frac{a}{2}}^{\frac{a}{2}} \int_{-\frac{b}{2}}^{\frac{b}{2}} \phi^2(x,y)\,\mathrm{d}y\,\mathrm{d}x = \tfrac{1}{9}\,m \qquad (14)$$

and vanishing off-diagonal coupling terms. Neglecting quadratic terms in the thickness $h$ and the flexible coordinate $q_{\mathrm{f}}$, the inertia tensor of the undeformed plate $\boldsymbol{\Theta}_1 = \mathrm{diag}\left(\tfrac{1}{12}mb^2,\ \tfrac{1}{12}ma^2,\ \tfrac{1}{12}m(a^2+b^2)\right)$, represented in $K_1$, is obtained.

The vector of generalised centrifugal and CORIOLIS forces $\mathbf{k}^{\mathrm{c}}$ is calculated by

$$\mathbf{k}^{\mathrm{c}} = \rho\,h \int_A \mathbf{J}(\hat{\mathbf{r}})^{\mathrm{T}}\,\bar{\mathbf{a}}(\hat{\mathbf{r}},\hat{\mathbf{v}})\,\mathrm{d}A. \qquad (15)$$

Neclecting coupling terms between rigid body and elastic coordinates, the gyro torque of the plate in the undeformed position remains,

$$\mathbf{k}^{\mathrm{c}} = \begin{bmatrix} \mathbf{0} \\ -\tilde{\boldsymbol{\omega}}\,\boldsymbol{\Theta}_1\,\boldsymbol{\omega} \\ 0 \end{bmatrix}. \qquad (16)$$

The vector of generalised applied forces $\mathbf{k}^{\mathrm{a}}$ is obtained from the gravity force $\mathbf{f}_{\mathrm{G}} = m\,g\,\mathbf{e}_{z0}$ at the mass center $O_1$ and the four suspension forces $\mathbf{f}_{ij}$ at the attachment points $Q_{ij}$ for $i = 1,2$ and $j = 1,2$. With the position vectors

$$\mathbf{r}_{Pij} = y_i\,\mathbf{e}_{y0} + x_{ij}\,\mathbf{e}_{x0}, \qquad \mathbf{r}_{Qij} = \mathbf{r}_1 + \mathbf{d}_{ij} \qquad (17)$$

of points $P_{ij}$ and $Q_{ij}$, see Fig. 3, and under consideration of linear attachment springs (stiffness $c$, elongations $s_{ij}$), the suspension forces are

$$\mathbf{f}_{ij} = c\,s_{ij}\,\mathbf{e}_{ij} \quad \text{with} \quad \mathbf{e}_{ij} = \frac{\mathbf{r}_{Pij} - \mathbf{r}_{Qij}}{|\mathbf{r}_{Pij} - \mathbf{r}_{Qij}|}, \quad s_{ij} = |\mathbf{r}_{Pij} - \mathbf{r}_{Qij}| - l_{ij}. \qquad (18)$$

With the JACOBI matrices $\mathbf{J}_1$ of point $O_1$ and the JACOBI matrices $\mathbf{J}_{ij}$ of points $Q_{ij}$ defined in analogy to (9), the generalised applied forces are calculated by

$$\mathbf{k}^{\mathrm{a}} = \mathbf{J}_1^{\mathrm{T}}\,\mathbf{f}_{\mathrm{G}} + \sum_{i=1}^{2}\sum_{j=1}^{2} \mathbf{J}_{ij}^{\mathrm{T}}\,\mathbf{f}_{ij} \quad \text{with} \quad \mathbf{J}_1 = \begin{bmatrix} \mathbf{E} & \mathbf{0} & \mathbf{0} \end{bmatrix},\ \mathbf{J}_{ij} = \begin{bmatrix} \mathbf{E} & \tilde{\mathbf{d}}_{ij}^{\mathrm{T}} & \phi_{ij}\mathbf{e}_{z1} \end{bmatrix}$$

$$(19)$$

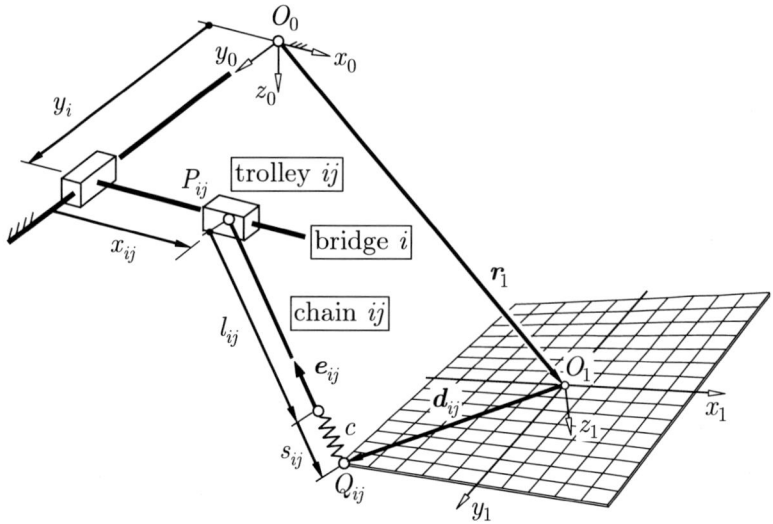

**Fig. 3.** Geometrical relations for calculation of the suspension forces.

($\phi_{ij} = 1$ for $i = j$ and $\phi_{ij} = -1$ for $i \neq j$) yielding

$$\mathbf{k}^{\mathrm{a}} = \begin{bmatrix} \mathbf{f}_{\mathrm{G}} + \mathbf{f}_{11} + \mathbf{f}_{12} + \mathbf{f}_{21} + \mathbf{f}_{22} \\ \tilde{\mathbf{d}}_{11}\,\mathbf{f}_{11} + \tilde{\mathbf{d}}_{12}\,\mathbf{f}_{12} + \tilde{\mathbf{d}}_{21}\,\mathbf{f}_{21} + \tilde{\mathbf{d}}_{22}\,\mathbf{f}_{22} \\ \mathbf{e}_{z1}^{\mathrm{T}}\,(\mathbf{f}_{11} - \mathbf{f}_{12} - \mathbf{f}_{21} + \mathbf{f}_{22}) \end{bmatrix}. \tag{20}$$

The generalised potential forces $\mathbf{k}^{\mathrm{P}}$ are related to the strain energy of the plate [1]

$$U = K \int_A \left[ \tfrac{1}{2}(w_{,xx}^2 + w_{,yy}^2)^2 - (1 - \nu)(w_{,xx}\,w_{,yy} - w_{,xy}^2) \right] \mathrm{d}A \tag{21}$$

with the plate bending stiffness (YOUNG's modulus $E$ and POISSON's ratio $\nu$)

$$K = \frac{E\,h^3}{12\,(1 - \nu^2)} \tag{22}$$

and the partial derivatives of the elastic displacements $w(x, y, t)$ from (5) with respect to $x$ and $y$, for example, $w_{,xy} = \frac{\partial^2 w}{\partial x\,\partial y} = \frac{4}{a\,b}\,q_{\mathrm{f}}$. With $w_{,xx} = w_{,yy} = 0$ the strain energy becomes

$$U(q_{\mathrm{f}}) = \frac{16\,K\,(1 - \nu)}{a\,b}\,q_{\mathrm{f}}^2. \tag{23}$$

The generalised potential forces are then obtained as

$$\mathbf{k}^{\mathrm{P}} = \begin{bmatrix} \mathbf{0} \\ \mathbf{0} \\ k_{\mathrm{f}} \end{bmatrix} \quad \text{with} \quad k_{\mathrm{f}} = -\frac{\partial U(q_{\mathrm{f}})}{\partial q_{\mathrm{f}}} = -\frac{32\,K\,(1 - \nu)}{a\,b}\,q_{\mathrm{f}}. \tag{24}$$

## 3    Flatness-Based Feedforward Control

To move the platform along a desired trajectory of the plate, given in terms of position $\hat{\mathbf{r}}(t)$, velocity $\hat{\mathbf{v}}(t)$ and acceleration $\dot{\hat{\mathbf{v}}}(t)$, the actuator positions $\mathbf{u}(t)$ according to (1) can be algebraically calculated from the equations of motion (11). By this, the system is flat [2]. The equations of motion (11) represent a system of seven nonlinear equations for the ten actuator positions $\mathbf{u}$,

$$\boldsymbol{\Gamma}(\mathbf{u}, \hat{\mathbf{r}}, \hat{\mathbf{v}}, \dot{\hat{\mathbf{v}}}) \equiv \mathbf{M}\dot{\hat{\mathbf{v}}} - \mathbf{k}^{\mathrm{c}} - \mathbf{k}^{\mathrm{a}} - \mathbf{k}^{\mathrm{P}} = \mathbf{0}. \tag{25}$$

Thus the problem is kinematically triple redundant. Appropriate solutions can be found by optimisation. A technically meaningful objective criterion are minimal chain lengths $l_{ij}$. This leads to solutions where the inclination angles of the four suspension chains are kept as small as possible during a platform motion. The constrained optimisation problem then is

$$Z(\mathbf{u}) \equiv \sum_{i=1}^{2} \sum_{j=1}^{2} l_{ij}^{2} = \min_{\mathbf{u}} \quad \text{with} \quad \boldsymbol{\Gamma}(\mathbf{u}, \hat{\mathbf{r}}, \hat{\mathbf{v}}, \dot{\hat{\mathbf{v}}}) = \mathbf{0}. \tag{26}$$

Solutions are obtained numerically by means of the Matlab function `fmincon`. The gradients can be analytically calculated by linearising the equations of motion (25). The linearised equations of motion additionally enable to get the actuator velocities $\dot{\mathbf{u}}$ in order to use $\mathbf{u}$ and $\dot{\mathbf{u}}$ as inputs for individual actuator position controllers.

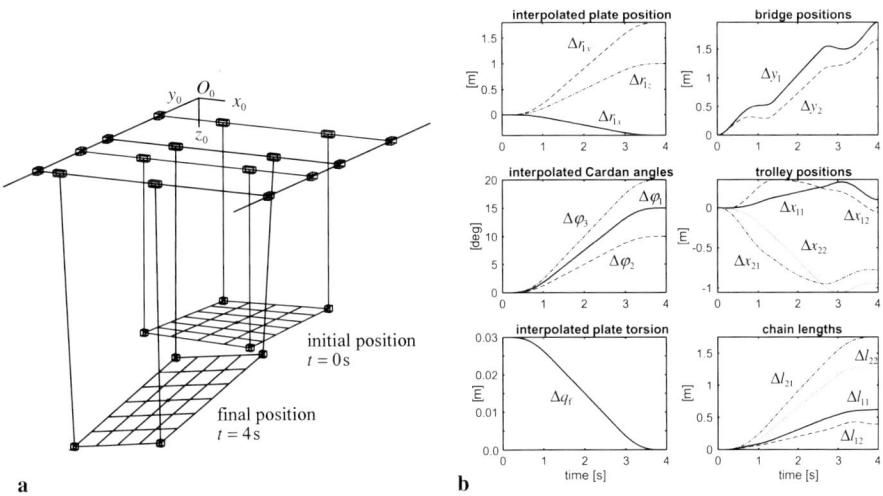

**Fig. 4.** Motion between rest positions. **a** Prescribed positions. **b** Time trajectories of the interpolated position variables $\Delta\hat{\mathbf{r}}(t) = \hat{\mathbf{r}}(t) - \hat{\mathbf{r}}(0)$ and of the actuator coordinates $\Delta\mathbf{u}(t) = \mathbf{u}(t) - \mathbf{u}(0)$.

## 4   Numerical Example

According to Fig. 4a a plate with $a = 2\,\mathrm{m}$, $b = 3\,\mathrm{m}$, $h = 0.01\,\mathrm{m}$ is moved from the initial position at $t = 0\,\mathrm{s}$ with $\mathbf{r}_1(0) = [2.2\ 3.0\ 3.0]^{\mathrm{T}}\,\mathrm{m}$, $\boldsymbol{\varphi}(0) = [0\ 0\ 0]^{\mathrm{T}}\,\mathrm{deg}$ and $q_{\mathrm{f}}(0) = 0.03\,\mathrm{m}$ into the final position at $T = 4\,\mathrm{s}$ with $\mathbf{r}_1(T) = [1.8\ 4.8\ 4.0]^{\mathrm{T}}\,\mathrm{m}$, $\boldsymbol{\varphi}(T) = [15\ 10\ 20]^{\mathrm{T}}\,\mathrm{deg}$ and $q_{\mathrm{f}}(T) = 0\,\mathrm{m}$. Physical parameters are $\rho = 7800\,\mathrm{kg\,m^{-3}}$, $E = 2.1 \cdot 10^{11}\,\mathrm{N\,m^{-2}}$, $\nu = 0.3$ and $c = 1.5 \cdot 10^4\,\mathrm{N\,m^{-1}}$. The seven position variables of the plate $\hat{\mathbf{r}}(t)$ according to (2) are interpolated over the time interval $[0, T]$ by functions being steady up to the fourth-order time derivative as shown in Fig. 4b together with the calculated trajectories of the ten actuator coordinates $\mathbf{u}(t)$ according to (1) and Fig. 1.

## 5   Conclusion and Outlook

The described feedforward control moves the flexible plate between rest positions along given trajectories of the position coordinates in the idealised disturbance-free case. While feedforward control is intended to contribute the major part of the control signals, model incertainties and external disturbances will have to be compensated by an additional feedback controller that actively dampens horizontal sway motions of the plate. This will be done by an additional linear feedback control using the bridge and trolley coordinates as control inputs similar to the procedure described in [5]. A prototype system being under construction will be used for experimental validations.

**Acknowledgements.** This research and development project was supported by the European Regional Development Fund (EFRE). Support was also provided by the lead partner Technologie-Beratungsinstitut (TBI) according to the directive for support, development and innovation of the Ministry of Economics, Construction and Tourism of Mecklenburg-Vorpommern.

## References

1. Bhaskar, K., Varadan, T.: Plates. Wiley, Chicester (2014)
2. Fliess, M., Lévine, J., Martin, P., Rouchon, P.: Flatness and defect of nonlinear systems: introductory theory and examples. Int. J. Control **51**, 1327–1361 (1995)
3. Heyden, T., Woernle, C.: Dynamics and flatness-based control of a kinematically undetermined cable suspension manipulator. Multibody Syst. Dyn. **16**, 155–177 (2006)
4. Shabana, A.A.: Dynamics of Multibody Systems, 4th edn. Cambridge University Press, Cambridge (2013)
5. Woernle, C.: Trajectory tracking for a three-cable suspension manipulator. In: Bruckmann, T., Pott, A. (eds.) Cable-Driven Parallel Robots, pp. 371–386. Springer, Berlin (2013)
6. Woernle, C.: Mehrkörpersysteme. Springer, Berlin (2016)

# Dynamics of Multi-body Systems

# Approach for Conformal Contact Detection for Wheel-Rail Interaction

Filipe Marques[1]([✉]), Hugo Magalhães[2], Jorge Ambrósio[2], and Paulo Flores[1]

[1] MIT-Portugal Program, CMEMS-UMinho,
Universidade do Minho, Guimarães, Portugal
{fmarques, pflores}@dem.uminho.pt
[2] LAETA, IDMEC, Instituto Superior Técnico,
Universidade de Lisboa, Lisbon, Portugal
{hugomagalhaes, jorge.ambrosio}@tecnico.ulisboa.pt

**Abstract.** No matter which methodology is used for the computational modelling and analysis, the wheel-rail interaction plays a fundamental role on the dynamic response of railway vehicles. For that, fast and accurate evaluation of the contact interaction is demanded. Bearing that in mind, realistic contact conditions must be taken into account to replicate as detailed as possible this interaction, namely in what concerns with the consideration of actual wheel and rail profiles. Often, parametric surfaces are used to describe their geometry; however, it is shown that the search for potential contact points may become troublesome when the contacting surfaces are conformal. In this work, a methodology to deal with contact detection between general wheel and rail profiles with conformal contact scenarios is presented. This method consists of the division of the wheel into strips together with a search approach to detect the contact between each strip and the rail surface. The static interaction between UIC54 rail and a wheel profiles is used as case study to demonstrate the effectiveness of the methodology described here. The results obtained show that the proposed approach is able to properly describe the contact zone and calculate the penetration along the patch.

**Keywords:** Contact detection · Wheel-rail interaction · Conformal contact Multibody dynamics · Strips · Parametric surfaces · Geometry

## 1 Introduction

The utilization of multibody systems methodologies to study the dynamics of railway vehicles has become a reliable tool for a wide variety of works, such as design of components, prevention of derailments, study of the vehicle's performance for a certain track, accident reconstruction, among others [8]. Having this issue in mind, the vehicle-track interaction plays a key role since it constrains the vehicle's motion and can be used on the study and prediction of damaging phenomena, namely wear and rolling contact fatigue. Therefore, it is of paramount importance to utilize accurate and efficient methodologies to deal with contact interaction between the wheel and the rail [10].

© Springer Nature Switzerland AG 2019
B. Corves et al. (Eds.): EuCoMeS 2018, MMS 59, pp. 71–78, 2019.
https://doi.org/10.1007/978-3-319-98020-1_9

The resolution of contact or impact problems between two or more bodies is quite frequent in the context of a multibody system dynamics. The accurate modelling of these events is of paramount importance, since the implementation of different methodologies has a significant impact on the obtained dynamic response of the system. During the resolution of the contact problems, two main phases must be considered. Firstly, the contact detection between both bodies needs to be efficiently performed, and, then, the contact forces must be accurately evaluated. The contact detection phase can be a very straightforward task and solved analytically if the colliding bodies have simple geometries. However, in the most general case the contacting surfaces can have an arbitrary shape, the search for the potential contact points is a quite demanding and complex task. Broadly, the contact evaluation methodologies can be divided into polygonal and non-polygonal models [5]. In turn, the former group can be classified in structured and Polygon soups, while the latter comprises constructive solid geometry, implicit surfaces and parametric surfaces.

Subsequently to the determination of the effective or potential contact points, the contact forces must be calculated, either on the normal and tangential directions. Although this task is not within the scope of the present work, the adopted contact detection methodology must be selected according to the contact model and bodies' formulation used, namely whether the bodies are flexible or rigid or penetration is allowed. In this work, an elastic approach is considered, therefore, it is fundamental not only to ensure an accurate contact detection in terms of spatial location, but also the precise time instant of the beginning of the contact to avoid unrealistic impulses on the colliding bodies [3].

Regarding the wheel-rail contact, in most of the cases, the contacting surfaces have a non-conformal geometry, which simplifies the search for contact points [12]. However, when the railway vehicle negotiates sharp curves or due to worn profiles, the wheel and rail geometries tend to interact in a conformal way. In this case, the uniqueness of solution is not ensured when searching for a local maximum of the contact penetration between the two surfaces [9].

Thus, the goal of this work is to analyze the problem of the conformal contact detection in the context of wheel-rail interaction, and to provide a reliable alternative to perform this task using parametric surfaces. The remaining of this paper is organized as follows. Section 2 describes the problem of searching for local maximum in the contact when dealing with conformal surfaces. An alternative methodology to handle this problem is described in Sect. 3. In Sect. 4, numerical cases are tested for this contact search algorithm. Finally, some concluding remarks are drawn in Sect. 5.

## 2   Problem Statement

The contact detection is crucial for the correctness of multibody dynamics analysis since it has an extreme impact on the evaluation of the contact forces. Regarding the definition of wheel and rail geometries, different methodologies can be employed. In the context of this work, two parametric surfaces are considered, which is a common approach to solve contact problems on a multibody dynamics framework [6, 12]. Both rail and wheel elements are defined by two parameters, as can be depicted in Fig. 1(a). The rail surface is given by the swept of the rail profile along its centerline, hence, $u_r$ is the lateral position on the rail, and $s_r$ is the position along the track. The wheel surface is described through the revolution of the wheel profile through its axis, thus, $u_w$ represents the lateral position on the wheel, and $s_w$ is its angular location. In order to find the potential contact points on both bodies, the minimum distance between surfaces is considered [4, 6, 12], which geometrically consists of ensuring the parallelism between the distance vector between two points, $\mathbf{d}$, and the normal vectors to the surfaces on those points (see Fig. 1(b)). These conditions can be mathematically described by a system of four nonlinear equations as follows

$$
\begin{aligned}
\mathbf{d} \times \mathbf{n}_j = \mathbf{0} &\equiv \begin{cases} \mathbf{d}^T \cdot \mathbf{t}_j = 0 \\ \mathbf{d}^T \cdot \mathbf{b}_j = 0 \end{cases} \\
\mathbf{n}_i \times \mathbf{n}_j = \mathbf{0} &\equiv \begin{cases} \mathbf{n}_i^T \cdot \mathbf{t}_j = 0 \\ \mathbf{n}_i^T \cdot \mathbf{b}_j = 0 \end{cases}
\end{aligned}
\tag{1}
$$

This system of equations can be solved by an iterative method, such as Newton-Raphson, in order to obtain the set of parameters, $s_r$, $u_r$, $s_w$ and $u_w$, that satisfies the minimum distance condition. In the presence of two convex surfaces and with a suitable initial approximation, this methodology has proven to be effective [12]. However, as it was aforementioned, the wheel and rail can interact in a conformal region, where one of the surfaces has negative curvature. In this case, the solution of the geometric problem given by (1) may not be possible. To better understand this aspect, a 2D scenario is considered, where the problem is simplified since the rail and wheel geometry only depend on $u_r$ and $u_w$, and the system of Eq. (1) can be reduced to two equations. The wheel profile of Light Rail Vehicle (LRV) [7] and the UIC54 rail profile are utilized in this demonstration. Figure 2 shows two tested cases, the top of the rail contacting the wheel tread (non-conformal) and the rail corner contacting the tread/flange concordance zone (conformal). For all the combinations of the curves' parameters ($u_r$ and $u_w$), the contact equations are evaluated, and the norm of the residual is calculated (see Fig. 2(e)–(f)). A null residual leads to the solution of the system of equations. From the analysis of Fig. 2, it is possible to verify that, for the non-conformal case, only one minimum exists (approximately zero) and it is well-defined. Moreover, for the conformal case, several combinations of the wheel and rail parameters spread through the domain are close to satisfy the equations although it is not possible to converge for an optimal solution. Therefore, alternative methodologies are required to solve the contact detection between wheel and rail surfaces for conformal situations.

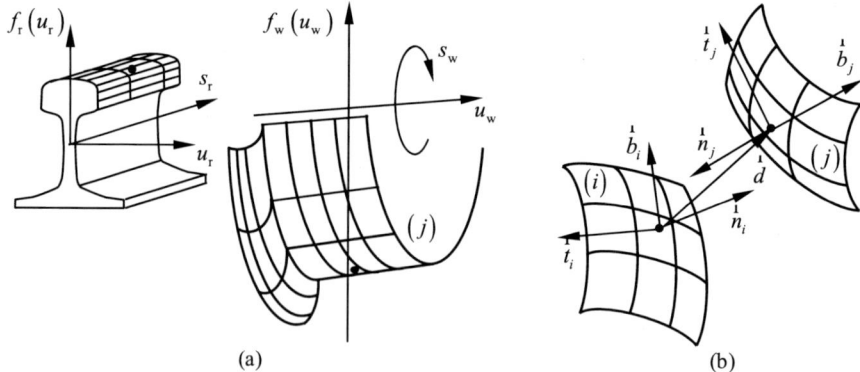

**Fig. 1.** (a) Parametrization of rail and wheel surfaces and (b) contact evaluation for two points of arbitrary parametric surfaces

## 3  Contact Detection for Conformal Wheel-Rail Profiles

In this section, a new approach to deal with contact detection between general wheel and rail profiles with conformal contact scenarios is described. The base of this methodology consists of taking advantage of the fact that the wheel is a solid of revolution. Thus, the wheel profile can be divided into strips or slices [1, 2], as it is represented in Fig. 3. In this way, the contact is not sought between the rail and wheel surfaces, but between the rail surface and each strip of the wheel instead. Bearing that in mind, and since $u_w$ is fixed for each strip, Eq. (1) must be replaced by a new system with only 3 equations with following form

$$\begin{cases} \mathbf{n}_i^T \cdot \mathbf{t}_{j,s} = 0 \\ \mathbf{d}_s^T \cdot \mathbf{t}_i = 0 \\ \mathbf{d}_s^T \cdot \mathbf{b}_i = 0 \end{cases} \tag{2}$$

Since the rail has a convex surface, there is a pair of points that is the solution of this nonlinear system of equations and satisfies the minimum distance condition between a circumference (wheel strip) and a surface (rail). This represents the penetration or distance between the profiles for each strip in the normal direction to the rail, as it can be depicted in Fig. 3(b).

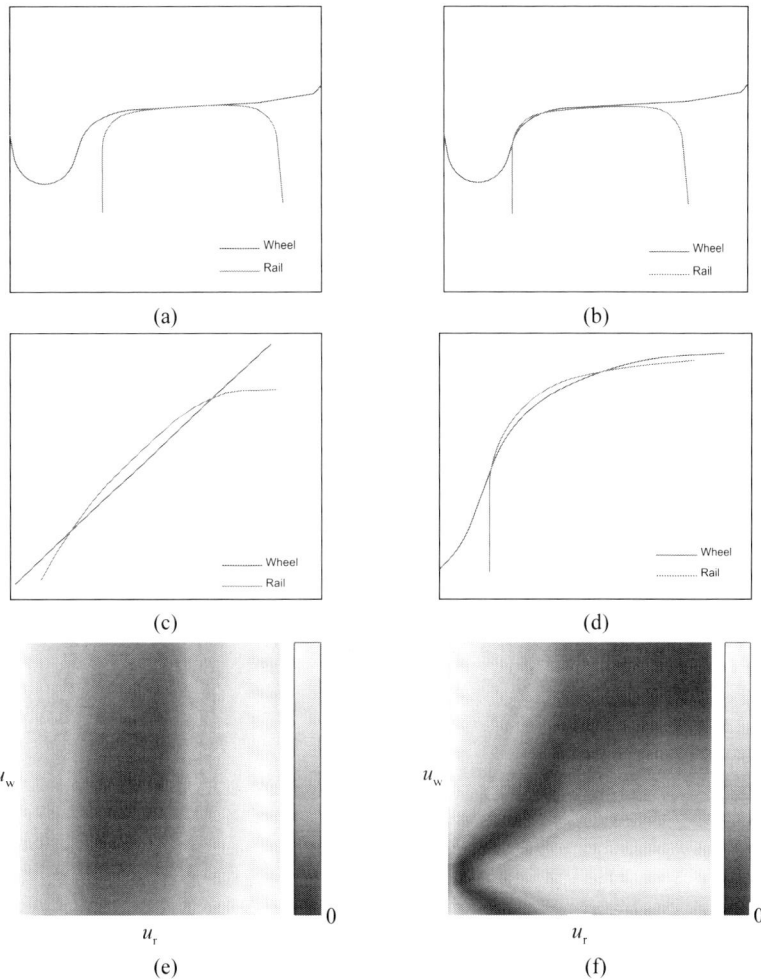

**Fig. 2.** Analysis of the existence of solution for the minimum distance problem for a non-conformal and conformal 2D cases: (a)-(b) wheel-rail relative position, (c)-(d) profiles relative position on the contact zone (distorted), and (e)-(f) norm of the residual of contact equations

## 4   Case Study

A static case of interference between the previous profiles is considered as example of application, (see Fig. 4). This interference state does not represent a real scenario since the wheel-rail contact area is typically smaller [10], however, to keep the visualization of results simple, this case study has been selected to assess the performance of the contact detection approach presented here.

Since the main goal of this study is to identify the wheel-rail contact accurately, the number of strips that represent the wheel surface has been analyzed. Figure 4 shows the case in which 10 strips are considered for sake of simplicity. Here, lines with circle

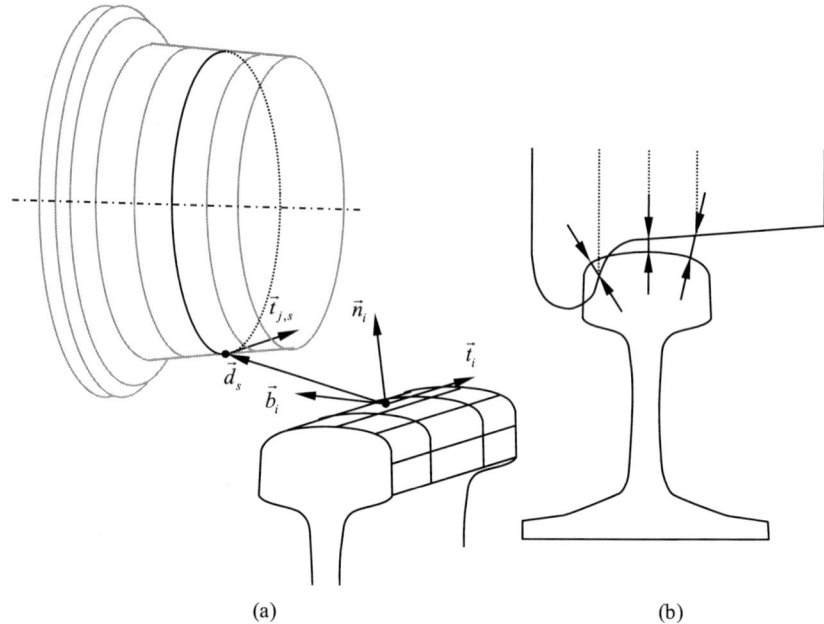

(a)                                          (b)

**Fig. 3.** (a) Representation of the wheel divided into strips and contact evaluation between two points; (b) Schematic representation of the resultant candidate points.

markers indicate points with no contact, while lines with asterisk markers indicate points in contact. It must be noted that these lines are perpendicular to the rail surface as expected. From this result, it is observed that more strips lead to a significant improvement of the contact detection quality. A convergence analysis has been performed being doubled consecutively the number of strips, starting from 10 up to 1280, and, hence, the strips' width has been reduced from 8 mm up to 0.0625 mm, respectively. Table 1 lists these cases, together with the percentage of strips in contact, the penetration of the strip at higher interference, $\|\mathbf{d}_s\|_{max}$, and the respective location of the contact points in the wheel and rail, that is, $u_w(\|\mathbf{d}_s\|_{max})$ and $u_r(\|\mathbf{d}_s\|_{max})$, respectively. From this analysis, it can be observed that the percentage of strips in contact, as well as the maximum penetration, tend to stabilize with the decreasing of the width of the strips. It must be highlighted also that, the location of the point of higher penetration, also called main point of contact, in the wheel and rail varied almost 3 mm, namely from the coarser to the finer mesh. An equally spaced division of the wheel strips was done, although it is important to mention that the refinement is only necessary in the contact zone and its neighborhood. Moreover, it must be noticed that main point of contact, satisfies Eq. (1) of the minimum distance between surfaces.

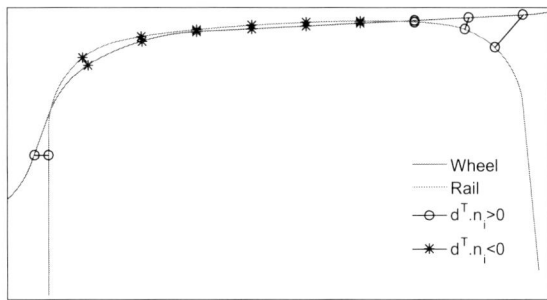

**Fig. 4.** Contact detection between the wheel and rail considering 10 strips

**Table 1.** Impact of number of strips on the accuracy of the contact detection

| Number of strips | Strip width [mm] | Contact strips [%] | $\|\mathbf{d}_s\|_{max}$ [mm] | $u_w(\|\mathbf{d}_s\|_{max})$ [mm] | $u_r(\|\mathbf{d}_s\|_{max})$ [mm] |
|---|---|---|---|---|---|
| 10 | 8.0000 | 70.0 | 1.058 | −24.9 | −23.6 |
| 20 | 4.0000 | 65.0 | 1.334 | −26.9 | −25.9 |
| 40 | 2.0000 | 62.5 | 1.379 | −27.9 | −27.0 |
| 80 | 1.0000 | 65.0 | 1.375 | −28.4 | −27.6 |
| 160 | 0.5000 | 64.4 | 1.379 | −28.2 | −27.3 |
| 320 | 0.2500 | 64.7 | 1.380 | −28.1 | −27.2 |
| 640 | 0.1250 | 64.4 | 1.380 | −28.1 | −27.3 |
| 1280 | 0.0625 | 64.4 | 1.380 | −28.1 | −27.2 |

## 5  Concluding Remarks

A contact detection methodology to model conformal contact in wheel-rail contact studies has been described in this work. The difficulties associated with the identification of the wheel-rail contact with a traditional approach have been investigated, and a modified strategy to deal with contact detection between conformal surfaces has been discussed. The proposed methodology considers the wheel as a set of circular strips while the rail is defined by the sweep of a generic profile. To assess the effectiveness of the proposed contact detection approach, the rail UIC54 and the wheel profile of the LRV at a given interference state have been considered as example. A convergence analysis of the strips size has been performed, which allows to observe that the increase of the number of strips may lead to significant variations on the size of contact area and on the location of the main point of contact where higher penetration is observed.

The current method involves performing the contact search per each strip of the wheel, which leads to higher computational cost. This is a quite significant aspect in a multibody dynamic simulation, because a railway vehicle has several pairs of contact and they have to be evaluated in each integration time step. Thus, as future work, the improvement of the discretization of the wheel by minimizing the number of strips without compromising the accuracy of the contact detection will be object of further investigation. The partitioning of the contact patch into strips for the contact detection may be advantageous in the measure that most of the relevant wheel-rail contact force models also require this type of discretization [10, 11].

**Acknowledgments.** The first and second authors are supported by the Portuguese Foundation for Science and Technology (FCT) under grant PD/BD/114154/2016, MIT Portugal Program, and SFRH/BD/96695/2013, respectively. This work has been supported by the FCT with the reference project UID/EEA/04436/2013, by FEDER funds through the COMPETE 2020 - *Programa Operacional Competitividade e Internacionalização* (POCI) with the reference project POCI-01-0145-FEDER-006941.

# References

1. Ambrósio, J., et al.: Multibody modelling and dynamic analysis of tapered roller bearings. In: Ambrósio, J., et al. (eds.) EUROMECH Colloquium 578 Rolling Contact Mechanics for Multibody System Dynamics, Funchal, Madeira, Portugal, pp. 1–29 (2017)
2. Ambrósio, J., Pombo, J.: A unified formulation for mechanical joints with and without clearances/bushings and/or stops in the framework of multibody systems. Multibody Syst. Dyn. **42**, 317–345 (2018)
3. Flores, P., Ambrósio, J.: On the contact detection for contact-impact analysis in multibody systems. Multibody Syst. Dyn. **24**(1), 103–122 (2010)
4. Gonçalves, A.A., et al.: A benchmark study on accuracy-controlled distance calculation between superellipsoid and superovoid contact geometries. Mech. Mach. Theory **115**, 77–96 (2017)
5. Lin, M., Gottschalk, S.: Collision detection between geometric models: a survey. In: Proceedings of IMA Conference on Mathematics of Surfaces, pp. 1–20 (1998)
6. Machado, M., et al.: A lookup-table-based approach for spatial analysis of contact problems. J. Comput. Nonlinear Dyn. **9**(4), 41010 (2014)
7. Magalhães, H., et al.: Railway vehicle modelling for the vehicle-track interaction compatibility analysis. Proc. Inst. Mech. Eng. Part K J. Multi-body Dyn. **230**(3), 251–267 (2016)
8. Magalhães, H., et al.: Railway vehicle performance optimisation using virtual homologation. Veh. Syst. Dyn. **54**(9), 1177–1207 (2016)
9. Malvezzi, M., et al.: Determination of wheel-rail contact points with semianalytic methods. Multibody Syst. Dyn. **20**(4), 327–358 (2008)
10. Meymand, S.Z., et al.: A survey of wheel–rail contact models for rail vehicles. Veh. Syst. Dyn. **54**(3), 386–428 (2016)
11. Piotrowski, J., Chollet, H.: Wheel–rail contact models for vehicle system dynamics including multi-point contact. Veh. Syst. Dyn. **43**(6–7), 455–483 (2005)
12. Pombo, J., et al.: A new wheel–rail contact model for railway dynamics. Veh. Syst. Dyn. **45**(2), 165–189 (2007)

# The Equations of Motion of a Four-Bar Linkage with Principal Vectors and Virtual Work

Jacob P. Meijaard$^{(\boxtimes)}$ and Volkert van der Wijk

Department of Precision and Microsystems Engineering,
Delft University of Technology, Delft, The Netherlands
{j.p.meijaard,v.vanderwijk}@tudelft.nl

**Abstract.** The motion of a four-bar linkage is considered with the goal to study the use of principal vectors to formulate the equations of motion and to get insight. Firstly, kinematic relations for the positions, velocities and accelerations are derived. Then, the motion of the centre of mass of the system is described with the aid of principal points and principal vectors, for which the mass of one link is replaced with equivalent masses. The condition of dynamic force balance is that the centre of mass is stationary. It is shown that the motion of the centres of mass of the links can be described in terms of the principal vectors. The equations of motion and the expressions for the force and moment on the base are derived with the aid of the principle of virtual work, which directly give conditions for dynamic force and moment balance. The equations of motion show a clear structure in their coefficients. The expression for the reaction force becomes simple, but the expression for the reaction moment remains rather complicated.

**Keywords:** Four-bar mechanism · Principal point · Principal vector
Virtual work · Equations of motion

## 1 Introduction

We consider a general planar four-bar mechanism $A_0A_1A_2A_3$ with ideal revolute joints and rigid links that have a general mass distribution, as shown in Fig. 1. The link $A_3A_0$ is considered the frame that is fixed to the ground; so for the dynamics, the mass of this link can be ignored. The system has one degree of freedom: if the orientation of the link $A_0A_1$ is given, the orientations of the other two moving links can be determined from the loop-closure conditions and then the positions of the joints can be found. These relations can be derived explicitly, as shown in the next section. Furthermore, the angular velocities and angular accelerations of the two dependent links can be expressed in the angular velocity and angular acceleration of the link $A_0A_1$.

The aim of this study is to describe the dynamics of the system with the aid of principal points and principal vectors, as introduced by Fischer [3,5].

© Springer Nature Switzerland AG 2019
B. Corves et al. (Eds.): EuCoMeS 2018, MMS 59, pp. 79–87, 2019.
https://doi.org/10.1007/978-3-319-98020-1_10

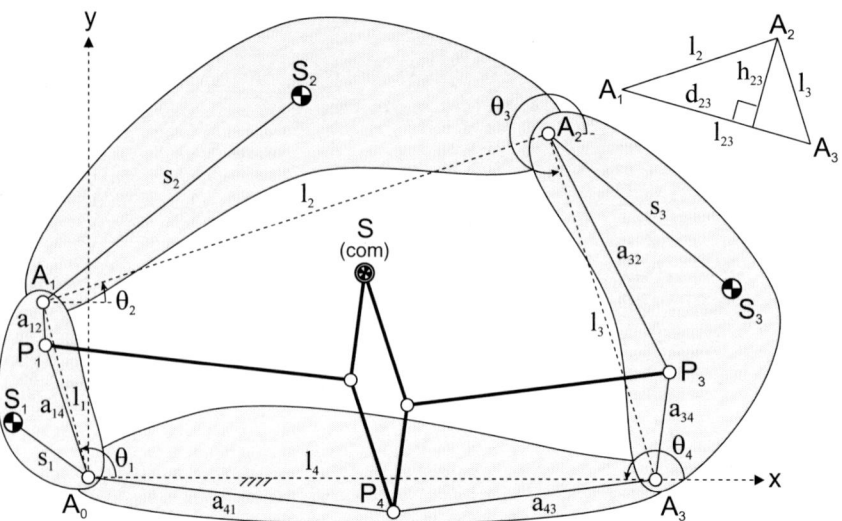

**Fig. 1.** A four-bar mechanism with the centres of mass of the moving links, the principal points $P_1$, $P_3$ and $P_4$ and principal vectors together with an auxiliary mechanism of parallelograms to determine the position of the centre of mass S of the system

With these points and vectors, the centre of mass of the system is easily found. The contribution of the linear acceleration of the centre of mass of one of the links can be expressed in the rotation angles of the other two moving links and their time derivatives. The equations of motion are derived with the aid of the principle of virtual work, where the terms due to accelerations are included. This contrasts with traditional methods in the literature, where Lagrange's equations are used. The reaction forces, that is, the shaking forces and moment, are also obtained with the principle of virtual work by releasing the constraints. We will show that the equations of motion have a clear structure, the expression for the reaction force becomes simple, but the expression for the reaction moment is rather complicated, even for a force-balanced system.

## 2   Kinematic Relations

A global system of coordinates $Oxy$ is introduced, where the origin coincides with the fixed point $A_0$, the global $x$-direction is along $A_0A_3$ and the global $y$-direction is found by rotating the $x$-axis counterclockwise by a right angle. The links $A_0A_1$, $A_1A_2$, $A_2A_3$ and $A_3A_0$ are numbered 1, 2, 3 and 4, respectively. The length of link $k$ ($k = 1, \ldots, 4$) is denoted by $l_k$. The relative and absolute positions can be represented by complex numbers, which are indicated by an overbar; they can also be seen as two-dimensional vectors. The link lengths and their orientations are denoted by $\bar{l}_k = l_k \exp(i\theta_k)$, where $\theta_k$ is the angle the line connecting the two joints of link $k$ makes with the global $x$-direction;

therefore, $\theta_4 = \pi$. The loop-closure condition can then be directly expressed as $\bar{l}_1 + \bar{l}_2 + \bar{l}_3 + \bar{l}_4 = \bar{0}$, where $\bar{0}$ is the complex zero or the zero vector. The angle $\theta_1$ is chosen as the independent angle and $\theta_2$ and $\theta_3$ are dependent variables. Several ways to determine the dependent coordinates can be found in the literature, see for instance [4]. Here, we present a way that appears to be convenient. With the vector $\bar{l}_{23} = \bar{l}_2 + \bar{l}_3 = -\bar{l}_1 - \bar{l}_4$, we introduce its length $l_{23}$ and the additional lengths $d_{23}$, $l_{23} - d_{23}$ and $h_{23}$ (see inset in Fig. 1) as

$$l_{23} = \sqrt{l_1^2 + l_4^2 - 2l_1l_4\cos\theta_1}, \quad d_{23} = \frac{l_{23}^2 + l_2^2 - l_3^2}{2l_{23}},$$

$$l_{23} - d_{23} = \frac{l_{23}^2 + l_3^2 - l_2^2}{2l_{23}}, \quad h_{23} = \pm\sqrt{l_2^2 - d_{23}^2} = \pm\sqrt{l_3^2 - (l_{23} - d_{23})^2}. \tag{1}$$

The combination of signs comes from the two possible configurations: the plus sign corresponds to the case in which $A_2$ is at the left side of the directed line $A_1A_3$, as drawn in Fig. 1, and the minus sign corresponds to the case in which this point is at the right side of this line. If it is assumed that the Grashof condition is satisfied and link 1 is the crank, no branching between these two configurations can occur. The dependent link orientation angles $\theta_2$ and $\theta_3$ are obtained from

$$\bar{l}_2 = (d_{23} + ih_{23})\frac{\bar{l}_{23}}{l_{23}}, \quad \bar{l}_3 = (l_{23} - d_{23} - ih_{23})\frac{\bar{l}_{23}}{l_{23}} \tag{2}$$

as the arguments of $\bar{l}_2$ and $\bar{l}_3$.

The velocities of the dependent angles are obtained from the loop-closure condition differentiated with respect to time, $i(\bar{l}_1\dot\theta_1 + \bar{l}_2\dot\theta_2 + \bar{l}_3\dot\theta_3) = \bar{0}$. The dependent angular velocities are obtained by taking the inner products of this expression with $\bar{l}_3$ and $\bar{l}_2$. The inner product is defined as the usual inner product of two vectors, which can be written with angle brackets, if $\bar{x}$ and $\bar{y}$ are two complex numbers, as $\langle \bar{x}, \bar{y} \rangle = \frac{1}{2}(\bar{x}^*\bar{y} + \bar{y}^*\bar{x})$, where a superscript asterisk denotes a complex conjugate. This results in

$$\dot\theta_2 = \frac{l_1\sin(\theta_1 - \theta_3)}{l_2\sin(\theta_3 - \theta_2)}\dot\theta_1, \quad \dot\theta_3 = \frac{l_1\sin(\theta_1 - \theta_2)}{l_3\sin(\theta_2 - \theta_3)}\dot\theta_1. \tag{3}$$

Another differentiation of the loop-closure condition with respect to time yields

$$i(\bar{l}_1\ddot\theta_1 + \bar{l}_2\ddot\theta_2 + \bar{l}_3\ddot\theta_3) - (\bar{l}_1\dot\theta_1^2 + \bar{l}_2\dot\theta_2^2 + \bar{l}_3\dot\theta_3^2) = \bar{0}. \tag{4}$$

Again taking the inner product of this expression with $\bar{l}_3$ and $\bar{l}_2$ yields

$$\ddot\theta_2 = \frac{l_1\sin(\theta_1 - \theta_3)}{l_2\sin(\theta_3 - \theta_2)}\ddot\theta_1 + \frac{l_1\cos(\theta_1 - \theta_3)\dot\theta_1^2 + l_2\cos(\theta_2 - \theta_3)\dot\theta_2^2 + l_3\dot\theta_3^2}{l_2\sin(\theta_3 - \theta_2)}, \tag{5}$$

$$\ddot\theta_3 = \frac{l_1\sin(\theta_1 - \theta_2)}{l_3\sin(\theta_2 - \theta_3)}\ddot\theta_1 + \frac{l_1\cos(\theta_1 - \theta_2)\dot\theta_1^2 + l_2\dot\theta_2^2 + l_3\cos(\theta_2 - \theta_3)\dot\theta_3^2}{l_3\sin(\theta_2 - \theta_3)}. \tag{6}$$

The expressions for the dependent angles and angular velocities can be substituted in the right-hand sides to obtain explicit results. In particular, $\cos(\theta_2 - \theta_3) = (l_{23}^2 - l_2^2 - l_3^2)/(2l_2l_3)$ and $\sin(\theta_2 - \theta_3) = \pm\sqrt{1 - \cos^2(\theta_2 - \theta_3)}$. In most cases, the dependent angles and their time derivatives will be kept in the equations, where it has to be understood that they are functions of the independent angle and its time derivatives. Another form to write the angular velocity and acceleration of link 2 that only contains the angles $\theta_1$ and $\theta_3$ and their derivatives is

$$l_2^2\dot{\theta}_2 = l_1[l_1 + l_3\cos(\theta_1 - \theta_3) - l_4\cos\theta_1]\dot{\theta}_1 + l_3[l_3 + l_1\cos(\theta_1 - \theta_3) - l_4\cos\theta_3]\dot{\theta}_3,$$

$$l_2^2\ddot{\theta}_2 = l_1[l_1 + l_3\cos(\theta_1 - \theta_3) - l_4\cos\theta_1]\ddot{\theta}_1 - l_1[l_3\sin(\theta_1 - \theta_3) - l_4\sin\theta_1]\dot{\theta}_1^2 \qquad (7)$$

$$+ l_3[l_3 + l_1\cos(\theta_1 - \theta_3) - l_4\cos\theta_3]\ddot{\theta}_3 + l_3[l_1\sin(\theta_1 - \theta_3) + l_4\sin\theta_3]\dot{\theta}_3^2 .$$

## 3   Mass Kinematics

The position of the centre of mass of link $k$ ($k = 1, 2, 3$) with respect to the line connecting its two joints, measured from the joint $A_{k-1}$, can be expressed by a complex number with the dimension of a length, $\bar{s}_k = s_k \exp(i\sigma_k)$. The absolute position of the centre of mass, $\bar{r}_k$, is then given by $\bar{r}_k = \bar{r}_{A_{k-1}} + \bar{s}_k \exp(i\theta_k) = \bar{r}_{A_{k-1}} + \bar{s}_k\bar{l}_k/l_k$, where $\bar{r}_{A_{k-1}}$ is the absolute position of the point $A_{k-1}$. Principal points are now defined for the case that link 2 is removed, its mass is redistributed by equivalent real masses at the joints $A_1$ and $A_2$ and so-called virtual equivalent masses as shown in [6]. The principal points of these three links are defined as the points $P_k$ that have the property that a rotation about this point, where the rotations of the other links are zero, does not change the position of the centre of mass of the reduced system. The positions of these points in link $k$ ($k = 1, 3, 4$) are given by the principal vectors, expressed as complex numbers, $\bar{a}_{kl}$, pointing from the principal point $P_k$ toward the joint connecting it to link $l$. These vectors are given by the equations

$$m_{\text{tot}}\bar{a}_{14} = -[m_1\bar{s}_1 + l_1m_2(1 - \bar{s}_2/l_2)]\bar{l}_1/l_1, \qquad (8)$$

$$m_{\text{tot}}\bar{a}_{32} = -\{m_3\bar{s}_3 + l_3[m_1 + m_2(1 - \bar{s}_2/l_2)]\}\bar{l}_3/l_3, \qquad (9)$$

$$m_{\text{tot}}\bar{a}_{43} = -[m_1 + m_2(1 - \bar{s}_2/l_2)]\bar{l}_4 , \qquad (10)$$

where $m_{\text{tot}} = m_1 + m_2 + m_3$ is the total mass of the system. Their complementary principal vectors are

$$\bar{a}_{12} = \bar{a}_{14} + \bar{l}_1, \quad \bar{a}_{34} = \bar{a}_{32} + \bar{l}_3, \quad \bar{a}_{41} = \bar{a}_{43} + \bar{l}_4 . \qquad (11)$$

The centre of mass of the system, denoted by S, can be found from the parallelogram construction shown in Fig. 1. The correctness of this construction can easily be seen by writing the position of the centre of mass of the system starting from one of the joints or principal points.

The positions of the centres of mass of the three links multiplied by their respective masses can be expressed in the coordinates of the centre of mass of the system, denoted by $\bar{r}_0$, the position vector of the centre of mass of link 2, $\bar{s}_2$, and the principal vectors as

$$m_1\bar{r}_1 = m_1(\bar{r}_0 + \bar{a}_{34} + \bar{a}_{41}) - (m_2\bar{s}_2/l_2 + m_3)\bar{a}_{14} - m_2(1 - \bar{s}_2/l_2)\bar{a}_{12}, \quad (12)$$

$$m_2\bar{r}_2 = m_2\big[\bar{r}_0 + (\bar{a}_{34} + \bar{a}_{41} + \bar{a}_{12})(1 - \bar{s}_2/l_2) + (\bar{a}_{14} + \bar{a}_{43} + \bar{a}_{32})\bar{s}_2/l_2\big], \quad (13)$$

$$m_3\bar{r}_3 = m_3(\bar{r}_0 + \bar{a}_{14} + \bar{a}_{43}) - \big[m_1 + m_2(1 - \bar{s}_2/l_2)\big]\bar{a}_{34} - m_2(\bar{s}_2/l_2)\bar{a}_{32}. \quad (14)$$

It is found by adding these three terms that the centre of mass is indeed given by $\bar{r}_0$. The constraints, some of which are dependent, can now also be expressed as

$$\bar{r}_{A_0} = \bar{0} = \bar{r}_0 + \bar{a}_{34} + \bar{a}_{14} + \bar{a}_{41}, \quad \bar{r}_{P_4} = -\bar{a}_{41} = \bar{r}_0 + \bar{a}_{34} + \bar{a}_{14}, \quad \theta_4 = \pi, \quad (15)$$

which relations can be used to eliminate $\bar{r}_0$ from Eqs. (12)–(14) as

$$m_1\bar{r}_1 = -(m_1 + m_2\bar{s}_2/l_2 + m_3)\bar{a}_{14} - m_2(1 - \bar{s}_2/l_2)\bar{a}_{12}, \quad (16)$$

$$m_2\bar{r}_2 = m_2\big[(\bar{a}_{12} - \bar{a}_{14})(1 - \bar{s}_2/l_2) + (\bar{a}_{43} - \bar{a}_{41} + \bar{a}_{32} - \bar{a}_{34})\bar{s}_2/l_2\big], \quad (17)$$

$$m_3\bar{r}_3 = m_3(\bar{a}_{43} - \bar{a}_{41}) - \big[m_1 + m_2(1 - \bar{s}_2/l_2) + m_3\big]\bar{a}_{34} - m_2(\bar{s}_2/l_2)\bar{a}_{32}. \quad (18)$$

The velocities are found by differentiation as

$$m_1\dot{\bar{r}}_1 = -\big[(m_1 + m_2\bar{s}_2/l_2 + m_3)\bar{a}_{14} + m_2(1 - \bar{s}_2/l_2)\bar{a}_{12}\big]i\dot{\theta}_1, \quad (19)$$

$$m_2\dot{\bar{r}}_2 = m_2\big[(\bar{a}_{12} - \bar{a}_{14})(1 - \bar{s}_2/l_2)i\dot{\theta}_1 + (\bar{a}_{32} - \bar{a}_{34})(\bar{s}_2/l_2)i\dot{\theta}_3\big], \quad (20)$$

$$m_3\dot{\bar{r}}_3 = -\big\{\big[m_1 + m_2(1 - \bar{s}_2/l_2) + m_3\big]\bar{a}_{34} + m_2(\bar{s}_2/l_2)\bar{a}_{32}\big\}i\dot{\theta}_3. \quad (21)$$

The virtual displacements are obtained by replacing the time derivatives by the corresponding virtual quantities denoted by a prefixed $\delta$, namely $\delta r_1$, $\delta r_2$, $\delta r_3$, $\delta\theta_1$, $\delta\theta_2$ and $\delta\theta_3$. Another differentiation gives the accelerations,

$$m_1\ddot{\bar{r}}_1 = -\big[(m_1 + m_2\bar{s}_2/l_2 + m_3)\bar{a}_{14} + m_2(1 - \bar{s}_2/l_2)\bar{a}_{12}\big](i\ddot{\theta}_1 - \dot{\theta}_1^2), \quad (22)$$

$$m_2\ddot{\bar{r}}_2 = m_2\big[(\bar{a}_{12} - \bar{a}_{14})(1 - \bar{s}_2/l_2)(i\ddot{\theta}_1 - \dot{\theta}_1^2) + (\bar{a}_{32} - \bar{a}_{34})(\bar{s}_2/l_2)(i\ddot{\theta}_3 - \dot{\theta}_3^2)\big], \quad (23)$$

$$m_3\ddot{\bar{r}}_3 = -\big\{\big[m_1 + m_2(1 - \bar{s}_2/l_2) + m_3\big]\bar{a}_{34} + m_2(\bar{s}_2/l_2)\bar{a}_{32}\big\}(i\ddot{\theta}_3 - \dot{\theta}_3^2). \quad (24)$$

The virtual displacements used for the calculation of the reactions are chosen as a virtual displacement of the centre of mass, $\delta \bar{r}_0$, and a rotation of the whole system, without changing its configuration, about the point S, denoted by $\delta \theta_0$. The corresponding virtual displacements of the centres of mass of the bodies are

$$m_1 \delta \bar{r}_1 = m_1 \delta \bar{r}_0 + \left[ m_1 (\bar{a}_{34} + \bar{a}_{41}) - (m_2 \bar{s}_2 / l_2 + m_3) \bar{a}_{14} - m_2 (1 - \bar{s}_2 / l_2) \bar{a}_{12} \right] i \delta \theta_0, \quad (25)$$

$$m_2 \delta \bar{r}_2 = m_2 \delta \bar{r}_0 + m_2 \left[ (\bar{a}_{34} + \bar{a}_{41} + \bar{a}_{12})(1 - \bar{s}_2 / l_2) + (\bar{a}_{14} + \bar{a}_{43} + \bar{a}_{32}) \bar{s}_2 / l_2 \right] i \delta \theta_0, \quad (26)$$

$$m_3 \delta \bar{r}_3 = m_3 \delta \bar{r}_0 + \left\{ m_3 (\bar{a}_{14} + \bar{a}_{43}) - \left[ m_1 + m_2 (1 - \bar{s}_2 / l_2) \right] \bar{a}_{34} - m_2 (\bar{s}_2 / l_2) \bar{a}_{32} \right\} i \delta \theta_0. \quad (27)$$

## 4   Equations of Motion

The virtual work expression for the system becomes

$$\delta W = \sum_{k=1}^{3} \left[ \delta \theta_k (M_k - I_k \ddot{\theta}_k) + \langle \delta \bar{r}_k, (\bar{F}_k - m_k \ddot{\bar{r}}_k) \rangle \right], \quad (28)$$

where $M_k$ is the resultant applied moment with respect to the centre of mass of link $k$, $\bar{F}_k$ is the resultant applied force, represented as a complex number, and $I_k$ is the moment of inertia with respect to the centre of mass. Substituting the quantities in this equation, while keeping the link angles provisionally as independent variables, one obtains

$$\delta W = \delta \theta_1 Q_1 + \delta \theta_2 Q_2 + \delta \theta_3 Q_3, \quad (29)$$

where $Q_k$ are generalized forces including inertia terms, which are given by

$$Q_1 = Q_1^c + M_1 - \langle i \left[ (m_1 + m_2 \bar{s}_2 / l_2 + m_3) \bar{a}_{14} + m_2 (1 - \bar{s}_2 / l_2) \bar{a}_{12} \right], \bar{F}_1 / m_1 \rangle$$
$$+ \langle i (\bar{a}_{12} - \bar{a}_{14})(1 - \bar{s}_2 / l_2), \bar{F}_2 \rangle - M_{11} \ddot{\theta}_1 - M_{13} \ddot{\theta}_3 - C_{13} \dot{\theta}_3^2, \quad (30)$$

$$Q_2 = Q_2^c + M_2 - I_2 \ddot{\theta}_2, \quad (31)$$

$$Q_3 = Q_3^c + M_3 + \langle i (\bar{a}_{32} - \bar{a}_{34})(\bar{s}_2 / l_2), \bar{F}_2 \rangle$$
$$- \langle i \left[ m_1 + m_2 (1 - \bar{s}_2 / l_2) + m_3 \right] \bar{a}_{34} + m_2 (\bar{s}_2 / l_2) \bar{a}_{32}, \bar{F}_3 / m_3 \rangle \quad (32)$$
$$- M_{13} \ddot{\theta}_1 - M_{33} \ddot{\theta}_3 + C_{13} \dot{\theta}_1^2,$$

where $Q_k^c$ are the generalized constraint forces and the mass and convective inertia terms are given by

$$M_{11} = I_1 + (a_{14}^2 / m_1) \left[ (m_1 + m_2 s_2 \cos \sigma_2 / l_2 + m_3)^2 + (m_2 s_2 \sin \sigma_2 / l_2)^2 \right]$$
$$+ \left[ a_{12}^2 m_2^2 / m_1 + m_2 (a_{12}^2 + a_{14}^2 - 2 \langle \bar{a}_{14}, \bar{a}_{12} \rangle) \right] (1 + s_2^2 / l_2^2 - 2 s_2 \cos \sigma_2 / l_2)$$
$$+ 2(m_2 / m_1) \langle (m_1 + m_2 \bar{s}_2 / l_2 + m_3) \bar{a}_{14}, (1 - \bar{s}_2 / l_2) \bar{a}_{12} \rangle, \quad (33)$$

$$M_{13} = m_2 \langle (1 - \bar{s}_2/l_2)(\bar{a}_{12} - \bar{a}_{14}), (\bar{s}_2/l_2)(\bar{a}_{32} - \bar{a}_{34}) \rangle, \tag{34}$$

$$M_{33} = I_3 + (a_{34}^2/m_3)\left[(m_{\text{tot}} - m_2 s_2 \cos\sigma_2/l_2)^2 + (m_2 s_2 \sin\sigma_2/l_2)^2\right]$$
$$+ \left[a_{32}^2 m_2^2/m_3 + m_2(a_{32}^2 + a_{34}^2 - 2\langle\bar{a}_{34}, \bar{a}_{32}\rangle)\right]s_2^2/l_2^2 \tag{35}$$
$$+ 2(m_2/m_3)\langle (m_{\text{tot}} - m_2\bar{s}_2/l_2)\bar{a}_{34}, (\bar{s}_2/l_2)\bar{a}_{32} \rangle,$$

$$C_{13} = m_2 \langle (1 - \bar{s}_2/l_2)(\bar{a}_{12} - \bar{a}_{14}), i(\bar{s}_2/l_2)(\bar{a}_{32} - \bar{a}_{34}) \rangle. \tag{36}$$

The velocity-dependent inertia terms have the structure that they show only squares of the angular velocities and a skew symmetry for its coefficients: the terms with $C_{13}$ in Eqs. (30) and (32) have opposite signs.

The unreduced equations of motion, which still contain the unknown constraint forces, are given by

$$Q_1 = 0, \quad Q_2 = 0, \quad Q_3 = 0. \tag{37}$$

The reduced equations are found by the incorporation of the relations between the virtual rotations, which eliminates the constraint forces as

$$Q_1 + \frac{l_1 \sin(\theta_1 - \theta_3)}{l_2 \sin(\theta_3 - \theta_2)} Q_2 + \frac{l_1 \sin(\theta_1 - \theta_2)}{l_3 \sin(\theta_2 - \theta_3)} Q_3 = 0 . \tag{38}$$

To get the explicit equation of motion, the dependent variables and their derivatives have to be expressed as functions of the independent variable and its derivatives. If this is explicitly written out, one gets a very long expression, not shown here.

## 5   Application to Dynamic Balancing

The resultant reaction forces can be obtained by applying the virtual displacement $\delta\bar{r}_0$ to the system, which yields the shaking forces

$$\bar{F}_{\text{sh}} = \bar{F}_1 + \bar{F}_2 + \bar{F}_3 + m_{\text{tot}}\left[\bar{a}_{14}(i\ddot{\theta}_1 - \dot{\theta}_1^2) + \bar{a}_{34}(i\ddot{\theta}_3 - \dot{\theta}_3^2)\right] . \tag{39}$$

It may be observed that the term within square brackets is just minus the acceleration of the centre of mass of the system. The conditions for force balancing are that the external applied forces are constant and the term in square brackets is zero. The last condition implies that the principal vectors $\bar{a}_{14}$ and $\bar{a}_{34}$ are zero, so the principal point $P_1$ coincides with $A_0$ and the principal point $P_3$ coincides with $A_3$. Written out in terms of lengths and angles, the conditions for force balance become

$$m_1(s_1/l_1)\cos\sigma_1 + m_2[1 - (s_2/l_2)\cos\sigma_2] = 0,$$
$$m_1(s_1/l_1)\sin\sigma_1 - m_2(s_2/l_2)\sin\sigma_2 = 0,$$
$$m_2(s_2/l_2)\cos\sigma_2 + m_3[1 - (s_3/l_3)\cos\sigma_3] = 0, \tag{40}$$
$$m_2(s_2/l_2)\sin\sigma_2 - m_3(s_3/l_3)\sin\sigma_3 = 0,$$

which agree with the conditions given in [2].

For the case of a force-balanced system, the shaking moment on the foundation is determined. In that case, $\bar{a}_{12} = \bar{l}_1$ and $\bar{a}_{32} = -\bar{l}_3$. By applying the principle of virtual work with the variations $\delta\theta_0$, the shaking moment, $M_{\text{sh}} = M_{\text{sh}}^{\text{a}} + M_{\text{sh}}^{\text{i}}$, is found as the sum of the resultant moment from the applied forces and moments,

$$
\begin{aligned}
M_{\text{sh}}^{\text{a}} = M_1 + M_2 + M_3 &+ \langle i[\bar{a}_{41} - (m_2/m_1)(1 - \bar{s}_2/l_2)\bar{a}_{12}], \bar{F}_1\rangle \\
&+ \langle i[(\bar{a}_{41} + \bar{a}_{12})(1 - \bar{s}_2/l_2) + (\bar{a}_{43} + \bar{a}_{32})\bar{s}_2/l_2], \bar{F}_2\rangle \quad (41) \\
&+ \langle i[\bar{a}_{43} - (m_2/m_3)(\bar{s}_2/l_2)\bar{a}_{32}], \bar{F}_3\rangle
\end{aligned}
$$

and a resultant moment from the inertia terms,

$$
\begin{aligned}
M_{\text{sh}}^{\text{i}} = -I_2\ddot{\theta}_2 &- \{I_1 + (m_2^2 l_1^2/m_1 + m_2 l_1^2 + m_2\langle\bar{a}_{41}, \bar{a}_{12}\rangle)(1 + s_2^2/l_2^2 - 2s_2\cos\sigma_2/l_2) \\
&\quad -m_2\langle\bar{a}_{41}, (1 - \bar{s}_2/l_2)\bar{a}_{12}\rangle + m_2\langle(\bar{a}_{43} + \bar{a}_{32})(\bar{s}_2/l_2), (1 - \bar{s}_2/l_2)\bar{a}_{12}\rangle\}\ddot{\theta}_1 \\
&- \{m_2\langle\bar{a}_{41}, i\bar{a}_{12}\rangle(1 + s_2^2/l_2^2 - 2s_2\cos\sigma_2/l_2) - m_2\langle\bar{a}_{41}, i(1 - \bar{s}_2/l_2)\bar{a}_{12}\rangle \\
&\quad +m_2\langle(\bar{a}_{43} + \bar{a}_{32})(\bar{s}_2/l_2), i(1 - \bar{s}_2/l_2)\bar{a}_{12}\rangle\}\dot{\theta}_1^2 \\
&- \{I_3 + (m_2^2 l_3^2/m_3 + m_2 l_3^2 + m_2\langle\bar{a}_{43}, \bar{a}_{32}\rangle)(s_2^2/l_2^2) \\
&\quad -m_2\langle\bar{a}_{43}, (\bar{s}_2/l_2)\bar{a}_{32}\rangle + m_2\langle(\bar{a}_{41} + \bar{a}_{12})(1 - \bar{s}_2/l_2), (\bar{s}_2/l_2)\bar{a}_{32}\rangle\}\ddot{\theta}_3 \\
&- \{m_2\langle\bar{a}_{43}, i\bar{a}_{32}\rangle(s_2^2/l_2^2) - m_2\langle\bar{a}_{43}, i(\bar{s}_2/l_2)\bar{a}_{32}\rangle \\
&\quad +m_2\langle(\bar{a}_{41} + \bar{a}_{12})(1 - \bar{s}_2/l_2), i(\bar{s}_2/l_2)\bar{a}_{32}\rangle\}\dot{\theta}_3^2 \, .
\end{aligned}
\quad (42)
$$

Internal driving moments do not give a contribution, because reactions are equal to actions with an opposed sign according to Newton's third law. Gravity forces have a constant value and direction, so their moment with respect to the centre of mass is zero. Moreover, if also $\sigma_2 = 0$, so $\sigma_1 = \sigma_3 = 0$, and if $I_2 = m_2 s_2(l_2 - s_2)$, the shaking moment depends linearly on the angular acceleration of the links 1 and 3, so these can be compensated by counterrotating masses in the base [1].

# 6   Conclusions

The derivation of the equation of motion of a four-bar linkage with the aid of principal points, principal vectors, and a replacement of the mass of one link by equivalent masses has been shown. The use of the principle of virtual work clearly reveals the structure of the velocity-dependent inertia terms, which show only squares of the angular velocities and a skew symmetry for its coefficients. The expressions for the reaction forces become simple and the condition for dynamic force balance is simply that two of the principal points coincide with the locations of the fixed joints. The expression for the reaction moment, however, remains complicated.

# References

1. Berkof, R.S.: Complete force and moment balancing of inline four-bar linkages. Mech. Mach. Theory **8**, 397–410 (1973)
2. Berkof, R.S., Lowen, G.G.: A new method for complete force balancing simple linkages. Trans. ASME J. Eng. Ind. **91**, 21–26 (1969)
3. Fischer, O.: Theoretische Grundlagen für eine Mechanik der lebenden Körper. B. G. Teubner, Leipzig (1906)
4. Paul, B.: Kinematics and Dynamics of Planar Machinery. Prentice-Hall, Englewood Cliffs (1979)
5. van der Wijk, V.: Methodology for analysis and synthesis of inherently force and moment-balanced mechanisms. Dissertation, University of Twente, Enschede (2014)
6. van der Wijk, V.: Design and analysis of closed-chain principal vector linkages for dynamic balance with a new method for mass equivalent modeling. Mech. Mach. Theory **107**, 283–304 (2017)

# Motion Analysis of Planar Flexible Mechanisms Using Vector Form Method

Mien-Li Wang[1], Tung-Yueh Wu[2], and Jyh-Jone Lee[1(✉)]

[1] National Taiwan University, Taipei, Taiwan
jjlee@ntu.edu.tw
[2] Atomic Energy Council, New Taipei City, Taiwan
tywuda@gmail.com

**Abstract.** In this paper, a procedure based on vector form intrinsic finite element method is developed for the dynamic analysis of flexible multi-body systems. Different from analytical mechanics such as variations or energy method, in this procedure, a structure is discretized to a set of particles and then the equations of motions of particle set via Newton's law is established. Evaluation of deformations and internal forces are subsequently calculated. Modeling of constraints are presented as well. Finally, the time integration using central difference is applied to solve the equations of motion. A numerical example of the planar flexible slider-crank is used to illustrate the procedure.

**Keywords:** Vector form analysis · Flexible multibody system dynamics
Computational dynamics

## 1 Introduction

The analysis of flexible multibody systems (FMS) has drawn much attention of researchers for the last forty years. Flexible multibody dynamics (FMD) are usually nonlinear can be formulated by different methods [9]. There are three essential constituents to be processed in a FMD problem, namely, (1) the modeling of flexible components including determination of reference frames, formulation of equation of motions, and treatment of deformed and displaced motions, (2) establishment of the constraints, and (3) solution method. As the finite element method became a powerful tool for analyzing solid mechanics, it has been widely applied to the FMD. Since 1970s, some efforts have been made by Bathe et al. [1]. In their work, both the updated Lagrangian (UL) formulation and total Lagrangian (TL) formulation were derived and examined [2]. Although the TL formulation has advantages in analyzing the problems with soft materials, it is not adequate for the analysis of FMS because the problems of negative Jacobian matrix may occur. The UL formulation improves the deficiency of TL, but it might lead to inaccurate modeling of rigid body dynamics. Shabana [3–5] proposed an absolute nodal coordinate formulation, where a set of global shape functions capable of describing an arbitrary rigid body translational and rotational displacement was developed. Recently, Ting et al. [7, 8] proposed a method called "vector form intrinsic finite element" (VFIFE) for the analysis of FMS with large deformation and large displacement. In their paper, a deformable body is represented

© Springer Nature Switzerland AG 2019
B. Corves et al. (Eds.): EuCoMeS 2018, MMS 59, pp. 88–96, 2019.
https://doi.org/10.1007/978-3-319-98020-1_11

by a set of particles where the motion of each particle is formulated by Newton equation without invoking continuum governing equations. The methodology has been successfully applied to the 2D frame structure [11], 3D membrane structures [10, 12] and 3D solids [13]. The presented examples shows that both large rotations and large deformation can be handled by VFIFE. Nonetheless, the constraints dealing with the mechanical joints such as revolute, prismatic and spherical joints are not well developed. In this paper, a procedure based on the VFIFE methodology is established for the analysis of FMD. A model of hinge joint is developed so that it fits for the VFIFE and is capable of handling a large number of joints with sufficient stability and accuracy.

## 2   Mathematic Background

### 2.1   Point Value Description

The VFIFE algorithm describes a continuous body by using finite number of particles rather than by a mathematical function. In this work, only 2D beam element is considered. A beam element consists of two mass points called nodal mass. For sake of simplicity, a beam element and each nodal mass are considered as independent variables, where they also provide internal force evaluation. The advantage of using particles to describe the object is the use of Newton's equation, resulting in equation of motions in ordinary differential equation. The mass and moment of inertia evaluation of the nodes adjacent to the element can be written as in Eq. (1) [7, 8]:

$$m_a = \frac{1}{2}m_{beam}, \; m_b = \frac{1}{2}m_{beam}, \; I_a = m_a r^2, \; I_b = m_b r^2 \tag{1}$$

where r is the radius of gyration.

### 2.2   Discrete Path and Governing Equation

Assume there is an arbitrary particle which underwent a motion from time $t_a$ to $t$, and its position vector $\mathbf{x}_a(t_a)$, has also been shifted to $\mathbf{x}(t)$. This motion path can be modeled via a set of discrete paths, as shown in Fig. 1. Within each path unit, the whole computational procedures can be simplified as a governing equation.

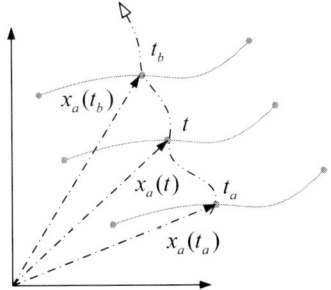

**Fig. 1.** A particle path

Considering the free body of a nodal mass, the equation of motion of the particle can be obtained via Newton's law, as in Eq. (2), where $\mathbf{d}$ is the displacement vector, $\mathbf{F}_{ext}$ is the concentrated external forces at the particle, $\mathbf{f}_{int}$ is the equivalent nodal forces due to deformations in element, and $n$ is the number of elements connected with particle $m$ at time $t_a$.

$$m\ddot{\mathbf{d}} = \mathbf{F}_{ext} - \sum_{i=1}^{n} \mathbf{f}_{int} \tag{2}$$

### 2.3    Reverse Rigid Body Motion

In order to calculate pure deformation, an element at time $t$ undergoes a fictitious reverse motion which includes a rigid body translation and rotation. To extract element deformation, a method for measuring the rigid body rotation is designed. As shown in Fig. 2, the rotation angle $\phi$ can be calculated by comparing the orientation of vector $\mathbf{e}_a$ with that of $\mathbf{e}_t$ as:

$$\varphi = \Delta\theta = \cos^{-1}(\mathbf{e}_a \cdot \mathbf{e}_t) \tag{3}$$

Further, $\beta^1$ and $\beta^2$ are nodal rotation angles caused by deformation and rigid body motion for nodes 1 and 2, respectively. When adding a reverse rigid body motion angle $\varphi$ to $\beta^1$ (and $\beta^2$), one obtains the angle $\theta_1$ (and $\theta_2$) due to deformation. Therefore, the pure deformation $\Delta_e$ and pure rotation angle $\theta_1$ and $\theta_2$ can be evaluated as:

$$\Delta_e = l - l_a \tag{4}$$

$$\theta_1 = \beta^1 + (-\varphi) \text{ and } \theta_2 = \beta^2 + (-\varphi) \tag{5}$$

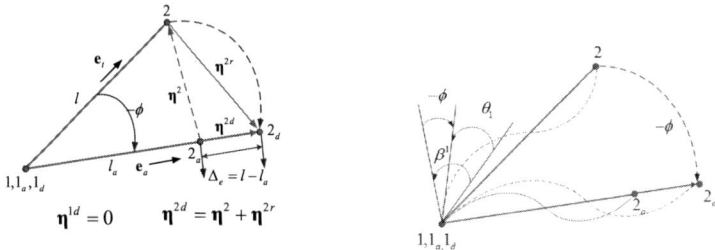

**Fig. 2.** Reverse rotation of a beam element

### 2.4    Internal Force Evaluation

The incremental internal force can be obtained as shown in Eq. (6). For detail derivation, it can be referred to Ting et al. [8].

$$\Delta \hat{\mathbf{f}} = \frac{E_a}{l_a} \begin{bmatrix} A_a & 0 & 0 \\ 0 & 4I_a & 2I_a \\ 0 & 2I_a & 4I_a \end{bmatrix} \begin{bmatrix} \Delta \\ \theta_1 \\ \theta_2 \end{bmatrix} \tag{6}$$

$$\hat{\mathbf{f}} = \hat{\mathbf{f}}_a + \Delta \hat{\mathbf{f}} \tag{7}$$

Three unknown components are calculated by the static equilibrium equation.

$$\left. \begin{array}{lll} \sum F_{\hat{x}} = 0 & : & \hat{f}_{1x} = -\hat{f}_{2x} \\ \sum F_{\hat{y}} = 0 & : & \hat{f}_{1y} = -\hat{f}_{2y} \\ \sum M_{1a} = 0 & : & \hat{f}_{2y} = -(m_{1z} + m_{2z})/l_a \end{array} \right\} \tag{8}$$

where the internal forces, $(\hat{f}_{jx}, \hat{f}_{jy})$ and moment, $m_{jz}$, j = 1, 2, are calculated in principal coordinate $(\hat{x}, \hat{y})$, and the nodal forces in the fictitious states. Thus, the internal force are then transformed to global coordinate system as

$$\mathbf{f}_j = \begin{bmatrix} f_{jx} \\ f_{jy} \\ m_{jz} \end{bmatrix} = \begin{bmatrix} \mathbf{R} & 0 \\ 0 & 1 \end{bmatrix} \begin{bmatrix} \hat{\mathbf{Q}} & 0 \\ 0 & 1 \end{bmatrix} \hat{\mathbf{f}}_j = \begin{bmatrix} \mathbf{R}\hat{\mathbf{Q}} & 0 \\ 0 & 1 \end{bmatrix} \begin{bmatrix} \hat{f}_{jx} \\ \hat{f}_{jy} \\ m_{jz} \end{bmatrix} \tag{9}$$

where $\hat{\mathbf{Q}}$ is the global coordinate transformation matrix and $\mathbf{R}$ is the rotation matrix.

## 3  Modelling of Revolute Joint Constraint

### 3.1  Local Coordinate System

As shown in Fig. 3(a), define a unit vector $\bar{e}_1$ by a reference point 5 and node 2 along the rotation axis of revolute joint J. Similarly, define the unit vector $\bar{e}_2$ node 2 and node 1. Therefore, a local coordinate system can be established as $\hat{x} = \bar{e}_1$, $\hat{y} = \bar{e}_2$, and $\hat{z} = \hat{x} \times \hat{y}$. Figure 3(b) shows the revolute joint where node s and node m are both mass points located on each side of J. The two nodes can be described by two independent equations of motion.

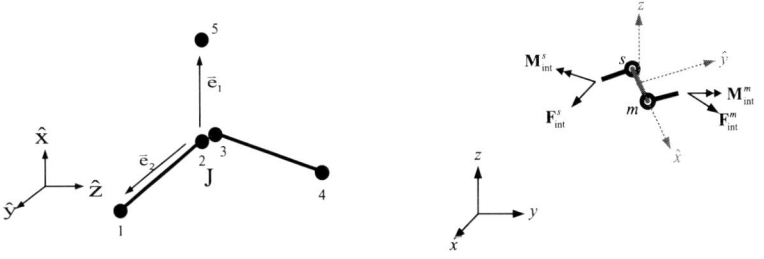

(a) Establishment of local coordinate system        (b) R-joint in local coordinate

**Fig. 3.**  Schematic of revolute joint.

## 3.2    Constraint of Revolute Joint

First, the internal moment and moment of inertia need to be transferred to local coordinate as shown by Eq. (10).

$$\hat{\mathbf{I}}^m = \mathbf{Q}\mathbf{I}^m\mathbf{Q}^T, \hat{\mathbf{I}}^s = \mathbf{Q}\mathbf{I}^s\mathbf{Q}^T, \hat{\mathbf{M}}^m_{int} = \mathbf{Q}\mathbf{M}^m_{int}, \hat{\mathbf{M}}^s_{int} = \mathbf{Q}\mathbf{M}^s_{int} \tag{10}$$

The equations of motion for nodes s and m are organized into matrix form, as shown by Eqs. (11) and (12), respectively.

$$\begin{bmatrix} \mathbf{I}^m_{ff} & \mathbf{I}^m_{fc} \\ \mathbf{I}^m_{cf} & \mathbf{I}^m_{cc} \end{bmatrix} \begin{Bmatrix} \ddot{\boldsymbol{\theta}}^m_f \\ \ddot{\boldsymbol{\theta}}^m_c \end{Bmatrix} = \begin{Bmatrix} \mathbf{M}^m_f \\ \mathbf{M}^m_c \end{Bmatrix} \tag{11}$$

where $\ddot{\boldsymbol{\theta}}^m_f = \left\{\ddot{\theta}^m_{\hat{x}}\right\}^T$, $\ddot{\boldsymbol{\theta}}^m_c = \left\{\ddot{\theta}^m_{\hat{y}} \quad \ddot{\theta}^m_{\hat{z}}\right\}^T$, and $\mathbf{I}^m_{fc} = (\mathbf{I}^m_{cf})^T$.

$$\begin{bmatrix} \mathbf{I}^s_{ff} & \mathbf{I}^s_{fc} \\ \mathbf{I}^s_{cf} & \mathbf{I}^s_{cc} \end{bmatrix} \begin{Bmatrix} \ddot{\boldsymbol{\theta}}^s_f \\ \ddot{\boldsymbol{\theta}}^s_c \end{Bmatrix} = \begin{Bmatrix} \mathbf{M}^s_f \\ \mathbf{M}^s_c \end{Bmatrix} \tag{12}$$

where $\ddot{\boldsymbol{\theta}}^s_f = \left\{\ddot{\theta}^s_{\hat{x}}\right\}^T$, $\ddot{\boldsymbol{\theta}}^s_c = \left\{\ddot{\theta}^s_{\hat{y}} \quad \ddot{\theta}^s_{\hat{z}}\right\}^T$, and $\mathbf{I}^s_{fc} = (\mathbf{I}^s_{cf})^T$. Deploying the second row of Eqs. (11) and (12), and noting $\ddot{\boldsymbol{\theta}}^s_c = \ddot{\boldsymbol{\theta}}^m_c$, yields

$$\mathbf{I}^m_{cc}\ddot{\boldsymbol{\theta}}^m_c = \mathbf{M}^m_c - \mathbf{I}^m_{cf}\ddot{\boldsymbol{\theta}}^m_f \tag{13}$$

$$\mathbf{I}^{cc}_s\ddot{\boldsymbol{\theta}}^m_c = \mathbf{M}^s_c - \mathbf{I}^s_{cf}\ddot{\boldsymbol{\theta}}^s_f \tag{14}$$

Adding Eqs. (13) and (15) side by side, one obtains

$$(\mathbf{I}^m_{cc} + \mathbf{I}^s_{cc})\ddot{\boldsymbol{\theta}}^m_c = (\mathbf{M}^m_c + \mathbf{M}^s_c) - \mathbf{I}^m_{cf}\ddot{\boldsymbol{\theta}}^m_f - \mathbf{I}^s_{cf}\ddot{\boldsymbol{\theta}}^s_f \tag{15}$$

Let $\mathbf{I}^m_{cc} + \mathbf{I}^s_{cc} = \mathbf{I}^{sm}_{cc}$ and $\mathbf{M}^m_c + \mathbf{M}^s_c = \mathbf{M}^{sm}_c$. This may lead to:

$$\ddot{\boldsymbol{\theta}}^m_c = (\mathbf{I}^{sm}_{cc})^{-1}\mathbf{M}^{sm}_c - (\mathbf{I}^{sm}_{cc})^{-1}\mathbf{I}^m_{cf}\ddot{\boldsymbol{\theta}}^m_f - (\mathbf{I}^{sm}_{cc})^{-1}\mathbf{I}^s_{cf}\ddot{\boldsymbol{\theta}}^s_f \tag{16}$$

Similarly, deploying the first row of Eqs. (11) and (12), yields

$$\mathbf{I}^m_{ff}\ddot{\boldsymbol{\theta}}^m_f + \mathbf{I}^m_{fc}\ddot{\boldsymbol{\theta}}^m_c = \mathbf{M}^m_f \tag{17}$$

$$\mathbf{I}^s_{ff}\ddot{\boldsymbol{\theta}}^s_f + \mathbf{I}^s_{fc}\ddot{\boldsymbol{\theta}}^m_c = \mathbf{M}^s_f \tag{18}$$

Substitute Eq. (16) into Eqs. (17) and (18), one obtains:

$$\left[\mathbf{I}_{ff}^m - \mathbf{I}_{fc}^m(\mathbf{I}_{cc}^{sm})^{-1}\mathbf{I}_{cf}^m \quad -\mathbf{I}_{fc}^m(\mathbf{I}_{cc}^{sm})^{-1}\mathbf{I}_{cf}^s\right]\left\{\begin{matrix}\ddot{\boldsymbol{\theta}}_f^m \\ \ddot{\boldsymbol{\theta}}_f^s\end{matrix}\right\} = \mathbf{M}_f^m - \mathbf{I}_{fc}^m(\mathbf{I}_{cc}^{sm})^{-1}\mathbf{M}_c^{sm} \qquad (19)$$

$$\left[-\mathbf{I}_{fc}^s(\mathbf{I}_{cc}^{sm})^{-1}\mathbf{I}_{cf}^m \quad \mathbf{I}_{ff}^s - \mathbf{I}_{fc}^s(\mathbf{I}_{cc}^{sm})^{-1}\mathbf{I}_{cf}^s\right]\left\{\begin{matrix}\ddot{\boldsymbol{\theta}}_f^m \\ \ddot{\boldsymbol{\theta}}_f^s\end{matrix}\right\} = \mathbf{M}_f^s - \mathbf{I}_{fc}^s(\mathbf{I}_{cc}^{sm})^{-1}\mathbf{M}_c^{sm} \qquad (20)$$

Combining Eqs. (19) and (20), yields the equations of motion for nodes m and s:

$$\begin{bmatrix}\mathbf{I}_{ff}^m - \mathbf{I}_{fc}^m(\mathbf{I}_{cc}^{sm})^{-1}\mathbf{I}_{cf}^m & -\mathbf{I}_{fc}^m(\mathbf{I}_{cc}^{sm})^{-1}\mathbf{I}_{cf}^s \\ -\mathbf{I}_{fc}^s(\mathbf{I}_{cc}^{sm})^{-1}\mathbf{I}_{cf}^m & \mathbf{I}_{ff}^s - \mathbf{I}_{fc}^s(\mathbf{I}_{cc}^{sm})^{-1}\mathbf{I}_{cf}^s\end{bmatrix}\left\{\begin{matrix}\ddot{\boldsymbol{\theta}}_f^m \\ \ddot{\boldsymbol{\theta}}_f^s\end{matrix}\right\} = \left\{\begin{matrix}\mathbf{M}_f^m - \mathbf{I}_{fc}^m(\mathbf{I}_{cc}^{sm})^{-1}\mathbf{M}_c^{sm} \\ \mathbf{M}_f^s - \mathbf{I}_{fc}^s(\mathbf{I}_{cc}^{sm})^{-1}\mathbf{M}_c^{sm}\end{matrix}\right\} \qquad (21)$$

The angular acceleration of mass points, $\ddot{\boldsymbol{\theta}}_f^m$ and $\ddot{\boldsymbol{\theta}}_f^s$, can now be solved in local coordinate. After obtaining $\ddot{\boldsymbol{\theta}}_f^m$ and $\ddot{\boldsymbol{\theta}}_f^s$, $\ddot{\boldsymbol{\theta}}_c^m$ can be solved via Eq. (16).

### 3.3 Time Integration

To solve the equations of motion of particles within a path unit, for instance, $t_a \leq t \leq t_b$, it requires the techniques of time integration. In this work, the explicit central difference time integration schemes is applied. The next step of nodal position can be written as:

$$d_{t+\Delta t} = \frac{(\Delta t)^2(\mathbf{f}_t^{ext} - \mathbf{f}_t^{int})}{m} + 2d_t - d_{t-\Delta t} \qquad (22)$$

Likewise, the revolute joint angular displacement, at time $(t + \Delta t)$ can be written as:

$$\hat{\boldsymbol{\theta}}_{t+\Delta t}^m = 2\hat{\boldsymbol{\theta}}_t^m + (\Delta t)^2\ddot{\hat{\boldsymbol{\theta}}}_t^m - \hat{\boldsymbol{\theta}}_{t-\Delta t}^m$$
$$\hat{\boldsymbol{\theta}}_{t+\Delta t}^s = 2\hat{\boldsymbol{\theta}}_t^s + (\Delta t)^2\ddot{\hat{\boldsymbol{\theta}}}_t^s - \hat{\boldsymbol{\theta}}_{t-\Delta t}^s \qquad (23)$$

Figure 4 also shows the flowchart of the proposed algorithm.

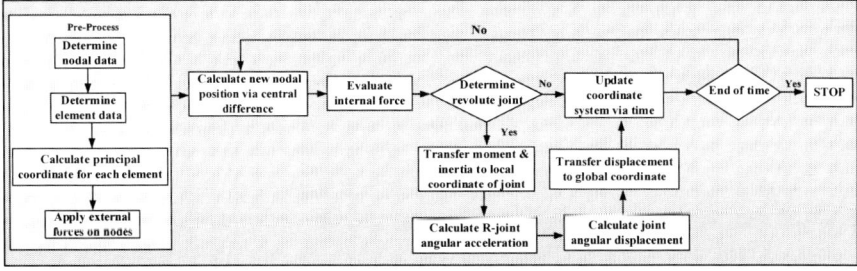

**Fig. 4.** Flowchart of VFIFE

## 4  Numerical Simulation

A slider-crank mechanism shown in Fig. 5 is used to illustrate the procedure developed. The connecting rod is assumed more flexible than the crankshaft, and the slider block is massless. The material and geometric properties are shown in Table 1. A constant time step $\Delta t = 1 \times 10^{-5}$ sec is used, and two element lengths, 0.019 m and 0.076 m are respectively assumed. A torque given by Eq. (24) is applied at the crankshaft O.

$$T_1(t) = \begin{cases} 0.01 \left[1 - \exp(\frac{-t}{0.167})\right] & t \leq 0.7 \text{ sec} \\ 0 & t \geq 0.7 \text{ sec} \end{cases} \tag{24}$$

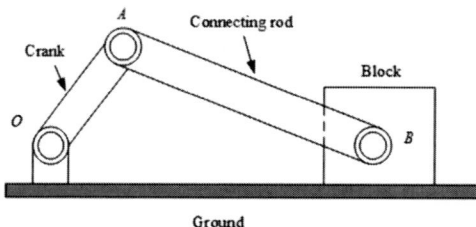

**Fig. 5.** Slider-Crank mechanism

Figure 6(a) shows the resulting response of the slider while Fig. 6(b) is the response of transverse deformation of the midpoint on the connecting rod. It can be observed from Fig. 6(b). These results show good agreement with those presented in reference [5]. Accordance with the above, not only verify as the revolute joint couples with flexible body which can analyzed by VFIFE, but also still maintains calculation accuracy.

**Table 1.**  Parameters of the slider-crank

|  | Crank | Connecting rod |
|---|---|---|
| Length (m) | 0.152 | 0.304 |
| Area (m$^2$) | 7.85E−05 | 7.85E−05 |
| Density (kg/m$^3$) | 2.77E +03 | 2.77E+03 |
| Area moment of inertia (m$^4$) | 4.91E−10 | 4.91E−10 |
| Young's modulus (N/m$^2$) | 1.00E+09 | 5.00E+07 |

Discussion

The Lagrange multiplier method together with Newmark implicit time integration has been widely used to solve the elastodynamic problem with constraints. In the method, the time step may be set larger than that in the explicit method; however, velocity and displacement have to be corrected to remain the accuracy in each step. Here, VFIFE uses explicit time integration schemes which provides better accuracy and stability. Another advantage is the proposed method avoids dealing with the matrix inverse as encountered in the Lagrange multiplier method.

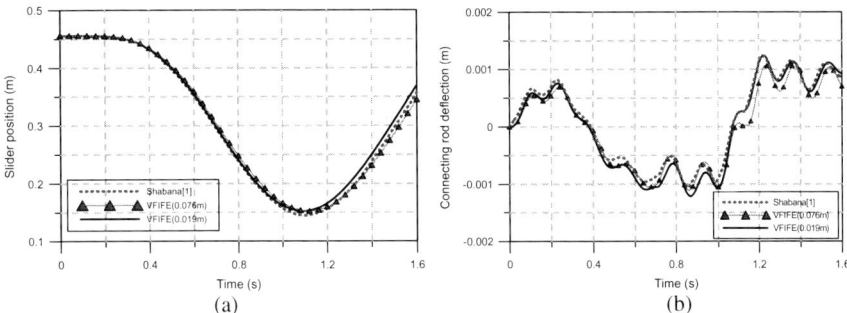

**Fig. 6.** Response of flexible slider crank. a Slider positon, b Transverse deformation of the mid-point

## 5   Summary

A procedure for planar flexible multibody system has been established in this work. Based on VFIFE, a method for formulating the constraint by revolute joint was also presented. This method avoids matrix inverse and is able to handle the flexible multibody systems with large deformation and large rotation effectively.

## References

1. Bathe, K.J., Ramm, E., Wilson, E.L.: Finite element formulations for large deformation dynamic analysis. Int. J. Numer. Methods Eng. **9**, 353–386 (1975)
2. Bathe, K.J.: Finite Element Procedures. Prentice-Hall, New York (1996)
3. Shabana, A.A.: Flexible multibody dynamics: review of past and recent developments. Multibody Syst. Dyn. **1**, 189–222 (1997)
4. Shabana, A.A.: Dynamics of Multibody Systems, 3rd edn. Wiley, New York (1998)
5. Escalona, J.L., Hussien, H.A., Shabana, A.A.: Application of the absolute nodal co-ordinate formulation to multibody system dynamics. J. Sound Vib. **214**(5), 833–851 (1998)
6. Lee, M.G., Chen, C.I.: Numerical solution of a flexible structure by differential algebraic equations. In: International Conference of Computational Method in Sciences and Engineering, vol. 3, pp. 350–359 (2003)
7. Ting, E.C., Shi, C., Wang, Y.K.: Fundamentals of a vector form intrinsic finite element: part I. Basic procedure and a planar frame element. J. Mech. **20**, 113–122 (2004)
8. Ting, E.C., Shi, C., Wang, Y.K.: Fundamentals of a vector form intrinsic finite element: part II. Plane solid element. J. Mech. **20**, 123–132 (2004)
9. Wasfy, T.M., Noor, A.K.: Computational strategies for flexible multibody systems. Appl. Mech. Rev. **56**(6), 553–613 (2003)

10. Wu, T.Y., Ting, E.C.: Large deflection analysis of 3D membrane structures by a 4-node quadrilateral intrinsic element. Thin-Walled Struct. **46**(3), 261–275 (2008)
11. Wu, T.Y., Wang, R.Z., Wang, C.Y.: Large deflection analysis of flexible planar frame. J. Chin. Inst. Eng. **29**, 593–606 (2006)
12. Wu, T.Y., Wang, C.Y., Chuang, C.C., Ting, E.C.: Motion analysis of 3D membrane structures by a vector form intrinsic finite element. J. Chin. Inst. Eng. **30**, 961–976 (2007)
13. Wu, T.Y., Wu, J.H., Ho, J.M., Chuang, C.C., Wang, R.Z., Wang, C.Y.: A study on motion analysis of 3D solids by a vector form intrinsic finite element. J. Chin. Inst. Civ. Hydraul. Eng. **19**, 79–89 (2007)

# On Classical Newmark Integration
# of Multibody Dynamics

Haritz Uriarte$^{(\boxtimes)}$, Igor Fernández de Bustos, Gorka Urkullu,
and Ander Olabarrieta

Department of Mechanical Engineering, School of Engineering of Bilbao,
University of the Basque Country (UPV/EHU), Paseo Rafael Moreno "Pitxitxi",
3, 48013 Bilbao, Vizcaya, Spain
{haritz.uriarte,igor.fernandezdebustos,
gorka.urkullu}@ehu.es, anderolabarrieta@gmail.com

**Abstract.** The use of the classical Newmark method for the integration of
Multibody System Dynamics (MBSD) is presented. This approach has the
advantage of directly integrating the second order differential equations which
appear in MBSD and, thus, does not require duplication of the variables,
reducing the computational cost. The resolution of each step is performed in a
full Newton approach, instead of using less efficient quasi-Newton approaches.
This requires the analytic computation of derivatives, but improves in convergence and precision. An implementation for 2D problems using Newton-Euler
formalism and cartesian coordinates has been developed to test the system. An
example is included.

**Keywords:** Multibody dynamics · Newmark integration
Newton-Euler formalism · Cartesian coordinates

## 1 Introduction

Rigid Multibody Dynamics is one of the most important types of analysis in
mechanics. Taking into account the application, one can classify the methods into two
big categories: Those which are used to simulate a determinate system and, thus, can
introduce particular expressions which can reduce computational cost and those which
are general and are formulated in a way that is independent of the topology. The first
kind of methods are often necessary in real time simulation because, for general
methods, the computational cost is high [1]. Another classification can be made taking
into account the formalism employed to derive the equilibrium equations. Thus, one
can use Newton-Euler equations, Lagrange's equations of the first or second Kind,
d'Alembert's principle of virtual work or Kane's equations. As exposed in [2], all of
these formalisms have its advantages and disadvantages. Another classification can be
done regarding the coordinate system employed to define the pose of the elements in
the space, thus, one can use cartesian coordinates, relative coordinates or the so called
natural coordinates [1, 3]. Once the problem is formulated, the integration process
begins. In order to do so, the usual approach in dynamics is to use a reduction of the
system to a first order of differential equations which is afterwards integrated by means
of a first order integration algorithm, such as Runge Kutta, Euler, or any other [3, 4].

© Springer Nature Switzerland AG 2019
B. Corves et al. (Eds.): EuCoMeS 2018, MMS 59, pp. 97–105, 2019.
https://doi.org/10.1007/978-3-319-98020-1_12

The problem with this approach is that the use of these kind of methods leads to the duplication of the amount of variables, which means a severe damage to performance. This can be somehow alleviated by the use of recursive formulations [6]. An alternative to this is the use of second order methods which do not require the system to be reduced to first order, thus not duplicating the equations. Although these methods (central differences, Newmark, Wilson-θ) have been intensively used in structural mechanics (specially in finite element analysis) it is quite seldom found in the multi-body dynamic field. An example of this application can be found in [5], but the presentation does not give any information on how was implemented, stating only that Newmark was used along with a quasi-Newton approach. In this presentation a full Newmark formulation is presented in 2D, employing a full Newton approach. This formulation has been verified with several examples (one is shown in the presentation) and has the advantages of simplicity, stability and, thus, it is of good use for teaching multibody dynamics. Furthermore, although currently it has been implemented in Octave, with little profiling, and thus the real computational efficiency cannot be experimentally studied, it should perform quite well in a well tailored implementation.

## 2   Newton-Euler Equation Integration with Newmark

We consider the Newton-Euler equation for 2D for a single solid:

$$\begin{Bmatrix} f_x \\ f_y \\ t_z \end{Bmatrix} = \begin{bmatrix} m & 0 & 0 \\ 0 & m & 0 \\ 0 & 0 & I_z \end{bmatrix} \begin{Bmatrix} \ddot{x} \\ \ddot{y} \\ \ddot{\theta} \end{Bmatrix} \tag{1}$$

In order to use it as a basis for a multi-body dynamic solver, one can expand and assembly that equation for the N solids in the system:

$$\begin{Bmatrix} f_{1x} \\ f_{1y} \\ t_{1z} \\ . \\ . \\ . \\ f_{nx} \\ f_{ny} \\ t_{nz} \end{Bmatrix} = \begin{bmatrix} m_1 & 0 & 0 & . & . & . & 0 & 0 & 0 \\ 0 & m_1 & 0 & . & . & . & 0 & 0 & 0 \\ 0 & 0 & I_{z1} & . & . & . & 0 & 0 & 0 \\ . & . & . & . & . & . & . & . & . \\ . & . & . & . & . & . & . & . & . \\ . & . & . & . & . & . & . & . & . \\ 0 & 0 & 0 & . & . & . & m_n & 0 & 0 \\ 0 & 0 & 0 & . & . & . & 0 & m_n & 0 \\ 0 & 0 & 0 & . & . & . & 0 & 0 & I_{zn} \end{bmatrix} \begin{Bmatrix} \ddot{x}_1 \\ \ddot{y}_1 \\ \ddot{\theta}_1 \\ . \\ . \\ . \\ \ddot{x}_n \\ \ddot{y}_n \\ \ddot{\theta}_n \end{Bmatrix} \tag{2}$$

In the force vector one must take into account the existence of elastic, dissipative forces and/or forces generated by joints. If one is to take advantage of the Newmark algorithm, these forces should be expressed in the form:

$$f_{total} = f + f_K + f_C + f_j = f - Kx - C\dot{x} + H^T \lambda \tag{3}$$

Being:

$K$ :    stiffness matrix of the elastic elements
$C$ :    damping matrix of the dissipative forces
$X$ :    displacement vector
$H^T$ :  restriction matrix (see later)
$\lambda$ :    generalized reaction vector (see later)

Which leads to:

$$M\ddot{x} + C\dot{x} + Kx = f + H^T \lambda \tag{4}$$

Newmark is an implicit method, and, thus, in the recursive formula one considers equilibrium in $t + \Delta t$, reaching eq.

$$M\ddot{x}(t + \Delta t) + C\dot{x}(t + \Delta t) + Kx(t + \Delta t) = f(t + \Delta t) + H^T \lambda \tag{5}$$

Applying Newmark algorithm and derivatives (12) and (13), one reaches:

$$Ax(t + \Delta t) = f(t + \Delta t) + M(a_0 x(t) + a_1 \dot{x}(t) + a_2 \ddot{x}(t)) + C(a_5 x(t) + a_6 \dot{x}(t) + a_7 \ddot{x}(t)) + H^T \lambda \tag{6}$$

being:

$$A = a_0 M + a_5 C + K, \quad a_0 = \frac{1}{\beta (\Delta t)^2}, \quad a_1 = \frac{1}{\beta \Delta t}, \quad a_2 = \frac{1}{2\beta} - 1, \quad a_3 = \Delta t(1 - \alpha),$$

$$a_4 = \alpha \Delta t, \quad a_5 = \frac{\alpha}{\beta \Delta t}, \quad a_6 = \frac{\alpha}{\beta} - 1, \quad a_7 = \frac{\Delta t}{2}\left(\frac{\alpha}{\beta} - 2\right)$$

One must include the restrictions imposed by the joints of the mechanism. These restrictions can be always written in the form:

$$H(x, t)x = b(x, t) \tag{7}$$

Although, in most cases, one can write:

$$H(x)x = b(x, t) \tag{8}$$

One of the possible methods to introduce these restrictions in the equilibrium formula is using the null-subspace method. One can obtain, from (7), evaluated in

$$t + \Delta t :$$

$$x(t + \Delta t) = x_p(t + \Delta t) + N_h(t + \Delta t)\alpha \tag{9}$$

being:
$x_p(t + \Delta t)$ :    a particular solution of (7)
$N_h(t + \Delta t)$ :    the null subspace of $H(x(t + \Delta t), t + \Delta t)$

It is advisable to use the minimal least squares solution for $x_p$. One can reach:

$$AN_h(t+\Delta t)\alpha = f(t+\Delta t) + M(a_0 x(t) + a_1 \dot{x}(t) + a_2 \ddot{x}(t)) + C(a_5 x(t) + a_6 \dot{x}(t) + a_7 \ddot{x}(t)) + H^T\lambda - Ax_p(t+\Delta t)$$

(10)

In order to get rid of $\lambda$, one multiplies this equation by $N_h^T$:

$$N_h^T(t+\Delta t)AN_h(t+\Delta t)\alpha$$
$$= N_h^T(t+\Delta t)\left[f(t+\Delta t) + M(a_0 x(t) + a_1 \dot{x}(t) + a_2 \ddot{x}(t)) + C(a_5 x(t) + a_6 \dot{x}(t) + a_7 \ddot{x}(t)) - Ax_p(t+\Delta t)\right]$$

(11)

And one obtains $x(t + \Delta t)$ by means of solving (9). Obviously, the resolution of Eq. (11) is an iterative process. In this paper a full Newton approach will be presented. Once obtained $x(t + \Delta t)$, one can obtain its derivatives by using eq.

$$\ddot{x}(t + \Delta t) = a_0(x(t + \Delta t) - x(t)) - a_1 \dot{x}(t) - a_2 \ddot{x}(t)$$

(12)

$$\dot{x}(t + \Delta t) = \dot{x}(t) + a_3 \ddot{x}(t) + a_4 \ddot{x}(t + \Delta t)$$

(13)

## 3    Resolution of the Non-linear System by Means of a Newton Method

In order to use a full Newton approach to solve the system composed by Eqs. (11) and (7), one needs to perform a linear approximation of both the restrictions and the matrices. To keep this presentation short, this paper will be limited to torsional springs and dampers, where the matrices do not depend on $x$, and, thus, linearization is not needed. The R joint restriction is introduced as described in Fig. 1.

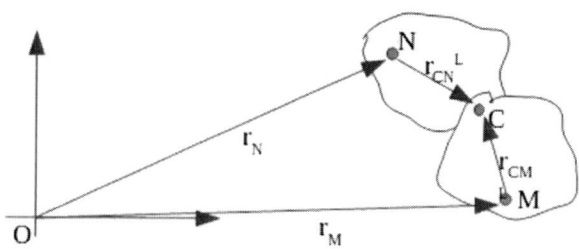

**Fig. 1.** R joint modelization

The restriction introduced by the R joint can be written as:

$$r_c = r_N + R_N r_{CN}^L = r_M + R_M r_{CM}^L \qquad (14)$$

Being:

$r_c$ :    coordinates of the R joint
$r_N$ :    coordinates of the gravity centre of the solid N
$r_M$ :    coordinates of the gravity centre of the solid M
$R_N$ :    rotation matrix of the solid N
$R_M$ :    rotation matrix of the solid M
$r_{CN}^L$ :    coordinates of the CN vector in N's local coordinate system
$r_{CM}^L$ :    coordinates of the CM vector in M's local coordinate system

One can linealize the left part of Eq. (14) as:

$$r_N + R_N(\theta_N)r_{CN}^L = r_N + R_N^j r_{CN}^L + \left| \frac{\partial R_N(\theta_N)}{\partial \theta_N} \right|_{\theta_N^j} \left( \theta_{N r_{CN}^L} - \theta_N^j \right) \qquad (15)$$

Being:

$R_N^j$ :    the rotation matrix of N in the previous time step
$\theta_N^j$ :    angular position of N in the previous time step

The same can be done for the right part of Eq. (14). This leads to an equation of the form:

$$H_c x = b_c$$

where:

$$H_c = \begin{bmatrix} \dots & 1 & 0 & -x_{CN}^L \sin\theta_{CN}^j - y_{CN}^L \cos\theta_{CN}^j & \dots & -1 & 0 & x_{CM}^L \sin\theta_{CM}^j + y_{CM}^L \cos\theta_{CM}^j & \dots \\ \dots & 0 & 1 & x_{CN}^L \cos\theta_{CN}^j - y_{CN}^L \sin\theta_{CN}^j & \dots & 0 & -1 & -x_{CM}^L \cos\theta_{CM}^j + y_{CM}^L \sin\theta_{CM}^j & \dots \end{bmatrix} \qquad (16)$$

and:

$$b_c = \begin{bmatrix} \cos\theta_{CM}^j & -\sin\theta_{CM}^j \\ \sin\theta_{CM}^j & \cos\theta_{CM}^j \end{bmatrix} \begin{Bmatrix} x_{CM}^L \\ y_{CM}^L \end{Bmatrix} - \begin{bmatrix} \cos\theta_{CN}^j & -\sin\theta_{CN}^j \\ \sin\theta_{CN}^j & \cos\theta_{CN}^j \end{bmatrix} \begin{Bmatrix} x_{CN}^L \\ y_{CN}^L \end{Bmatrix} + \begin{Bmatrix} -x_{CN}^L \sin\theta_{CN}^j - y_{CN}^L \cos\theta_{CN}^j \\ x_{CN}^L \cos\theta_{CN}^j - y_{CN}^L \sin\theta_{CN}^j \end{Bmatrix} \theta_{CN}^j$$
$$- \begin{Bmatrix} -x_{CM}^L \sin\theta_{CM}^j - y_{CM}^L \cos\theta_{CM}^j \\ x_{CM}^L \cos\theta_{CM}^j - y_{CM}^L \sin\theta_{CM}^j \end{Bmatrix} \theta_{CM}^j \qquad (17)$$

One can add as many joints as needed by stacking the corresponding restrictions. Similar developments can be performed for other joints, such as P-joints. The restrictions of fixed elements can be introduced by means of a restriction in the form (18), which is also stacked into the global restriction matrix $H$ and the global restriction vector $b$.

$$[.. \quad I \quad ...] = x_f \tag{18}$$

being $x_f$ the coordinates of the fixed element.

## 4  Initial Conditions

In the beginning of the simulation, one has $x(0)$ and $\dot{x}(0)$. $\ddot{x}(0)$ is also needed, and it can be derived from the equilibrium equation evaluated in $t = 0$.

$$M\ddot{x}(0) + C\dot{x}(0) + Kx(0) = f(0) + H(0)^T \lambda(0) \tag{19}$$

The problem here is that, as one needs to obtain accelerations, the restrictions must be formulated in terms of accelerations, which is achieved by taking the derivative of (14) twice:

$$\ddot{x}_N + \ddot{\theta}_N \left|\frac{\partial R_N(\theta_N)}{\partial \theta_N}\right|_{\theta_N^j} r_{CN}^L - \dot{\theta}_N^2 [R_N] r_{CN}^L = \ddot{x}_M + \ddot{\theta}_M \left|\frac{\partial R_M(\theta_M)}{\partial \theta_M}\right|_{\theta_M^j} r_{CM}^L - \dot{\theta}_M^2 [R_M] r_{CM}^L \tag{20}$$

Rearranging items, an equation with the following form is achieved:

$$H_{ac}\ddot{x}_{ac} = b_{ac} \tag{21}$$

Being:

$$H_{ac} = H_c \tag{22}$$

$$\ddot{x}_{ac} = \left\{ \ddot{x}_N \quad \ddot{y}_N \quad \ddot{\theta}_N \quad . \quad . \quad . \quad \ddot{x}_M \quad \ddot{y}_M \quad \ddot{\theta}_M \right\}^T \tag{23}$$

$$b_{ac} = \dot{\theta}_N^2 [R_N] \left\{ \begin{matrix} x_N^L \\ y_N^L \end{matrix} \right\} - \dot{\theta}_N^2 [R_M] \left\{ \begin{matrix} x_M^L \\ y_M^L \end{matrix} \right\} \tag{24}$$

From Eq. (21) a particular solution is obtained and, after introducing the restrictions in the equilibrium formula (19) with the null subspace method, leading to Eq. (25). The initial acceleration is thus obtained with Eq. (26).

$$N_{ha}^T M (\ddot{x}_{Pa} + N_{ha}\alpha_a) = N_{ha}^T (f - C\dot{x}(0) - Kx(0)) \tag{25}$$

$$\ddot{x} = x_{Pa} + N_H^T \alpha_a \tag{26}$$

## 5   Example of Application

As an example, the following system (taken from [5]) will be solved. It's a double pendulum, with a solid body hanging from a fixed point, and a second body hanging from the other side of the first one. It can be represented with the following image:

The full information about the problem is presented in Tables 1 and 2. Both solids are simulated as cylindrical bars, whose length is the separation between both rotation constraints (Fig. 2).

**Table 1.**  Initial conditions and mass properties

| | Initial conditions | | | | | | | |
|---|---|---|---|---|---|---|---|---|
| | x (m) | y (m) | $\theta$(rad) | $v_x$(m/s) | $v_y$(m/s) | $\omega$(rad/s) | Mass (kg) | Length (m) |
| n | 1 | 0 | 0 | 0 | 0 | 0 | 3 | 2 |
| m | 3,4488887 | −0,388228 | | 3,88228 | 14,488887 | 10 | 0,3 | 3 |

**Table 2.**  Stiffness and damping properties

| | K (N/m) | C (N*s/m) |
|---|---|---|
| F joint | 400 | 15 |
| R joint | $310^5$ | $510^4$ |

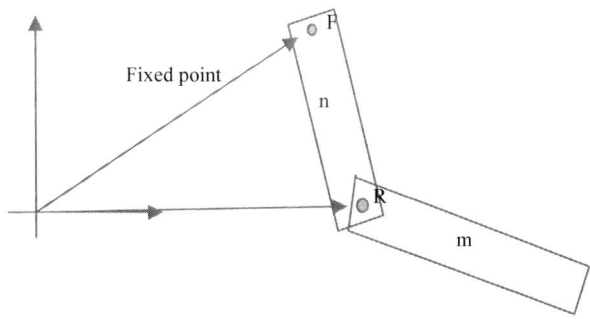

**Fig. 2.**  Example diagram

For the simulation, the following values have been used:

$$\Delta t = 0,001 seg$$

$$t = 10 seg$$

$$\alpha = 0,501$$

$$\beta = 0,2507$$

The obtained angular position and velocity of both centres of mass are shown in Figs. 3 and 4. They match the results presented in [5].

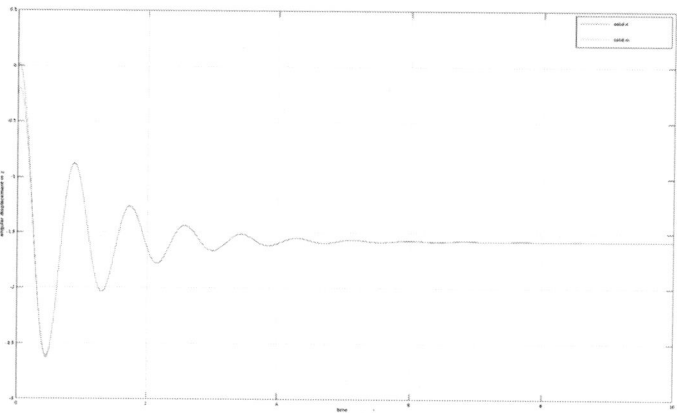

**Fig. 3.** Angular position over time

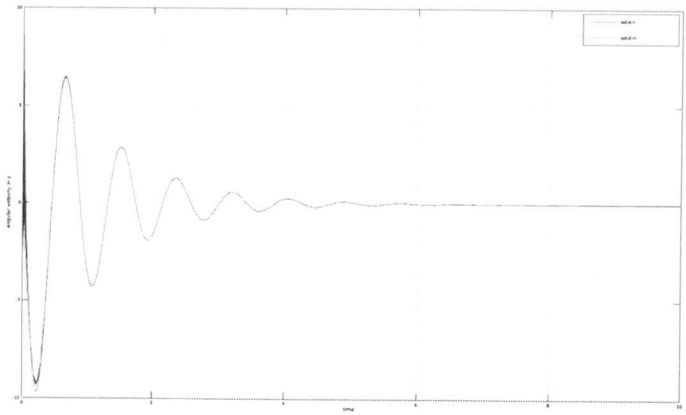

**Fig. 4.** Angular velocity over time

## 6   Conclusions

The use of second order methods for the integration of Multibody Dynamics, although not quite explored, is a useful option to be considered. The particular case of the Newmark method usually employed in structural dynamics has shown not only to be applicable, but also a reliable approach due to its known stability. The use of a full Newton scheme to solve the equilibrium in each iteration improves convergence and speed. Thanks to its simplicity, when applied to bidimensional problems this approach is quite didactic, not only for teaching dynamics, but also for a general introduction to non linearities.

**Acknowledgments.** The authors wish to acknowledge the Spanish and the Basque Goverment for the funding of this project through project DPI2016-80372-R, and IT947-16. The authors also want to thank Prof. J. Cuadrado for his advices in the development of this work.

# References

1. Cuadrado, J., Cardenal, J., Bayo, E.: Modeling and solution methods for efficient real-time simulation of multibody dynamics. Multibody Syst. Dyn. **1**, 259 (1997). https://doi.org/10. 1023/A:1009754006096
2. Slaats, P.M.A.: Recursive Formulations in Multibody Dynamics. Technische Universiteit Eindhoven, Eindhoven (1991)
3. Shabana, A.A.: Dynamics of Multibody Systems, 4th edn. Springer, Berlin (2013)
4. Haugh, E.J.: Computer-Aided Kinematics and Dynamics of Mechanical Systems. Allyn and Bacon, Boston (1989)
5. Gavrea, B., Negrut, D., Potra, F.A.: The Newmark integration method for simulation of multibody systems: analytical considerations. In: Proceedings of IMECE 2005. ASME International Mechanical Engineering Congress and Exposition 2005, 5–11 November, Orlando, USA (2005)
6. Müller, A.: Screw and Lie group theory in multibody dynamics: recursive algorithms and equations of motion of tree-topology systems. Multibody Syst. Dyn. (2017). https://doi.org/ 10.1007/s11044-017-9583-6

# Experimental Mechanics

# Velocity Characteristics of Active Omni Wheel Considering Transmitting Mechanism

Masaharu Komori[✉] and Kippei Matsuda

Kyoto University, Kyoto, Japan
komorim@me.kyoto-u.ac.jp, matsuda_kippei@khi.co.jp

**Abstract.** Transportation vehicles and mobile robots that can move in an arbitrary direction on a floor are currently in high demand. The authors proposed a novel mechanism called the active omni wheel, which is able to actively move in an arbitrary direction. The active omni wheel is composed of a main body and outer rollers. Both the main body and the outer rollers can be rotated actively by using a differential gear mechanism. The moving velocity of the active omni wheel was analyzed. It was clarified that the maximum velocity are different under different moving directions and that the wheel can output the maximum velocity in the forward–backward and transverse directions. An active omni wheel was produced. Experiments on the wheel showed that the main body and the outer rollers can be rotated actively and that the rotation agrees with the theoretical formulas.

**Keywords:** Omni wheel · Active motion · Arbitrary direction
Velocity

## 1 Introduction

Vehicles that can move in any direction are required for efficient process of transportation. However, transverse movement is impossible for conventional vehicles. Therefore, a novel mechanism to move in an arbitrary direction is needed. The omni wheel is an example of an omnidirectional wheel [1–6]. It is possible for free rollers along the outer circumference of the omni wheel to rotate passively. A motor can drive the main body of the omni wheel. Passive transverse movement is possible because free rollers can rotate passively. Therefore, two omni wheels enable active omnidirectional movement. Researches on the omni wheel in motion analysis, modeling [1–6], or control [7, 8] have been reported. However, the omni wheel has rotational resistance of the free rollers and has difficulty to move precisely. The omni wheel has another problem in that two wheels are necessary to travel in an arbitrary direction.

To solve these problems, we proposed the active omni wheel [9]. The active omni wheel can travel in any direction by itself. The main body of the wheel and the outer rollers can rotate actively due to a differential gear mechanism. In this study, the velocity characteristics of the active omni wheel are analyzed theoretically and clarified. Moreover, a prototype of the active omni wheel is produced. Experiments are conducted on the wheel, and the effectiveness of the active omni wheel is verified.

© Springer Nature Switzerland AG 2019
B. Corves et al. (Eds.): EuCoMeS 2018, MMS 59, pp. 109–116, 2019.
https://doi.org/10.1007/978-3-319-98020-1_13

## 2 Structure and Motion of Active Omni Wheel

The structure of the active omni wheel is shown in Fig. 1. The main body of the wheel and outer rollers are used for the wheel. The wheel has the drive mechanism with input shafts, a differential gear mechanism, and outer roller drive system. The outer roller is driven by the outer roller drive system and the rotation is transmitted by couplings between the outer rollers. The main body of the wheel supports the output shaft through a bearing.

The differential gear mechanism is shown in Fig. 2. When the rotational speed and direction of the two input shafts is the same, revolution of the output shaft around axis A occurs. The main body of the wheel rotates by the revolution of the output shaft. This means forward or backward movement of the active omni wheel. Conversely, when the rotational direction of the two input shafts is opposite but the rotational speed is the same, rotation of the output shaft around axis B occurs. The outer rollers rotate due to the rotation of the output shaft. This results in left or right movement of the active omni wheel. When the two input shafts rotate at different absolute value of the rotational speeds, the rotation of the output shaft around both axes A and B occurs. This rotation results in the rotation of both the main body of the wheel and the outer rollers and a diagonal movement of the active omni wheel is caused. It is possible for the active omni wheel to move in any direction by appropriate rotation of input shafts.

In the case of the conventional omni wheel, two omni wheels are necessary to move actively in an arbitrary direction. Three omni wheels are necessary for rotation. Conversely, a single active omni wheel can move actively in an arbitrary direction without other wheels. Rotation is possible by using two active omni wheels. With a conventional omni wheel, it is difficult to move to a precise target location because of the uncertainty of the rotational resistance of the free rollers rotating passively. The proposed active omni wheel can accurately control the motion because the motors control not only the rotation of the main body of the wheel but also that of the outer rollers. This makes it possible to accurately reach the target position.

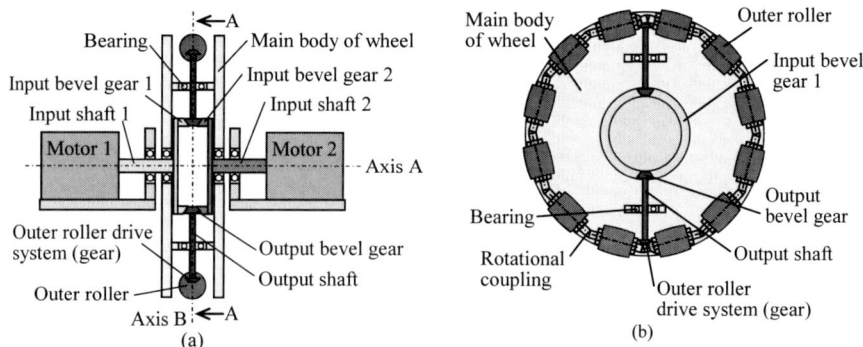

**Fig. 1.** Active omni wheel: (a) cross section including the rotation axis of the main body of the wheel, (b) cross section AA labeled in (a) [9]

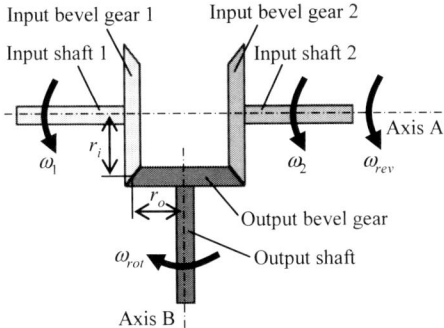

**Fig. 2.** Differential gear mechanism of active omni wheel [9]

## 3   Analysis of Characteristics of Moving Velocity

The motion characteristics of the active omni wheel are analyzed theoretically in this section. $\omega_1$ and $\omega_2$ are the angular velocities of input shafts 1 and 2, respectively. $r_i$ and $r_o$ are the representative radii of the input and output bevel gear, respectively. $\omega_{rev}$ and $\omega_{rot}$ are the angular velocities of the output shaft around axes A and B, respectively as shown in Fig. 2. If the radius of the main body is $R$, $R\omega_{rev}$ is the moving velocity of the main body. If the radius of the outer roller is $r$ and the reduction ratio of the outer roller drive system is $k$, $r\omega_{rot}/k$ is the moving velocity of the outer roller. $v_x$ is the forward–backward moving velocity due to the rotation of the main body and $v_y$ is the transverse moving velocity due to the rotation of the outer roller. $v_x$ and $v_y$ are given by

$$v_x = R/2(\omega_1 + \omega_2) \tag{1}$$

$$v_y = rr_i/2kr_o(\omega_1 - \omega_2). \tag{2}$$

Here, the condition $kr_oR/r_ir = 1$ is applied. In this case, the transverse moving velocity $v_y$ of the active omni wheel can be given as

$$v_y = R/2(\omega_1 - \omega_2). \tag{3}$$

In general, the motors driving input shafts 1 and 2 have an upper limit for the rotational velocity. The maximum value of the angular velocities $\omega_1$ and $\omega_2$ of input shafts 1 and 2 is assumed to be $\pm\omega_{imax}$. The active omni wheel moves forward when $v_x$ is positive and it moves to the left when $v_y$ is positive. The moving direction of the active omni wheel is analyzed. The moving direction $\phi$ is defined such that a value of $\phi = 0$ is the forward direction of the motion of the active omni wheel, and it is expressed as $\tan\phi = v_y/v_x$. Figure 3 shows the relationship between $\omega_1$, $\omega_2$, and $\phi$. From Fig. 3, it is understood that the values of $\omega_1$ and $\omega_2$ that yield a constant moving direction $\phi$ have a proportional relationship because the contour lines corresponding to constant $\phi$ values are straight lines that pass through the origin. That is, for a constant ratio of the values of $\omega_1$ and $\omega_2$, the moving direction $\phi$ of the active omni wheel remains constant. The equation $\tan\phi = (1 - \omega_2/\omega_1)/(1 + \omega_2/\omega_1)$ is obtained. From this expression, it can be confirmed that $\phi$ is constant if $\omega_2/\omega_1$ is constant.

The active omni wheel can move in any direction, but the characteristics of the velocity may differ in each direction. The moving velocity of the active omni wheel in each direction is analyzed here. The maximum moving velocity in each direction is shown in Fig. 4. A vector from the origin to a point on this graph is considered. In this graph, the direction of the vector indicates the moving direction $\phi$ and the length of the vector represents the moving velocity. The edges of the diamond in the graph indicate the maximum velocity that the wheel can achieve in each direction. At the upper right and lower left lines of the diamond, $\omega_1$ takes the maximum value ($\pm\omega_{imax}$). Conversely, at the lower right and upper left lines of the diamond, $\omega_2$ takes the maximum value. The wheel can maximize the moving velocity ($R\omega_{imax}$) if the moving direction is forward–backward or transverse. However, the moving velocity is limited to a lower value in the other directions. In particular, the possible velocity is the lowest in the directions of $\pm45°$ and $\pm135°$.

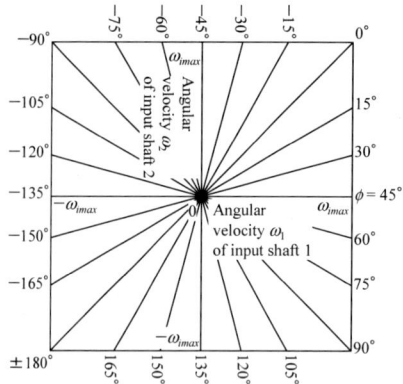

**Fig. 3.** Relationship between angular velocities of input shafts and the moving direction $\phi$ of the active omni wheel (contour lines correspond to constant moving directions)

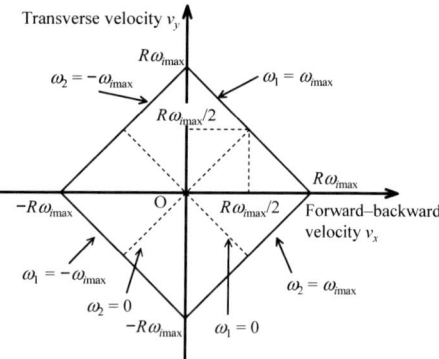

**Fig. 4.** Maximum moving velocity of active omni wheel under different moving directions

## 4   Production of Active Omni Wheel and Experiment

In this section, an active omni wheel is produced, and experiments are conducted. Figure 5 shows a photograph of the produced active omni wheel. Figure 6 shows the internal structure of the wheel. The output shaft of the differential gear mechanism is supported by the main body of the wheel via bearing 1. The rotation of the output shaft around axis A (Fig. 1(a)) is transmitted to the main body of the wheel via bearing 1, causing the main body to rotate. The outer roller drive system that connects the output shaft to the outer roller is composed of a gear set, transmission shaft, toothed belt, and pulley. Two sets of outer roller drive systems are incorporated into the wheel. Figure 7 shows the internal structure of the outer roller. There are two types of outer rollers: one is driven directly by the toothed belt of the outer roller drive system (2 rollers), and the other is not (10 rollers). Both types of outer rollers are supported by bearings and bearing support parts at both ends of the roller. The outer roller driven directly by the toothed belt is composed of two main roller parts, and a pulley is set between them. For the other outer rollers, which are not directly driven by the toothed belt, is almost the same as the outer roller mentioned above, although the outer roller body is not divided in two.

This outer roller is driven through universal joints by the outer roller that is driven directly by the toothed belt. The universal joint used in this wheel is composed of two joint sections. Therefore constant velocity is transmitted. The numbers of teeth in the input and output bevel gears and spur gears 1 and 2 are 45, 18, 60, and 30, respectively. The diameter of the wheel is 300 mm, and the diameter of the outer roller is 60 mm. The parameters are selected to satisfy $kr_oR/r_ir = 1$.

In the motion experiment, the relationship between the rotation of input shafts 1 and 2 and the rotation of the active omni wheel is examined by driving the two input shafts by hand. The rotation angles $\theta_m$ and $\theta_o$ of the main body and outer roller are given using the input shaft rotation angles $\theta_1$ and $\theta_2$ as $\theta_m = (\theta_1 + \theta_2)/2$ and $\theta_o = R(\theta_1 - \theta_2)/2r$. Because $R/r$ is 5 in the produced active omni wheel, $\theta_o = 5(\theta_1 - \theta_2)/2$ is obtained.

Figure 8 shows the experimental conditions. White labels are adhered to the outer rollers, and the state of rotation is recorded by observing these labels. Figure 8(a) shows photographs of the experiment in which the two input shafts are rotated in the same direction by the same angle. It was confirmed that the outer roller does not rotate, though the main body does. When the two input shafts are rotated by 360°, the main body rotates by the same angle. Figure 8(b) shows the experiment in which the two input shafts are rotated in the opposite direction by the same angle. A triangular label is adhered to the outer roller to allow the observation of the rotation of the outer roller. It is confirmed that the outer roller rotates and the main body does not. When the two input shafts are rotated by 72°, the outer roller rotates by 360°. In Fig. 8(c), one of the

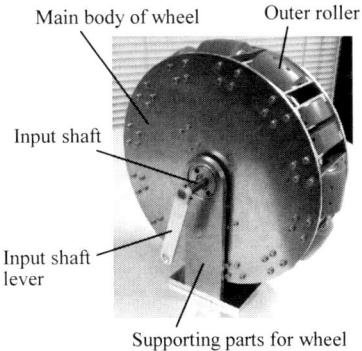

**Fig. 5.** Photograph of produced active omni wheel

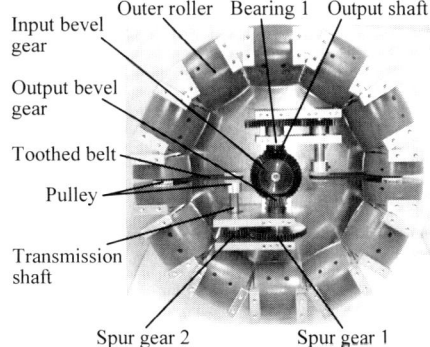

**Fig. 6.** Internal structure of produced active omni wheel

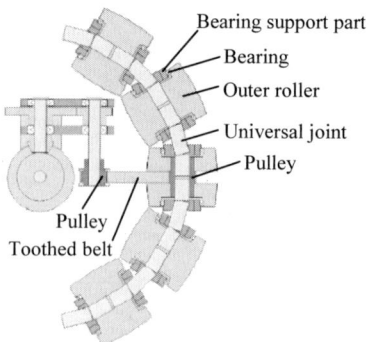

**Fig. 7.** Structure of active omni wheel

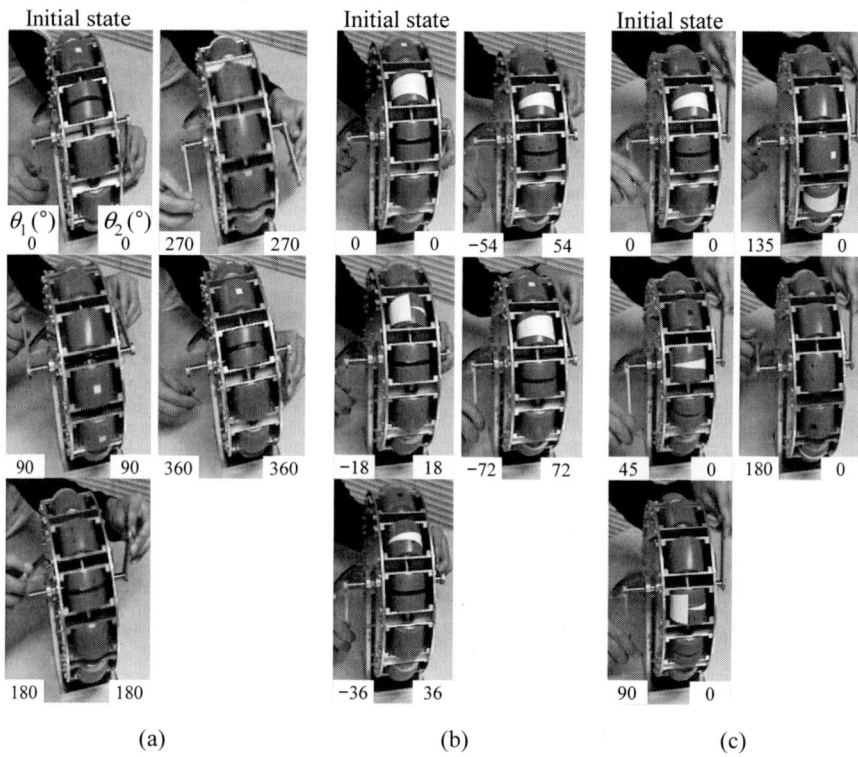

(a)    (b)    (c)

**Fig. 8.** Motion experiment of the active omni wheel, where $\theta_1$ and $\theta_2$ are rotation angles of input shafts: (a) Rotation in same direction, (b) Rotation in opposite direction, (c) Rotation with one side fixed

input shafts is rotated without rotating the other input shaft. It is confirmed that both the main body and the outer roller rotate. These experimental results confirm that the produced active omni wheel is able to make both the main body and the outer roller rotate actively. Moreover, the experimental relationship between the rotation angles of the main body, outer roller, and input shaft is shown to agree with the theoretical formulas. As a result, it was confirmed that the proposed active omni wheel can actively move in an arbitrary direction.

## 5    Conclusion

A transportation vehicle or mobile robot with high mobility performance that can move to a target location in an arbitrary direction on the floor is important to ensure work can proceed quickly and effectively in a limited space. We proposed a novel mechanism called the active omni wheel, which is able to actively move in an arbitrary direction. In this study, analysis and experiments on the active omni wheel were conducted. The moving velocity of the active omni wheel was analyzed. It was shown that the maximum velocity that the active omni wheel can realize are different under different moving directions and that the wheel can output the maximum velocity in the forward–backward or transverse directions. An active omni wheel was produced, and experiments were conducted. It was experimentally confirmed that it is possible to actively rotate the main body of the wheel and the outer roller by rotating the input shafts.

**Acknowledgment.** The authors are grateful to Mr. Hiroo Ohashi of Kyoto University for his contribution to this study.

## References

1. Leow, Y.P., Low, K.H., Loh, W.K.: Kinematic modelling and analysis of mobile robots with omni-directional wheels. In: Proceedings of ICARCV, Singapore, pp. 820–825. IEEE (2002)
2. Loh, W.K., Low, K.H., Leow, Y.P.: Mechatronics design and kinematic modelling of a singularityless omni-directional wheeled mobile robot. In: Proceedings of IEEE ICRA, Taipei, pp. 3237–3242. IEEE (2003)
3. Fisette, P., Ferriere, L., Raucent, B., Vaneghem, B.: A multibody approach for modelling universal wheels of mobile robots. Mech. Mach. Theory **35**(3), 329–351 (2000)
4. Williams II, R.L., Carter, B.E., Gallina, P., Rosati, G.: Dynamic model with slip for wheeled omnidirectional robots. IEEE Trans. Robot. Autom. **18**(3), 285–293 (2002)
5. Conceição, A.S., Moreira, A.P., Costa, P.J.: Practical approach of modeling and parameters estimation for omnidirectional mobile robots. IEEE/ASME Trans. Mechatron. **14**(3), 377–381 (2009)
6. Huang, H.C.: SoPC-based parallel ACO algorithm and its application to optimal motion controller design for intelligent omnidirectional mobile robots. IEEE Trans. Ind. Inform. **9**(4), 1828–1835 (2013)
7. Kim, H., Kim, B.K.: Online minimum-energy trajectory planning and control on a straight-line path for three-wheeled omnidirectional mobile robots. IEEE Trans. Ind. Electron. **61**(9), 4771–4779 (2014)

8. Sharbafi, M.A., Lucas, C., Daneshvar, R.: Motion control of omni-directional three-wheel robots by brain-emotional-learning-based intelligent controller. IEEE Trans. Syst. Man Cybern. C Appl. Rev. **40**(6), 630–638 (2010)
9. Komori, M., Matsuda, K., Terakawa, T., Takeoka, F., Nishihara, H., Ohashi, H.: Active omni wheel capable of active motion in arbitrary direction and omnidirectional vehicle. JAMDSM **10**(6), JAMDSM0086 (2016)

# Experimental Analysis of the Dynamic Behavior of a Non-stationary Two Stage Planetary Gearbox

Claudia Aide González-Cruz[1]($\boxtimes$), Marco Ceccarelli[2],
and Juan Carlos Jáuregui-Correa[1]

[1] Universidad Autónoma de Querétaro, 74010 Querétaro, Mexico
claudia.aide.gonzalez@gmail.com, jc.jauregui@uaq.mx
[2] LARM, University of Cassino and South Latium, 03043 Cassino, Italy
ceccarelli@unicas.it

**Abstract.** This paper presents a methodology to analyze the dynamic behavior of a two stages planetary gearbox working under non-stationary operational conditions. Experimental tests are carried out on a test bed, which consists in a motor drive that is coupled to the input shaft of a gearbox and a break system on the output shaft. The system is instrumented to measure the power consumption of the motor drive, the vibrations on the bearing next to the output shaft, the torques in both shafts, and the speed of the motor shaft. The analysis method consists in a combined use of statistical descriptors and mathematical tools in frequency and time-frequency domains, such as FFT and the continuous wavelet transform.

**Keywords:** Two-stage planetary gearbox · Non-stationary analysis
Continuous wavelet transform · Dynamic analysis

## 1 Introduction

Planetary gearboxes are widely used as power transmission elements in heavy industry such as automotive, wind turbines, aerospace and marine applications. Since they are complex systems, they have many internal and external excitation sources, such as gear mesh and bearing stiffness, eccentricity and manufacturing errors, variation operational speed and load, to mention a few, which can increase the modal interaction in nonlinear regimes and produce nonlinear jump, bifurcation and chaotic behavior [1, 2].

Dynamic modelling helps for good understanding of dynamic characteristics of the structure. It is suitable during the design stage in order to improve the dynamic behavior of the whole system by the simulation of different combinations of design parameters and the variation of the external excitation sources. Lumped parameter model (LPM) has been the most used approach in this task and it has proven to be an effectiveness and robust method [3]. On the other hand, analysis of the dynamic response helps for developing monitoring and analysis tools, which can identify failures and nonlinear and transient behaviors even in early stages. Different data processing techniques are used to analyze and characterize the dynamic response of

© Springer Nature Switzerland AG 2019
B. Corves et al. (Eds.): EuCoMeS 2018, MMS 59, pp. 117–125, 2019.
https://doi.org/10.1007/978-3-319-98020-1_14

planetary gearboxes through numerical and experimental data. Fast Fourier transform (FFT) is the most commonly used technique [4]. It let know the harmonic frequency content of a time domain signal. However, it does not shown the evolution of the frequency content over time. On the other hand, continuous wavelet transform (CWT) plots time-frequency maps that display the behavior of each frequency as a function of time [5, 6]. When a system is linear, the amplitude of each frequency remains constant at any time; otherwise, it varies and, in the time-frequency map, it shows changes in the amplitude. The fundamentals of this analysis tool can be referred to [7]. Other methods, are also used and they have demonstrated effectiveness to identify dynamic behavior and nonlinear characteristics [8].

In this paper a methodology based in the combined use of frequency and time-frequency analysis is proposed to analyze the dynamic behavior of planetary gearboxes. Experiments were carried out on a two-stage planetary gearbox that was designed in LARM [9]. It was operated at different operational speeds and the torque and vibrations on the output shaft were measured. Post-processing is worked by means of frequency spectrum and CWT in order to analyze the dynamic behavior of the system. The organization of the article is as follows. Kinematic modelling is presented in Sect. 2, experimental setup is thoroughly described in Sect. 3, results of signal processing are discussed in Sect. 4 and conclusions are given in Sect. 5.

## 2   Design Modeling

The planetary gearbox under examination was designed at the Laboratory of Robotics and Mechatronics (LARM) in the University of Cassino and South Latium [9]. A kinematic scheme of the planetary gearbox is shown in Fig. 1.

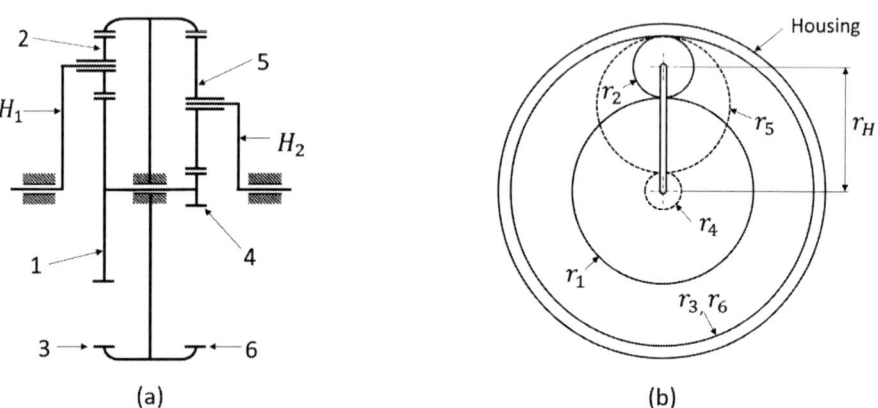

**Fig. 1.** Kinematic scheme of the planetary gearbox: (a) Longitudinal view, (b) Cross-section view with $r_i$ radius of $i$-th component ($H_1$-Input carrier, $H_2$-Output carrier, 1-Input sun gear, 2-Input planet gear, 3-Epicyclic input gear, 4-Output sun gear, 5-Output planet gear, 6-Epicyclic output gear).

The mechanism consists of one input carrier ($H_1$), one output carrier ($H_2$), two central sun gears (1, 4), two planet gears (2, 5) and two ring gears (3, 6). The input carrier $H_1$ applies the driving force to the closed gear chain, while the output carrier $H_2$ generates resistance force due to the connected load on its shaft. The gears 2-3-4-5-6-1 form a closed gear chain that is operated differentially. At the beginning of the gearbox operation, $H_1$ moves the gear 2, which transmits forces to the gears 1 and 3. Then, gears 4 and 6 are moved too. However, the friction forces that are generated by the gear contact forces are small and the gear 5 runs like the output link while holding the output carrier in a fixed position. When the friction forces between gears increases, gear 5 is locked to the gears 4 and 6 and the movement is transmitted from $H_1$ to $H_2$.

A kinematic characterization of the planetary gearbox can be expressed as function of the design parameters, the internal torques and the angular velocity [9]. Thus, the relationships between the angular velocities can be expressed as

$$\frac{\omega_1 - \omega_{H1}}{\omega_3 - \omega_{H1}} = -\frac{z_3}{z_1} = u_{13} \tag{1}$$

$$\frac{\omega_1 - \omega_{H2}}{\omega_3 - \omega_{H2}} = -\frac{z_6}{z_4} = u_{46} \tag{2}$$

where $\omega_i$ are the angular velocity of the corresponding elements in Fig. 1 ($i = 1 \ldots 6, H_1, H_2$), $z_i$ are the teeth number of the gears and $u_{ij}$ are the gear ratios.

From Eqs. (1) and (2), the angular velocities of gears 3 and 1 can be obtained as

$$\omega_3 = \frac{(u_{13} - 1)\omega_{H1} - (u_{46} - 1)\omega_{H2}}{u_{13} - u_{46}} \tag{3}$$

$$\omega_1 = u_{13}(\omega_3 - \omega_{H1}) + \omega_{H1}. \tag{4}$$

The internal forces on gears can be expressed from a static equilibrium analysis. When an external force $F_{H1}$ is applied on the input carrier, reaction forces and torques are produced internally as in Fig. 2.

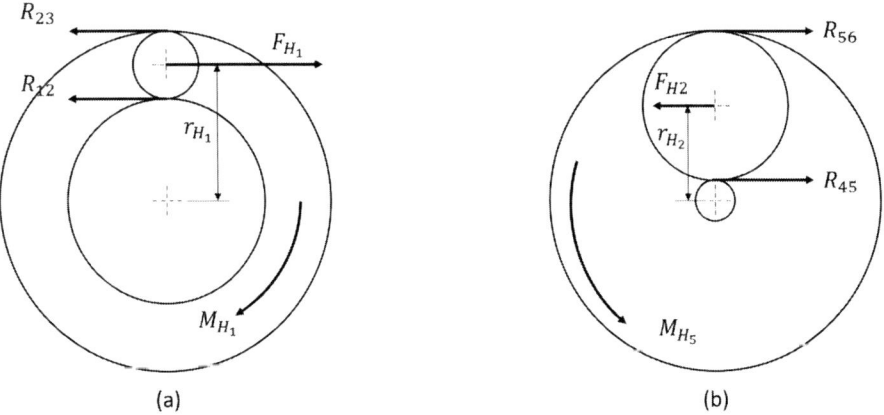

**Fig. 2.** A free body diagram for static equilibrium: (a) Input link; (b) Output link.

The input and output mechanical power can be expressed in terms of the torques and the angular velocity as

$$W_{in} = M_{H1}\omega_{H1} \tag{5}$$

$$W_{out} = M_{H2}\omega_{H2} \tag{6}$$

where,

$$M_{H1} = F_{H1}r_{H1} \tag{7}$$

$$M_{H2} = F_{H2}r_{H2}. \tag{8}$$

Thus, the efficiency of the mechanism is given by

$$\eta = \frac{W_{in}}{W_{out}} \tag{9}$$

## 3  Experimental Setup

A laboratory setup of the test bed is shown in Fig. 3 at LARM in Cassino. It consists of an AC motor drive Cantoni Sh-80-2B (1) that is coupled to the planetary gearbox prototype (2) to drive the attached load (3). The test bed is instrumented with an accelerometer Kistler 8305B2 (5), which is attached on the bearing next to the output shaft of the gearbox to measure the vibrations and a torque sensor CD1050 (6), which measures the mechanical torque in the same shaft. Since the output signal of the torque sensor has low values, a voltage amplifier (7) is included. The data acquisition is carried on by the NI USB 6009 DAQ system (8).

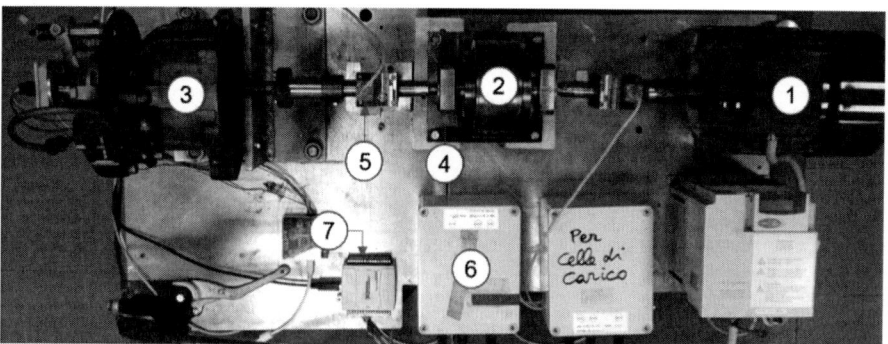

**Fig. 3.** The laboratory set up of test bed at LARM (1-AC motor drive, 2-Planetary gearbox, 3-Load, 4-Accelerometer, 5-Torque sensor, 6-Voltage amplifier, 7-NI USB 6009 DAQ system).

Tests were carried out at two different speeds: 500 and 1,112 rpm by means varying the frequency of the power supply at 9 and 20 Hz, respectively. Those speeds were selected since the gearbox was designed to operate at those frequencies. A LabVIEW user interface in a PC was used to monitor the tests and to save the acquired data for post-processing. The sampling frequency was set as 1k samples per second. The design characteristics of the planetary gearbox in Fig. 3 are listed in Table 1. The gears are designed with a module equal to 0.5 mm and a pressure angle equal to 20°.

**Table 1.** Design parameters of the LARM's planetary gearbox.

| Stage | Element | Sun | Planet | Ring | Carrier |
|---|---|---|---|---|---|
| 1st | Number of teeth | 60 | 20 | 100 | - |
| | Base circle radius (mm) | 30 | 10 | 50 | 40 |
| 2nd | Number of teeth | 12 | 44 | 100 | - |
| | Base circle radius (mm) | 6 | 22 | 50 | 28 |

The time series of measured data are processed first by the FFT in order to get the frequency content of the system; later, the CWT is used to extract time-frequency characteristics and analyze dynamic behavior. Signal processing is performed in MatLab and Autosignal software. The results are given in the next section.

## 4    Results and Discussion

Results of two tests are reported from an experimental campaign as to indicate both, the utility of the analysis procedure and the characterization of the examined gearbox. Time domain signals were recorded for five seconds of the motor operation, see Fig. 4.

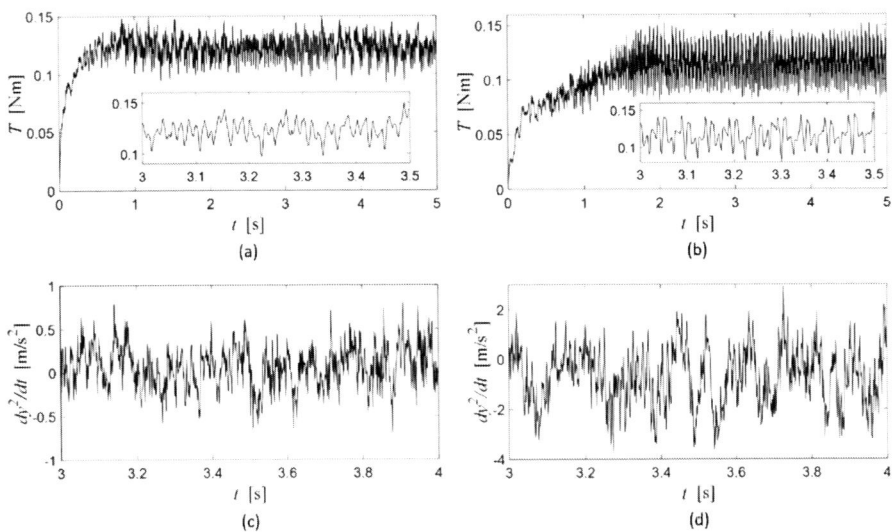

**Fig. 4.** Measured data on the output shaft: (a) torque at 500 rpm, (b) torque at 1,112 rpm; (c) vibrations at 500 rpm, (d) vibrations at 1,112 rpm.

The torque on the output shaft when the motor is operated at 500 rpm and 1,112 rpm are shown in Fig. 4(a) and (b), respectively; inner graphs show a zoomed view of the torque signals for a short time period. Radial vibrations on the bearing close to the output shaft at 500 rpm and 1,112 rpm are shown in Fig. 4(c) and (d), respectively.

Statistic for the measured data are listed in Table 2. It can be seen the range of the amplitudes increases at higher operational speed for both measured variables, torque and vibrations signals, while the mean value decreases and the standard deviation increases. The difference in the time response of the torque signals is related to the time that the motor takes to reach the nominal speed.

**Table 2.** Statistic of the measured data for the tests in Fig. 4.

| Statistical descriptor | Torque [Nm] | | Vibrations [m/s²] | |
|---|---|---|---|---|
| | 500 rpm | 1,112 rpm | 500 rpm | 1,112 rpm |
| Mean | 0.123 | 0.116 | 0.063 | −0.721 |
| Range | 0.055 | 0.064 | 1.495 | 6.634 |
| Standard deviation | 0.009 | 0.014 | 0.245 | 1.162 |

Frequency spectrums for measured torque and vibrations are shown in Fig. 5. It can be seen there are higher amplitudes when the system is operated at 1,112 rpm. Furthermore, the torque spectrums have lower amplitudes and lower bandwidth detection than the vibration ones.

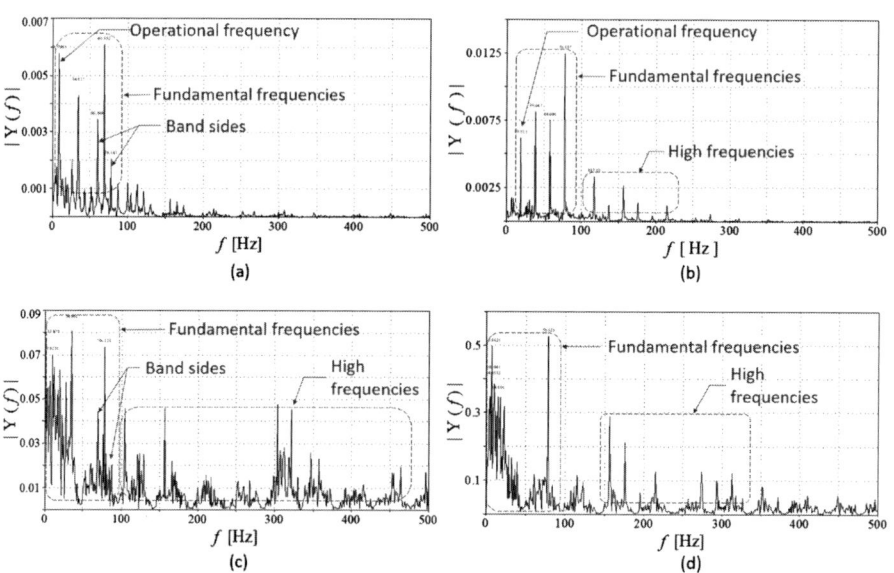

**Fig. 5.** Frequency spectrums of the measured data on the output shaft: (a) torque at 500 rpm, (b) torque at 1,112 rpm; (c) vibrations at 500 rpm, (d) vibrations at 1,112 rpm.

The difference between the bandwidth detection ranges of both spectrums, torques and vibrations, Fig. 5(a–b) and (c–d), respectively, are related to the fact that the accelerometer is placed closer to the gear train than the torque meter, making it possible to detect the complex operational frequencies of the gear train elements. In the other hand, disturbances in the frequency content in Fig. 5 are related to poor assembly, the band sides that appear around the fundamental frequencies suggest misalignment and clearance between the shafts and the couplings. Fundamental frequencies are computed from kinematic model and they can be found in the frequency spectrums in Fig. 5.

Time-frequency spectrograms are presented in Fig. 6, from which it is possible to see the changes in the frequency of the signals throughout time, as there is further interaction within frequencies the system has stronger nonlinear behavior.

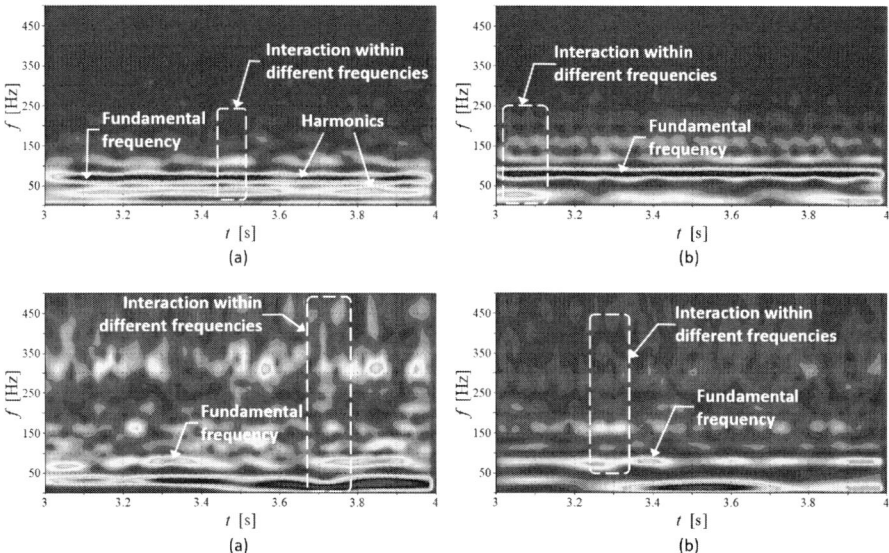

**Fig. 6.** Time-frequency spectrograms from the CWT of the measured data: (a) torque at 500 rpm, (b) torque at 1,112 rpm; (c) vibrations at 500 rpm, (d) vibrations at 1,112 rpm.

Finally, power spectral density, $P(f)$, of the time-frequency spectrograms is computed and the results are shown in Fig. 7. It can be seen that $P(f)$ varies throughout time and strong variations occurs at higher operational speed.

In order to measure the dispersion of $P(f)$, its standard deviation is computed: $\sigma_{T@500rpm} = 2.3 \times 10^{-8}$, $\sigma_{T@1112rpm} = 6.9 \times 10^{-8}$, $\sigma_{Acc@500rpm} = 1.7 \times 10^{-5}$ and $\sigma_{Acc@1112rpm} = 69.74 \times 10^{-5}$. Results show that $\sigma$ has smaller values in the case of torque than in the case of vibrations; this is related to the lower amplitudes that the former have compared with the later in time domain. Furthermore, comparing the different tests for the same variable, $\sigma$ takes higher values as operational speed is increased.

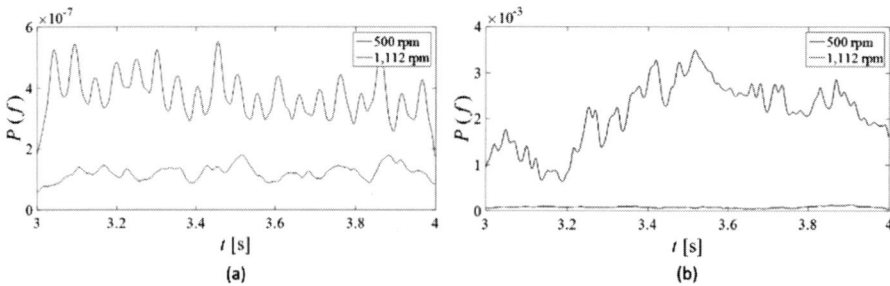

**Fig. 7.** Power spectral density, $P(f)$, from the results of the CWT of the measured data: (a) torque at 500 rpm, (b) torque at 1,112 rpm; (c) vibrations at 500 rpm, (d) vibrations at 1,112 rpm.

## 5   Conclusions

This paper presents a methodology for the analysis of the dynamic behavior of a two-stage planetary gearbox. Tests are performed in a test bend at LARM at two different speeds, 500 and 1,112 rpm, and torque and vibrations are measured on the output shaft. Significant dynamic disturbances at low frequencies can be related to misalignment and the clearance that there is between shafts and the coupling elements. Spectrograms from the CWT are computed to analyze the dynamic behavior of the system and their power spectral density are computed in order to measure the variation of the frequency content throughout time. Stronger nonlinear behavior can be seen in the spectrograms of the CWT as the interaction within frequencies increases. The standard deviation of the power spectral density is useful to measure the variation in the frequency content of the system throughout time. It shown an increment of three times in the case of the torque signals, while it was $\approx 41$ times in the case of vibrations.

**Acknowledgments.** The first author would like to gratefully acknowledge the Mexican Government Foundation CONACYT for the fellowship to pursue her postdoctoral studies at LARM in the academic year 2017–2018.

## References

1. Zhu, C., Chen, S., Liu, H., Huang, H., Li, G., Ma, F.: Dynamic analysis of the drive train of a wind turbine based upon the measured load spectrum. J. Mech. Sci. Technol. **28**(6), 2033–2040 (2014)
2. Li, S., Wu, Q., Zhang, Z.: Bifurcation and chaos analysis of multistage planetary gear train. Nonlinear Dyn. **75**(1–2), 217–233 (2014)
3. Masoumi, A., Pellicano, F., Samani, F.S., Barbieri, M.: Symmetry breaking and chaos-induced imbalance in planetary gears. Nonlinear Dyn. **80**(1–2), 561–582 (2015)
4. Chaari, F., Abbes, M., Rueda, F.V., del Rincón, A.F., Haddar, M.: Analysis of planetary gear transmission in non-stationary operations. Front. Mech. Eng. **8**(1), 88–94 (2013)
5. Chen, X., Cheng, G., Li, H., Li, Y.: Fault identification method for planetary gear based on DT-CWT threshold denoising and LE. J. Mech. Sci. Technol. **31**(3), 1035–1047 (2017)

6. Janeliukstis, R., Rucevskis, S., Wesolowski, M., Chate, A.: Experimental structural damage localization in beam structure using spatial continuous wavelet transform and mode shape curvature methods. Measurement **102**, 253–270 (2017)
7. Gao, R.X., Yan, R.: Wavelets: Theory and applications for manufacturing, 1st edn. Springer, Boston (2010)
8. Lei, Y., Lin, J., Zuo, M.J., He, Z.: Condition monitoring and fault diagnosis of planetary gearboxes: a review. Measurement **48**, 292–305 (2014)
9. Balbayev, G., Ceccarelli, M., Carbone, G.: Design and numerical characterization of a new planetary transmission. IJITR **2**(1), 735–739 (2014)

# Haptic Systems

# A Critical Review of Unpowered Performance Metrics of Impedance-Type Haptic Devices

İbrahimcan Görgülü, Gökhan Kiper, and Mehmet Ismet Can Dede[(✉)]

Izmir Institute of Technology, İzmir, Turkey
{ibrahimcangorgulu,gokhankiper,candede}@iyte.edu.tr

**Abstract.** A kinesthetic haptic device's performance relies on unpowered, powered and controlled system characteristics. In this paper, a critical review is carried out for the well-known metrics for kinematics, stiffness and dynamic aspects of robots that can be applied in evaluating the unpowered system performance of kinesthetic haptic devices. The physical meanings of these metrics are discussed and the important factors that affect the unpowered system performance of a kinesthetic haptic device are revealed.

**Keywords:** Haptics · Performance metrics · Stiffness · Kinematics
Dynamics

## 1  Introduction

Haptic devices are interfaces that enable the interaction of the human with a slave environment either virtual or a real one and they can be categorized with respect to their targeted touch sensation as kinesthetic and cutaneous. The devices constructed to stimulate kinesthetic sensation can be further categorized as impedance and admittance type devices. Impedance-type devices acquire the human motion and transmit this information to a slave environment while the interaction forces measured or calculated in the slave environment is displayed by these devices to the human. The quality of the haptic interaction depends on unpowered, powered and controlled system characteristics of this haptic device as explained in [16].

In [16], powered and controlled system performances were investigated in detail but, in terms of the unpowered system characteristics, some performance metrics were listed but not critically reviewed. The focus of this work is the evaluation of impedance-type haptic devices based on the unpowered system performance, which is only related with the mechanical properties of the device. The unpowered system performance is investigated in three main topics: Kinematics, Stiffness, and Dynamics, which are among the performance metrics of robot manipulators. There are some studies that investigate these metrics with

© Springer Nature Switzerland AG 2019
B. Corves et al. (Eds.): EuCoMeS 2018, MMS 59, pp. 129–136, 2019.
https://doi.org/10.1007/978-3-319-98020-1_15

case studies [4] however, this work reviews these metrics and relates them to the haptic device performance metrics.

In the optimal design of impedance-type haptic device manipulators, the desired objective is to minimize the minimum mechanical impedance, maximize the maximum mechanical impedance and enlarge the operational frequency range, the bandwidth. Minimum impedance is achieved when the back-drivability is at its best conditions, which require the minimization of dynamic effects by constructing a lightweight manipulator and by maximizing the manipulability. Maximum impedance, on the other hand, can be achieved when the manipulator is the least back-drivable, which states that the requirements of the minimum impedance are reversed. There are also other critical constraints in the design optimization of a haptic device such as workspace and isotropy of the force and velocity ellipsoids. The objective function construction becomes highly nonlinear due to the competitive relationship between these constraints and performance metrics. Therefore, an optimization algorithm such as a genetic algorithm can be used to solve the optimization problem. However, the critical issue in this solution is the composition of the objective function in a meaningful way. This work undertakes this challenge.

Manipulator kinematics is the backbone of the design procedure since the dimensions of the links affect the kinematic, stiffness and dynamic properties. On the other hand, mechanical impedance characteristics of a device are driven by its stiffness and dynamics properties. In order to understand the relation between the kinematics and impedance, all the design parameters affecting the impedance performance should be considered simultaneously. Therefore, the next sections are organized to review the kinematics, stiffness, and dynamics related performance metrics and finally conclude the paper by providing a discussion on the adaptation of these metrics in the optimal design of haptic device manipulators.

## 2   Kinematics Performance Metrics

Kinematic performance of the manipulators is usually evaluated depending on their dexterity measure. Dexterity is evaluated in terms of manipulability and condition number. Both metrics use the Jacobian matrix but evaluate its different properties. Thus, as a first step, Jacobian matrix properties should be understood.

The Jacobian matrix is the mapping matrix of the velocities between the joint space and task space in which its elements are called the velocity influence coefficients. Eigenvalues and eigenvectors of the Jacobian matrix reveal the physical meaning of Jacobian matrix for a unit change in joint space velocities [6]. Eigenvectors of the Jacobian matrix of a 3 degree of freedom (DoF) translational manipulator, for instance, represent the orientation of an ellipsoid's axes for a specific pose of the manipulator. Eigenvalues of the matrix define the semi-axis dimensions of this ellipsoid. This ellipsoid is named as velocity ellipsoid and the boundary of the ellipsoid shows the motion capability of the manipulator in an arbitrary direction.

Higher eigenvalues correspond to higher motion capability which is desired for minimizing the minimum impedance of a haptic device. The manipulability concept is introduced in [18] to describe the easiness of motion of the end effector. Here, the determinant of the Jacobian is used as the performance measure. The manipulability measure is shown in Eq. (1) for both redundant and non-redundant manipulators.

$$\mu_v = \sqrt{\det\left(\hat{J}\hat{J}^T\right)} \tag{1}$$

Here, $\hat{J}$ is the Jacobian matrix, $\mu_v$ is the manipulability measure. This metric is a value that is proportional to the volume of the velocity ellipsoid. If the Jacobian matrix is square, the absolute value of its determinant is used for evaluation [12].

Condition number calculates the magnitude of ratio between the maximum and minimum eigenvalues of Jacobian matrix. The ratio is evaluated in order to understand the motion resolution of the mechanism [15]. The formulation is shown as:

$$c_v = \|\hat{J}\|\|\hat{J}^{-1}\|, \tag{2}$$

where $c_v$ is the condition number calculated by using the Euclidean norm of the Jacobian matrix and its inverse. If Jacobian matrix is a non square matrix, psuedo inverse can be used. The resolution of the acquired motion is important in haptic devices since the acquired motion is the demand for the slave system that might be employed in a critical-precision operation such as telesurgery.

A key activity of a haptic device is reflecting forces back to the user. Therefore, in contrast to the previous motion-related metrics, force-related metrics should be evaluated. The relation between the actuator inputs, $\bar{\tau}$, and the end-effector forcing, $\bar{F}$, is established by using the Jacobian matrix as shown in Eq. (3).

$$\bar{F} = (\hat{J}^{-1})^T \bar{\tau} \tag{3}$$

The force capability, $\mu_f$, of the manipulator can be written as a scalar index as:

$$\mu_f = \sqrt{\det\left(\hat{J}\hat{J}^T\right)^{-1}} = 1/\mu_v. \tag{4}$$

In haptic device design, it is favorable to maximize the force capability, which results in receiving the most force output for unit input to the actuators. This would maximize the maximum impedance of the device. On the other hand, maximization of this metric means the minimization of the manipulability. Therefore, minimizing the minimum impedance and maximizing the maximum impedance by using these metrics are contradictory goals and a trade-off should be adjusted by the designer.

The abovementioned metrics are generalized and normalized for a fair comparison between different manipulators. If the design problem of the manipulator consists of the selection of manipulators which have different degree-of-freedom (DoF), the comparison between the performance metrics will be impossible since they have different physical meaning (area, volume). This order dependency can

be solved with the manipulation in Eqs. (1) and (4) as proposed in [11], which
is:

$$\mu_v^n = \sqrt[n]{\det(\hat{J}\hat{J}^T)} \text{ and } \mu_f^n = \sqrt[n]{\det(\hat{J}\hat{J}^T)^{-1}} = 1/\mu_v^n, \tag{5}$$

where $n$ is the rank of the Jacobian matrix (instantaneous DoF of the manipulator).

Another problem stated in [11] is the scaling problem. In order to evaluate
the different size manipulators in a common framework, the characteristic length
is introduced in [1]. This length is used to normalize the Jacobian matrix. The
characteristic length $L$ is defined as the ratio between the maximum desired
reach, $R_d$, and the maximum actual reach, $R_a$, which can be achieved by synthesized links. For a normalized Jacobian matrix, $\hat{J}_n$, the desired reach is 1.

$$\hat{J}_n = \hat{J}L \text{ for } R_d = 1, \text{ where } L = \frac{R_d}{R_a} \tag{6}$$

Another problem in using kinematic performance indices arises if the manipulator has both translational and rotational DoFs. In this case, Eqs. (1) and
(3) lose their physical meanings. A solution is devised by using homogeneous
coordinates [10].

## 3   Stiffness Performance Metrics

The stiffness of a haptic device manipulator determines the maximum impedance
along with the damping and inertial properties of the system. This property
directly determines the quality and the limits of rendered forces in a haptics
scenario. The stiffness of the manipulator can be analyzed using finite element
methods (FEM).

Stiffness matrix can be obtained by using Hooke's law [7] or virtual work
method [14]. In Eq. (7), the most general form of force-stiffness relation is presented where $\hat{K}_C$ is the Cartesian stiffness matrix, and $\delta\bar{r}$ denotes the deflection
at the tip point. Equation (8) represents the joint reactions ($\bar{F}_\theta$)-task space force
relation and Eq. (9) represents the joint reaction-joint space deflections ($\delta\bar{\theta}$) relation. Here, $\hat{K}_\theta$ denotes the structural stiffness of the link of interest.

$$\bar{F} = \hat{K}_C \delta\bar{r} \tag{7}$$

$$\bar{F}_\theta = \hat{J}_\theta^T \bar{F} \tag{8}$$

$$\bar{F}_\theta = \hat{K}_\theta \delta\bar{\theta} \tag{9}$$

Using the Conservative Congruency Transformation [5], the mapping between
the joint space stiffness matrix $\hat{K}_\theta$ and Cartesian space stiffness matrix is as
follows:

$$\hat{K}_C = (\hat{J}_\theta \hat{K}_\theta^{-1} \hat{J}_\theta^T)^{-1}, \tag{10}$$

where the $\hat{J}_\theta$ denotes the homogeneous Jacobian matrix developed for virtual
joint variables. Here, $\hat{K}_\theta$ can be a diagonal matrix for a simplified model approach

or non-diagonal matrix which represents the real case. Cartesian space stiffness is critical in design since it directly relates the external forces and tip point deflections.

Kinematic evaluation methods can be extended for stiffness matrix [3]. Singular value decomposition (SVD) of $\hat{K}_C$ reveals directional stiffness properties of the manipulator. Similar to force and velocity ellipsoids, SVD can be used for graphical illustration of stiffness. Frobenius norm of $\hat{K}_C$ is used to evaluate overall stiffness of the manipulator, $S_f$, ib Eq. (11). On the other hand, the Euclidean norm of $\hat{K}_C$ and $(\hat{K}_C)^{-1}$ denoted by $S_e$ exposes the stiffest and the most compliant axes of the manipulator which is similar to condition number, Eq. (11).

$$S_f = \sqrt{tr(\hat{K}_C \hat{K}_C^T)} \text{ and } S_e = \|\hat{K}_C\|\|(\hat{K}_C)^{-1}\| \tag{11}$$

If $S_e$ value is equal to 1, it can be stated that the manipulator is in an isotropic pose in terms of stiffness. This index is useful when the link weights and dynamic effects are included to the calculation of stiffness matrix. Hence, manipulator can be designed to be more stiff along the axes affected by gravitational and dynamic forces.

Another scalar index can be defined as the determinant of $\hat{K}_C$ shown in Eq. (12). $S_d$ is a value that is proportional to the volume of stiffness ellipsoid. Naturally, higher volume indicates higher stiffness.

$$S_d = \det(\hat{K}_C) = \det(\hat{J}_\theta \hat{K}_\theta^{-1} \hat{J}_\theta^T)^{-1} \tag{12}$$

Notice that, $S_d$ can be increased by increasing the determinant of $\hat{K}_\theta$ and/or by decreasing determinant of $\hat{J}_\theta$. If the manipulator is in a singular pose $S_d$ becomes infinite. Therefore, the designer should be careful while using stiffness oriented performance metrics as a design objective and should make use of global indices to resolve this problem. If the design parameters consist of link lengths, which is the general case, in order to avoid such problems, kinematic synthesis should be concluded before designing the link geometry for desired stiffness.

## 4   Dynamics Performance Metrics

Dynamic properties are hard to implement to design procedure since, dimensional parameters must be solved first. Therefore, dynamics oriented designs are iterative. Acceleration mapping between the task and joint space in Eq. (13) and dynamic equation of motion in Eq. (14) are manipulated to define $\ddot{r}^*$ and $\bar{\tau}^*$, respectively.

$$\ddot{r} = \hat{J}\ddot{\theta} + \dot{\hat{J}}\dot{\theta} \Rightarrow \ddot{r}^* = \ddot{r} - \dot{\hat{J}}\dot{\theta} \tag{13}$$

$$\bar{\tau} = \hat{M}\ddot{\theta} + \hat{V}\dot{\theta} \Rightarrow \bar{\tau}^* = \bar{\tau} - \hat{V}\dot{\theta} \tag{14}$$

where $\bar{\tau}$ contains the actuator inputs, $\hat{M}$ is the generalized inertia matrix, $\hat{V}$ contains the Coriolis and centripetal effects and $\bar{\theta}$ is the vector of generalized coordinates (joint variables). Further manipulation can be issued to transform

the Eq. (14) from joint space to Cartesian space by using Eqs. (3) and (15) can
be obtained.

$$\bar{F}^* = \hat{M}_C \ddot{\bar{r}}^* \text{ where } \hat{M}_C = \hat{J}^{-T} \hat{M} \hat{J}^{-1} \tag{15}$$

Asada referred to the $\hat{M}_C$ matrix as generalized inertia matrix (GIM), however,
this GIM is derived by using the generalized coordinates as Cartesian space pose
components [2]. He used the eigenvalues and eigenvectors in order to graphically
illustrate the properties of GIM and named them as generalized inertia ellipsoids
(GIE). In terms of haptics, $\hat{M}_C$ is the dominating factor of minimum impedance.
If a user jiggles the end effector with $\ddot{\bar{r}}^*$ input, the felt force due to the dynamic
effects is the $\bar{F}^*$ and the relationship between the input and output is established
by the $\hat{M}_C$ matrix. Therefore, for minimization of the minimum impedance of a
haptic device, the $\hat{M}_C$ matrix should be minimized.

Similar evaluations conducted for stiffness performance can be repeated for
GIM. Average inertia of the manipulator, $I_f$, and inertia condition number, $I_e$,
as the Euclidean norm of inertia matrix are defined in Eq. (16).

$$I_f = \sqrt{tr(\hat{M}_C \hat{M}_C^T)} \text{ and } I_e = \|\hat{M}_C\| \|(\hat{M}_C)^{-1}\| \tag{16}$$

By using Eqs. (13) and (14), (17) is derived. By using this, in [17], the
dynamic manipulability measure is introduced (Eq. (17)). This index indicates
the amplification rate between the actuator inputs and output acceleration of
the end-effector. Better dynamic manipulability can be achieved by increasing
the manipulability and/or decreasing the determinant of inertia matrix.

$$I_d = \det(\hat{J}\hat{M}^{-1}) = \det(\hat{J})/\det(\hat{M}) \tag{17}$$

Another evaluation approach in [9] proposed the acceleration radius measure
to emphasize the acceleration capability of the end-effector for any arbitrary
direction. However, $I_e$ and $I_d$ are more suitable as performance indices since
they already consider the acceleration radius.

Final evaluation can be carried out by computing the natural frequency of the
manipulator. Natural frequency can be computed by evaluating the eigenvalues
of the dynamic matrix $\hat{D}$ which is defined in Eq. (18) in joint space and in
Cartesian space. Modal vectors are the eigenvectors of the dynamic matrix.

$$\hat{D} = -\hat{M}^{-1}\hat{K} \text{ and } \hat{D}_C = \hat{J}\hat{D}\hat{J}^{-1} \tag{18}$$

Increasing the natural frequency corresponds to increasing bandwidth and
decreasing the response time of the mechanism. This also maximizes the
maximum impedance by increasing gains of $\hat{K}$ and minimizes the minimum
impedance by decreasing gains of $\hat{M}$. It should be noted that $\hat{K}_C$ and $\hat{M}_C$ are
functions of link lengths and joint variables. Thus, natural frequency can be
intuitively optimized in kinematic level. Matrix $\hat{D}$ or $\hat{D}_C$ should be simplified as
a scalar performance index. The easiest way is by evaluating the Frobenius norm
of the dynamic matrix (Eq. (19)), which takes in account the value of natural

frequency for Cartesian space, $\omega_{nc}$, and for joint space, $\omega_n$. This way an average value of natural frequency is obtained.

$$\omega_{nc} = \sqrt{tr(\hat{D}_C \hat{D}_C^T)} \text{ and } \omega_n = \sqrt{tr(\hat{D}\hat{D}^T)} \tag{19}$$

## 5    Globalization of Indicies

All indices, mentioned above, are pose depended. A manipulator cannot be designed for a single pose. It should be evaluated at each discrete pose. The global performance index is proposed by Gosselin and Angeles [8] to address this problem:

$$k_i = \left( \int_W idw \right) \Big/ \left( \int_W dw \right), \tag{20}$$

where $k_i$ is the average of the globalized index, $i$ is the index which will be globalized and $W$ denotes the workspace. As a scalar value, $k_i$ enables evaluating the performance of related index for all discrete positions of the workspace. If there are no singular poses within the workspace, this method can be simplified by considering only the critical poses which are generally at the boundaries of the workspace.

Evaluation of $k_i$ alone may mislead the designer. If the manipulator is close to a singular pose, the effect of the index at that pose will be reduced when the average value is calculated. In [13] the uniformity of the performance index is introduced as

$$U = I_{min}/I_{max}, \tag{21}$$

where $U$ is the uniformity measure, $I_{min}$ and $I_{max}$ are the minimum and maximum values of the observed index through the workspace. As an example, the uniformity of the force capability is especially important for haptic devices so that same amount of impedance can be displayed to the user throughout the workspace.

## 6    Conclusions

The critical metrics to optimize the unpowered system characteristics of a haptic device are reviewed and discussed in terms of kinematics, stiffness and dynamics indices, and their physical meanings. All indices include the Jacobian matrix or share the same parameters with the Jacobian matrix. Therefore, these metrics can be narrowed down as a function of Jacobian matrix entries and it is possible to optimize all performance indices in kinematic level up to a certain point. This is computationally faster and it handles most of the design parameters, simultaneously. However, Jacobian matrices should be normalized and made homogeneous to remove the scale and unit dependency to compare different manipulator topologies in the same framework. The metrics investigated in this paper will be applied in the optimal design of a new haptic device.

**Acknowledgements.** This work is supported in part by The Scientific and Technological Research Council of Turkey via grant number 117M405.

# References

1. Angeles, J.: Fundamentals of Robotic Mechanical Systems, vol. 2. Springer, New York (2002)
2. Asada, H.: A geometrical representation of manipulator dynamics and its application to arm design. J. Dyn. Syst. Meas. Contr. **105**(3), 131–142 (1983)
3. Carbone, G., Ceccarelli, M.: Comparison of indices for stiffness performance evaluation. Front. Mech. Eng. China **5**(3), 270–278 (2010)
4. Ceccarelli, M., Carbone, G., Ottaviano, E.: Multi criteria optimum design of manipulators. Bull. Polish Acad. Sci. Tech. Sci. **53**, 9–18 (2005)
5. Chen, S.F., Kao, I.: Conservative congruence transformation for joint and cartesian stiffness matrices of robotic hands and fingers. Int. J. Robot. Res. **19**(9), 835–847 (2000)
6. Chiu, S.L.: Kinematic characterization of manipulators: an approach to defining optimality. In: Proceedings of the 1988 IEEE International Conference on Robotics and Automation, pp. 828–833. IEEE (1988)
7. Gosselin, C.: Stiffness mapping for parallel manipulators. IEEE Trans. Robot. Autom. **6**(3), 377–382 (1990)
8. Gosselin, C., Angeles, J.: A global performance index for the kinematic optimization of robotic manipulators. J. Mech. Des. **113**(3), 220–226 (1991)
9. Graettinger, T.J., Krogh, B.H.: The acceleration radius: a global performance measure for robotic manipulators. IEEE J. Robot. Autom. **4**(1), 60–69 (1988)
10. Khan, W.A., Angeles, J.: The kinetostatic optimization of robotic manipulators: the inverse and the direct problems. J. Mech. Des. **128**(1), 168–178 (2006)
11. Kim, J.O., Khosla, K.: Dexterity measures for design and control of manipulators. In: IEEE/RSJ International Workshop on Intelligent Robots and Systems' 91. Intelligence for Mechanical Systems, Proceedings IROS 1991, pp. 758–763. IEEE (1991)
12. Paul, R.P., Stevenson, C.N.: Kinematics of robot wrists. Int. J. Robot. Res. **2**(1), 31–38 (1983)
13. Pham, H.H., Chen, I.M.: Optimal synthesis for workspace and manipulability of parallel flexure mechanism. In: 11th World Congress in Mechanism and Machine Science, Tianjin, China, pp. 18–21, August 2003
14. Quennouelle, C., Gosselin, C.: Stiffness matrix of compliant parallel mehanisms. In: ASME 2008 International Design Engineering Technical Conferences and Computers and Information in Engineering Conference, pp. 151–161. American Society of Mechanical Engineers (2008)
15. Salisbury, J.K., Craig, J.J.: Articulated hands: force control and kinematic issues. Int. J. Robot. Res. **1**(1), 4–17 (1982)
16. Samur, E.: Performance Metrics for Haptic Interfaces. Springer, London (2012)
17. Yoshikawa, T.: Dynamic manipulability of robot manipulators. Trans. Soc. Instrum. Control Eng. **21**(9), 970–975 (1985)
18. Yoshikawa, T.: Manipulability of robotic mechanisms. Int. J. Robot. Res. **4**(2), 3–9 (1985)

# Experimental Evaluation of Actuation and Sensing Capabilities of a Haptic Device

Emir Mobedi, İbrahimcan Görgülü, and Mehmet Ismet Can Dede[✉]

Izmir Institute of Technology, İzmir, Turkey
{emirmobedi,ibrahimcangorgulu,candede}@iyte.edu.tr

**Abstract.** Haptic devices are used to increase the telepresence level by providing the sense of touch to the human operator. Simultaneously, they capture the targeted motion of the human operator to generate a motion demand for the teleoperated slave system. Considering a scenario where the slave system's end-effector is handled by the human operator at the master side, which is attached to the haptic device, an ideal haptic interaction involves the feeling of only the end-effector dynamics and the accurate sensation of the end-effector pose. The performance of a haptic device is based on these two functionalities. In this paper, the experimental evaluation of the actuation and sensing capabilities of a haptic device, HIPHAD v1.0 kinesthetic haptic device, is presented.

**Keywords:** Haptic device performance · Experimental performance evaluation
System identification

## 1 Introduction

Haptic devices are used for a variety of tasks such as surgical robotics [11], robotic rehabilitation [3]. These devices are categorized into two groups in terms of feedback types as kinesthetic and tactile. Kinesthetic haptic devices are developed to stimulate bodily position, weight, or movement of the muscles, tendons, and joints [9] while measuring the pose of the human's targeted motion. Therefore, a kinesthetic haptic device is used both as a sensory and an actuation system.

Back-drivability is defined in the literature as the level of easiness of the transmission of movement from the end-effector to the input axes [6]. In this definition, user handles the end-effector and the actuators are coupled with the input axes. If back-drivability can be achieved without any power input to the device, this type of a haptic device is called impedance-type haptic device [7]. In an ideal case, if the haptic device is intended to act as a lumped mass of the tool, the force felt by the user should be proportional to the acceleration of this lumped mass. In the case of a physical interaction, interaction forces are calculated based on the impact velocity, mass, damping and stiffness of the objects and slave system [2].

Mass, damping and stiffness properties of a haptic device manipulator are defined as the impedance of the device. If the inertial, damping and stiffness properties of the device are identified with some accuracy, a feedforward term can be utilized in order to

© Springer Nature Switzerland AG 2019
B. Corves et al. (Eds.): EuCoMeS 2018, MMS 59, pp. 137–144, 2019.
https://doi.org/10.1007/978-3-319-98020-1_16

cancel the effect of these dynamics [11]. Hence, when there is no contact, the user feels almost only the resistance due to the tool inertia and this situation increases the backdrivability performance of the system or in other words, minimizes the minimum impedance. The maximum impedance that a haptic device can display is related to its actuation capacity which depends on the inertial, damping and stiffness properties of the device and actuation system located at the input axes. Therefore, actuation performance evaluation of a haptic device is important in order to characterize the system and also to enlarge the impedance range by control.

The measured pose of the targeted human motion by the haptic device is used to transmit motion demand to a slave system. In precise operation the resolution and the repeatability of this sensory acquisition system become important.

A number of methods have been proposed for evaluation of actuation and sensing capabilities of haptic devices [4, 10, 11]. The aim of this study is to evaluate actuation and sensing capabilities of a haptic device, HIPHAD v1.0, by determining its impedance parameters, and workspace position resolution and repeatability.

## 2  Kinematics and Static Force Analysis of HIPHAD

In this section, the results of the kinematics and static force analysis of HIPHAD v1.0 are presented. The device was designed and produced in IzTech Robotics Laboratory as an open-loop impedance-type structure with the hybrid mechanism. The translational parallel mechanism of the device has three actuated degrees-of-freedom (DoF). The mechanism is a modified version of the R-CUBE mechanism and the modifications are described in [1]. Forward kinematics equation of the mechanism is shown in Eq. (1).

$$W_{ri} = S + l_{i1} \cdot \sin \theta_i; i = 1, 2, 3 \tag{1}$$

Here, $W_{ri}$ indicates the end-effector location with respect to the origin along the $\vec{u}_i$ direction, $S$ parameter is the distance of the actuation axis from the origin, $\theta_i$ represents the joint variable, $l_{i1}$ is the effective link length and $i$ represents the $i^{th}$ serial chains and $\vec{g}$ is the gravity vector as indicated in Fig. 1. The map of the end-effector force to the actuator inputs is shown as follows;

$$T_i = F_i \cdot l_{i1} \cdot \cos \theta_i; i = 1, 2, 3 \tag{2}$$

where $F_i$ represents the force applied by the device in the task space along the $\vec{u}_i$. direction and $T_i$ indicates the torque generated by the actuator connected to the $i^{th}$ serial kinematic chain. Detailed analyses of the kinematics and quasi-static force are presented in [5].

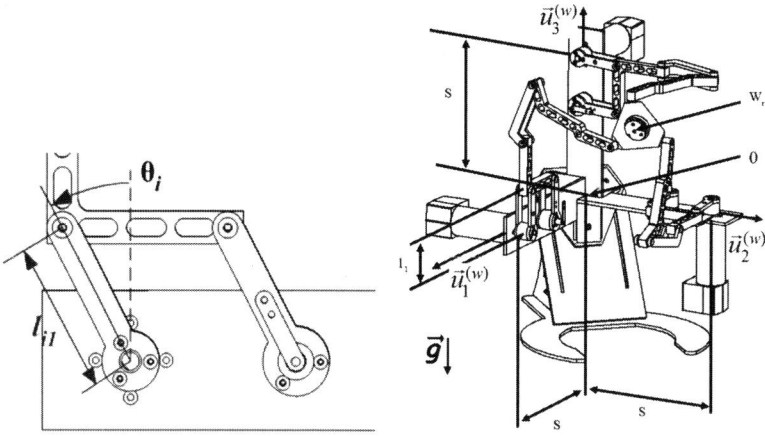

**Fig. 1.** Kinematic parameters of HIPHAD v1.0

## 3   Actuation Capabilities

In this section, frequency response method is explained to identify the actuation capability of HIPHAD v1.0. For an impedance-type haptic interface, a sweep force signal is applied to the actuators over a certain range of frequencies at a certain amplitude and velocity and force responses are measured [11]. Afterwards, the behavior of the system is identified by making use of the experimentally obtained Bode diagrams. Then, a transfer function is fitted to determine the model of the system. In other words, actuation performance of a haptic device is determined. During the tests, no control strategy is employed on the system. However, in our setup, in order to cancel the nonlinear effects of gravity, a feedforward term is used for gravity compensation. The test was carri out only for $\vec{u}_1$ axis since all axes shares the same structure and the actuation of the system is uncoupled along the three axes. The steps of the experiment are presented as follows:

**1st step:** A gravity compensation algorithm is employed as explained in [5].
**2nd step:** The input axis' joint position is set to $\theta_1 = 0°$ position before each test and the other input axes are mechanically fixed at the same angular position.
**3rd step:** An accelerometer (Type: CXL10GP3, MEMSIC Brand) is attached to the end-effector as presented in Fig. 2 by ① and sinusoidal end-effector force ranging from 0.1 to 50 Hz is modeled and then translated as joint torque inputs by using Eq. 2. The amplitude of the end-effector force is kept at 0.17 N between 0.1–0.9 Hz and 0.5 N for the rest. During the tests, the accelerometer data were collected after the system has reaches a steady-state condition in the operating frequency.
**4th step:** The accelerometer data is processed and then integrated to calculate the end-effector velocity. By calculating the ratio between the output velocity and input force, modified admittance Bode diagram ($V_{out}/F_{in}$) is obtained as shown in Fig. 3. This technique is defined in the literature as the open-end experiment [11].

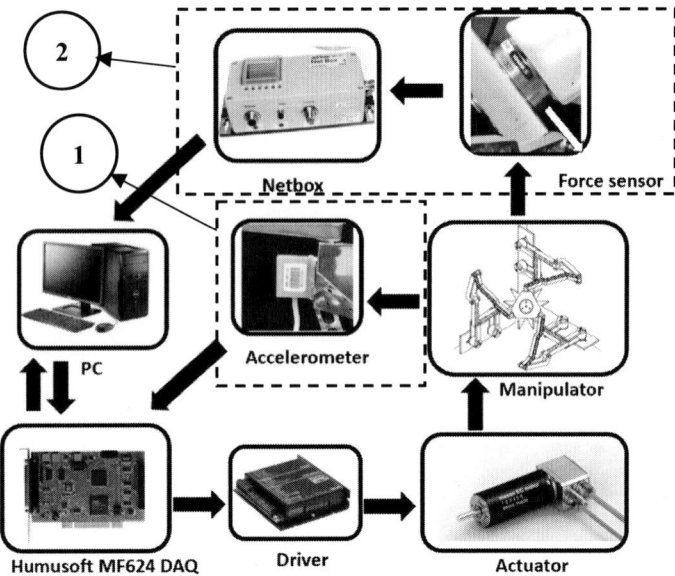

**Fig. 2.** Information flow of the experimental setups for frequency response method: (1) Fixed-end experiment (2) Open-end experiment

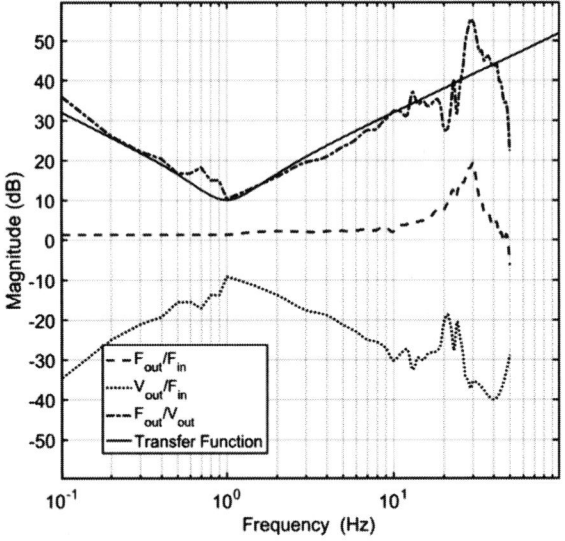

**Fig. 3.** Bode diagram of obtained data and fitted transfer function.

**5$^{th}$ step:** In this step, force response of the system is evaluated. A 6-DOF force/torque sensor system (type: Mini45, calibration: SI-580-20, with a DAQ called Netbox from ATI Industrial Automation, Inc.) indicated as ② in Fig. 2 is used to measure the output force. First, the end-effector is fixed mechanically and the force sensor is attached between the end-effector and the fixture (Fig. 2(1)). Then, the similar procedure expressed in the 3$^{rd}$ step is carried out. However, the force amplitude for the whole frequency range is set to 3 N and the output forces are acquired through the force sensor. Finally, the Bode diagram of force response is plotted for $F_{out}/F_{in}$ as illustrated in Fig. 3. This method is indicated in the literature as the fixed-end experiment [11].

**6$^{th}$ step:** A final calculation is made to find the ratio of the force response of the fixed-end frequency experiment to the velocity response of the open-end experiment. Thus, Bode diagram for $F_{out}/V_{out}$ of the device is obtained.

**7$^{th}$ step:** Finally, a second order mass-spring-damper in parallel connection model is fitted to the impedance Bode diagram ($F_{out}/V_{out}$) as shown in Fig. 3.

According to Fig. 3; fixed-end experiment results revealed that the applicable force range (bandwidth) is up to 10 Hz and a natural frequency is observed at 30 Hz which is caused due to the dynamics of the actuator (Type: EC 45, Maxon motor).

$$Z_{device} = \frac{F_{out}}{V_{out}} = \frac{ms^2 + bs + k}{s} = K_P \cdot \frac{s^2 + 2\zeta w_n s + w_n^2}{w_n^2 s} = \frac{0.63s^2 + 3.184s + 25}{s} \quad (3)$$

Investigating the impedance plot of the total system, it is clear that the corner frequency of the system is found to be around 1 Hz. Comparing the transfer functions presented in Eq. 3, the system's equivalent mass is calculated to be 0.63 kg, its damping is 3.184 Ns/m and stiffness is 25 N/m. The mass of the model is compared with the CAD model of the manipulator and the error is found to be 3% between the real mechanism and the modelled one.

## 4   Sensing Capabilities

### 4.1   Workspace Position Resolution

In this section, the position resolution of the $\vec{u}_1$ axis is studied. Position resolution is indicated as a distance which is obtained through the minimum applicable encoder resolution [8]. Therefore, the experiment is conducted by following this criterion and the details of the procedure are described as follows:

**1$^{st}$ step:** Theoretical workspace resolution is calculated through the forward kinematics in Eq. (1) for the encoder which has 0.087890625° resolution per step.

**2$^{nd}$ step:** Five test points are chosen along the test axis by examining the most critical points from the theoretical workspace resolution and the end-effector is situated at these points, respectively. In addition, the dial indicator is located at these points, for external measurement of end-effector location.

**3$^{rd}$ step:** At the first test point, the motor is driven for two counts change in encoder by using a position controller. Afterward, the end-effector position variation is measured by the help of dial indicator, which has 0.01 resolution. The information flow of the experiment is presented in Fig. 4. The reason for the two counts change is that single step count could not be followed by the controller without overshoot.

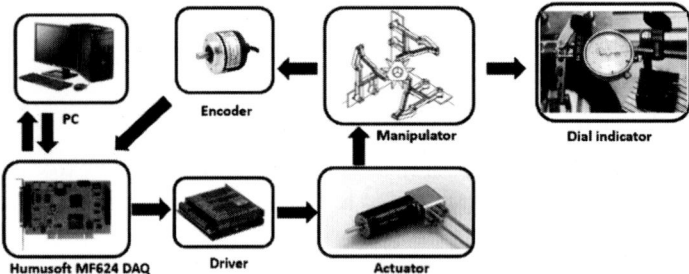

**Fig. 4.** Information flow of the experimental setup of the position resolution test

**Table 1.** Results of the workspace resolution experiment

| Test range (degree) | | Measured resolution (mm) | Theoretical resolution (mm) | Error value (mm) |
|---|---|---|---|---|
| 59.94 | 59.77 | 0.11 | 0.1 | 0.01 |
| 41.46 | 41.20 | 0.17 | 0.15 | 0.02 |
| 0.04 | 0.021 | 0.2 | 0.19 | 0.01 |
| −39.57 | −39.74 | 0.17 | 0.15 | 0.02 |
| −60.20 | −60.28 | 0.13 | 0.1 | 0.03 |

During the experiment, the dial indicator was in contact with the end-effector so that there was no gap between the end-effector and the dial indicator.

**4th step:** Finally, the differences between the measured and calculated end-effector positions are calculated and presented in Table 1.

According to the Table 1, it is obvious that there is a relatively large difference in terms of resolution between the 0° and 59.94° test ranges. The reason arises from the nonlinear kinematic relationship between the joint space and task space which is indicated in Eq. 1. Therefore, the mechanism has the capacity to achieve precise operations if the links are located away from 0° position. On the other hand, the errors are observed because of joint clearance as well as the compliance of the manipulator. Moreover, the spring of the dial indicator may have affected the results.

## 4.2  Workspace Position Repeatability

In this section, position repeatability of the $\vec{u}_1$ axis is presented. The repeatability term is expressed as the numerical position error which occurs when the end-effector is moved to a predefined point for several times [12]. The same test setup explained in Fig. 4 is used to measure the position differences at the end-effector. Moreover, the "Statistical testing of the operational and positional accuracy of machine tools basis-VDI/DGQ 3441" is implemented as follows:

**1ˢᵗ step:** Three test points are specified with unequal distance from each other so that recurrent errors are detected. Also, test points are chosen by evaluating the most critical points of the workspace.

**2ⁿᵈ step:** The positive initiation point, which is the starting point of the end-effector, is located at +60 mm along the workspace for (−) direction measurement. Afterwards, the dial indicator and the end-effector are positioned at the first test point (+1.22 mm) so that the dial indicator is set to zero by contacting the probe of the dial indicator with the end-effector.

**3ʳᵈ step:** The end-effector is moved back to the positive initiation point and then moved to the first test point. This step is repeated five times by recording the position differences that are observed at the dial indicator in Table 2.

**Table 2.** The repeatability test results of $\vec{u}_1$ axis for both directions

| # of tests | 1.22 mm (−) | 1.22 mm (+) | 14 mm (−) | 14 mm (+) | 50.2 mm (−) | 50.2 mm (+) |
|---|---|---|---|---|---|---|
| 1 | 0.03 | 0 | −0.02 | 0.01 | 0.07 | 0.05 |
| 2 | 0.04 | 0.01 | 0 | 0 | −0.04 | 0.04 |
| 3 | 0.03 | 0.01 | 0.06 | −0.08 | 0.03 | −0.03 |
| 4 | 0.03 | 0.01 | 0.03 | 0.03 | 0.02 | 0.03 |
| 5 | 0.02 | 0 | 0.04 | 0.02 | −0.05 | 0.04 |
| Standard deviation | 0.006325 | 0.0054 | 0.028566 | 0.043 | 0.044989 | 0.028705 |

**4ᵗʰ step:** The same procedure expressed in the previous step is performed for the second (+14 mm) and the third (+50.2 mm) test points.

**5ᵗʰ step:** For opposite direction measurement, the end-effector is located at the negative initiation point (−10 mm). In addition, the process described in 3ʳᵈ and 4ᵗʰ step is repeated for negative direction approach.

**6ᵗʰ step:** Finally, the standard deviation is calculated for each test point.

During the experiment, a position controller was used and the measurements were performed when the joint is moved to and stopped at the required test position.

From Table 2, it is observed that the test point at 1.22 mm has the minimum standard deviation in both directions among the overall set of test points. Since the manipulator is operating close to its nominal position, which is also the symmetric pose, required input torque to approach this test point from both directions is almost identical. On the other hand, for the other test points, depending on the approach direction, the required input torque is differed more.

# 5   Conclusions

In this study, actuation and sensing capabilities of the HIPHAD v1.0 haptic device were experimentally evaluated. Making use of obtained system model, the minimum impedance of the device can be decreased by canceling the dynamic effects as required. In this way the impedance width of the device would be enlarged. In addition, the obtained experimental data will be used to guide us in developing the new version of the device.

**Acknowledgments.** This work is supported in part by The Scientific and Technological Research Council of Turkey via grant number 117M405.

# References

1. Bilgincan, T., Gezgin, E., Dede, M.I.C.: Integration of the hybrid-structure haptic interface HIPHAD v1.0. In: Proceedings of the International Symposium of Mechanism and Machine Theory AzCIFToMM, Izmir, Turkey (2010)
2. Colgate, J.E., Brown J.B.: Factors affecting the z-width of a haptic display. In: Proceedings of the IEEE International Conference on Robotics and Automation, pp. 3205–3210 (1994)
3. Görgülü, I., Maaroof, O.W., Taner, B., Dede, M.I.C., Ceccarelli, M.: Experimental verification of quasi-static equilibrium analysis of a haptic device. In: Proceedings of the International Symposium of Mechanism and Machine Science AzCIFToMM, Baku, Azerbaijan, pp. 57–64 (2017)
4. Gupta, A., O'Malley, M.K.: Design of a haptic arm exoskeleton for training and rehabilitation. IEEE/ASME Trans. Mechatron. **11**, 280–289 (2006)
5. Hayward, V., Astley, O.R.: Performance measures for haptic interfaces. In: Robotics Research, pp. 195–206 (1996)
6. Ishida, T., Atsuo, T.: A robot actuator development with high backdrivability. In: IEEE Conference on Robotics Automation and Mechatronics, pp. 1–6 (2006)
7. Kern, T.A.: Engineering Haptic Devices A Beginner's Guide for Engineers. Springer, London (2009)
8. Martin, S., Hillier, N.: Characterisation of the novint falcon haptic device for application as a robot manipulator. In: Australasian Conference on Robotics and Automation (ACRA), pp. 1–9 (2009)
9. Martinez, M.O., Campion, J., Gholami, T., Rittikaidachar, M.K., Barron, A.C., Okamura, A.M.: Open source, modular, customizable, 3-D printed kinesthetic haptic devices. In: IEEE World Haptics Conference (WHC 2017), pp. 142–147 (2017)
10. Morrell, J.B., Salisbury, J.K.: Parallel-coupled micro-macro actuators. Int. J. Robot. Res. **17**, 773–791 (1998)
11. Samur, E.: Performance Metrics for Haptic Interfaces. Springer, London (2012)
12. Taner, B.: Development and Experimental Verification of the Haptic Device, Izmir Institute of Technology (2015)

# History of Mechanism Science

# Forgotten Facility: The Pneumatic Tube Mail System in Russian State Library

Andrei Vukolov[(✉)]

Institute of Modern Educational Technologies,
Bauman Moscow State Technical University, Moscow, Russian Federation
twdragon@bmstu.ru

**Abstract.** This paper describes history of development, installation and usage of pneumatic mail system in Russian State Library. Main peculiarity of this single-pipe system is ability to transfer several carriers simultaneously through the building which it is installed within. Technical solutions which were used to develop the system practically turned it to forgotten relic with both mechanical and electronic transport control techniques combined.

**Keywords:** Pneumatic tube transport · Russian State Library
Pipeline transport · Library service · Documents delivery

## 1   Introduction

The first known work on pneumatic transport was done by British engineer George Medhurst near 1810. Now pneumatic mail is in limited use. Pneumatic tubes are typical transport solution [2,3] for currency and paper documents in banks, hospitals, large factories, libraries etc. This type of transport leaves loads practically intact and allows to transfer such materials as radioactive samples [3–5]. In Russia the one of the most unusual applications of the pneumatic mail was made: the pneumatic pipeline for commands installed on the famous "Ilya Murometz" aircraft. Below the description of not so famous pneumatic mail system will be given. This system provides main part of functionality of user requests' delivery process in Russian State Library.

Russian State Library (RSL) is one of the largest libraries in the world. Its main building (Fig. 1[1]) placed on Vozdvizhenka street near Moscow Kremlin at the center of Moscow.

RSL was founded in 1862 based on Rumyantzev's books collection [13,15]. In 1918 after transfer of capital from St. Petersburg to Moscow it became the main library of the whole country. Creation of special department that work with classified documents also can be referred to the same period. From 1922 sending two copies of all books issued in USSR to Russian State Library had become

---

[1] Image produced by V. Tokarev (Wikimedia Commons) and published under conditions and constraints of Creative Commons Attribution License 3.0.

© Springer Nature Switzerland AG 2019
B. Corves et al. (Eds.): EuCoMeS 2018, MMS 59, pp. 147–154, 2019.
https://doi.org/10.1007/978-3-319-98020-1_17

mandatory. In 1925 the library was finally divided from museum department and renamed to "Russian State Library named after V. I. Lenin" [1]. Today Russian State Library is largest public library of Russia and Europe [1,14].

RSL uses paper-based orders placement system with paper duplication of electronic reader's requests. Such system requires to develop effective delivery solution for paper requests. Figure 2 shows simplified schema of request transport streams within the library. It contains storage facilities as collection points for requests. Each storage should have addressing system to distribute requests by service desks. At the endings of 1960s the decision was made to use pneumatic tubes. However, pneu-

**Fig. 1.** Russian State Library

matic transport systems that were being developed that time in USSR used 65 mm master pipe diameter. It was not possible to set up such system in the library building [8,12]. Development of new system with 50 mm master pipe diameter had begun near 1968–1970, but it is not possible to determine the date more precisely due to ambiguity between documents of different years [9,12].

## 2  Facility Development and Construction

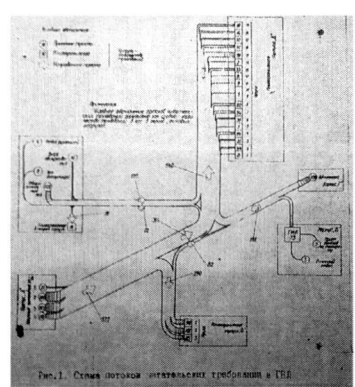

**Fig. 2.** Request transport schema (original facsimile)

In 1971 the contract was officially established between Central Bureau of Design, Mechanization and Automation and Russian State Library on installation of pneumatic mail facility. Main project was prepared in 1970. The main choice for developers was an alternative between double-pipe and single-pipe design [9]. The first variant of the project included 6 double-pipe pneumatic mail systems within the facility. It was declined immediately because of absence of reservation possibility for documents departure frequency. Singled reversible system supports this possibility, so this type of design was chosen. According to Fig. 2 after analysis of collecting points workflow the final schema (Fig. 3) was developed. It includes the following peculiarities: the facility is divided into four subsystems with independent control: two master (*all-over*) subsystems are connecting all requests collection points and two slave (*bypass*) subsystems are connecting especially heavily loaded service desks for reservation purposes.

Master subsystems include 38 endpoints each. Bypass subsystems include 29 and 30 endpoints respectively. Thus each endpoint with more than one transceiver installed have possibility to double their transfer rate. Each subsystem include one master pipe and from one to five slave pipes [9] connected to endpoints through switches [10]. Each slave pipe can be connected to not more than 25 endpoint transceivers. Each pipe has own reversible blower on the end, so the subsystem can transfer one carrier in each slave pipe plus one in the master pipe simultaneously. Endpoint

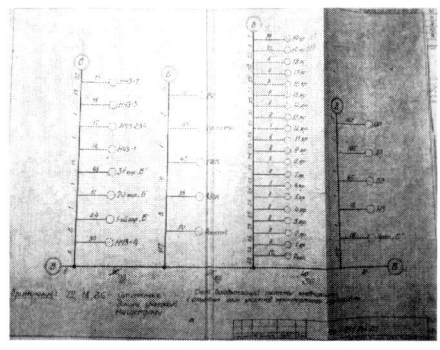

**Fig. 3.** Facility schema with blowers and subsystems, 1970 (photograph of original facsimile)

address consists of master pipe (subsystem) code and endpoint code defined within destination subsystem.

Main facility was built near 1971 is presented on Fig. 4 where I and II are master subsystems, III and IV are bypass subsystems.

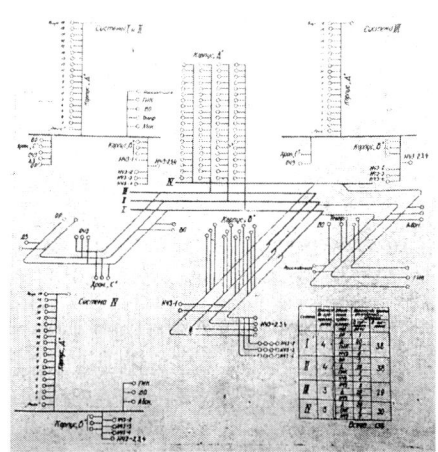

**Fig. 4.** Main facility schema with all requests collection points (original facsimile)

The schema (Fig. 4) contains all requests collection points snapped to their slave pipes and all master pipes enumerated. In general main project included five subsystems [9] but the fifth one was converted into experimental facility in the design bureau. On that facility the endpoint transceiver design decisions (Fig. 5) were checked and the experiments for transfer rate determination were performed.

Also on that stage the decisions were made about installation of busy pipe detectors directly into the endpoint transceivers what allowed to avoid mistaken departure of multiple carriers by single address and/or orphan carriers in the subsystem [6]. Due to low weight of load it was not needed to solve problems of inappropriate accelerations [11] and the announced speed of the carrier was set (overriding the initial value of 6 m/s) near 10 m/s for master pipe. During experiments also reversing time was measured for single carrier (0.46 s). Timing constant for control system was set to 1 s [9].

## 2.1    Addressing and Address Encoding

Now it is not possible to reconstruct true addressing algorithm due to absence of documentation. But according to [9] each master pipe of the facility can address from 1 to 5 slave pipes and from 1 to 25 endpoint transceivers on each slave pipe respectively, as it was described above. But the real endpoint transceivers uses button blocks with only one button allowed in pressed state at the moment [10]. Cause all facility uses binary logical control blocks this fact allows to make an assumption about addresses encoding (please see the example below):

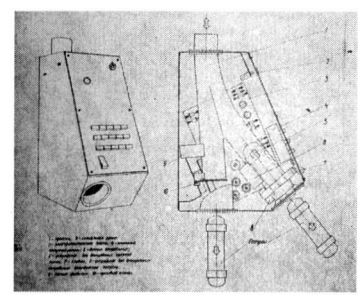

**Fig. 5.** Design sketch for endpoint transceiver (original facsimile)

010             01001
slave pipe      endpoint address

Thus each address in the subsystem contains single byte (8 bits), big-endian, where first 3 bits represent slave pipe code with range from 0 to 7. The next 5 bits represent the endpoint transceiver number with range from 0 to 31. In this case construction of the transceiver can use three 5-button blocks on the keyboard where the top one is for entering of slave pipe code (Fig. 5) what complies to real situation. This type of address makes subsystems scalable on control level without necessity to change construction of electronic blocks. Between the subsystems manual transfer of carrier is required.

## 3    Installation Workflow

The first installed subsystem consists of master pipe and three slave pipes with 30 endpoint transceivers (Fig. 6a) was installed in Russian State Library between 1974 and 1975. It was the master subsystem (marked as I on Fig. 4) configured for test usage [10, 12]. According to specification (Fig. 6b) installation of parts was performed.

During the installation some problems were revealed. One of them was existing compression blowers which were not reversible. To solve this problem new blowers were developed based on ordinary high-speed motors for vacuum cleaners. It became needed to change location of the blowers to make airflow reversible. On the new schema (Fig. 7) master pipe of the subsystem has two blowers on endings which are working only in pressurization mode. Only one blower can be run at the moment. This is an interesting fact that practically the same solution for obtaining of reversible airflow was patented in United States in 1973 as carrier slower [16]. But it is not possible to determine interconnection between these solutions now.

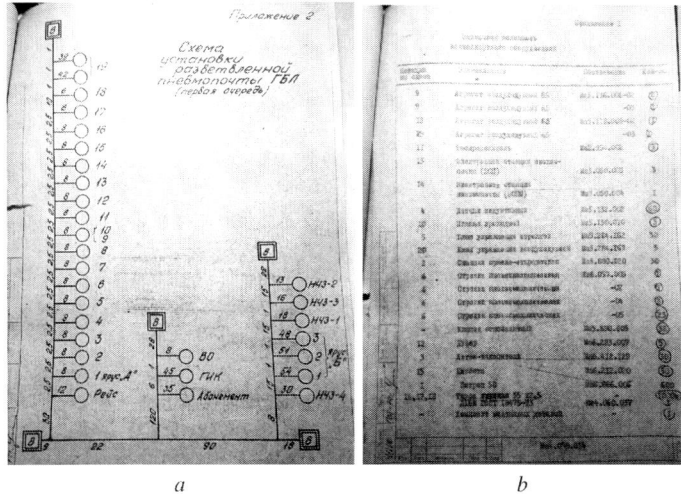

*a*                                   *b*

**Fig. 6.** Installation schema and specification of first pneumatic mail subsystem (photographs of original facsimile)

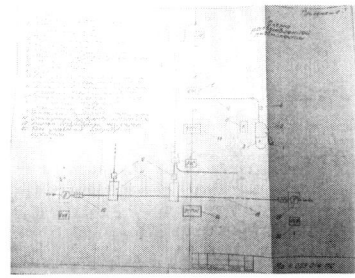

**Fig. 7.** Changed installation schema with relocated blowers (photograph of original facsimile)

For slave pipe that has one ending connected to the switch the blower was installed connected to ringed pipes with turning valves that changes direction of an airflow through the pipe. The valves are driven by electric actuators.

An another problem was connected with construction of switches. Traditional electro-pneumatic switches [3,5,7] require too high air pressure to work, so the decision was made to replace them with switches driven by electromagnetic actuators. They allowed to use exhaust mode of one blower for slave pipe to deliver the carrier on large distances. Carriers themselves were redesigned with thin plastic bodies and narrower sealing to reduce weight. Lightweight carriers became able to work on the switches and carry announced load of 50 request sheets.

All facility is equipped with scalable modular control system. Each block of the system (Fig. 8) contains logical elements which translate binary addressing commands from endpoint transceivers to electric signals.

All facility is powered by three power supplies. The main one provides standard 220 V, 50 Hz alternating current for all mechanisms, endpoint transceivers and blowers. Also it provides power for two another supplies.

The second one transforms standard current into 60 V direct current for electromagnetic switches. The transformer in each control rack provides 12 V direct current for control devices. The facility uses electronic transport control together with mechanical solutions. At the moment of beginning of everyday usage (1975) the subsystem I (Fig. 4) contained 5 endpoint transceivers for first slave pipe, 8 for the second, 20 for third and 5 for fourth one. After entering of an address using keyboard (Fig. 5) the librarian places loaded carrier into front

**Fig. 8.** Control block

hole of the transceiver and then internal sensor immediately sends a signal to the control block. This signal turns the switches to prepare them for delivery.

The second (II on Fig. 4) subsystem was installed after tests in 1978 [12]. Its endpoint transceivers work simultaneously with the identical ones from the subsystem I to increase transfer rate on service desks (Fig. 9).

### 3.1  Handling of Exceptions

**Fig. 9.** Service desk and pneumatic mail station with endpoint transceivers of first (I) and second (II) subsystems

The facility uses predefined set of methods to handle exceptions. Its documentation [10] contains the list of typical errors. Carrier dismantling is considered as common mechanical issue. Disconnected cup of the carrier may damage the pipe and jam switches. In case of this library staff can try to push the dismantled carrier through the pipe to nearest switch (the switch must be turned off to do that) with another empty carrier.

User mistake while addressing is considered as most common control error. When user enters non-existent destination address the design of endpoint transceiver and slave pipe does not allow to handle this issue automatically and to return the carrier, it must be sent anyway. The handling algorithm contains three branches:

- If the slave pipe code is the same than the code of origin slave pipe but the destination code points to non-existent endpoint: sending transceiver is became blocked and transferred to waiting mode until the correct destination address will be entered;
- If the destination address points to non-existent endpoint:
  - If the non-existent endpoint located on existent slave pipe: the carrier will be sent from origin slave pipe to destination slave pipe switch. After that the carrier will be dropped back into master pipe. Master pipe controller sends dropped carriers to specially marked endpoint transceiver of the first slave pipe;

– If destination address is fully invalid the sending transceiver switches to emergency mode. Now user must change master pipe code to make it valid.

As the most dangerous malfunction of the facility the fire (or flame throw) on one of the blowers is considered. High speed motors used in blowers are liable to such issue. Air flow can heathen the pipes and throw the flame and smoke trough them very fast. To prevent that the special fire-proof gaskets placed between pipes and walls. For additional protection against fire each blower is installed into metal locker equipped with air filters and earthing buses to prevent short circuit.

## 4    Current Status

At present the pneumatic pipe mail facility in Russian State Library works as main transport system for readers' requests. It transfers printed requests from printing racks and reception desks on the first floor to storage enclosure [14]. After completion of each order empty pneumatic mail carriers are reassembled and sent back to reception desks. In fact, each endpoint transceiver of each subsystem has an another transceiver from another subsystem nearby. Installation of the facility remained incomplete. Due to unknown reasons no attempts to finish the work were made in between 1980 and 1993. Two successfully installed subsystems now are working as

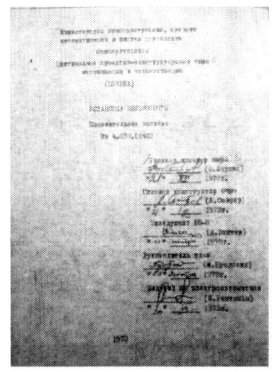

**Fig. 10.** Title page of technical project (original cyanotype)

replacement of the whole facility announced in the project which exists (Fig. 10). Due to destruction of main equipment manufacturers it is not possible to renovate or replace endpoint transceivers, switches and controllers. All of this allow to make an assumption that any serious injury will lead all facility to complete destruction.

## 5    Conclusion

The Russian State Library pneumatic pipe mail facility is now turned into forgotten relic of technologies. It is the main instrument which is transferring readers' requests from reception desks to administration and books storage enclosure. In fact construction of the facility demonstrates high level of reliability and efficiency due to usage of most advanced solutions of its time.

**Acknowledgements.** Author wants to kindly acknowledge **Arkady Igorevich Aksenoff** the Chief Engineer of subsystem II for his assistance and advice.

Mr. Aksenoff granted a permission to visit the facility and to access unique documents. Author also wants to acknowledge BMSTU ordinary professor **Olga Egorova** for her advice dedicated to earlier history of pneumatics and Library.

# References

1. History of Russian State Library [CD-ROM] (in Russian). Russian State Library, Moscow. Fund of Electronic Library, 16-1/152 (2006)
2. Capsule pipelines — Mainland Europe. Capsu.org website (2010). Accessed 12 Feb 2010
3. Rohrpost — pneumatic city mail. BUISPOST.EU Website. http://buispost.eu/pneumatic-city-mail/. Accessed 10 June 2016
4. Pneumatic networks (updated 23.07.2008). Douglas-self website. http://www.douglas-self.com/MUSEUM/COMMS/pneumess/pneumess.htm#con. Accessed 24 Mar 2016
5. Adámek, A., Severa, F.: A dual pneumatic tube transfer system for the analysis of fast neutron activated short-lived isotopes. J. Radioanal. Chem. **7**(1), 119–125 (1971). https://doi.org/10.1007/BF02520882
6. Antonenko, V.F., Borovikov, V.K.: Novel signalling and control element for pneumatic mail systems. Metallurgist **13**(6), 184–385 (1969)
7. Batcheller, B.: Pneumatic-tube system. US Patent 840,194 (1907). https://www.google.com/patents/US840194
8. Borisenko, G.P., Belyaev, A.E.: About determining of pneumatic mail tube carrier dimensions (in Russian). Izvestia Tomsk Polytech. Univ. **224**, 51–56 (1976)
9. Central Bureau of Design, Mechanization and Automation, Moscow: Pneumatic Mail Facility. Detailed Technical Description (in Russian) (1973). Item no. 40371, stored in Russian State Library
10. Central Bureau of Design, Mechanization and Automation, Moscow: Pneumatic Mail Facility. Technical Passport (in Russian) (1974). Stored in Russian State Library, fund 51, item 19
11. Davydov, S.Y., Kosyrev, N.P., Valiev, N.G., Simisinov, D.I., Kurochkin, V.A., Zamuraev, A.E.: Theoretical studies of the unloading of containers in the pneumatic transport systems of today and tomorrow. Refract. Ind. Ceram **54**(3), 178–187 (2013). https://doi.org/10.1007/s11148-013-9572-0
12. Dvorkina, M.Y.: Library service: theoretical aspect (in Russian). MGIK (1993)
13. Koval, L.M.: For the Good Education: from the History of Russian State Library (in Russian). Pashkov's Mansion, Moscow — Russian State Library (2012)
14. Lukyanov, D.: The labyrinths of 'Leninka' (in Russian). COLTA.RU Website. http://m.colta.ru/articles/specials/3824?page=9. Accessed 10 June 2016
15. Tatarinova, E.A.: The exhibition activities of Moscow Public Museums and Rumyantzev's museum of Russian State Library in 1861–2000: Tendencies and succession (in Russian). Ph.D. thesis, Russian State Library, Moscow (2009)
16. Van Otteren, W.: Pneumatic tube system. US Patent 3,711,038 (1973). https://www.google.com/patents/US3711038

# A Vehicle Driven Upwind by the Horizontal Axis Wind Turbine

Marat Dosaev[(✉)], Liubov Klimina, and Yury Selyutskiy

Institute of Mechanics of LMSU, Moscow, Russia
{dosayev,klimina}@imec.msu.ru

**Abstract.** A mathematical model of a wind powered vehicle is constructed. The vehicle is driven by a horizontal axis wind turbine. It is supposed that the velocity of the center of mass of the vehicle is directed upwind. Conditions of existence and stability of the steady upwind motion are discussed. The maximum speed of the upwind motion is estimated. The maximum speed is compared with that obtained for different types of wind powered vehicles.

**Keywords:** Wind powered vehicle · Horizontal axis wind turbine
Dynamical system · Steady motion

## 1 Introduction

For ages people use the power of wind or water stream for motion of boats. Similar approach sometimes is used for the overland traveling when a wheeled vehicle is equipped with a sail. It's a normal situation to wait for the favorable wind for a long time or to use tacking for the upwind motion. At first sight, it seems that the wind never causes an upwind motion, as well as a river never brings a boat from the estuary to the source.

In spite of seeming impossibility, a principle of a self-sustained upstream motion was proposed by Italian engineers of the early Renaissance ([1, 2], Fig. 1). Very similar idea was suggested by Claude François Milliet Dechales (1621–1678) in his book [3]. The same principle was independently reinvented by Ivan Kulibin (1735–1818). The first official full-scale test of his self-moving boat took place on the Neva River in 1782; the boat was moving against the stream and against the wind [4]. Such boat was used for some time to transport goods upriver.

However, the idea of self-sustained upstream motion was forgotten until the second half of the XX century. Then, within a short period, several prototypes of wind powered cars moving upwind or "downwind faster than the wind" were designed using very similar general principle, and corresponding mathematical models were discussed [5–7].

For instance, a vehicle powered by a horizontal axis wind turbine (HAWT) was studied in [7], where the possibility of the upwind motion, as well as of the motion "downwind faster than the wind" was shown in the context of the Betz theory.

Maybe, the most popular schemes of wind powered vehicles are those driven by horizontal axis wind turbines (HAWT) and those driven by Savonius wind turbines [8, 9].

© Springer Nature Switzerland AG 2019
B. Corves et al. (Eds.): EuCoMeS 2018, MMS 59, pp. 155–161, 2019.
https://doi.org/10.1007/978-3-319-98020-1_18

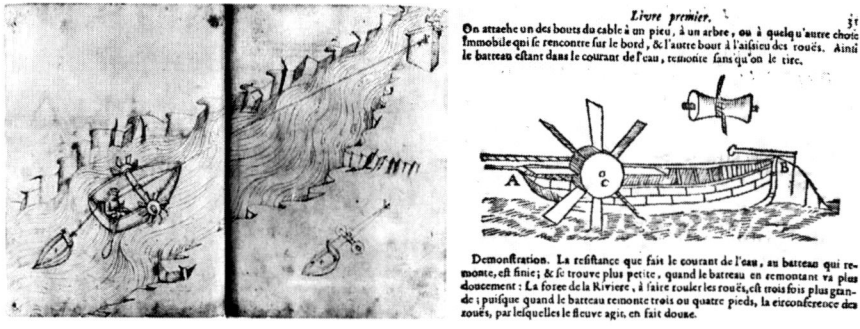

**Fig. 1.** The fragments from books of Taccola and Dechales describing the upriver motion

In the same time, novel types of wind turbines are also used for wind powered cars (e.g., [10]). It is important to compare different types of wind turbines with respect to their efficiency for flow powered vehicles.

In the present paper, the following criteria of such efficiency is discussed: the maximum upwind speed. The quasi-steady approach is applied to the HAWT-type wind car; the mathematical model is constructed. The parametrical analysis of steady motions of the HAWT type wind car is carried out. The maximum speed of the upwind motion is estimated. Its value is compared with that obtained earlier for Savonius and slider-crank types of wind vehicles in the context of the similar model of an aerodynamic load [9, 10].

## 2  Description of the System

The wind powered vehicle is equipped with the HAWT of the radius $r$ (Fig. 2). Denote the central point of the HAWT as $O$. The shaft of the HAWT is rotating around the axis $Oz$. Suppose that the system is located in a steady wind flow with the wind velocity $\mathbf{V}$ parallel to the $Oz$ axis. Notations: $J$ is the moment of inertia of the HAWT around the axis $Oz$; $S$ is the characteristic swept area of the HAWT; $\Omega$ is the angular speed of the HAWT; $m$ is the mass of the vehicle.

Suppose that the turbine shaft is connected to the axis of driving wheels of the vehicle by means of a reduction gear. Energy losses in the gear mechanism are neglected. Thus, the mechanical torque produced by the turbine is used to drive the vehicle.

It is supposed that the vehicle can move only along the $Oz$ axis (i.e., directly upwind or downwind). In what follows, positive values of the absolute speed $V_O$ of the point $O$ correspond to the upwind motion.

It is supposed that there is no slipping between the wheels and the ground (hence, the system has one degree of freedom). The gear ratio and the radius of wheels provide the following relation between the speed $V_O$ of the point $O$ and the tip speed $V_{tip} = r\Omega$ of the HAWT:

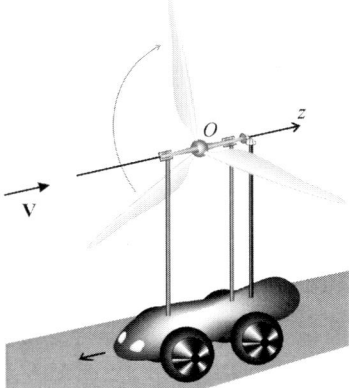

**Fig. 2.** The scheme of the HAWT-type wind vehicle

$$V_O = kV_{tip}. \tag{1}$$

Here $k$ is the kinematic "reduction coefficient". Further, it is supposed that $k$ is positive. The configuration of the mechanism is intended for the upwind motion. This case is interesting because it allows using the energy of the wind flow for the direct upwind motion (without tacks).

Assume that the flow interacts only with the turbine blades. Suppose that the aerodynamic load that influences the motion of the vehicle along the $Oz$ axis can be represented as the torque $T$ about the axis $Oz$ and drag force **D** directed along the wind speed and applied at the point $O$. In the context of the quasi-steady approach [11–13], when the aerodynamic load is supposed to depend only on the instantaneous state of the system, these values are described as follows:

$$T = \frac{1}{2}\rho SU^2 rC_T(\lambda), \quad D = \frac{1}{2}\rho SU^2 C_D(\lambda), \tag{2}$$

Here $\rho$ is the density of the air; $\lambda = r\Omega/|U|$ is the tip speed ratio of the HAWT; $C_T(\lambda)$, $C_D(\lambda)$ are the dimensionless coefficients of the torque and the drag force, respectively, that are taken from the experiments. For further calculations, these functions are interpolated (using splines) basing on the experimental data [14]: Fig. 3. Dots show experimental results.

The experimental data for negative tip speed ratios and for large tip speed ratio (for which $C_T(\lambda) < 0$ that is $\lambda > \lambda_0 \approx 11$) are not available.

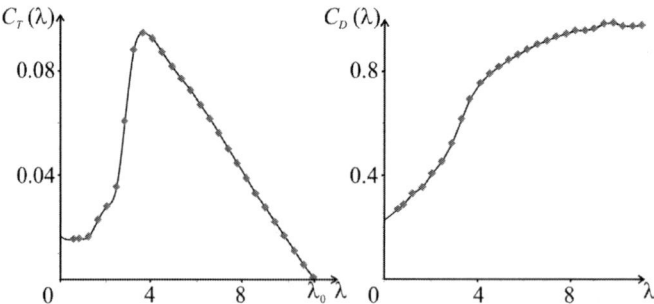

**Fig. 3.** Aerodynamic coefficients approximated basing on the experimental data [14].

## 3   Equations of Motion

Due to the relations (1) and (2), equations of motion of the system can be represented in the following form:

$$(J + mr^2k^2)\dot{\Omega} = \frac{1}{2}\rho SU^2 r(C_T(\lambda) - kC_D(\lambda)\text{sign}(U)). \tag{3}$$

Equations (3) can be reduced to the following dimensionless form:

$$(a^2 + k^2)\omega' = (1 + k\omega)^2(C_T(\lambda) - kC_D(\lambda)\text{sign}(1 + k\omega)),$$
$$\lambda = \frac{\omega}{|1 + k\omega|},$$
$$a = \frac{J}{mr^2} > 0. \tag{4}$$

Here the prime denotes the derivative with respect to the dimensionless time $\tau = 0.5\rho SVt/m$; $\omega = r\Omega/V$ is the dimensionless angular speed of the HAWT.

Stable fixed point of Eq. (4) corresponds to the steady motion of the vehicle: the center of mass moves with the constant speed, and the angular speed of the HAWT is also constant.

The equation of the steady motion is as follows:

$$C_T\left(\frac{\omega}{|1 + k\omega|}\right) - kC_D\left(\frac{\omega}{|1 + k\omega|}\right)\text{sign}(1 + k\omega) = 0. \tag{5}$$

Let $v$ be the speed $V_O$ normalized by the wind speed. So, from (1), $v = k\omega$. The efficiency of the mechanism is characterized by the value $v$ at steady motion. The case of positive $v$ corresponds to the upwind motion of the vehicle. Notice, that for another significant case of motion – "downwind faster than the wind", – propellers with special aerodynamic characteristics are used that significantly different from those for wind turbine rotors [5–7].

## 4   Steady Upwind Motion

In the case of the upwind motion, $k$ is positive, the value $U$ of the air speed of the point $O$ is positive (i.e. $1 + k\omega > 0$). Denote the solution of (5) as $\omega_*$.

Condition of stability of the steady motion takes the following form:

$$C_T' - kC_D' < 0, \qquad C_T' = \left.\frac{dC_T(\lambda)}{d\lambda}\right|_{\omega=\omega_*}, \qquad C_D' = \left.\frac{dC_D(\lambda)}{d\lambda}\right|_{\omega=\omega_*}. \tag{6}$$

The curve $v(k)$ corresponding to solutions of (5) can be represented explicitly in the parameterized form $k(\lambda), v(\lambda)$ using the following formula:

$$k(\lambda) = \frac{C_T(\lambda)}{C_D(\lambda)}, \qquad v(\lambda) = \frac{\lambda C_T(\lambda)}{C_D(\lambda) - C_T(\lambda)\lambda}. \tag{7}$$

Taking into account (7), one obtains that the condition (6) of stability of the fixed point can be rewritten as follows:

$$\left.\frac{d}{d\lambda}(k(\lambda))\right|_{\omega=\omega_*} < 0. \tag{8}$$

The points of the bifurcation curve $v(k)$ where the tangent is vertical correspond to the merging of stable and unstable fixed points. The bifurcation curve $v(k)$ is represented in the Fig. 4. Solid branches (*AB, CD, EF, FG*) correspond to stable steady motions; dashed branches (*BC, DE*) correspond to unstable steady motions. White dots (*B, C, D, E*) represent points where the linear approximation is not enough to check stability. The red part of the curve (*PQ*) corresponds to the case when the absolute speed of the vehicle is large than the wind speed. Magenta dotted lines show an example of a transition process that is discussed further.

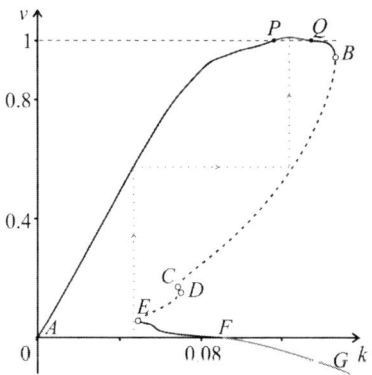

**Fig. 4.** Speed of the vehicle normalized by the wind speed for the case of the upwind motion. (Color figure online)

The blue part of the bifurcation curve (*FG...*) corresponds to the stable downwind motion ($v<0$). This part of the curve $v(k)$ is constructed qualitatively, because experimental data for negative $\lambda$ are not available, as mentioned before. Notice that the downwind motion is not a desired regime for $k > 0$.

## 5   Discussion

In the case of the chosen aerodynamic characteristics (Fig. 3), the maximum upwind speed is about 101% of the wind speed and is achieved for $k \approx 0.12$. Notice that the result predicted in [7] by the Betz theory was that maximum upwind speed of the vehicle tends to infinity if energy losses in the mechanism are neglected.

Compare the maximum speed of the HAWT-type wind car with that obtained in the context of the same aerodynamic model for other types of wind powered vehicles. The maximum upwind speed of the Savonius-type wind powered vehicle is about 15% of the wind speed as can be concluded from [9]. The maximum upwind speed of the slider-crank wind car is about 60% of the wind speed as shown in [10]. Thus, the HAWT-type wind car can be treated as the most effective from these types for the purpose of the upwind motion.

For a certain range of $k$, the domain of attraction of the "fast" stable steady upwind motion (the upper branch of the curve $v(k)$) is limited by the unstable steady regime (unstable steady regimes correspond to the branches *BC* and *DE* in the Fig. 4). For such $k$, if the initial speed of the vehicle is below the value that corresponds to the unstable motion, then the system tends to the "slow" steady upwind motion. The corresponding "slow" stable steady regimes are represented by the branches *CD* and *EFG* in the Fig. 4 (*FG* corresponds to slow downwind motion).

In order to bring the system to the upwind motion with the maximum speed (supposing that the initial speed is zero), one could organize switching of the reduction coefficient $k$. An example of such switching is illustrated by magenta dotted lines in the Fig. 4. At least two different values of $k$ have to be involved in this transition process. It is noticeable that rather slight change of the bifurcation curve *AB...ED* can lead to the situation when at least one more switching of $k$ will be needed to bring the system to the domain of attraction of the optimal regime. It can be concluded that the control of the value $k$ is enough to reach the regime of the upwind motion with the maximum speed.

## 6   Conclusions

The mathematical model of a wind car driven by the horizontal axis wind turbine is constructed and studied. The car is intended for the upwind motion. The parametrical analysis of steady motions is performed. It is shown that in the case of the chosen aerodynamic characteristics of the HAWT, the maximum upwind speed can be larger than the wind speed (101% of the wind speed). The strategy of control of the reduction coefficient is suggested that allows transferring the system to the fastest steady upwind motion.

**Acknowledgments.** This work was partially supported by the Russian Foundation for Basic Research, projects NN 17-08-01366, 18-01-00538.

# References

1. Feldhaus, F.M.: Ruhmesblätter der Technik von den Urerfindungen bis zur Gegenwart, pp. 399–401. Verlag F. Brandstetter, Leipzig (1910). (in Deutsch)
2. Nanni, R.: Il Badalone di Filippo Brunelleschi e l'iconografia del «navigium» tra Guido da Vigevano e Leonardo da Vinci. Annali di Storia di Firenze **6**, 65–119 (2011). (in Italian)
3. Dechales, C.F.M.: L'art de naviger demontré par principes & confirmé par plusieurs observations tirées de l'expérience, Paris (1677). (in French)
4. Tatarenkov, V.I.: The history of ship means of traffic. Saint Petersburg, Gallery Print (2006). (in Russian)
5. Lysenko, G.P., Grigoriev, B.V., Karpin, K.B.: Wind motor applications for transportation. In: Proceedings of the 31st Intersociety Energy Conversion Engineering Conference, IECEC 96, pp. 1783–1785. IEEE (1996)
6. Sannikov, V.: Downwind faster than the wind. Pop. Mech. **9**, 84–87 (2010). (in Russian)
7. Gaunaa, M., Øye, S., Mikkelsen, R.F.: Theory and design of flow driven vehicles using rotors for energy conversion. In: 2009 European Wind Energy Conference and Exhibition (2009)
8. Kassem, Y., Hüseyin, Ç.: Wind turbine powered car uses 3 single big C-section blades. In: International Conference on Aeronautical & Manufacturing Engineering (ICAAME 2015), pp. 42–45 (2015)
9. Selyutskiy, Yu., Klimina, L., Masterova, A., Hwang, Sh.-Sh., Lin, Ch.-H.: On dynamics of a Savonius rotor-based wind power generator. In: Proceedings of 14th Conference on Dynamical Systems: Theory and Applications (DSTA 2017), vol. 3, pp. 275–284. The Technical University of Lodz (Poland) (2017)
10. Klimina, L., Dosaev, M., Selyutskiy, Yu.: Asymptotic analysis of the mathematical model of a wind-powered vehicle. Appl. Math. Model. **46**, 691–697 (2017)
11. Dosaev, M.Z., Samsonov, V.A., Seliutski, Y.D.: On the dynamics of a small-scale wind power generator. Dokl. Phys. **52**(9), 493–495 (2007)
12. Dosaev, M.Z., Lin, Ch.-H, Lu, W.-L., Samsonov, V.A., Selyutskii, Yu.D: A qualitative analysis of the steady modes of operation of small wind power generator. J. Appl. Math. Mech. **73**(3), 259–263 (2009)
13. Samsonov, V.A., Dosaev, M.Z., Selyutskiy, Y.D.: Methods of qualitative analysis in the problem of rigid body motion in medium. Int. J. Bifurcat. Chaos **21**(10), 2955–2961 (2011)
14. Adaramola, M.S., Krogstad, P.A.: Experimental investigation of wake effects on wind turbine performance. Renew. Energy **36**(8), 2078–2086 (2011)

# Industrial and Non-industrial Applications

# Electrical Torque Addition Mechanism for Engines with High Levels of EGR

Madhusudan Raghavan[1(✉)] and Andrew Balhoff[2]

[1] General Motors R&D Center, Pontiac, USA
madhu.raghavan@gm.com
[2] (Intern) General Motors R&D Center, Pontiac, USA
abalhoff@gmail.com

**Abstract.** Gasoline engines that employ high levels of exhaust gas recirculation (EGR) may suffer from slow torque transients during tip-in and tip-out maneuvers. This is due to the residual EGR present in the engine intake manifold. In order to fill these torque holes, we explore the use of a belted alternator starter (BAS) mechanism. The BAS supplements the torque provided by the engine. In the present work, we investigate the operating modes and show that the power and torque levels of the BAS are adequate to add the necessary driveline torque for acceptable tip-in response.

**Keywords:** BAS · EGR · Mild electrification

## 1 Introduction

Fuel economy and energy consumption are key drivers for innovations in automotive propulsion systems. The International Energy Outlook [1] offers insights into potential future scenarios. World population is expected to grow by approximately 25% from 2015 to 2040. In the same time period, world marketed energy is expected to grow by nearly 40%. So there is an urgent need for energy saving breakthroughs in the transport sector, the power sector, agriculture and industry. Emissions and air quality are also areas of concern in future scenarios.

Automakers are exploring the use of exhaust gas recirculation (EGR) as a means to improving fuel economy [2, 3]. External EGR removes a portion of the air from the exhaust manifold, and re-routes it through a throttle back to the intake manifold of the engine. EGR helps reduce emissions by lowering combustion temperatures and reducing excess oxygen in the combustion mixture, preventing NOx formation. Due to the potential benefits of EGR, it is currently in use in many modern-day vehicles, and is being considered for broad implementation. The use of EGR may pose challenges when power is required. As power is demanded from the engine, increased levels of fuel mixed with the high levels of EGR may lead to cylinder knock. In addition the introduction of the burned emissions displaces potentially useable oxygen for combustion. Therefore, the combusted gases must flow through the intake manifold and cylinders before full engine output can be realized. This physical process can lead to delays in excess of a third of a second from throttle demand to output, which may be unacceptable to the consumer. Therefore, some other mechanism is required to provide

© Springer Nature Switzerland AG 2019
B. Corves et al. (Eds.): EuCoMeS 2018, MMS 59, pp. 165–172, 2019.
https://doi.org/10.1007/978-3-319-98020-1_19

nearly instantaneous power while the EGR flows from the intake through to the exhaust manifold.

A belted alternator starter (BAS) mechanism could satisfy this requirement. The implementation of this system necessitates very little modification to the engine, as it is in effect a "bolt on" assembly. Additionally, electric motors are noted for their rapid power delivery capability. This solution to the EGR flow problem shows promise, as little redesign is necessary to the original powerplant.

Prior work on belted alternator starters is as follows. Wezenbeek et al. [4], investigated combustion-assisted starting as a potential method of aiding a 12 V accessory drive belted alternator starter in the starting process on larger engines. Canova et al. [5], use simulation tools to design a closed-loop controller for starting and stopping a 1.9L diesel engine equipped with a 10.6 kW BAS system, following a prescribed speed trajectory. Hawkins et al. [6], describe the GM eAssist system which includes a water-cooled induction motor-generator BAS, an accessory drive with a coupled dual tensioner system, air cooled power electronics integrated with a 115 V lithium-ion battery pack, a direct-injection 2.4 liter 4-cylinder gasoline engine, and a modified 6-speed automatic transmission. Fulks et al. [7], detail a 2.4 liter 4-cylinder GDI application of a non-boosted 14 V BAS system with integrated battery state of health monitoring. A key enabler for the application is a high (3.24:1) pulley ratio which reduced cranking current by 43%.

We differentiate our present work from these cited prior work references by investigating the coupling of the BAS functionality with the characteristics of an EGR-equipped engine to achieve acceptable drive quality and response.

## 2   Engine Model

Our simulations were conducted on a 1.5L gasoline engine. This is a four cylinder SI engine modeled in GT-Power with intake and exhaust flow to allow for EGR flow modeling. The EGR was controlled through a throttle, and passed through a cooler assembly before reintroduction in the intake manifold. Care was taken to include the full cranktrain dynamics. Inertias and masses were used to model the individual cranktrain components and were then assembled into a working model.

A number of cases were then chosen to test the engine dynamics. The torque requirements studied tip-in and tip-out maneuvers varying from 10 Nm to 107 Nm, which is the engine's maximum torque at wide open throttle (WOT). Points at 36 Nm and 70 Nm were also studied to provide finer detail. The engine speeds chosen were 1360, 2100, and 2600 RPM. Lower speeds were deemed more important, as manifold turnover time is inversely related to engine speed. Valve timing was then adjusted to prevent cylinder knock spikes in the worst possible case. Upon investigation, the case with the greatest torque lag was the 10 Nm to WOT case at 1360 RPM. The phaser rate was set to 0.2 s, the throttle tip-in rate to 0.02 s, the throttle tip-out rate to 0.2 s, and the EGR tip-in/tip-out rate to 0.2 s. These values account for a worst case scenario approach to the engine valve timing, and were set to be consistent across all investigated cases.

Once the timing values were set, the torque and RPM cases were again swept to estimate response times. The values corresponding to the 1360 RPM cases can be seen in Table 1. The $3\tau$ response times (time to reach within 5% of target) were found to range between 0.26 s to 0.37 s. This trend holds over all engine RPMs considered.

**Table 1.** Response times for varying torque demands at 1360 RPM

| Torque range [Nm] (at 1360 RPM) | $3\tau$ response time [sec] |
| --- | --- |
| 10–70 | 0.34 |
| 10–107 | 0.37 |
| 36–70 | 0.26 |
| 36–107 | 0.27 |
| 70–107 | 0.32 |

As engine RPM is increased, the $3\tau$ response times decrease. This follows from the premise that the EGR exits the engine more quickly at higher engine speeds. This is because the response times are directly tied to the turnover rate of the engine, the volume of the cylinders, and the volume of the intake manifold. The turnover rate of the manifold is approximately 2/3 of the response time in each case, or 6 combustion cycles, and the rest of the time is taken by approximately two full combustion cycles per cylinder.

**Table 2.** Time response over varying engine speeds

| Engine speed [RPM] (10–70 Nm) | $3\tau$ response time [sec] |
| --- | --- |
| 1360 | 0.34 |
| 2100 | 0.17 |
| 2600 | 0.15 |

Table 2 clearly shows the response time decreasing as engine speed increases, indicating the study should be focused on the 1360 RPM cases. This sweep indicates that the worst case was the 10–107 Nm, 1360 RPM scenario, with a $3\tau$ response time of 0.37 s. This time is 0.12 s longer than the desired response time, chosen to be 0.25 s. This 0.25 s requirement set the standard for the BAS system performance demands. From the 0.25 s response time choice, the cases at engine speeds above the 1360 RPM were discarded, as their response times were all satisfactory.

In order to test the validity of the chosen BAS system, the power requirement for the insufficient cases was calculated. Power is calculated from the formula

$$Power = RPM\left(\frac{1}{60}\right)2\pi(Torque) \tag{1}$$

and the resulting power requirements in various scenarios can be seen in Table 3.

The BAS motor chosen has an output of 4 kW, and is capable of satisfying the power requirements of the system.

**Table 3.** Power requirement

| Engine speed [RPM] | Torque range [Nm] | Torque@0.25 s[Nm] | Power [W] |
|---|---|---|---|
| 1360 | 10–70 | 13.9 | 1981.5 |
| 1360 | 36–70 | 2.0 | 290.3 |
| 1360 | 70–107 | 2.4 | 348.6 |
| 1360 | 10–107 | 19.8 | 2839.4 |

## 3   BAS Mechanism Model

The next step to validate the potential of the BAS mechanism was to create a simulation model to study the dynamic effects on the engine, as well as the potential issues related to belt slip. First a GT-Power model was created to model the BAS system. The physical design of the system as seen in Fig. 1, consists of an electric motor marked, a dual tensioner assembly, an air-conditioner compressor and a pulley connected to the crankshaft. These rotary devices are all connected by a serpentine belt. The dual tensioner ensures that bi-directional torque can be exerted by the motor/generator on the remaining systems via the belt.

The electric motor was modeled as an ideal motor applying up to 4 kW of power. The dual tensioner assembly was modeled as two individual rigid tensioning arms, due to limitations in GT-Power. The alternator was modeled as a pulley with inertia. The crankshaft pulley was then attached to the GT-Power model. The belt to run the system is a 7-rib serpentine belt with a pre-tension of 170 N. The belt was modeled as a cold belt, as this would be the case with the least friction, and therefore provide a worst-case baseline for estimation. The nominal maximum pulley torque, or the torque required to pull the belt free, is calculated using the following formula,

$$M_{max} = 2R\left(T_0 - \rho v^2\right)\frac{e^{\mu\beta} - 1}{e^{\mu\beta} + 1} \tag{2}$$

where $M_{max}$ is the nominal maximum pulley torque, R is the pulley radius, $T_0$ is the nominal static tension, $\rho$ is the belt density per unit length, v is the belt initial linear speed, $\mu$ is the friction coefficient, and $\beta$ is the belt wrapping angle. By iterating along the potential belt speeds, a plot comparing torque and belt speed was generated, as seen in Fig. 2. Due to the squared belt speed in the formula, the maximum torque available decreases in a non-linear fashion as speed of the crankshaft increases.

By associating the crankshaft pulley with the inertias of the engine and running the BAS assembly at maximum torque, the engine speed slew rate was calculated. Then, by multiplying the speed by the torque available, the power at any given time is determined and is nearly linear with time. This is due to the small speed range operated on in 0.25 s. From this, it is evident that even with the effects of belt slip, the electric motor of the BAS can provide approximately 2800 Watts of power at 0.25 s, capable of handling all of the cases presented. The only potential issue could be at a low engine speed transition from low torque to WOT, with a cold belt. This power delivery would be greater with a warm belt, which should be present in most operating conditions.

Additionally, the torque available to the crankshaft is nearly constant, decreasing from 22.59 Nm to 22.57 Nm over the initial 0.25 s.

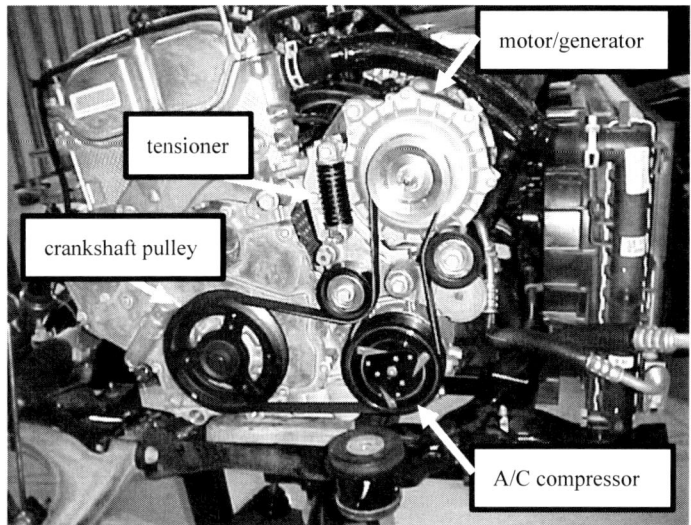

**Fig. 1.** BAS assembly

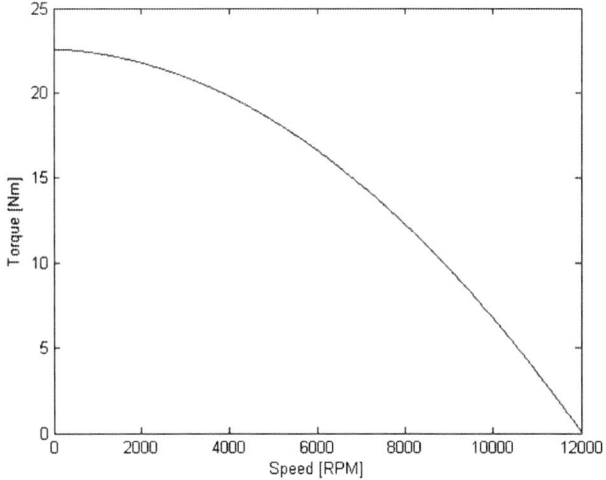

**Fig. 2.** Torque vs. Crankshaft Speed

## 4 Combined System

Based on this estimated performance, the system model was operated with the BAS providing extra torque to the crankshaft in GT-Power. This was implemented by establishing a rigid connection from the pulley object to the crankshaft block. The control for the BAS system provided the maximum amount of power until the desired torque was achieved, and then maintained that torque until the engine matched it, and then shut off. Four cases were studied, and are pictured below in Fig. 3.

The profiles all significantly outperformed their unadjusted counterparts. The assistance was more significant in the lower torque range tip-ins, as improvements were seen in excess of 50%. The adjusted $3\tau$ response times can be seen in Table 4.

**Table 4.** Adjusted torque response times

| Torque range [Nm] (at 1360 RPM) | $3\tau$ response time [sec] | New $3\tau$ response time [sec] | % Improvement |
|---|---|---|---|
| 10–70 | 0.34 | 0.21 | 37% |
| 10–107 | 0.37 | 0.23 | 38% |
| 36–70 | 0.26 | 0.13 | 51% |
| 70–107 | 0.32 | 0.14 | 56% |

Additionally, the energy requirement was investigated for each of the cases considered. These values are calculated using

$$Energy = \int \left( RPM.T.\frac{\pi}{30} \right) dt \tag{3}$$

where T is the torque required from the BAS. The results can be seen in Table 5.

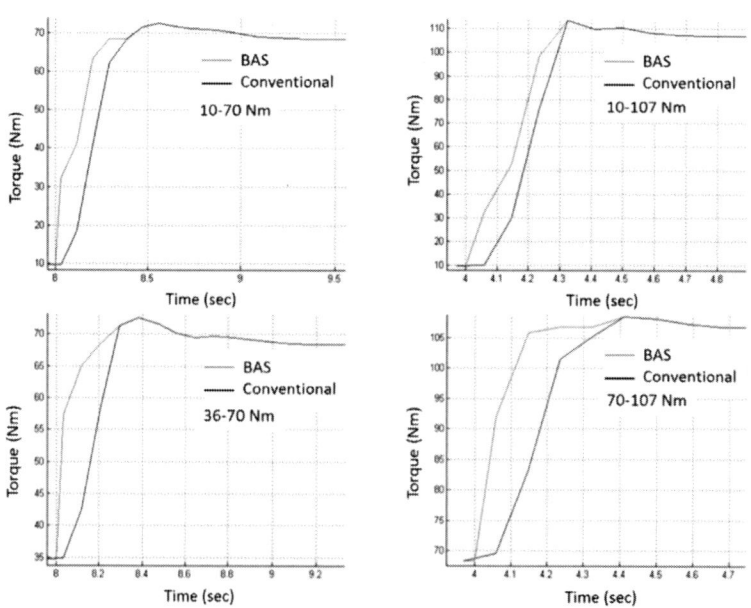

**Fig. 3.** BAS assisted torque profiles

**Table 5.** Energy consumption for 1360 RPM cases

| Torque range [Nm] (at 1360 RPM) | Energy [J] |
|---|---|
| 10–70 | 908 |
| 10–107 | 906 |
| 36–70 | 662 |
| 70–107 | 707 |

The energy consumption at higher engine speeds were also considered. These values are similar to the ones in Table 5, as the increase in operating speed is balanced by the decreased run time of the BAS system. Subsequent to these GT-Power simulations, the BAS mechanism and EGR engine models were inserted into a Matlab program for computing fuel consumption on certification drive cycles. These MATLAB simulations confirmed our initial projections from GT Power, regarding the adequacy of the BAS torque to improve tip-in response on the high EGR engines, The use of the BAS system in regen/boost mode with a Li-ion battery enables a 6–7% improvement in fuel economy over the baseline (non-BAS) gasoline propulsion system. These fuel economy estimates include the effect of the increased mass of the BAS system.

## 5   Summary and Conclusions

Detailed modeling of a 1.5L gasoline engine was established in GT-Power. This dynamic model proved important for later implementation of the BAS model, and belt-slip modeling of the BAS system. The increased EGR effected the greatest delay in low engine speed cases. These effects were especially pronounced in cases with high torque demands. These cases had response times greater than ¼ s, the chosen maximum response time. A 4 kW BAS system was deemed capable of satisfying these requirements, as it is capable of providing satisfactory power and torque. Further work in MATLAB was conducted, establishing a pathway for dynamic EGR simulation. In addition to improving engine response time, the 4 kW BAS system offers a 6–7% fuel economy improvement over baseline on various drive cycles.

## References

1. International Energy Outlook 2016, DOE/EIA-0484 (2016), May 2016. https://www.eia.gov/outlooks/ieo/pdf/0484(2016).pdf
2. Przastek, J., Dabkowski, A., Teodorczyk, A.: The study of exhaust gas recirculation on efficiency and NOx emission in spark ignition engine. SAE (1999). https://doi.org/10.4271/1999-01-3514
3. Cairns, A., Blaxill, H.: The effects of combined internal and external exhaust gas recirculation on gasoline controlled auto-ignition. SAE (2005). https://doi.org/10.4271/2005-01-0133
4. Wezenbeek, P., Evans, D., Sczomak, D., Absmeier, J., Fattic, G.: Combustion assisted belt-cranking of a V-8 engine at 12-Volts. SAE (2004). https://doi.org/10.4271/2004-01-0569

5. Canova, M., Sevel, K., Guezennec, Y., Yurkovich, S.: Control of the start/stop of a diesel engine in a parallel HEV with a belted starter/alternator. SAE (2007) https://doi.org/10.4271/2007-24-0076
6. Hawkins, S., et al.: Development of general motors' eAssist powertrain. SAE (2012) https://doi.org/10.4271/2012-01-1039
7. Fulks, G., Roth, G., Fedewa, A.: High performance stop-start system with 14 volt belt alternator starter. SAE (2012). https://doi.org/10.4271/2012-01-1041

# Motion Planning of a Rotation Type Peach Fruit Moth Inspection System

Koji Makino(✉), Kazuyoshi Ishida, Hiromi Watanabe, Yutaka Suzuki, Shinji Kotani, and Hidetsugu Terada

University of Yamanashi, Kofu, Japan
{kohjim,isawa,hwatanabe,yutakas,kotani,terada}@yamanashi.ac.jp

**Abstract.** A peach fruit moth is very serious problem while exporting the peach, because the peach fruit moth affects the ecological system of the destination country where it does not live in. This paper aims to develop a new type of a peach fruit moth inspection system and to evaluate its performance in terms of motion planning theoretically. The validity of the equation for system evaluation is confirmed by numerical computer simulation. Finally, the rotation type of the real system is developed, and the validity of the theoretical values is verified.

**Keywords:** Motion planning · Detection system · Real system
Numerical simulation

## 1 Introduction

Since the items with the free trade or tariff elimination have increased in Japan, it is important to export not only the industrial products but also the agricultural products [8]. Peaches among the agricultural products in Japan are very popular in foreign countries [1]. In foreign countries, peaches are often used for processing. The Japanese peaches can be eaten raw and are much larger than the ones found in foreign countries. Among all the fruits that are found in Japan, peach is the third largest exported fruit, and it has increased in the past 10 years. The export of the peach fruit moth along with the peach fruit [4] is a very serious problem. If a peach containing a larva of the peach fruit moth is export to a country where the moth is not endemic, it damages the ecological system of that country. Even worst is that the peach fruit moth lays on the rosaceous fruits such as apples and pears, and the larva enters and feeds on the peach pulp. Therefore, the peach fruit moth inspection system has been developed [3]. This system has various desirable features, but is not time cost efficient. And it is important to compare the success rate of the detection of the moth between human operator and the inspection system. However it is difficult to investigate the success rate of the detection of the human, since it is necessary to cut the peaches in order to confirm the inner of the peach. The human operator could not detect the moth one time in 5 years ago.

© Springer Nature Switzerland AG 2019
B. Corves et al. (Eds.): EuCoMeS 2018, MMS 59, pp. 173–180, 2019.
https://doi.org/10.1007/978-3-319-98020-1_20

This paper aims to develop a new rotation type peach fruit moth inspection system, and evaluates it in terms of motion planning. Initially, the characteristics are described. Then, the evaluation of the system is theoretically considered from the viewpoint of time cost and operating time ratio. The validity of the evaluation is confirmed by numerical computer simulation. Finally, it is verified using the real system.

## 2    Peach Fruit Moth Problems and Search Methods

First, the peach fruit moth is described briefly [5]. The moth lays eggs on the peach surface. The hatching larva enters the peach. The size of the hole that the larva digs is approximately 0.2 mm. The larva grows in the peach pulp till the time the size of the larva becomes is 5 mm. If the peach containing the larva is exported from Japan to a foreign country where the moth does not live in, for example Taiwan, the ecological system in the country will be severely damaged. In Japan, examiners search for an extremely small hole and eggs in all the peaches that are being exported by performing a 1 min visual inspection per peach. However, searching them is very difficult.

Second, each problem of the three morphologies (egg, larva and adult moth) is described. The adult moth is not a serious problem. It is easy to detect the adult moth, since the shape is similar to a butterfly and the size is greater than 5 mm. Larva is the most serious problem. It cannot be seen on the outer surface of the peach, because the larva does not appear on the surface of the peach till the time it reaches the adult stage. This study adopted the X-ray analysis to detect the larva that enters the peach. Some methods, such as infrared and magnetic resonance imaging (MRI), exist to check the inner part of the peach. The detection rate using the infrared is more than 90% [9]. However more than 40% of the peaches that does not contain the larva is mistakenly identified as peaches with the larva. In other study, the larva of size approximately 1.5–2.0 mm can be detected by using the MRI [2], however this method is necessary to process the image for 2 h. The X-ray image that is obtained by the proposed system is shown in Fig. 1(a) [6]. The space that is generated by the larva is searched in this system, because the transmission rate of the larva is similar to one of the pulps of the peach. The two image conversions, brightness gradient vector image (Fig. 1(b)) and gradient concentration vector image (Fig. 1(c)), are performed to detect the space more easily. The time for both conversions is less than 10 s. Moreover, the space is not often detected in cases where the space overlaps the shade of the seed. The peach is rotated to take X-ray images from various directions [7]. On the other hand, the eggs are also a serious problem. Egg detection and its removal are also important. The former is used to directly detect the egg using a camera. Our group is developing the detection system using the partial division method of the photo obtained using the high-resolution camera. The latter removes the egg by air blow. If the short and thin trichome of the peach is removed, the almost eggs are removed. Additionally, the automatic air blow machine is being developed, too.

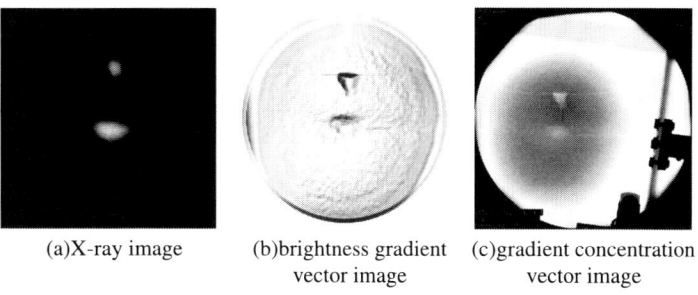

| (a)X-ray image | (b)brightness gradient vector image | (c)gradient concentration vector image |

**Fig. 1.** The images of the peach containing a space

## 3   Peach Fruit Moth Inspection System

The linear type peach fruit moth inspection system has been developed. The linear type system has a desirable characteristic such as the deviation of the area into a "safety area" and "risk area". In the safety area, only peaches are seen after the checking process. In the risk area, peaches are seen before the checking process and peaches, in which the larva is detected, are also seen. On the other hand, if the some sensors and machines are connected, some problems from the viewpoint of time cost and operating time ration exist. This study aims to propose a new rotation type system that is expanded from the linear type system.

The rotation type system is illustrated in Fig. 2. This system consists of some linear sliders that are circularly connected, handling units that grab the peach and carries it along the slider, and the X-ray detection box. In the rotation type system, all the handling units move toward one direction along the circularly disposed linear sliders. One examiner places the peach on the handling unit and the other examiner picks it up at another position. The differences in the linear type system are the arrangement of the sliders and movement of the handling unit. The slider of the linear type system is straight, and the handling unit moves forward and back in the same slider. Therefore, only one handling unit is used in the linear type system. On the other hand, many handling units can be placed on the rotation type system, because collision within the handling units does not occur.

The structure of the rotation type system (Fig. 2) is illustrated is Fig. 3. To evaluate the system, the time for checking the peach is important. The time is called as the time cost ($TC$) in this paper. It is desirable that the operating time ration of the sensors and machines are high. The average operating time ratio ($OT$) is evaluated.

First, $TC$ and the average ratio of $OT$ are calculated for the simple structure shown in Fig. 3. The time from the setting position to the picking-up position is defined as $t_1$. $t_1$ includes the moving and checking times and $t_2$ is the moving time on the return path. In fact, the time that the examiner sets the peach and picks it up have to be considered. These times are ignored in this study

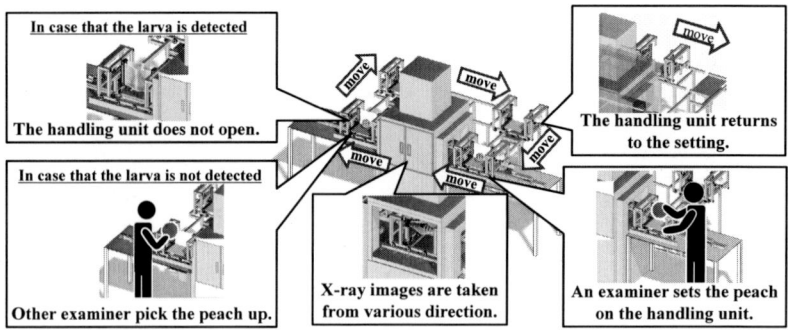

**Fig. 2.** Rotation type system consisting the X-ray sensors and circularly disposed linear slider

to evaluate the system performance. In this study, the setting and picking-up process are planned by the robot hand using the deep Q network that is a kind of the deep learning. $N$ is the number of the units on the system, which is taken as three in the case of the Fig. 3. $m_1$ is the number of the units that are checked at the same time. For example, $m_1$ becomes two if the X-ray boxes are connected in parallel. $m_2$ is the number of the units on the return path at the same time, which is three in this figure. Two kinds of time cost are considered (when the bottleneck occurs and does not occur). Each $TC$ is expressed in Eq. (1). The bottleneck occurs when the value obtained using the upper equation in Eq. (1) is larger than the value obtained in the lower equation.

$$TC = \begin{cases} max(t_1/m_1, t_2/m_2) & \text{if bottleneck occurs} \\ (t_1 + t_2)/N & \text{else} \end{cases} . \tag{1}$$

where the $max$ function, in this paper, is the maximum input values.

Second, the average operating time ratio is considered. The condition shown in Fig. 3 is described in Eq. (2) because the sensor is one. If the bottleneck occurs, $OT$ becomes one.

$$OT = \frac{t_x}{TC}. \tag{2}$$

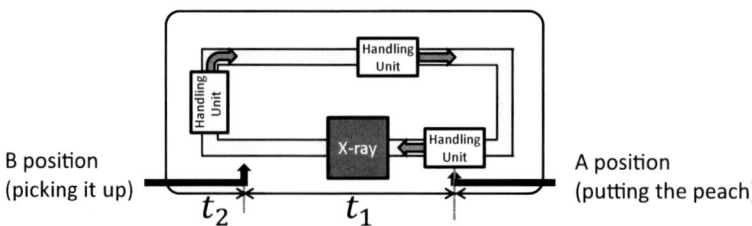

**Fig. 3.** The structure of the rotation type system shown in Fig. 2

Next, the system is expanded, and the various sensors and machines are connected. Figure 4 shows an example of the system. This system consists of three kinds of sensors and a machine (X-ray sensor, air blow machine, camera sensor and three-dimensional sensor). The three-dimensional sensor can observe the shape of the peach and is useful for quality guarantee. The sensor is also developed by our group. The sensors and the machines are connected with four junctions that are expressed with cross-marks in this figure. Two X-ray sensors and three air blow machines are connected in parallel. $TC$ and average ratio of $OT$ are calculated using Eqs. (3) and (4), respectively.

$$TC = \begin{cases} max(t_1/m_1, t_2/m_2 \cdots, t_M/m_M) & \text{if bottleneck occurs} \\ \dfrac{1}{N}\displaystyle\sum_{i=1}^{M} t_i & \text{else} \end{cases} \tag{3}$$

$$OP = \frac{\dfrac{\sum_{i=1}^{M}(t_i/m_i)}{M}}{TC} \tag{4}$$

Equation (4) is explained. The $\displaystyle\sum_{i=1}^{n}(t_i/m_i)$ is the fastest $TC$ by one handling unit. $M$ is the sum of the kinds of sensors and machines. Dividing it by $M$ is equal to the average of the $TC$ of the all sensors and machines. The average ratio of $OT$ is obtained by dividing it by $TC$.

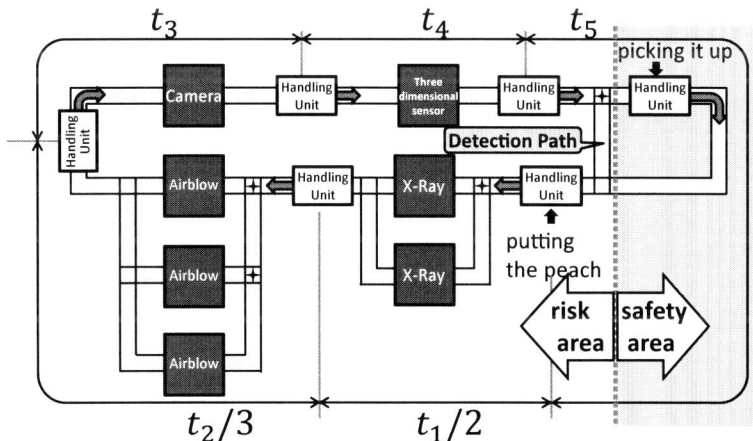

**Fig. 4.** An example of the structure of the extended system

In addition, the system enables the area to be divided into the "safety area" and the "risk area" using the detection path. The handling unit in the peach where the larva is detected moves along the detection path. Using the detection path, the peach containing the larva does not enter the safety area.

Final, The TC and average ratio of OT of the linear type inspection system proposed by [3] are considered. The structure of the linear type corresponds with the structure of the rotation type shown in Fig. 2 using only one handling unit. Therefore, they can be calculated by Eqs. (1) and (2) in case that the number of the handling unit is set to one.

## 4    Arrangement Sensors and Machines Using Simulation

In this section, the effectiveness using various sensors and machines is discussed and the validity of the theoretical values is confirmed by numerical computer simulation. The $TC$ of each sensor and machine was assumed as follows; X-ray sensor is 60, air blow machine is 120, camera sensor is 40, three-dimensional sensor is 20, and return path is 10. $TC$ is measured in seconds. The simulation is coded using the processing software which is a language similar to C language and easily realizes visualization. Each simulation is performed up to 10,000 s. The stopped time of the units on each sensor and machine is counted. The units in the picking-up position are also counted. $TC$ and average ratio of $OT$ are calculated using the counted numbers.

The results of the simulation are shown in Figs. 5 and 6. In both figures, the horizontal axis represents the number of the handling unit and the name of sensors and machines. X is expressed as X-ray sensor. A, C, T, and R are the air blow machine, camera sensor, three-dimensional sensor, and return path, respectively. For example, 5(XXAAACTR) means five handling units, two X-ray sensors, three air blow machines, one camera sensor, one three-dimensional sensor, and the return path. Each vertical axis in Figs. 5 and 6 is $TC$ and average ratio of $OT$, respectively. The bar graphs depict the values obtained using the simulation, and the ×mark shows the calculated value using the Eqs. (3) and (4). The value obtained using the simulation is very similar to the calculated values. As a result, the validity of the theoretical value is confirmed.

First, $TC$ shown in Fig. 5 is addressed. The $TC$s of 1(XACTR) and 2(XACTR) are compared. The $TC$ becomes half in case where two handling units are used. However, the $TC$ does not become lower though the larger number of the handling unit (for example, 3(XACTR)) is employed because the bottleneck occurs. In this theoretical equation, the bottleneck can be represented too.

Second, the kind of sensors and machines is considered. The $TC$ of 5(XAACTR) is lower than that of the 5(XACTR). The $TC$ of 5(XXAAACTR) is the lowest and is one-fifth of 1(XACTR). The bottleneck is resolved and the condition can be calculated using the equation. The $TC$ becomes low in proportion to the number of handling units if the sensors and machines increase.

Third, the average ratio of $OT$, as shown in Fig. 6, is focused on. The average ratio of $OT$ using 5(XXAACTR) is the highest though the $TC$ is not the lowest. It is confirmed that there is an optimal value of the average ratio of $OT$. It is clear that the balance between the $TC$ and $OT$ has to be considered.

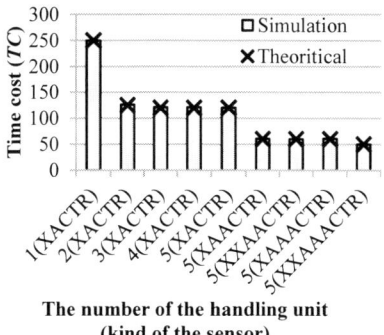

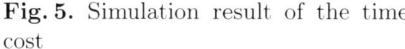

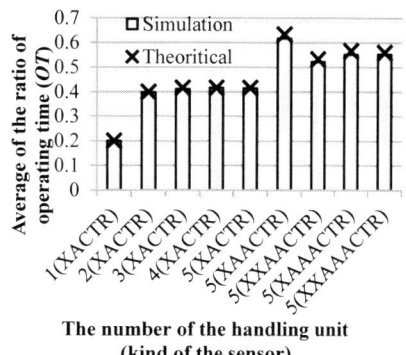

**Fig. 5.** Simulation result of the time cost

**Fig. 6.** Simulation result of the average of the ratio of the operating time

## 5   Experiment and Demonstration

The real system of the rotation type of the peach fruit moth inspection system consisting of the X-ray sensor shown in Fig. 2, and the three handling units used were developed. The real system is shown in Fig. 7. This section describes the result of the experiment and demonstration of the rotation type system.

First, the basic performance is measured. Both $TC$ of the X-ray sensor and the return path are 44 s that are average values, when the trials using a real system are ten times. The $TC$ of the X-ray sensor (44 s) is longer than the $TC$ of the return path (44/3 s). Therefore, the $TC$ of the system ($TC$) is 44 s and the $OT$ ratio is 1. In this setting, the handling unit reaches the setting position before X-ray checking using the theoretical equation, as confirmed in the experiment.

Second, to investigate whether this rotation type system can be operated on the real environment, the system was demonstrated from 19 to 21 October 2016 at the International Exhibition (Japan Robot Week) and checked greater than 100 peaches that are imitation and are not fresh. No problems exist through

**Fig. 7.** Demonstration at Japan Robot Week

the demonstration. It is confirmed that the system can be operated in the real environment.

## 6    Conclusion

This paper describes the rotation type peach fruit moth inspection system that is expanded from the linear type system that has been developed. First, the equations for obtaining the theoretical value to evaluate the time cost ($TC$) and the average ratio of the operating time ($OT$) in terms of motion planning are shown. The validities of the equation are confirmed using numerical computer simulation. It is clear that the system has an optimal size. Finally, the rotation type of the real system is developed, and the validities of the theoretical values are verified. It is confirmed that the system can be operated in the real environment through the demonstration.

**Acknowledgements.** This research was supported by grants from the Project of the NARO Bio-oriented Technology Research Advancement Institution (the special scheme project on regional developing strategy).

## References

1. Fruits and Vegetables Grown in Japan Are Safe and of the Finest Quality, WEB site on the Ministry of Agriculture, Forestry and Fisheries. http://www.maff.go.jp/j/export/e_info/vege_fruit/pdf/fruit_e_all.pdf. Accessed Jan 2018
2. Ihara, F., et al.: Non-destructive observation of peach fruit moth, carposina sasakii matsumura (lepidoptera: carposinidae), in young apple fruits by MRI. Jpn. J. Appl. Entomol. Zool. **52**(3), 123–128 (2008)
3. Ishida, K., et al.: Development of a soft material handling mechanism with the concavo-convex sheet for the peach fruit moth inspection system. In: ICMDT (2017)
4. Japan Fruit Association: Introduction to Fruit Export and Trade, Agriculture Study Group Report (2017)
5. Kawashima, K.: Bionomics of the peach fruit moth. Bull. Appl. Exp. Stat. Aomori Prefect. Agric. For. Res. Center **35**, 1–51 (2008)
6. Kotani, S., Suzuki, Y.: Development of an X-ray penetration imaging system for detecting peaches injured by the peach fruit moth. J. Jpn. Soc. Precis. Eng. **79**(11), 995–998 (2013)
7. Makino, K., et al.: Arrangement Method of a Peach for Export to Taiwan in a Peach Moth Inspection System. SII WeD.4.2 (2017)
8. Summary of the Basic Plan for Food, Agriculture and Rural Areas, WEB site on the Ministry of Agriculture, Forestry and Fisheries. http://www.maff.go.jp/e/policies/law_plan/attach/pdf/index-2.pdf. Accessed Jan 2018
9. Toyoshima, S., et al.: The ability of a double-sensored NIR device to detect apples ('Fuji' cultivar) injured by the peach fruit moth, Carposina sasakii Matsumura (Lepidoptera: Carposinidae). Bull. Nat. Inst. Fruit Tree Sci. **7**, 13–20 (2008)

# Access Systems to Marine Energy Production Units. Review and New Challenges

Pablo García$^{(\boxtimes)}$, Javier Sanchez-Espiga,
Alfonso Fernandez-del-Rincon, Ana De-Juan, Miguel Iglesias,
Alberto Diez-Ibarbia, and Fernando Viadero

ETSIIT, University of Cantabria, Santander, Spain
{garciafp,sanchezespij,viaderof}@unican.es

**Abstract.** The search of new energy production forms in order to fulfil the raising energetic demand and diminish the environmental issues derived from the production of energy by using non-renewable resources results in the development of new technologies like the ones related to the offshore marine renewable energy. The interest in the exploitation of this energy source is tangible according to the fact that different projects are going to be developed in the following years, both in Europe and worldwide.

Despite having a huge potential [13] as a very efficient way of energy production, there are several challenges regarding the economic viability of this kind of infrastructure. In this way, the cost reduction associated to maintenance and installation fixings is crucial. The disposal of staff transfer systems between a ship and the generator is one of the main factors that could contribute to the cost reduction. Simultaneously, this system should guarantee the security and being effective in the widest range of sea conditions possible.

In this work a review of the main options available in the market is done. Taking into account for each of them the work ranges and the kind of installation, but also security and other important factors that will affect the economic viability of this energy production way. Likewise, new ideas, in ongoing development, will be analysed, which are related to the staff transferring onto floating infrastructures, being this a non-appropriately solved problem these days.

**Keywords:** Offshore energy · Movement compensation · Transfer system
Safety

## 1  Introduction

Several problems are being faced by humankind in terms of sustainable energy production and rising energy consumption. First of all, the raise in the energy demand due to the increasing worldwide population and also because of the need of more energy per capita. At the same time, there are different environmental issues directly related to the greenhouse gases emission, consequence of energy production, which makes necessary the development and implantation of new ways to produce "clean" energy. Also, countries committed to the recent Paris agreement and the previous Kyoto protocol have to fulfil their renewable energy targets.

© Springer Nature Switzerland AG 2019
B. Corves et al. (Eds.): EuCoMeS 2018, MMS 59, pp. 181–188, 2019.
https://doi.org/10.1007/978-3-319-98020-1_21

As a result, along the last decades, different renewable energies have experienced a rapid growth in terms of power installed and production. Out of those possibilities wind energy has been the renewable energy mostly used until now, reaching a global cumulative installed capacity of 486,749 MW at the end of 2016 [3]. This is because of the high wind resource and the reached maturity in the development of the necessary technology.

However, appropriate places to set this kind of technology on-shore are running short; hence wind energy production in the last two decades has set towards sea. Offshore wind energy production has steeply grown, reaching a cumulative installed capacity total of 12,631 MW in ten different European countries [1, 2, 4] and a global cumulative offshore wind capacity of 14,384 MW [3] by the end of 2016. Besides, several different new projects are in ongoing development and construction worldwide in spite of the high costs associated to the construction and maintenance of these infrastructures.

Many diverse enterprises and research projects have been studying different possible systems and manoeuvres to reduce maintenance cost providing forms of reaching the fixed marine structures in challenging sea conditions. By assuring the access to these structures the overall operation time will be diminished and therefore their economic viability.

In terms of safety and work conditions in this kind of manoeuvres, the systems used for these operations normally follow "walk to work" technology. This technology is currently applied to the design of many staff transfer systems to fixed marine structures trying to guarantee safety for the individuals walking between the elements involved in the manoeuvre. To this aim there should be a system that connects physically both origin and destination and, at the same time, dispose a surface on which operators will step in order to complete those metres between both locations. Apart from the elements involved in the fulfilment of these requirements, this kind of equipment incorporates other systems, mainly movement compensations.

## 2   Main Existent Systems

Normally, "walk to work" systems follow a morphology according to which there is a gangway, a movement compensation system (MCS) and a control system to rule all the operation. Basically, the gangway is the only element mandatory in systems based on this technology. From this point onwards, these components will be described in detail from different points of view taking into account the different solutions currently offered in the market.

### 2.1   Gangway

A gangway is a common element to any staff transfer system based on "walk to work" technology. It allows the personnel walk from the service ship to the fixed marine structure. Even though this kind of system could have a totally passive gangway, it is not enough to obtain a proper behaviour of the system. As a result, the control of some of its degrees of freedom will be necessary.

Related to the gangway actuators, in order to obtain a proper functioning; firstly, each gangway needs actuators that enable to modify its orientation and tilt as well as its total length; first to attach it properly and them to keep the contact with the destination area. Given that, a telescopic section is included, as well as actuators to control the orientation and the tilt allowing overcoming misalignment, height differences and increases or decreases in the distance always between the vessel and the landing area in the marine structure. At the same time, the gangway incorporates a device in its tip to assure contact between itself and the landing area. All this controls will not always be active depending on the solution chose and the phase throughout the manoeuvre in which the system is.

## 2.2 Passive

In terms of "walk to work" technology systems, the simplest solutions, not always the best, consist on gangways that are totally passive once attached to the marine structure. From a theoretical point of view, once the physical connection between the gangway and the unloading area is accomplished, all the actuators are disconnected and the system drifts by the movement of the ship, leaving the joints free. On the other hand, drifting could result in too much movement and therefore increased danger during the transfer (Fig. 1).

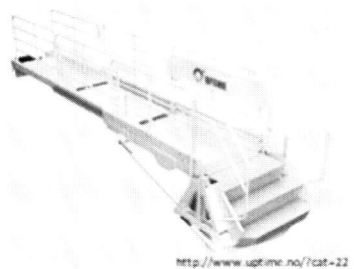

http://www.uptime.no/?cat=22

**Fig. 1.** UPTIME 4–12 m gangway [7]

These are systems in which the gangway is controlled to achieve a physical connection to the unloading area in the fixed structure and once this is done, the joints are let free. These are completely open chain solutions, which enable continuous crossing from the origin to the destination.

Despite the existing options seen above in the majority of solutions, the main component, which is the gangway, is jointed to an MCS, which works in parallel to enhance its workability. From the compensation system point of view some solutions are considered to be active. Despite the existing options seen above in the majority of solutions the main component, which is the gangway, is jointed to a MCS, which works in parallel to enhance its workability. From the compensation system point of view, some of the solutions are considered to be active.

## 2.3  Active

Calling "active" this kind MCS refers to its impossibility of disconnecting any or the whole DOF actuators during the manoeuvre and the fact that doing such a thing would compromise the entire operation.

Ampelmann is the greatest exponent in the production of active movement compensation systems in order to compensate ship movements while transferring operators to a fixed marine structure. This enterprise disposes a gamut of products based on the "walk to work" technology. Typically, these gangways incorporate configurations that allow compensating the vessel movements caused by swell. In the vast majority of solutions commercialized by Ampelmann the MCS consist on a Stewart-Gough platform, Fig. 2. By changing the orientation of the mechanism the end effector and the base change as well. So, the fixed element is the upper platform and the end effector is the lower platform. The gangway lays on top of the upper platform. By this mean, the compensation of the vessel movements is achieved by activating the hexapod's DOF in such a way that it reproduces the vessel movements but oppositely. Consequently, summing up movements taking into consideration the boat movements and compensation system movements it results in a null movement in the upper platform. This way the gangway stays steady (Fig. 3).

**Fig. 2.** Ampelmann gangway in service [4]

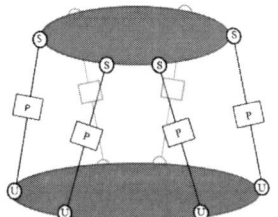

**Fig. 3.** 6-UPS schematic view

In terms of its construction, Ampelmann system kinematic chains present a modification from 6-UPS to 6-UCU [1, 9]. Another point of interest in the construction is the combination of a parallel manipulator and an open chain system, always enabling continuous crossing. Kinematically, in order to accomplish the proper position in the end-effector its total movement is composed by each chain movement. This allows keeping the gangway steady without shortening or lengthening it or changing its orientation. Whereas, this movement composition makes active DOF mandatory if any disconnects, the whole operation is compromised.

However, these active solutions need from a continuous energy consumption and are moving during the entire manoeuvre. In search of a simpler solution and energetically more efficient there are semi-passive solutions.

## 2.4   Semi-passive

In contrast to active systems, in semi-passive ones once the gangway is coupled to a fixed marine structure, DOF disconnection is allowed. Actually, it is a must so as to obtain a proper behaviour by the transfer system simplifying the control system and reducing its energy consumption. In this category many of the transfer systems employed for this kind of manoeuvre are involved. Normally, in this kind of systems the gangway is disposed on a movement compensation pedestal. Typically, this pedestal possesses the vertical translation separated from the rest of DOF, given that this is the most determining DOF. As a matter of fact, in different semi-passive systems (UPTIME 23,4 m, UPTIME 26 m, MACGREGOR OFFSHORE GANGWAY) [7, 12] the MCS strictly compensates this translation, Figs. 4 and 5.

**Fig. 4.**  MACGREGOR offshore gangway [12]

**Fig. 5.**  UPTIME 23,4 m

However, there exist more complicated systems (BM GANGWAY 3.0, BM GANGWAY 4.5 and SAFEWAY SEAGULL) that incorporate in the lower part of the pedestal kinematic chains to compensate the vessel desk rotations (roll and pitch). So as to ensure that effect, normally a couple of cylinders is introduced apart from a couple of rotational joints. Consequently, the pedestal is kept vertical assuring the translation parallel to the Z axis, but experiencing residual translations in the X and Y axis due to the positioning of the rotational joints with regard to the vessel pivot point, Fig. 6.

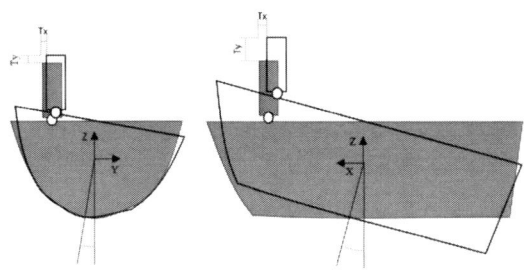

**Fig. 6.**  Residual translations in the pedestal due to desk rotations

As far as its construction is concerned, this kind of systems are a parallel manipulator, the pedestal, and an open chain manipulator, the vertical translation and the gangway.

## 3   Control System and Manoeuvre

So as to ensure movement compensation the unpredictable sea behaviour makes impossible its modelling in order to base the control system on an accurate enough model. As a result, these control systems function by using measurements of the different movements induced in the vessel by swell. To measure the movements generated by swell, vessels equip an IMU (Inertial Measurement Unit) [1] composed by three accelerometers and three gyroscopes that measure movements in the six DOF. The actuators in the MCS use these measurements.

Besides, systems that incorporate "push" technology in the tip of the gangway have the necessary sensors to guarantee the contact, such as a load cell. This technology base contact keeping on obtaining a steady load level in the joint by shortening or lengthening the gangway in order to augment or relieve pressure in the tip.

## 4   Apart from "Walk to Work"

The previous solutions were based on "walk to work" technology, but even though the vast majority of new systems follow this trend, this is not unique. In the market there are other solutions in which open chain robots or cranes on stabilized platforms are used. Normally, these solutions consists in a cage or a booth as end-effector. Then, for its positioning an arm is used and a wrist is used to orientate it. This configuration makes impossible the continuous staff crossing [4].

There exist other solutions related to form closure, therefore the transfer system grasps the fixed structure by using different configurations normally positioned in the tip of the boat.

## 5   New Challenges

In search of a higher energy production, new technologies related to marine renewable energy production are on ongoing development nowadays. These consist on floating structures that allow exploiting wind resource in deeper seas and the energy from waves. This sets a new challenge in personnel transfer, which is being able to guarantee a high workability in the new systems by being capable of transferring personnel between two floating bodies moved by swell.

# 6  Conclusions

All the new transfer systems described above in this review were sorted out into active, semi-passive and passive based on the actuators functioning once the gangway is already attached and in each option there are different upsides and downsides.

The fact that passive systems do not compensate movement makes risks associated to this kind of manoeuvre much higher. Besides, it requires more proximity between the structure and the vessel, which affects to the vessel stability. On the other hand, these are much simpler systems with a smaller size allowing to equip them on smaller boats and reducing the cost of using a vessel.

In Ampelmann products, a redundancy in the DOF actuators is visible taking into account the possibility of activating some of them in the gangway itself. With this, the only thing achieved is an augmentation in the system's workspace, but making the control system more complicated. However, systems that incorporate a hexapod as movement compensation system are capable of maintaining the gangway tip position without being affected by the difference in the pivot point between the compensation system and the vessel. In all the solutions mentioned based on "walk to work" technology, the continuous crossing is possible. However, just in Ampelmann systems there is no accelerations in the gangway, making the manoeuvre safer.

Semi-passive systems are a more efficient solution kinematically and energetically speaking compared to the active ones. There is no visible redundancy in the DOF actuators. These systems achieve total compensation by combining pedestal's and gangway's actuators. This allows disconnecting some DOF when the coupling is accomplished. In this moment, the system works to assure the connection and the safety of the operators during crossing.

As far as non-"walk to work" solutions are concerned the main downside consists in the impossibility of having continuous crossing. Therefore, manoeuvre duration is increased and its associated cost either. At the same time, this solutions suppose a more dangerous option taking into account the fact that if any eventuality occurred, operators would have no way to receive any assistance or quickly return to the boat desk given the fact that there is no continued connection between origin and destination.

# References

1. Cerda, D.-J.: Ampelmann, Development of the Access System for Offshore Wind Turbines. Ph.D. thesis, TU Delft (2010)
2. Floating Offshore Wind: Market and Technology Review. Carbon Trust, June 2015
3. GWEC Global Wind Statistics 2016
4. Katsouris, G., Savenije, L.-B.: Offshore Wind Access 2017. ECN (2016)
5. Merlet, J.-P.: Parallel Robots. Solid Mechanics and Its Applications. Springer, Dordrecht (2006)
6. Siciliano, B., Khatib, O.: Handbook of Robotics. Springer, Heidelberg (2008)
7. http://www.uptimc.no/. Acccssccd 11 Jan 2017
8. https://www.barge-master.com/products/. Accessed 11 Jan 2017
9. http://www.ampelmann.nl/systems. Accessed 11 Jan 2017

10. http://www.vanaalstmarine.com/safeway. Accessed 12 Jan 2017
11. http://m-a.no/wp-content/uploads/2015/03/MA-Gangway2016_LOW.pdf. Accessed 15 Jan 2018
12. https://www.macgregor.com/Products-solutions/Offshore-oil-and-gas-and-renewables/. Accessed 12 Jan 2017
13. http://www.owjonline.com/news/view,norway-recognises-offshore-winds-huge-potential_50452.htm. Accessed 17 Jan 2017

# Radial Cam Grinder with a Camera Inspection System

Pavel Fišer[✉], Petr Jirásko[✉], and Miroslav Václavík[✉]

VÚTS, a.s., Liberec, Czech Republic
{pavel.fiser,petr.jirasko,miroslav.vaclavik}@vuts.cz

**Abstract.** The radial cam grinder of the here presented concept has not been implemented anywhere in the world. This is mainly the application of direct servo drives in interpolation axes, where the direct rotary servomotor is carried by a linear servomotor. They both work in electronic cam mode. The inspection camera system for checking cam contour accuracy has never been used for this purpose too. The issue of production data is the unification of the kinematic calculation of the conventional cam contour with the principle of the program realization of the displacement law in the PLC control system of the electronic cam. The paper deals with the development and the current state of the unique grinding machine.

**Keywords:** Cam · Electronic cam · Control system · Grinding machine

## 1   Introduction – General Characteristics of the Radial Cam Grinding Machine

It is a special single-purpose machine that processes hardened active surfaces or radial cam contours with grinding technology. The required geometric deviation from the theoretical course is within the tolerance of 0.01 up to 0.03 mm. The standard arrangement of the interpolation axes is in the horizontal or vertical plane with the perpendicular positioning sliding axis of the tool. The interpolation axes are one rotational (C) and one sliding (V). The sliding positioning axis (Z) carries the grinding spindle. The rotational axis C is in the form of a rotary table with a clamping plate on which the cam clamping fixtures are placed. The axis C is then firmly connected to the sliding axis V. The axis of the rotation of the spindle is identical to the rotational axis C for V = 0. The cams are clamped on the rotary table by means of the fixtures or are deposited on the magnetic plate. The zero position of the C axis relative to the cam blank is ensured by the index hole of the cam with the guide groove of the rotary table. The systems of aligning (matching) the workpiece coordinate system with the machine coordinate system are different. This is manual or automatic adjustment using external metering devices.

Grinding contours is essentially done in two ways. The first method is grinding with conventional grinding discs, which need to be dressed. The second method is grinding with CBN (Cubic Boron Nitride) tools, which are mostly not dressed. Both methods are applied in a balanced way, depending on the ground material and

B. Corves et al. (Eds.): EuCoMeS 2018, MMS 59, pp. 189–196, 2019.
https://doi.org/10.1007/978-3-319-98020-1_22

production economics. In both cases, the radial cam is ground in cycles with a varying size of removals, where the tool center paths are equidistant to the active surface or cam contour.

The issue of radial cam grinding can be divided into the following areas. It is the machine design concept itself, the production technology, the control system and the accuracy checking of the ground contour or the inspection system.

## 2    Design Concept of the Machine

A common NC axis solution on machine tools is with an inserted constant transmission between the working motion and the servo drive to reduce the force effect of machining that enters the servomotor. For relatively standard feed rates, different types of gears (planetary, worm-gear) and ball screws are available. Mechanical inaccuracies (dimensional, clearance) are addressed by direct measurement of working movements. The grinding spindle bearing concept is horizontal or vertical. The first development variant was the BRV-300 CNC grinder according to Fig. 1, where the rotational interpolation axis C was realized by a direct servomotor with a horizontal spindle bearing.

**Fig. 1.** BRV-300 CNC radial cam grinder

The results of the grinding with an excellent shape accuracy of the cam contour in the tolerance field of one hundredth of a millimeter led to the next developmental step, which is the use of a direct linear servomotor of the sliding interpolation axis V, see Fig. 2. In Fig. 2, it is a CAD model of a new grinder concept where the spindle is in the vertical direction of the fundamental portal tri-axial frame which allows the grinding of the cam positioned on the magnetic clip with the possibility of matching the coordinate systems of the machine and the workpiece without additional program corrections.

The trouble-free use of direct rotary drive is verified from the machine according to Fig. 1, where the standard values of the control parameters in the servo inverter have been set. The measured position error PERR of both axes C and V is according to Fig. 3 below the value of 0.002 [mm]. The PERR of the rotary axis C is a thick red line.

**Fig. 2.** Portal concept of the machine with interpolation axes C and V (CAD model and assembly of the machine)

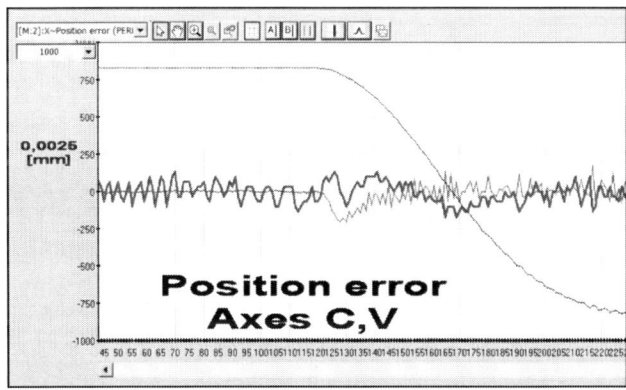

**Fig. 3.** PERR of interpolation axes C and V of the BRV-300 CNC grinder

The satisfactory size of the PERR error of the used linear servomotor was verified on the dynamic stand according to Fig. 4. The servomotor was loaded with simulated load which exceeded the technological power of grinding many times. The monitored position error is in a comparable range as in Fig. 3.

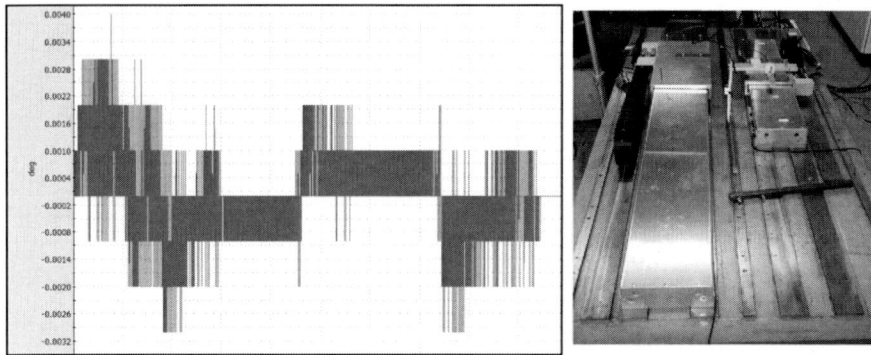

**Fig. 4.** PERR of the linear servo drive of the stand

## 3 Production Technology and Control System

The reasons for the development of the control system are in the ever-evolving technology of the shape-grinding itself of hardened materials, which is related to progressive tools (grinding and milling) and new cam materials. X-series tooling materials with a yield strength of up to 3000 [MPa] are a great hope for high speed cam mechanisms and high cam contour life. However, these materials are difficult to process and the development of grinding technology is necessary with the possibility of internal interventions in the machine control system. In principle, it is concerned the possible reactions of the system to *geometric* and *technological* errors of cam contours. Geometric errors arise from the mutual displacement and rotating of the workpiece coordinate system and that of the machine when the cam is clamped, and the technological errors are a general consequence of grinding the cam contour with a variable radius of curvature.

Another important reason for the development of the system is the dynamics of the inertial forces of the mass elements of the interpolation axes in the machining process. We are currently focusing on such grinding technology where the dynamics of inertia forces does not play a crucial role. With increasing shifts, in the case of small reductions and roughing of cam contours by milling, the dynamics of inertia forces is at the forefront. Position error control then requires the use of new own algorithms. Generally speaking, it can be stated that the development of modern production technology is not possible without the open control system of the machine that implements the technology.

The goal is therefore the concept of a shape 2D grinding machine open control system. The system will allow a flexible response to the development of the production technology, the used drives and the associated quality, productivity and production economics.

## 3.1    Geometric and Technologic Source of Errors

*Geometric* errors are those parts of manufacturing errors that are caused by the eccentricity of the clamp, inappropriate tolerance of the center hole of the cam and the diameter of the clamp, inappropriate clamping construction, and the rotation of the cam with respect to the defined coordinate system of production data. In Fig. 5 (left picture), they are visible from the measurement protocol.

Geometric errors can often be minimized by suitable bearing, but for many design reasons of the cam mechanism itself, they cannot be completely suppressed. It is well known that in the production of double cams, deviations occur in the relative rotation of the cams themselves when re-clamping the double cam for particular contour grinding. Therefore, it is an effort to grind this type of cams with one clamp, which is not always possible.

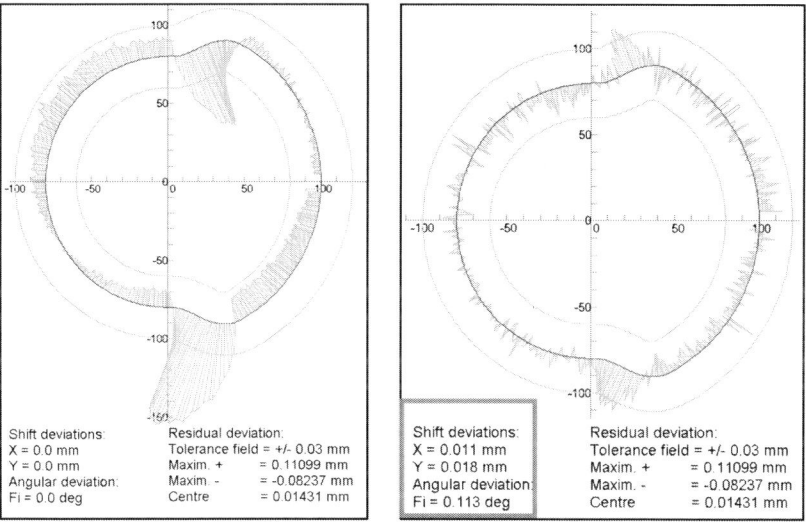

**Fig. 5.** Measurement protocol before/after determining correction deviations

When the profile of the cam is evaluated or the measuring protocol is additionally suitably adjusted by displacement and rotation of the coordinate system, geometric errors are suppressed, and technological errors are evident, as shown also in Fig. 5 (right picture). The absolute shift in the XY plane and the rotation of the cam by a *Fi* angle to the zero point can be deduced from the protocol.

The problem of correctly evaluating the geometric precision of the contour of the cam is that it is necessary to measure the contour on the machine without removing the cam from the fixture or the clamping magnetic plate. In Fig. 6, it is drawn a situation in a rectangular coordinate system. The non-dashed coordinates (x, y) are the coordinate system of the machine, the dashed coordinates (x′, y′) represent the cam coordinate system in the checking coordinate measurement situation with a defined beginning according to the center and index hole of the cam. In Fig. 6, it is a scanning of the contour of a cam clamped on the machine according to Fig. 1.

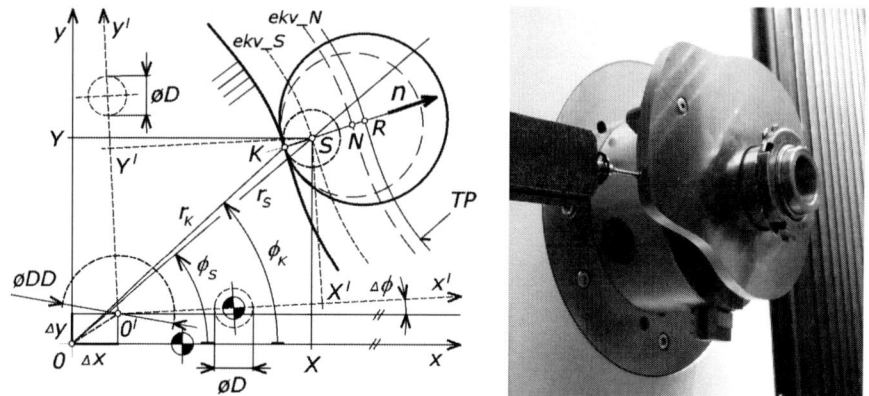

**Fig. 6.** Geometry of the coordinate systems machine/workpiece

The *technological* source of errors is often a principle reality. The contour of the cam has a variable radius of curvature (positive and negative), and the grinding wheel occupies a variable surface or a variable arc portion. In the areas of small positive radii of curvature, the cam is relieved, in places with a small negative curvature (it must be larger than the radius of the grinding wheel), the contour is, on the contrary, not ground (the wheel is pushed away). The source of these errors is also dependent on the possible usability and character of the oscillation motion of the Z positioning axis which carries the machine spindle.

### 3.2    Control System

The control system [1] is a complete development of the authors on the Yaskawa MP2000/3000 HW controller platform with Weintek HMI communications. The control system allows classic control in manual (JOG, STEP, POSI, HW, etc.) and automatic mode (CNC). CNC mode is the automatic workflow control according to production data in two interpolation axes C (rotation) and V (sliding). Implementation of interpolation motion is programmed in the conventional *Motion* area where the motion between two positions is defined by the coordinates with the given feed rate. *However, the uniqueness of the new concept of the radial cam grinder system is based on the fact that the second way of controlling the movements of the production axes is to use the function (PLC user function) of two electronic cams interconnected.*

The production data of the cam in Fig. 7 are counted on any equidistant and in the form of polar coordinates (angle and vector) with normal angle values are stored in the memory of the main controller. These parameters then control the production axes motions of the machine with the value of the tool correction. The system also processes the eccentricity (external measurement) values of the cam position on the C rotary axis and transforms it into the machine coordinate system. The values of the drive control output registers are generated in the PLC user function. The electronic cam's user function allows control with a number of correction parameters, forward speed and torque constraints, and functions that perform different grinding conditions. One of the conditions tested is such shifting when grinding the contour of the cam when the torque load on the rotary axis is theoretically constant. This motion is derived from the data of the normal angle and the radius of curvature of the ground contour. The importance of this condition stems from the used direct drive of the rotational axis (*Direct Drive*), where it is principally present electromagnetic compliance that is influenced by the PI regulation and that is therefore a source of a positional error depending on torque load.

The conditions for determining the motion functions of the C rotary axis are arbitrary. It can be ground with a constant feed along the contour of the cam or on the basis of other conditions given by the particular contour of the cam. The cam contour measurement is also used without the need of removing the cam workpiece and using the measured data to a direct correction of the production data generated by the user function.

*The production axes thus controlled can be said that two electronic cams are grinding the contour of a conventional radial cam. Axes control (realization of their displacement laws) is essentially identical to the computational synthesis of the basic cam mechanism of a radial cam with a central follower and a roller. The axes of the grinder are really this basic cam mechanism (C-axis rotation and V-axis displacement).*

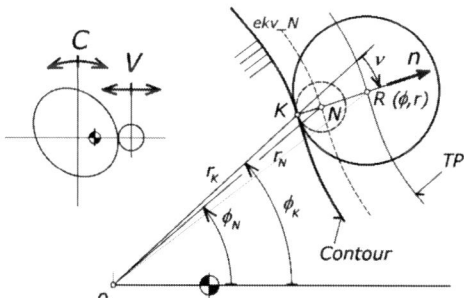

**Fig. 7.** Interpolation axes C and V, computational and production coordinates

## 4  Camera Inspection System

The camera inspection system for evaluating the geometric accuracy of the radial cam contour has not been used anywhere so far. In Fig. 8, it is the cam contour edge scanning with an error representation on the HMI screen. The projected cam is a standard with a chamfered edge that will be sharp and completely distinct during grinding. One pixel is in the order of thousandths of a millimeter.

**Fig. 8.** Scanned cam contour with error evaluation

## 5    Conclusions

The basic characteristic of the new concept of the radial cam grinder according to Fig. 2 is a three-axis vertical spindle guide with two interpolation axes C and V realized by direct drives. The control system is based on years of research into electronic cam applications [2], which are integrated in the system and enable efficient NC production axes control. The control system allows carrying out dimensional analysis of the contour of the cam without the need of removing the workpiece from the machine. The use of direct drives of interpolating NC axes in electronic cams mode and the camera inspection system in competing cam manufacturing machines is not known to us.

## References

1. Jirásko, P., Crhák, V., Bureš, P.: The conception of the control system of radial cam grinder. In: IFToMM, Liberec (2012)
2. Václavík, M., Jirásko, P.: The problems of electronic cam applications. In: 13th World Congress in Mechanism and Machine Science, Guanajuato, México, 19–25 June 2011

# Linkages and Cams

# Development of a Novel Linkage
# for Low-Profile Sickle Drive

Alexey Fomin[1,3(✉)], Sergey Kiselev[1], Andreas Jahr[2], and Hunn Sim[2]

[1] Siberian State Industrial University, Novokuznetsk, Russia
{alexey-nvkz,ksv01}@mail.ru
[2] Hochschule Düsseldorf, Düsseldorf, Germany
{andreas.jahr,hunn.sim}@hs-duesseldorf.de
[3] Mechanical Engineering Research Institute of the RAS, Moscow, Russia

**Abstract.** Currently agricultural complex is presented by diverse techniques with different manufacturing capabilities to perform various tasks. Efficiency of agricultural mechanisms and machines seriously depends on a design of their kinematic structures. In the presented study we carry out novel kinematic design of linkage-based sickle drive for actuation of combine harvester's blades. The developed sickle drive is designed as a symmetrical planar six-bar linkage with only rotational joints. The linkage is actuated by a single pneumatic drive and allows moving blades in opposite directions with equal translations. The optimal structure of the linkage provides lightweight, reliability, speedwork, stiffness and precision for the whole drive. The study presents structural and kinematic analysis as well as numerical calculations of the proposed sickle drive. The functional capabilities of novel drive allow reducing maintenance costs and saving power.

**Keywords:** Kinematic pair · Mechanism · Degree of freedom
Sickle drive · Combine harvester

## 1 Introduction

Agricultural industry forms the basis of the economy for many countries, it guarantees food supply and economic security. Even the most developed countries pay high attention to the development of agricultural industry and make significant investments in it. However, there is number of factors that thwart a progress of the industry. Among them there are weak technical equipment and significantly non-current facilities and resources. In this regard the actual problem lies in a development of agricultural machinery that is key part of whole agro-industrial complex. The creation of highly efficient agricultural equipment is possible only on a serious scientific basis.

In this study we aim the development of a novel configuration of a sickle drive for combine harvesters that will allow improving technical characteristics of existing models. Sickle drives of combine harvesters are variously designed with the use of diverse linkages and transmissions. Examples of drives with a combination of gear wheels and planar linkages are shown in [1–4], with the use of cam mechanisms are presented in [5], with flexible couplings in [6, 7] and with eccentric mechanisms in [8, 9]. Analysis of these kinematic schemes of sickle drives shown that main

© Springer Nature Switzerland AG 2019
B. Corves et al. (Eds.): EuCoMeS 2018, MMS 59, pp. 199–206, 2019.
https://doi.org/10.1007/978-3-319-98020-1_23

characteristics that lead to increase a combine harvester's efficiency are based on simultaneous actuation of blades, placement of all links in horizontal plane for reduction of dynamic loads in vertical directions, assembly with minimum of elements and joints to reach minimum weight, exclusion of flexible couplings as well as joints with degree-of-freedom (DoF) more than one for reliability growth. In this way we would like to present a novel configuration of low-profile sickle drive mechanism which is designed as a symmetrical six-bar linkage with all rotational joints and single drive.

## 2   Mobility Analysis

Figure 1 provides the low-profile linkage-based sickle drive [3] developed by CNH Industrial America LLC (New Holland, PA, USA). It includes pneumatic drive $P$, driving bevel wheel 1, driven bevel wheels 2 and $2'$, which are fixedly connected with cranks 3 and $3'$, couplers 4 and 6, rockers 5 and 7, body frame 8 that serves as a fixed link. Blades are fixedly connected with rockers 5 and 7 which serve as end-effectors. Links 4 and 6 as well as 5 and 7 are pairwise identical. This mechanism allows having inclinations of rockers 5 and 7 on identical angles and move blades on equal distances.

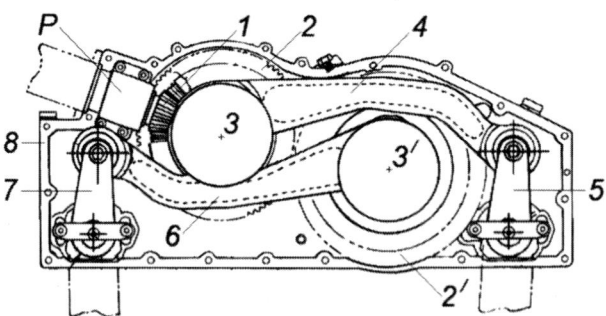

**Fig. 1.** Low-profile linkage-based sickle drive [3]

The linkage includes seven movable links ($n = 7$), nine one-DoF rotational joints ($p_5 = 9$) and two two-DoF gear joints ($p_4 = 2$). According to Chebishov's mobility formula written as follows

$$W_3 = 3n - 2p_5 - p_4, \tag{1}$$

where $W_3$ is a mobility, defining number of DoF of a planar kinematic chain with three imposed constraints, $n$ is a number of movable links of a kinematic chain, $p_5$ and $p_4$ are numbers of one- and two-DoF kinematic pairs [10, 11], the mobility of the drive mechanism is equal to one ($W_3 = 1$). It proves the mechanism's workability with single drive. The input rotational motion is applied to wheel 2. Mechanism is structurally separated out into driving link (bevel wheel 1), two dyads *RRR* (groups of links 4-5 and 6-7) and two monads (links 3 and $3'$), where each monad includes one rotational and one gear kinematic pairs. Structural formula for this mechanism can be written as I (1) → II(3) → III(4,5) → IV($3'$) → V(6,7), where I indicates driving link, II–V indicate groups having zero DoFs included in the mechanism.

The construction of this drive mechanism has some drawbacks, which can be improved. The availability of wheels 2 and $2'$ results to increase the overall weight and requires permanent lubrication to provide high frequency of end-effectors 5 and 7. The leftward drive causes strict requirements applicable to balancing of the whole mechanism. The centrally arranged drive would provide equal distribution of power for both sides of the mechanism and allow excluding additional transmission gears.

Figure 2 provides optimized kinematic scheme for the drive mechanism. It includes five movable links ($n = 5$) and seven rotational joints ($p_5 = 7$). According to (1) the mobility is equal to one ($W_3 = 1$). Predefined motions of rockers 5 and 7 are controlled by single actuator which rotates crank 3. Mechanism is structurally separated out into two dyads $RRR$ (groups of links 4-5 and 6-7) that connect to crank 3 in points $B$ and $E$. Structural formula for this linkage can be written as I(1) → II(2,3) → III(4,5).

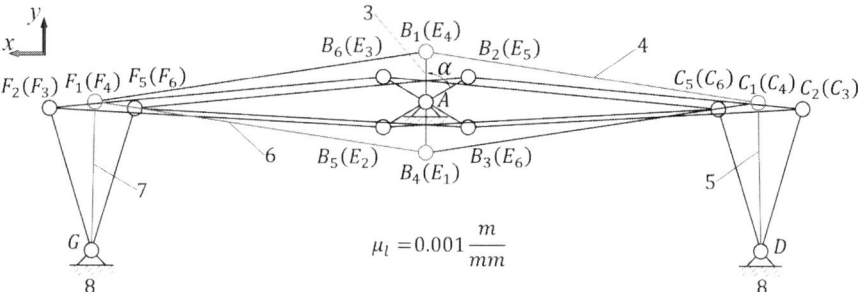

**Fig. 2.** Kinematic scheme of the proposed sickle drive linkage and its position plan

The optimized kinematic design of the proposed linkage provides an opportunity to set drive align the center of the body frame, to exclude additional gears for lightweightness and simplicity of its structure, at the same time the linkage provides equal inclinations of rockers 5 and 7 and is actuated from single drive.

## 3  Kinematic Analysis

Turn to the task on position analysis of the developed sickle drive linkage. Figure 2 shows the linkage in six positions. The first position is indicated with red color, when angle $\alpha = 0°$ (angle $\alpha$ defines a position of crank 3). Points $B_1$ and $E_4$, $B_2$ and $E_5$, $B_3$ and $E_6$, $B_4$ and $E_1$, $B_5$ and $E_2$, $B_6$ and $E_3$, $F_5$ and $F_6$, $F_1$ and $F_4$, $F_2$ and $F_3$, $C_5$ and $C_6$, $C_1$ and $C_4$, $C_2$ and $C_3$ coincide each other. Following link lengths are accepted: length of crank 3 is $l_3 = 0.068$ m, lengths of couplers 4 and 6 are $l_4 = l_6 = 0.230$ m, lengths of rockers 5 and 7 are $l_5 = l_7 = 0.100$ m. The kinematic scheme of the linkage is presented in Fig. 2 on a scale $\mu_l = 0.001$ m/mm.

Let's address to the velocity and acceleration calculation of the developed sickle drive. Accept rotation frequency of crank 3 as $n_3 = 450$ rpm. Then angular velocity of crank 3 is equal to $w_3 = 47.2$ s$^{-1}$. Velocities $V_B$ and $V_E$ of points $B$ and $E$ are equal to 1.6 m/s. Velocity vector diagrams for each of six positions of the linkage are shown in Fig. 3a–f. Vector $pb$ is equal to 40 mm there, so the scale of the diagrams is equal to $\mu_V = V_B/pb = 0.04$ (m/s)/mm. Velocities of points $C$ and $F$ can be calculated from the following equations

$$\begin{cases} \overline{V}_C = \overline{V}_B + \overline{V}_{CB}, \text{ where } \overline{V}_{CB} \perp CB, \\ \overline{V}_C = \overline{V}_D + \overline{V}_{CD}, \text{ where } \overline{V}_{CD} \perp CD, \end{cases} \begin{cases} \overline{V}_F = \overline{V}_E + \overline{V}_{FE}, \text{ where } \overline{V}_{FE} \perp EF, \\ \overline{V}_F = \overline{V}_G + \overline{V}_{FG}, \text{ where } \overline{V}_{FG} \perp FG. \end{cases} \quad (2)$$

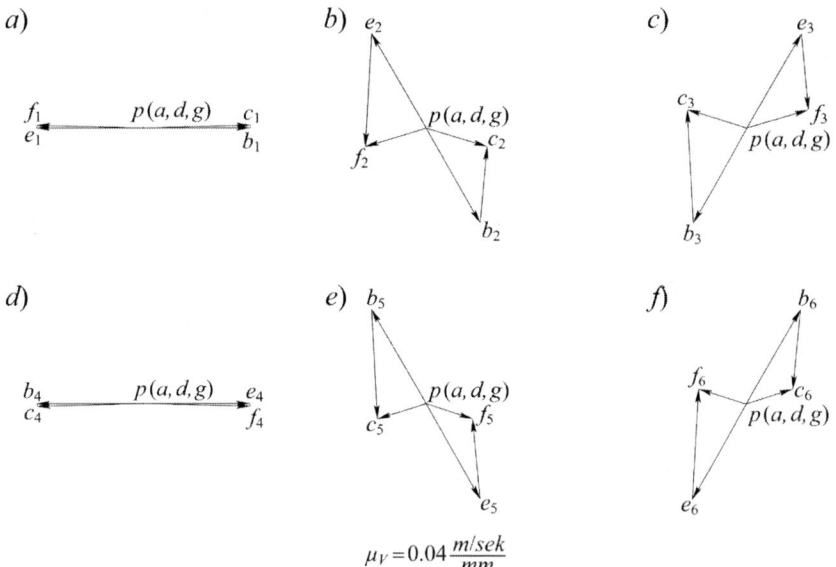

**Fig. 3.** Velocity vector diagrams for six positions of the sickle drive linkage when: (a) $\alpha = 0°$; (b) $\alpha = 60°$; (c) $\alpha = 120°$; (d) $\alpha = 180°$; (e) $\alpha = 240°$; (f) $\alpha = 300°$

Numerical values of velocities $V_{CB}$, $V_C$, $V_{FE}$ and $V_F$ indicated in (2) are calculated as multiplications ($bc \cdot \mu_V$), ($pc \cdot \mu_V$), ($ef \cdot \mu_V$) and ($pf \cdot \mu_V$), where $bc$, $pc$, $ef$ and $pf$ - are vector lengths from the velocity diagram shown in Fig. 3a–f. Angular velocities of links 4–7 are calculated as $\omega_4 = V_{CB}/l_4$, $\omega_5 = V_{CD}/l_5$, $\omega_6 = V_{FE}/l_6$, $\omega_7 = V_{FG}/l_7$. Show in Table 1 linear and angular velocities of the linkage's links in six positions depending on angle $\alpha$.

Accelerations $a_B$ and $a_E$ of points $B$ and $E$ are equal to 75.42 m/s$^2$. Acceleration vector diagrams for each of six positions of the linkage are shown in Fig. 4a–f. Draw vector $\pi b$ at the acceleration diagram that is equal to 150.84 mm, so the diagram scale is equal to $\mu_a = a_B/\pi b = 0.5$ (m/s$^2$)/mm. Accelerations of points $C$ and $F$ can be calculated as

**Table 1.** Variability of linear and angular velocities depending on angle $\alpha$

| $\alpha$, deg | 0 | 60 | 120 | 180 | 240 | 300 |
|---|---|---|---|---|---|---|
| $V_{CB}$, m/s | 0.040 | 1.123 | 1.659 | 0.040 | 1.610 | 1.171 |
| $V_{FE}$, m/s | 0.040 | 1.659 | 1.123 | 0.040 | 1.171 | 1.610 |
| $V_C(V_{CD})$, m/s | 1.606 | 0.942 | 0.932 | 1.594 | 0.746 | 0.724 |
| $V_F(V_{FG})$, m/s | 1.594 | 0.932 | 0.942 | 1.606 | 0.724 | 0.746 |
| $\omega_4$, $s^{-1}$ | 0.174 | 4.883 | 7.213 | 0.174 | 7.000 | 5.091 |
| $\omega_5$, $s^{-1}$ | 16.060 | 9.420 | 9.320 | 15.94 | 7.460 | 7.240 |
| $\omega_6$, $s^{-1}$ | 0.174 | 7.213 | 4.882 | 0.174 | 5.091 | 7.000 |
| $\omega_7$, $s^{-1}$ | 15.940 | 9.320 | 9.420 | 16.060 | 7.240 | 7.460 |

$$\begin{cases} \bar{a}_C = \bar{a}_B + \bar{a}_{CB}^n + \bar{a}_{CB}^\tau, \text{ where } \bar{a}_{CB}^n \parallel CB, \ \bar{a}_{CB}^\tau \perp CB, \\ \bar{a}_C = \bar{a}_D + \bar{a}_{CD}^n + \bar{a}_{CD}^\tau, \text{ where } \bar{a}_{CD}^n \parallel CD, \ \bar{a}_{CD}^\tau \perp CD, \end{cases}$$
$$\begin{cases} \bar{a}_F = \bar{a}_E + \bar{a}_{FE}^n + \bar{a}_{FE}^\tau, \text{ where } \bar{a}_{FE}^n \parallel EF, \ \bar{a}_{FE}^\tau \perp EF, \\ \bar{a}_F = \bar{a}_G + \bar{a}_{FG}^n + \bar{a}_{FG}^\tau, \text{ where } \bar{a}_{FG}^n \parallel FG, \ \bar{a}_{FG}^\tau \perp FG. \end{cases} \quad (3)$$

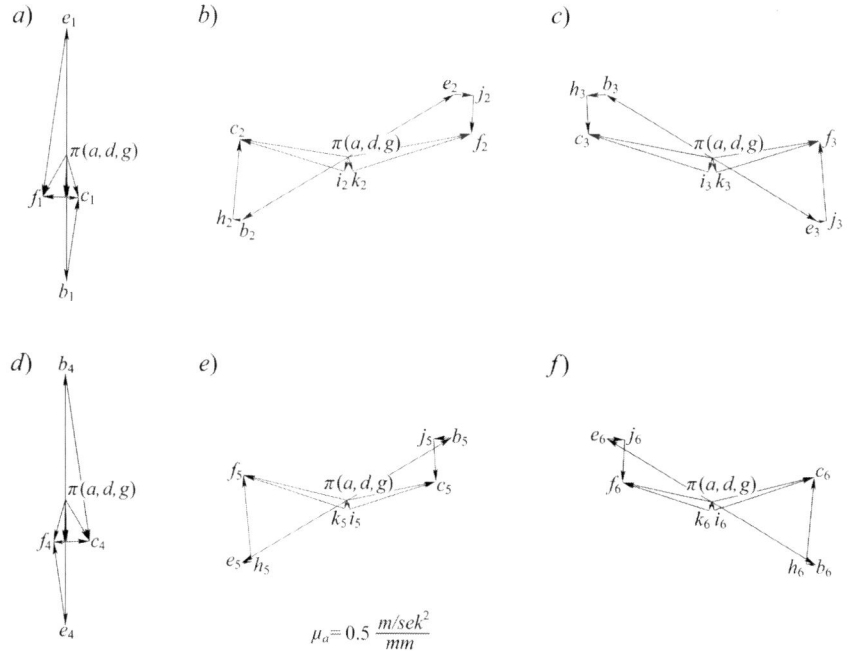

$\mu_a = 0.5 \ \frac{m/sek^2}{mm}$

**Fig. 4.** Acceleration vector diagrams for six positions of the sickle drive linkage when: (a) $\alpha = 0°$; (b) $\alpha = 60°$; (c) $\alpha = 120°$; (d) $\alpha = 180°$; (e) $\alpha = 240°$; (f) $\alpha = 300°$

Numerical values of normal components ($a_{CB}^n$, $a_{CD}^n$, $a_{FE}^n$ and $a_{FG}^n$) of relative accelerations for $BC$, $CD$, $EF$ и $GF$ indicated in (3) can be calculated as $a_{CB}^n = w_4^2 \cdot l_4$, $a_{CD}^n = w_5^2 \cdot l_5$, $a_{FE}^n = w_6^2 \cdot l_6$, $a_{FG}^n = w_7^2 \cdot l_7$. Accelerations of points $C$ and $F$ can be found from the acceleration vector diagram as ($\pi c \cdot \mu_a$) and ($\pi f \cdot \mu_a$). Tangential components ($a_{CB}^\tau$, $a_{CD}^\tau$, $a_{FE}^\tau$ and $a_{FG}^\tau$) of relative accelerations for $BC$, $CD$, $EF$ и $GF$ can be also calculated through acceleration diagram as ($hc \cdot \mu_a$), ($ic \cdot \mu_a$), ($jf \cdot \mu_a$), ($kf \cdot {}_a$). Angular accelerations of links 4–7 are calculated as $\varepsilon_4 = a_{CB}^\tau/l_4$, $\varepsilon_5 = a_{CD}^\tau/l_5$, $\varepsilon_6 = a_{FE}^\tau/l_6$, $\varepsilon_6 = a_{FG}^\tau/l_7$. Show in Table 2 linear and angular accelerations of the linkage's links in six positions depending on angle $\alpha$.

**Table 2.** Variability of linear and angular accelerations depending on angle $\alpha$

| $\alpha$, deg | 0 | 60 | 120 | 180 | 240 | 300 |
|---|---|---|---|---|---|---|
| $a_{CB}^n$, m/s$^2$ | 0.007 | 5.483 | 11.966 | 0.007 | 11.270 | 5.962 |
| $a_{CB}^\tau$, m/s$^2$ | 50.34 | 47.820 | 23.095 | 101.550 | 25.695 | 52.405 |
| $a_{FE}^n$, m/s$^2$ | 0.007 | 11.966 | 5.483 | 0.007 | 5.962 | 11.270 |
| $a_{EF}^\tau$, m/s$^2$ | 101.550 | 23.095 | 47.820 | 50.350 | 52.405 | 25.695 |
| $a_{CD}^n$, m/s$^2$ | 25.792 | 8.874 | 8.686 | 25.408 | 5.565 | 5.242 |
| $a_{CD}^\tau$, m/s$^2$ | 6.780 | 66.595 | 76.760 | 14.330 | 56.330 | 65.835 |
| $a_{FG}^n$, m/s$^2$ | 25.408 | 8.686 | 8.874 | 25.792 | 5.242 | 5.565 |
| $a_{FG}^\tau$, m/s$^2$ | 14.345 | 76.760 | 66.595 | 6.775 | 65.835 | 56.330 |
| $a_C(a_{CD})$, m/s$^2$ | 26.670 | 67.185 | 77.250 | 29.170 | 56.600 | 66.045 |
| $a_F(a_{FE})$, m/s$^2$ | 29.175 | 77.250 | 67.185 | 26.665 | 66.045 | 56.600 |
| $\varepsilon_4$, s$^{-2}$ | 218.869 | 207.913 | 100.413 | 441.523 | 111.717 | 227.848 |
| $\varepsilon_5$, s$^{-2}$ | 67.800 | 665.950 | 767.600 | 143.300 | 563.300 | 658.350 |
| $\varepsilon_6$, s$^{-2}$ | 441.522 | 100.413 | 207.913 | 218.913 | 227.848 | 111.717 |
| $\varepsilon_7$, s$^{-2}$ | 143.450 | 767.600 | 665.950 | 67.750 | 658.350 | 563.300 |

Thus kinematic analysis of novel low-profile linkage-based drive mechanism is completely solved. The results obtained in this section allow detecting velocities and accelerations of blades which are rigidly connected with end-effectors 5 and 7. The obtained data will be used for dynamic analysis when calculating forces and moments of inertia.

## 4  Numerical Simulations

In this section we present results on numerical analysis of the proposed linkage. The CAD model of the drive has been developed and coordinates of points $C$ and $F$ which belong to rockers 5 and 7, are identified and shown in Fig. 5a and b as coordinate-time diagrams. Velocities and accelerations of points $C$ and $F$ have been numerically calculated and presented in Fig. 6a and b as velocity-time and acceleration-time diagrams.

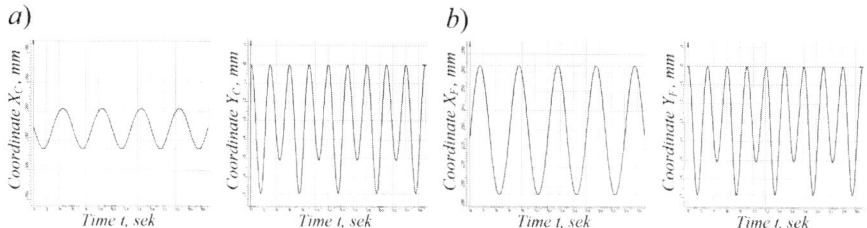

**Fig. 5.** Coordinate-time diagrams $X(t)$ and $Y(t)$: ($a$) for point $C$; ($b$) for point $F$

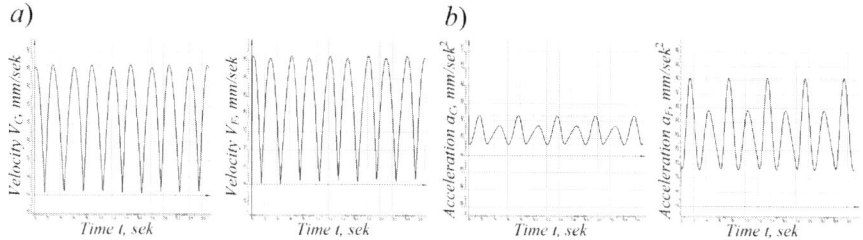

**Fig. 6.** ($a$) Velocity-time diagrams $V(t)$ for point $C$ (left) and point $F$ (right); ($b$) Acceleration-time diagrams $a(t)$ for point $C$ (left) and point $F$ (right)

## 5   Conclusions and Future Works

The study presents novel linkage for low-profile sickle drive used in combine harvester. Its unicity and novelty are based on advanced technical properties such as simultaneous actuation of both end-effectors from single drive, planar orientation of all movable links, compactness, an opportunity to get relatively big velocities and accelerations of end-effectors and minimum number of elements. The developed linkage has been structurally and kinematically investigated: mobility analysis has been carried out and velocities and accelerations of all links have been calculated for six positions of the linkage. Numerical simulations have been carried out. The obtained data will be used for further kinetostatic and dynamic analysis, where reactions in all kinematic pairs and counterbalance moment on the driving link will be found and then correct law motion will be identified. Future works will also include a more advanced kinematic and dynamic analysis which is necessary to obtain a posterior optimized mechanical elements dimension. Also singular points in the motion of the linkage will be identified to avoid possible critical situations.

## References

1. Cook, J.T., Rich, G.L., DeChristopher, D.M., Ungs, P.J.: Slot driven low profile sickle drive. US Patent 9,545,051, 17 January 2017
2. Klenin, N.I.: Agricultural and Meliorative Machines. Koloss, Moscow (2005)

3. Cook, J.T., Gahres, R.T., Bich, G.L., DeChristopher, D.M.: Low profile sickle drive. US Patent 9,532,502, 3 January 2017
4. Polk, G.C., Lou, B.: Reversing transfer drive for sickle cutting knives on a header of an agricultural combine. US Patent 6,889,492, 10 May 2005
5. Cook, J.T., Bich, G.L., DeChristopher, D.M.: Cam driven low profile sickle drive. US Patent 9,545,052, 17 January 2017
6. Naurizbaev, R.K., Kazikhanov, E.H.: The theory of self-aligning flexible cardans. Alma-Ata, Filim (1998)
7. Brimeyer, A., Weichholdt, D., Losa, L.L., Meyer-Hamme, F.: Drive assembly for an agricultural harvesting platform. US Patent, 8,973,345, 10 March 2015
8. Priepke, E.H.: Compact knife head assembly bearing and eccentric for a sickle drive for a header of an agricultural plant cutting machine. US Patent 7,810,304, 12 October 2010
9. Priepke, E.H.: Compact sickle drive for a header of an agricultural plant cutting machine. US Patent US 7,401,458, 22 July 2008
10. Fomin, A., Dvornikov, L., Paik, J.: Calculation of the general number of imposed constraints of kinematic chains. Procedia Eng. **206**, 1309–1315 (2017)
11. Dvornikov, L., Fomin, A.: Development of basic conditions for division of mechanisms into subfamilies. Procedia Eng. **150**, 882–888 (2016)

# Advanced Technique of Type Synthesis and Construction of Veritable Complete Atlases of Multiloop F-DOF Generalized Kinematic Chains

Vladimir Pozhbelko[✉]

South Ural State University (National Research University), Chelyabinsk, Russia
pozhbelkovi@susu.ru

**Abstract.** This paper proposes a new straightforward and advanced approach to structural synthesis, classification and construction of complete atlas databases for creative design of multiloop F-DOF planar closed mechanisms. First, a new simple and reliable synthesis method based on various compositions of the contracted graphs circles, the number of independent loops and the quantity of binary links as primary main structural parameters for generation and systematization of all feasible K-loop F-DOF topological graphs and generalized closed multiloop kinematic chains is presented. Then the complete atlas of all feasible 20 types of 1-DOF, 3-loop, 8-link, $V = 0$ planar closed kinematic chains without rigid subchains (including the 4 new kinematic chains) and the complete atlas of all feasible 59 types of 2-DOF, 3-loop, 9-link, $V = 0$ planar closed kinematic chains without rigid subchains (including the 19 new non-fractionated and link-fractionated kinematic chains) are synthesized and revealed for the first time. The results of this paper can be used for the construction of complete atlases of all feasible generalized kinematic chains with specified input parameters and the invention of F-DOF novel mechanisms in creative design of mechanisms.

**Keywords:** Kinematic chains · Mechanisms · Atlas database

## 1 Introduction

Structural synthesis of all possible non-isomorphic types of kinematic chains (KCs) in batch with given input parameters is one of the important challenging mathematical and topological problems in the field of conceptual design stage of mechanisms [1, 3, 7, 8]. So, structural synthesis is the first stage in the mechanical conceptual creative design [1–9], which can not only optimize existing mechanisms, but also invents novel mechanisms.

It is very important to synthesize all feasible topological structures of KCs for selection of multi-loop mechanisms with the best performance for a specified mechanical task. However, conceptual design of mechanisms based on traditionally designer's experience and intuition does not allow to obtain all feasible mechanisms with specified input parameters. Therefore, recently various researches use computer aided structural synthesis methods [1, 2, 4, 7, 8] for automatic generation of non-isomorphic simple-jointed KCs.

© Springer Nature Switzerland AG 2019
B. Corves et al. (Eds.): EuCoMeS 2018, MMS 59, pp. 207–214, 2019.
https://doi.org/10.1007/978-3-319-98020-1_24

Hence, until now, the structural synthesis results present maximum 16 types of 1-DOF, 8-link and 40 types of 2-DOF, 9-link planar simple joint KCs [1, 2, 4–9]. However, in [4, 8] is shown that the quantity of synthesized complex multiloop chains with large number of links ($\tilde{n} > 6$) and independent loops ($K > 2$) can be smaller than must be in reality. There is one possible reason: applied methods of isomorphic identification can be unreliable, so some non-isomorphic kinematic chains are regarded as isomorphic kinematic chains by mistake [4]. Thus, the development of more effective methods for synthesis of all the possible types of multiloop F-DOF kinematic chains is a difficult and important problem in mechanisms and machine theory and in creative design up to present time.

The main purpose of the paper is to present a new straightforward approach to structural synthesis, classification and construction of veritable complete atlases of all the possible multiloop F-DOF topological graphs and corresponding generalized kinematic chains. The new approach is based on various compositions of (1) possible multiloop contracted graphs kinds; (2) the number of independent loops and (3) the quantity of binary links as a primary structural parameter for creative design of multiloop mechanisms.

*Nomenclature*

| | |
|---|---|
| $\tilde{n}$ | total number of links in the kinematic chain (KC) ($\tilde{n} = \Sigma n_i$); |
| $n_i$ | the number of binary ($n_2$), ternary ($n_3$), quaternary ($n_4$), pentagonal ($n_5$) links, etc., in that order ($i = 1,2,3,4,5,\ldots, i_{max}$), having $i$ joints in the kinematic chain; |
| $i_{max}$ | the maximum number of joints incident to a link; |
| | $[LA] = [n_2, n_3, n_4, n_5,\ldots,n_i]$ – link assortment in the kinematic chain; |
| $j$ | multiplicity of joints ($j = n' - 1$), $n'$ is the number of adjacent links composed of simple ($n' = 2, j = 1$) and multiple ($n' \geq 3, j \geq 2$) joints; |
| $v_j$ | the number of simple ($j = 1$) and multiple joints with multiplicity $j \geq 2$ ($j = 2,3,4,5,\ldots, j_{max}$); |
| $j_{max}$ | the highest multiplicity of joints in the kinematic chain; |
| $V$ | total multiple joint factor of the kinematic chain, $V = 0$ – simple joint kinematic chain (SJKC), $V \geq 1$ – multiple joint kinematic chain (MJKC); |
| $K$ | the number of independent closed loops in the kinematic chain; |
| $L_i$ | denotes closed loop with $k$ –sides; |
| $F$ | degree of freedom (DOF) of KC with the fixed link; |
| $H$ | $= 1$ – one d.o.f. of lower kinematic pair; |
| $p_2$ | the number of 2-d.o.f. higher kinematic pairs in the kinematic chain. |

## 2   Basic Topological Dependences and Universal Table with All Feasible *K*-Loop Link Assortment Arrays

According to unified structure theory of any mechanical systems [5] for planar closed kinematic chains (KCs) intended to function in $h = 3$ mobile parameter of workspace, degree of freedom $F$ satisfies the following new link based mobility equation:

$$F = \left[ \sum_{i=2}^{i_{max}} (3-i)n_i - V - 3 \right] + p_2 = [n_2 - (n_4 + 2n_5 + 3n_6 + \ldots) - V - 3] + p_2,$$
(1)

where the number of binary links $(n_2)$, the value of total multiple joint factor $(V)$ and number of independent loops $(K)$ are constrained by the following topological dependencies [5]:

$$[3 + F + V - p_2] \leq n_2 \leq [\tilde{n} = F + 2K + 1 - p_2],$$
(2)

$$n_2 = (F + V + 3) + \sum (i - 3)n_{i \geq 4},$$
(3)

$$n_{i \geq 3} \leq \frac{2(K-1) - V}{i - 2}, \quad V = \sum_{j=1}^{j_{max}} (j-1)v_j, \quad 0 \leq V \leq 2(K-1)$$
(4)

$$K = 1 + \frac{1}{2}\left[ V + \sum_{i=3}^{i_{max}} (i-2)n_i \right] \Rightarrow K = 1 + \frac{1}{2}V \, if \, \tilde{n} = n_2 \Rightarrow K = 1 \, if \, V = 0, \tilde{n} = n_2$$
(5)

$$i_{max} = \begin{cases} K + F \, if \, F \leq K \\ 2K \, if \, F \geq K \end{cases}, \quad j_{max} = \begin{cases} K + F \, if \, F \leq K - 1 \\ 2K - 1 \, if \, F \geq K - 1 \end{cases}.$$
(6)

Note, that our new Eqs. (1) and (5) for calculation of $F$-DOF and number of independent loops $(K)$ can be useful for structural synthesis of all feasible kinematic chains and are alternative to well-known Gruebler-Kutzbach's equation $F$-DOF and Euler's formula for graphs [3].

According to general structural Eq. (1), we can:

(1) propose to choose the number of binary links $(n_2)$ as the main structural parameter for classification, analysis and synthesis of kinematic chains;
(2) derive the following "*Universal Table 1*" (for all feasible 1-DOF K-loop V-factor planar closed chains) by solving the structural Eqs. (1)–(6) in integers.

So, the "*Universal Table 1*" (Table 1) is calculated based on Eqs. (1)–(6). The table contains exact boundary quantities of binary links $(n_2)$ (marked in bold-italic) and all possible link assortments [LA] for 1–DOF up to 5-independent loop, 12-link planar KCs with various $n_2 \geq 4+V$ in full diapason $1 \leq V \leq 2(K-1)$. In total, the "*Universal Table 1*" contains all feasible $N_1 = 1$ [LA] for case $K = 1$, $N_2 = 3$ [LA] for case $K = 2$, $N_3 = 9$ [LA] for case $K = 3$, $N_4 = 23$ [LA] for case $K = 4$ and $N_5 = 53$ [LA] for case $K = 5$ independent loop KCs with simple $(V = 0)$ and multiple $(1 \leq V \leq 8)$ joints, i.e. with up to 8 multiple joints. In Table 1 the value $N_k$ $(k = K)$ denotes number of possible link assortments for KCs with given $K − 1, 2, 3, 4, 5$.

**Table 1.** All the possible link assortments for 1-DOF simple and multiple joint kinematic chains with up $K = 5$ and $V_{max} = 8$

$F = 1, h = 3, H = 1$

| | K = 1 | K = 2 | | | K = 3 | | | | | | | | |
|---|---|---|---|---|---|---|---|---|---|---|---|---|---|
| | ($\tilde{n}$=4) | ($\tilde{n}$=6) | | | ($\tilde{n}$=8, $N_3$=9) | | | | | | | | |
| $V$ | 0 | 0 | 1 | 2 | 0 | 0 | 0 | 1 | 1 | 2 | 2 | 3 | 4 |
| $n_2$ | 4 | 4 | 5 | 6 | 4 | 5 | 6 | 5 | 6 | 6 | 7 | 7 | 8 |
| $n_3$ | – | 2 | 1 | 0 | 4 | 2 | 0 | 3 | 1 | 2 | 0 | 1 | 0 |
| $n_4$ | – | – | – | – | 0 | 1 | 2 | 0 | 1 | 0 | 1 | 0 | 0 |

$h = 3$
$F = 1$
$i \le K + F$

$K = 4$ ($\tilde{n} = 10$, $N_4=23$)

| | | | | | | | | | | | | | | | | | | | | | | | |
|---|---|---|---|---|---|---|---|---|---|---|---|---|---|---|---|---|---|---|---|---|---|---|---|
| $V$ | 0 | 0 | 0 | 0 | 0 | 0 | 0 | 1 | 1 | 1 | 1 | 1 | 2 | 2 | 2 | 2 | 3 | 3 | 3 | 4 | 4 | 5 | 6 |
| $n_2$ | 4 | 5 | 6 | 6 | 7 | 7 | 8 | 5 | 6 | 7 | 7 | 8 | 6 | 7 | 8 | 8 | 7 | 8 | 9 | 8 | 9 | 9 | 10 |
| $n_3$ | 6 | 4 | 2 | 3 | 0 | 1 | 0 | 5 | 3 | 1 | 2 | 0 | 4 | 2 | 0 | 1 | 3 | 1 | 0 | 2 | 0 | 1 | 0 |
| $n_4$ | 0 | 1 | 2 | 0 | 3 | 1 | 0 | 0 | 1 | 2 | 0 | 1 | 0 | 1 | 2 | 0 | 0 | 1 | 0 | 0 | 1 | 0 | 0 |
| $n_5$ | 0 | 0 | 0 | 1 | 0 | 1 | 2 | 0 | 0 | 0 | 1 | 1 | 0 | 0 | 0 | 1 | 0 | 0 | 1 | 0 | 0 | 0 | 0 |

$F= 1, h = 3, H = 1$

$K = 5$ ($\tilde{n} = 12$, $N_5=53$)

| | | | | | | | | | | | | | | | | | | | | | | | | | | | |
|---|---|---|---|---|---|---|---|---|---|---|---|---|---|---|---|---|---|---|---|---|---|---|---|---|---|---|---|
| $V$ | 0 | 0 | 0 | 0 | 0 | 0 | 0 | 0 | 0 | 0 | 0 | 0 | 0 | 0 | 0 | 0 | 1 | 1 | 1 | 1 | 1 | 1 | 1 | 1 | 1 | 1 | 1 |
| $n_2$ | 4 | 5 | 6 | 6 | 7 | 7 | 7 | 7 | 8 | 8 | 8 | 8 | 9 | 9 | 9 | 10 | 5 | 6 | 7 | 7 | 8 | 8 | 8 | 9 | 9 | 9 | 10 |
| $n_3$ | 8 | 6 | 4 | 5 | 2 | 3 | 4 | 0 | 1 | 2 | 2 | 0 | 0 | 1 | 0 | 0 | 7 | 5 | 3 | 4 | 1 | 2 | 3 | 0 | 1 | 1 | 0 |
| $n_4$ | 0 | 1 | 2 | 0 | 3 | 1 | 0 | 4 | 2 | 0 | 1 | 1 | 2 | 0 | 0 | 0 | 1 | 2 | 0 | 3 | 1 | 0 | 2 | 0 | 1 | 0 | |
| $n_5$ | 0 | 0 | 0 | 1 | 0 | 1 | 0 | 0 | 1 | 2 | 0 | 2 | 0 | 1 | 0 | 0 | 0 | 0 | 1 | 0 | 1 | 0 | 1 | 2 | 0 | 1 | |
| $n_6$ | 0 | 0 | 0 | 0 | 0 | 0 | 1 | 0 | 0 | 0 | 1 | 0 | 1 | 1 | 2 | 0 | 0 | 0 | 0 | 0 | 1 | 0 | 0 | 1 | 1 | | |

$K = 5$ (continued)

| | | | | | | | | | | | | | | | | | | | | | | | | | | | |
|---|---|---|---|---|---|---|---|---|---|---|---|---|---|---|---|---|---|---|---|---|---|---|---|---|---|---|---|
| $V$ | 2 | 2 | 2 | 2 | 2 | 2 | 2 | 2 | 2 | 3 | 3 | 3 | 3 | 3 | 3 | 4 | 4 | 4 | 4 | 4 | 5 | 5 | 5 | 6 | 6 | 7 | 8 |
| $n_2$ | 6 | 7 | 8 | 8 | 9 | 9 | 9 | 10 | 10 | 7 | 8 | 9 | 9 | 10 | 10 | 8 | 9 | 10 | 10 | 11 | 9 | 10 | 11 | 10 | 11 | 11 | 12 |
| $n_3$ | 6 | 4 | 2 | 3 | 0 | 1 | 2 | 0 | 0 | 5 | 3 | 1 | 2 | 0 | 1 | 4 | 2 | 0 | 1 | 0 | 3 | 1 | 0 | 2 | 0 | 1 | 0 |
| $n_4$ | 0 | 1 | 2 | 0 | 3 | 1 | 0 | 0 | 1 | 0 | 1 | 2 | 0 | 1 | 0 | 0 | 1 | 2 | 0 | 0 | 1 | 0 | 0 | 1 | 0 | 0 | |
| $n_5$ | 0 | 0 | 0 | 1 | 0 | 1 | 0 | 2 | 0 | 0 | 0 | 0 | 1 | 1 | 0 | 0 | 0 | 0 | 1 | 0 | 0 | 1 | 0 | 0 | 1 | 0 | |
| $n_6$ | 0 | 0 | 0 | 0 | 0 | 0 | 1 | 0 | 1 | 0 | 0 | 0 | 0 | 0 | 1 | 0 | 0 | 0 | 0 | 1 | 0 | 0 | 1 | 0 | 0 | 0 | |

Hence the total sum of all feasible link assortments for 1-DOF up to 5-independent loop 12-link simple and multiple joint kinematic chains (with up to 8 multiple joints) is: $N_\Sigma = 1 + 3 + 9 + 23 + 53 = 89$.

## 3  Technique of Structural Synthesis and Construction of Veritable Complete Atlas of Closed Loop Kinematic Chains

In this paper we propose the structural synthesis method for the generation of all feasible generalized kinematic chains (KCs) with specified $K$-independent loops. We use the given quantity of independent loops (i.e. $K \geq 2$) and exactly calculated number of binary links in kinematic chain (i.e. $n_2 \geq 4$) for systematic enumeration of various multiloop F-DOF KCs.

The synthesis process (Fig. 1) is based on finding all the possible combinations of $n_2$ into various kinds of $K$-loop contracted graphs (CG). Then we convert it into corresponding kinematic chains by determining and combining the required number of binary links, satisfying to new basic degree of freedom equation Eq. (1) and "Universal Tables 1".

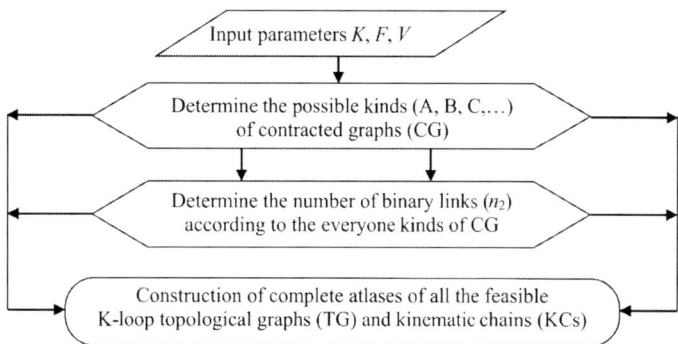

**Fig. 1.** The flow chart of the algorithm of type synthesis and the construction complete atlases

The constructed atlas of all possible contracted graphs with 3 independent loops is presented in Fig. 2, where symbolic notation ③ denotes ternary link and notation ④ denotes quaternary link. Accordingly, it will be used below for topological and structural synthesis of 3-loop TG and kinematic chains presented in Sects. 4 and 5.

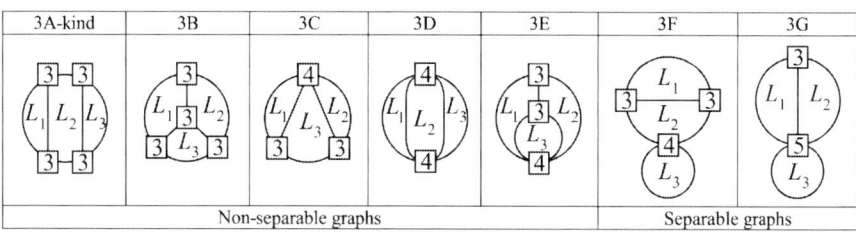

**Fig. 2.** Atlas of all the 7 possible kinds of simple joint contracted graphs with 3 loops ($L_1$, $L_2$, $L_3$)

## 4   Synthesis of 1-DOF 3-Loop 8-Link Kinematic Chains

The well-known synthesis results for creation of all feasible types of 1-DOF 3-loop 8-link planar simple joint kinematic chains (PSJKCs) contains only 16 types of PSJKCs [1–9].

According to "Universal Table 1", all feasible 1-DOF, 3-loop PSJKCs can have only 3 combinations of $n_2$ and [LA]: (a) $n_2 = 4$, [LA] = [4,4,0]; (b) $n_2 = 5$, [LA] = [5,2,1]; (c) $n_2 = 6$, [LA] = [6,0,2].

We use the algorithm (Fig. 1) to synthesize 1-DOF, 3-loop topological graphs and KCs. Then we classify it on the basis of the number of binary links (Figs. 3 and 4). The complete atlas of all feasible 20 types of PSJKCs (Fig. 4) includes the 4 new kinematic chains (No.4, No. 14, No. 17 and No. 20).

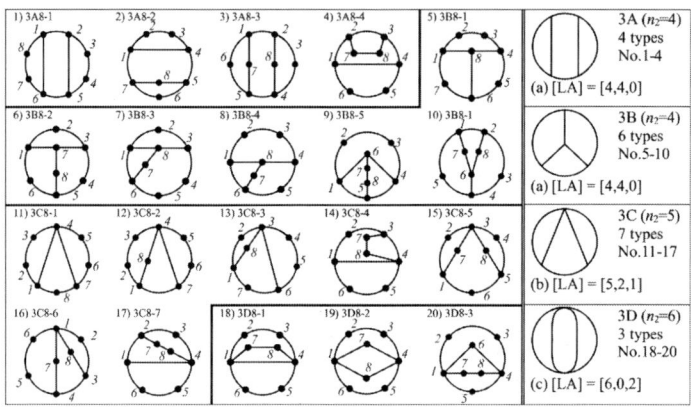

**Fig. 3.** Complete atlas of graphs for all feasible 20 types of 1-DOF 3-loop 8-link PSJKCs

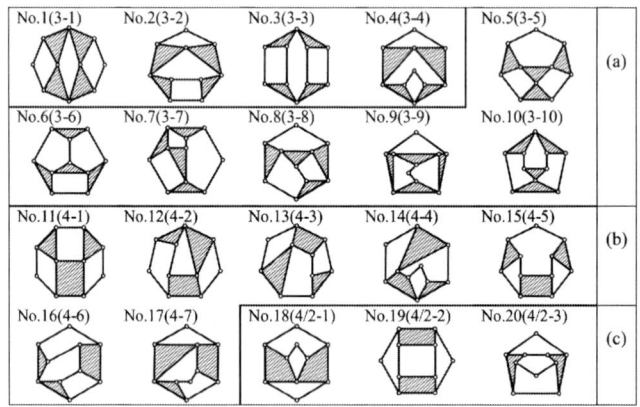

**Fig. 4.** Complete atlas of all feasible 20 types of 1-DOF 3-loop 8-link PSJKCs

## 5   Synthesis of 2-DOF 3-Loop 9-Link Kinematic Chains

The well-known synthesis results for creation of all feasible types of 2-DOF 3- loop 9-link planar simple joint kinematic chains (PSJKCs) contains only 40 types of PSJKCs [1, 2, 4–9].

According to new Eq. (1), all feasible 2-DOF, 3-loop PSJKCs can have only 4 combinations of $n_2$ and $[LA]$: (a) $n_2 = 5$, $[LA] = [5,4,0]$; (b) $n_2 = 6$, $[LA] = [6,2,1]$; (c) $n_2 = 7$, $[LA] = [7,0,2]$; (d) $n_2 = 7$, $n_5 = 1$, $[LA] = [7,1,0,1]$.

In such a way, based on general algorithm (Fig. 1) and after the classification of KCs on the basis of the number of binary links ($n_2 \geq 5$) and $[LA]$, the complete atlas of all feasible 59 types of 2-DOF, 3-loop closed PSJKCs (including the $59 - 40 = 19$ new non-fractionated and fractionated KCs) is synthesized and shown in Fig. 5, which can be used for creative design of novel mechanisms.

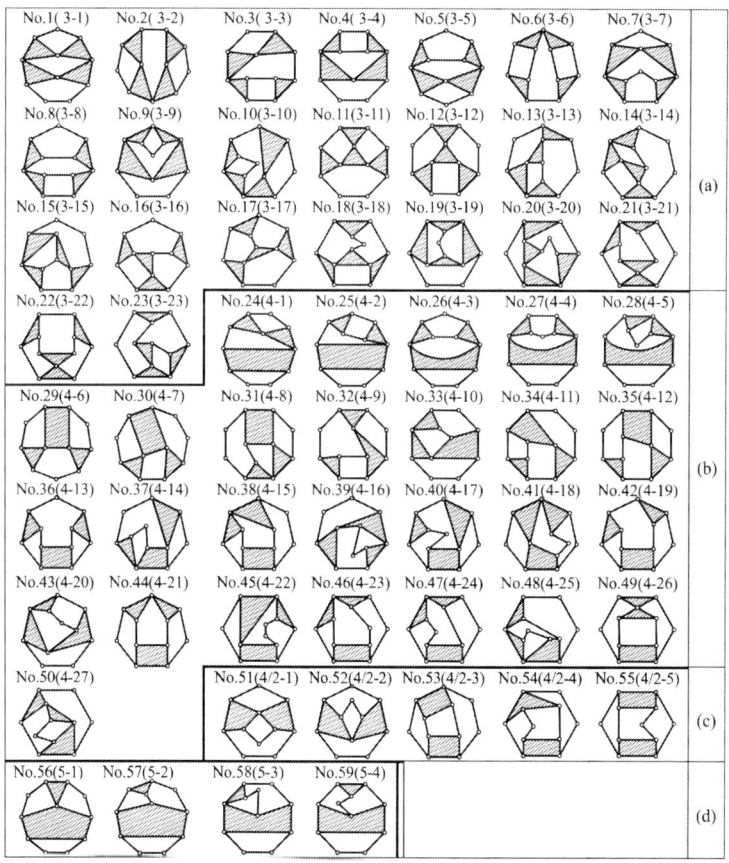

**Fig. 5.**   Complete atlas of all feasible 59 types of 2-DOF 3-loop 9-link planar PSJKCs

# 6 Conclusions

The proposed synthesis method allows to obtain all feasible F-DOF, K-loop kinematic chains (for a example, 20 types of 1-DOF and 59 types of 2-DOF 3-loop KCs) for the first time. We can easy prove non-isomorphism of synthesized novel kinematic chains using "Loop-base detection isomorphic method" [7]. The obtained KCs have different loop assortments $[L_\alpha]$ [5]:

(i) No.1/$[L_\alpha]$ = [4-4-4-8]  and  No.4/$[L_\alpha]$ = [4-4-6-6];  No.12/$[L_\alpha]$ = [4-4-5-7]  and No.17/$[L_\alpha]$ = [4-5-5-6] for 1-DOF 3-loop 8-link KCs (Fig. 4);
(ii) No.51/$[L_\alpha]$ = [4-4-7-7] and No.52/$[L_\alpha]$ = [4-5-6-7]; No.53/$[L_\alpha]$ = [4-5-6-7] and No.54/$[L_\alpha]$ = [5-5-6-6] for 2-DOF 3-loop 9-link KCs (Fig. 5).

Thus, total synthesis results present the extended necessary data bank for mechanical designers in the form of atlas of all feasible 20 types of 1-DOF $K = 3$ 8-link kinematic chains (with 4 new KCs) and atlas of all feasible 59 types of 2-DOF $K = 3$ 9-link kinematic chains (with 19 new KCs). It confirms the effectiveness of this method for conceptual creative mechanical design.

**Acknowledgments.** The work was supported by Act 211 Government of the Russian Federation, contract № 02.A03.21.0011.

# References

1. Alfattani, R., Lusk, C.: Lamina-emergent frustum using a bistable collapsible compliant mechanism (BCCM). In: Proceedings of ASME 2016 International Design Engineering Technical Conferences and Computers and Information in Engineering Conference, Charlotte, North Carolina, USA, 21–24 August 2016
2. Ding, H.: Automatic Structural Synthesis of Planar Mechanisms and its Application to Creative Design Duisburg. Universitätsbibliothek Duisburg-Essen, Essen (2016)
3. Grübler, M.: Allgemeine eigenschaften der zwangläufigen ebenen kinematische kette: I. Civilingenieur **29**, 167–200 (1883)
4. Nie, S.H., Liao, A.H., Qiu, A.H.: Addition method with 2 links and 3 pairs of type synthesis to planar closed kinematic chains. Mech. Mach. Theory **58**(12), 179–191 (2012)
5. Pozhbelko, V.: A unified structure theory of multibody open, closed loop and mixed mechanical systems with simple and multiple joint kinematic chains. Mech. Mach. Theory **100**(6), 1–16 (2016)
6. Pozhbelko, V., Ermoshina, E.: Number structural synthesis and enumeration process of all possible sets of multiple joints for 1-DOF up to 5-loop 12-link mechanisms on base of new mobility equation. Mech. Mach. Theory **90**(8), 108–127 (2015)
7. Rao, A.C., Prasad Raju Pathapati, V.R.: Loop based detection of isomorphism among chains, inversions and type of freedom in multi degree of freedom chain. ASME J. Mech. Des. **122** (1), 31–42 (1999)
8. Tsai, L.-W.: Mechanism Design: Enumeration of Kinematic Structures According to Function. CRC Press, Florida (2001)
9. Yan, H.S., Chiu, Y.T.: On the number synthesis of kinematic chains. Mech. Mach. Theory **89** (7), 128–144 (2015)

# Fundamentals for Web-Based Analysis and Simulation of Planar Mechanisms

Stefan Gössner[(✉)]

Dortmund University of Applied Sciences and Arts, Dortmund, Germany
stefan.goessner@fh-dortmund.de

**Abstract.** With the growing importance of distance education comes an increasing demand for web-based tools for analysis and simulation of mechanisms.

Current commercial and non-commercial software for simulating mechanisms is usually desktop-based and not web-enabled (GrafiCalc, Linkage, SAM, WinMeCC). With the intention of filling this gap, the fundamentals for web-based analysis and simulation of planar mechanisms are discussed and presented here first.

Of particular importance is a compact, redundancy-free data structure for describing a mechanism together with its pleasing visualization. The suitability of directed graphs is presented as well as an advantageous restriction to four basic vector types.

**Keywords:** Planar mechanisms · Web-based simulation · Directed graph
Basic vector types

## 1 Introduction

As distance education continues to evolve and moves gradually to the mainstream, new challenges will arise also in the fields of mechanisms and kinematics education. The theory of mechanisms is said to be hard to understand on a mathematical level. But learning success may be significantly improved by parallel use of descriptive and responsive simulation tools. Distance learning environments are usually based on web technologies (Moodle, Ilias, Canvas, Blackboard), so we need powerful web-based mechanism synthesis, analysis and simulation tools in future.

This paper presents concepts and work, still in its initial state of development. The discussion of an advantageous mechanism-notation as a data exchange format is of considerable importance in this context. The authors former work regarding a mechanical engine [4] as well as a proposal of a meta-programming language MML by Nagarajan and Bandyopadhyay [6] are considered here as starting points.

The vector algebra applied here is making use of an orthogonal operator. The orthogonal vector to $a$ is denoted here as $\tilde{a}$, while in other publications it is often written as $a^{\perp}$ [3].

© Springer Nature Switzerland AG 2019
B. Corves et al. (Eds.): EuCoMeS 2018, MMS 59, pp. 215–222, 2019.
https://doi.org/10.1007/978-3-319-98020-1_25

## 2  Loop-Closure Model

It is generally known, that mechanisms may be easily represented as directed or undirected graphs. So a central point in this paper is the description of mechanisms by one or more closed vector loops [5, 6] or a contour graph [7], which are then articulated in mathematical notation as loop-closure equations, on basis of which a mechanism's pose and kinematic state can be determined finally.

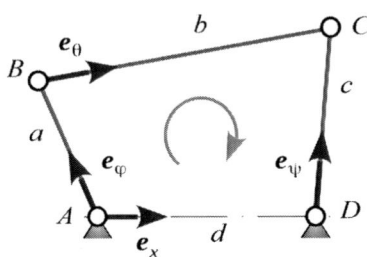

**Fig. 1.**  Closed loop of a four-bar mechanism

In contrast to the usual concept in building a graph with joints represented as nodes and links as edges, we prefer a more abstract model here. The graph edges correspond directly to the loop vectors, which not necessarily represent links. The more abstract nodes may or may not represent joints. Figure 1 illustrates a single loop mechanism with corresponding link length and direction vectors. Its loop-closure equation reads

$$a e_\varphi + b e_\theta - c e_\psi - d e_x = \mathbf{0}. \tag{1}$$

The polar vector representation $r e_\varphi$ used here clearly separates a scalar *length* $r$ from a *unit direction vector* $e_\varphi$. From these two values both, one of them or none are known.

With an abstract graph-centric point of view the nodes might be also interpreted as 2-dimensional 'bodies' without extension. The vectors function then as constraints by connecting exactly two nodes while restricting their mutual relative mobility. As each known value is taking away one single degree of freedom, the mobility $M$ and the number of loops $L$ of a graph, representing a planar mechanism, may then be calculated by

$$\begin{aligned} M &= 2(N-1) - E_1 - 2E_2 \\ L &= E_0 + E_1 + E_2 - N + 1, \end{aligned} \tag{2}$$

with $N$ nodes, $E_1$ edges restricting a single vector value, $E_2$ edges restricting both and $E_0$ none of them.

See [1, 2] for a more general discussion regarding these aspects. We will return to these vectors in Sect. 6.

## 3   Constraint Equations

Assuming a constraint equation $C$ of dimension $m$, composed of some closed vector loops and possibly some additional scalar equations. Then we introduce a set of generalized coordinates $q$ of dimension $n$. If their values are consistent with the constraints, the constraint equation is valid, i.e.

$$C(q) = 0. \tag{3}$$

The derivation of constraint equation $C$ with respect to time leads to the constraint velocities

$$\dot{C} = \frac{\partial C}{\partial q}\frac{dq}{dt} = J\dot{q} = 0, \tag{4}$$

where $J$ is the *Jacobian matrix* of $C$. Deriving the constraint velocities again with respect to time results in the constraint accelerations

$$\ddot{C} = J\ddot{q} + \frac{\partial(J\dot{q})}{\partial q}\dot{q} = J\ddot{q} + (J\dot{q})_q\dot{q} = 0. \tag{5}$$

Demanding clearly solvable kinematic conditions, $m < n$ must hold. The difference $k = n - m$ gives us the total *degree of freedom* (*DoF*) of the mechanism, which corresponds to the necessary number of input values.

So from $n$ coordinates in $q$ only $k$ coordinates are independent, which we would like to collect into vector $v$. The $m$ remaining dependent coordinates will go into $u$ then. After this process of coordinate partitioning [5] we have

$$C(q) = 0 \text{ with } q = \begin{pmatrix} u \\ v \end{pmatrix}. \tag{6}$$

The constraint velocities now read

$$\dot{C} = (J_u J_u)\begin{pmatrix} \dot{u} \\ \dot{v} \end{pmatrix} = J_u\dot{u} + J_v\dot{v} = 0. \tag{7}$$

and the constraint accelerations are

$$\ddot{C} = J_u\ddot{u} + J_v\ddot{v} + (J\dot{q})_q\dot{q} = 0. \tag{8}$$

## 4   Kinematic Analysis

Having composed the constraint equation from one or more loop-closure equations, a short outline may be presented, how to get to the associated kinematic and kinetostatic state.

### 4.1 Positions

We assume a positional state $q$ inconsistent with the constraints, i.e. $C(q) \neq 0$. So we are demanding a correction vector $\delta q$ for validating the constraint equation

$$C(q + \delta q) = 0. \tag{9}$$

Approximating the sufficiently differentiable function $C$ around the point $q$ according to *Taylor's Theorem* yields [5]

$$C(q + \delta q) = C(q) + J\delta q + higher\ order\ terms.$$

Neglecting the higher order terms and partitioning of coordinates gives

$$C + J_u\delta u + J_v\delta v = 0.$$

Since the independent coordinates are actuated, they are always intrinsically consistent with the constraints. So the corrective term $\delta v$ is ever zero. The corrective vector of dependent coordinates is then

$$\delta u = -J_u^{-1}C, \tag{10}$$

which is identical to applying *Newton-Raphson*'s method for numerically solving nonlinear function $C$ for $u$. We assume here that $J_u$ is a nonsingular matrix, so Eq. (10) can be used for position correction and even initial mechanism assembly.

### 4.2 Velocities

Velocity constraint equations are always linear in terms of velocity coordinates. Since independent velocity coordinates $\dot{v}$ are explicitly set, we can directly solve for dependent velocity coordinates $\dot{u}$ from Eq. (7).

$$\dot{u} = -J_u^{-1}J_v\dot{v} \tag{11}$$

The matrix $-J_u^{-1}J_v$ is called an *influence coefficient matrix*, which relates velocities $\dot{u}$ to velocities $\dot{v}$ [5].

### 4.3 Accelerations

In analogy to the velocities, independent acceleration coordinates $\ddot{v}$ are actuated, so we can again directly solve for the dependent acceleration coordinates $\ddot{u}$ from Eq. (8), which results in

$$\ddot{u} = -J_u^{-1}\left(J_v\ddot{v} + (J\dot{q})_q\dot{q}\right). \tag{12}$$

## 5 Kinetostatic Analysis

We start with *Newton's Law*

$$M\ddot{q} = Q_c + Q_a, \tag{13}$$

with mass matrix $M$ and forces $Q$ on the right side of the equation already separated in constraint forces $Q_c$ and applied forces $Q_a$. Observing the kinetic energy of the system $E_k = \frac{1}{2}(M\dot{q})^T\dot{q}$, we derive it with respect to time and yield by reusing Eq. (13)

$$\dot{E}_k = (M\ddot{q})^T\dot{q} = (Q_c + Q_a)^T\dot{q}. \tag{14}$$

Now we demand the term $Q_c^T\dot{q}$ being ever zero in order to satisfy the *principle of virtual work*. This dot product vanishes as both vectors $Q_c$ and $\dot{q}$ are always orthogonal to each other, which basically means: *Forces in frictionless constraints do no work.*

As the columns of the Jacobian $J$ show the directions perpendicular to the movability of the system, we can express the constraint forces as, $Q_c = J^T\lambda$ where $\lambda$ is the vector of the *Lagrangian Multipliers*. With these we can rewrite Eq. (14) while renaming the applied forces $Q_a$ as $Q$

$$M\ddot{q} = J^T\lambda + Q.$$

Partitioning of coordinates again results in

$$\begin{pmatrix} M_u & 0 \\ 0 & M_v \end{pmatrix}\begin{pmatrix} \ddot{u} \\ \ddot{v} \end{pmatrix} = \begin{pmatrix} J_u^T \\ J_v^T \end{pmatrix}\lambda + \begin{pmatrix} Q_u \\ Q_v \end{pmatrix}. \tag{15}$$

With kinetostatics the kinematic state of the system is completely known. We are rather interested in the Lagrange Multipliers $\lambda$ and the external (driving) forces $Q_v$. So we separate from Eq. (15) the first row for determining $\lambda$

$$\lambda = J_u^{-T}(M_u\ddot{u} - Q_u). \tag{16}$$

The second row can be used then for getting the possibly unknown actuating forces $Q_v$

$$Q_v = M_v\ddot{v} - J_v^T\lambda. \tag{17}$$

## 6 Vector Chains

Proposing a mechanism representation by not necessarily closed vector chains consisting of polar vectors, we distinguish four types of vectors.

**Table 1.** Vector types

| Type | bas | trn | rot | trn+rot |
|---|---|---|---|---|
| Symbol | | | | |
| Length | fixed | variable | fixed | variable |
| Orientation | fixed | fixed | variable | variable |
| $q$ | - | $r$ | $\varphi$ | $r, \varphi$ |
| $C$ | $r$ | $re_\varphi$ | $re_\varphi$ | $re_\varphi$ |
| $J$ | $0$ | $e_\varphi$ | $r\tilde{e}_\varphi$ | $e_\varphi, r\tilde{e}_\varphi$ |
| $\dfrac{\partial(J\dot{q})}{\partial q}$ | $0$ | $0$ | $-\dot{\varphi}re_\varphi$ | $\dot{\varphi}\tilde{e}_\varphi, \dot{r}\tilde{e}_\varphi - \dot{\varphi}re_\varphi$ |

Each vector in the chain can have fixed or variable *length* and/or *orientation*. Every fixed value reduces the overall degree of freedom by one and every variable value contributes to the list of generalized coordinates $q$ and the Jacobian $J$ (see Table 1).

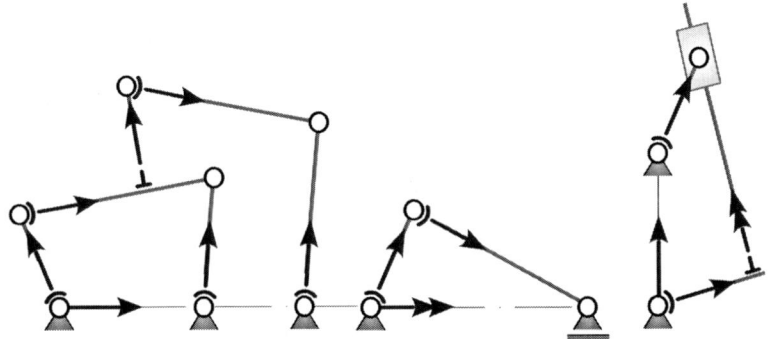

**Fig. 2.** Mechanism examples using basic vector types.

Please observe, that there is another vector type carrying a *perp* symbol. This is merely an associate of a fixed or rotational vector (*bas* or *rot*), with which it is always building an orthogonal base, so not introducing another variable (Fig. 2).

## 7   Implementation Details

The mechanism model is represented as a tree and written to a textual document. Its structure is inspired by the MML language [6]. Although in contrast to that it is not a programming language API, but a JSON based mechanism-notation.

JSON is a lightweight, text-based, language-independent data interchange format [8]. It was derived from *JavaScript Object Notation* of the ECMAScript Programming Language Standard and is heavily used for data exchange on the web.

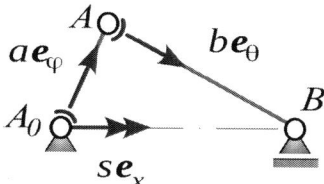

**Fig. 3.** Slider crank mechanism

The JSON document representing a mechanism is read and written by a web-based mechanism editor, which is then able to compile the graph structure into an incidence matrix as well as into one or more associated constraint equations, numerically calculate, graphically visualize and simulate different mechanism poses with their corresponding velocity and acceleration state. The mechanism editor is implemented as an open source web application using pure standard web technologies (HTML, CSS, JavaScript). Web applications traditionally are client-server software, but today modern webapps can run without the need for internet connectivity and even without a web browser, similar to native desktop applications.

Listing 1 shows the JSON based mechanism-notation describing the slider crank mechanism from Fig. 3.

```
{
    "id": "Slider Crank",
    "q": [0,300,1.047],   // values of theta, s, phi
    "v": [2],             // indices of independent coordinates in q
    "nodes": {            // named array of nodes
        "A0": {"base": true, "x":0, "y":0},
        "A": {},          // node coordinates ...
        "B": {}           // ... are optional
    },
    "edges": [            // unnamed array of directed edges/vectors
        {"from":"A0","to":"A","typ":"rot","r":50,"wq":0},
        {"from":"A0","to":"B","typ":"trn","w":0,"rq":1},
        {"from":"A","to":"B","typ":"rot","r":300,"wq":2}
    ]
}
```

**Listing 1.** JSON document of slider crank mechanism

The mechanism model for this example is sufficiently complete and redundancy free. Array 'q' contains initial values of the generalized coordinates $q$, not necessarily consistent with the constraints. Array 'v' contains the indices of the independent (actuated) coordinates in Array 'q'. The 'nodes' array is holding the nodes accessible by their names. At least one of them must be defined as *base node* including position coordinates with respect to an implicit global coordinate system. Finally, the 'edges' array contains the vectors pointing from one node to another. In fact, this edge list corresponds to the graph incidence list [1, 2]. The vectors are equipped with their type according to Table 1, as well as their polar coordinates either as const values ('r', 'w') or as indices pointing to the generalized coordinates ('rq', 'wq').

# 8  Conclusion

With this work, an attempt was made to provide the development of a web-based simulation and analysis system with an appropriate JSON based mechanism-notation. Future work may include the definition of a JSON schema for documenting and validating the structure of that JSON data [9].

This admittedly abstract model can, however, be made visually pleasing and accessible to the user by means of a small set of different vector symbols. On the other hand the model harmonizes perfectly with the mathematical formulation of loop-closure equations. These are at the same time constraint equations that act on a set of generalized coordinates.

The present work has the character of a proof of concept. The positive results so far encourage further action in this direction.

# References

1. Bang-Jensen, J., Gutin, G.: Directed Graphs: Theory, Algorithms and Applications. Springer, London (2000)
2. Diestel, R.: Graph Theory, 3rd edn. Springer, New York (2005)
3. Gössner, S.: Mechanismenanalyse im Vektorraum $R^2$, IFToMM D-A-CH Konferenz, Innsbruck (2016)
4. Gössner, S.: Eine Physik-Engine zur webbasierten Mechanismensimulation – Ergebnisse einer Studie. 9. In: Kolloquium Getriebetechnik - Conference Proceedings, pp. 333–354, Chemnitz (2011). ISBN 978-3-941003-40-8
5. Nikravesh, P.E.: Computer-Aided Analysis of Mechanical Systems. Prentice-Hall, Upper Saddle River (1988)
6. Nagarajan, A., Bandyopadhyay, S.: A framework for analysis and dynamic visualization of mechanisms (2011)
7. Drewniak, J., Garlicka, P., Borowik, B.: Analysis of the kinematics of planar link mechanism with non-stationary motion of crank. In: Mechanisms and Machine Science. Springer (2017)
8. The JavaScript Object Notation (JSON) Data Interchange Format. IETF, December 2017. https://tools.ietf.org/html/rfc8259
9. JSON Schema: A Media Type for Describing JSON Documents, 19 March 2018. https://tools.ietf.org/html/draft-handrews-json-schema-01

# Effect of the Roller Crown Shape
# on the Cam Stress

Jiří Ondrášek$^{(\boxtimes)}$

VÚTS Liberec, a.s., Liberec, Czech Republic
jiri.ondrasek@vuts.cz

**Abstract.** The aim of this paper is to outline the issues of the influence of the cam follower crown geometric shape in cam mechanisms on the stress of the surface layers of the general kinematic pair. The mentioned type of kinematic constraint in the practical embodiment is usually most often constituted by a straight roller and a cam or a crowned roller and a cam. Near the shape discontinuities, intersections of the cylindrical profile with the cam profile, there is a sharp increase in the size of the reduced stresses. This is one of the reasons for the practical application of crowned rollers. The disadvantage of this solution is that the greatest stress values will be reached in the middle of the contact area. With the logarithmic profile of the roller crown, a uniform stress distribution for different load levels of the general kinematic pair can be achieved. However, the production of such a general roller crown profile is a technologically demanding issue in the required precision and quality. A certain compromised solution is the design of a part-crown roller. In order to optimize the profile shape of the part-crown roller, the Finite Element Method - FEM can be used with some advantages.

**Keywords:** Cam roller · Contact region · Contact stress · Roller crown shape
General kinematic pair

## 1 Introduction

The geometrical shape of the roller crown itself has a significant effect on the stress distribution, due to the load and inertia effects in the contact areas of the general kinematic pair. The mentioned type of kinematic constraint in the practical embodiment is usually most often constituted by a cylindrical roller and a cam or a roller with a convex crown and a cam. In the case of the said type of kinematic pair, the contact load on the surface of the body contact areas in the contact and under the respective surface has a periodic course. The concluded stresses are transient, having a character of pulses with a period of $2\pi$. If a certain limit of this stress is exceeded at any point of the contact area, fatigue damage of the cam and follower contact surfaces may occur after a certain number of cycles in the operation of the cam mechanisms. One of the criteria for such damage may be the value of the largest pressure main stress in the contact area. This is subsequently brought into the relationship with the strength limit of the respective material, whether heat-treated or without heat treatment.

© Springer Nature Switzerland AG 2019
B. Corves et al. (Eds.): EuCoMeS 2018, MMS 59, pp. 223–230, 2019.
https://doi.org/10.1007/978-3-319-98020-1_26

In order to calculate the main stresses, the conclusions of the contact mechanics for the respective case of contact between two elastic bodies, but also the possibilities of the finite element method can be used. The finite element method is particularly applicable in the cases when it comes to contacting a cam and a roller with a different crown shape than was above mentioned. For both of these cases, the results of Hertz's theory of contact can be used simply to establish the distribution of the contact stress, but with some limitations.

## 2 The Shape of the Roller Crown

The publications [3–6] conclude that contact stresses tend to be significantly higher in the vicinity of certain discontinuities in the contact area than those achieved in the middle of this area. As a result of this, it can be stated that the shape of the roller crown profile affects the stress distribution in the contact surface and its vicinity of the general kinematic pair of the cam mechanism and thus radically affects the dynamic load and life of the cam and the roller themselves. For example, the cylindrical profile of the roller crown appears to be a key factor in achieving a longer life of the general kinematic pair of any cam mechanism. However, in the case of a conventional cylindrical roller, there are discontinuities in the intersections of the cylindrical profile with the cam profile, i.e. if one contact part is axially shorter than the other, as well as in the chamfer of the roller edges, see Fig. 1. In the vicinity of those profile discontinuities, the contact between the roller and the cam cannot be considered to be straightforward to which Hertz's theory of contact can be applied, but to be a more complex three-dimensional type of contact. These discontinuities cause a very rapid increase in the pressure distribution in the respective contact area of the bodies. In fact, these local increases in pressure distribution can exceed the strength limit of the given material, thereby causing plastic deformations, residual stresses in the material or hardening of steel. Furthermore, the area of question will be more susceptible to fatigue damage to the contact surfaces, i.e. to pitting or spalling of the material.

One of the possible variants of ensuring a more even distribution of contact stress is a structural change in the shape of the roller crown axial cross section. This is one of the reasons for the practical application of crowned rollers, where the radius $R$ of curvature of the crown profile is far greater than the radius $D/2$ of the roller, see Fig. 1. If the general kinematic pair load is predictable, the crown curvature radius $R$ can be dimensioned so as to achieve the distribution of the contact stress as even as possible over the entire axial length $l_e$ of the contact area. The disadvantage of this solution is that the greatest stress values will be reached in the middle of the contact area, and once the limit value is exceeded, the same undesirable phenomena can occur again as in the case of the cylindrical roller. Another reason of using a roller with a convex crown is that we are able to easily determine the possible misalignment between the roller and the cam, without fundamentally changing the distribution of the contact stress.

In order to reduce the excess edge stress in the case of cylindrical rollers, it is possible to achieve a crown axial cross section with such a shape that includes a straight line and a single circular arc or a combination of several circular arcs, see Fig. 2. However, such shape of a roller crown whose one segment is cylindrical and the

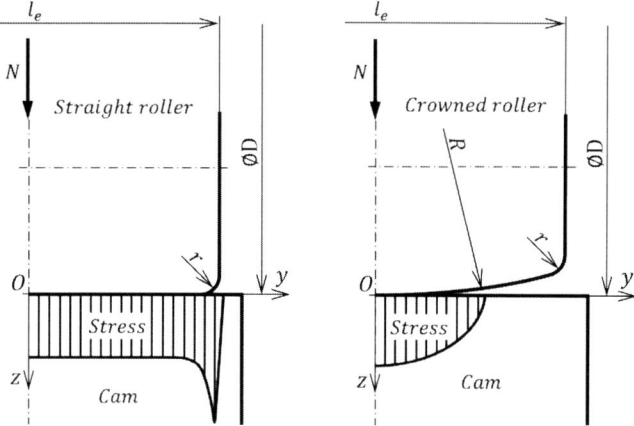

**Fig. 1.** Schematic drawing of a straight and crowned roller

subsequent segment thereof is convex leads to a certain concentration of stress in the transition from the cylindrical section to the convex one.

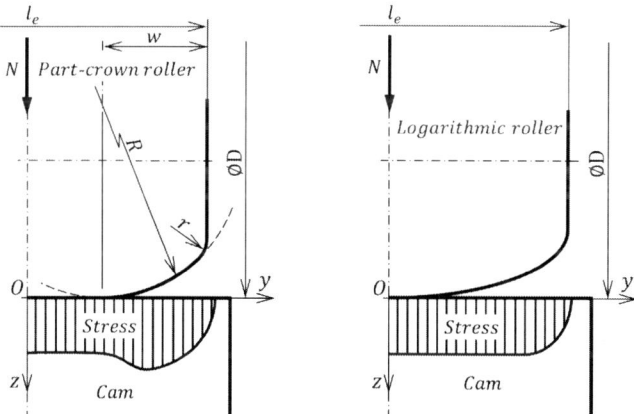

**Fig. 2.** Schematic drawing of a part-crown and logarithmic roller

According to [6, 7], the logarithmic profile of the roller crown we can achieve uniform stress distributions for different load levels of the general kinematic pair, see Fig. 2. A characteristic feature of this profile is that it descends monotonously from its center to the edge according to the logarithmic function. It should be noted here that the production of the general profile of the roller crown is a technologically demanding issue in required precision and quality.

## 3  The FEM Use When Designing the Roller Crown Shape

The logarithmic profile of the roller crown is advantageous both from the point of view of the uniform distribution of the contact stress in the contact area and also from the point of view of the determination of the roller and cam misalignment. However, this type of profile is very demanding as far as manufacture is concerned. A certain compromised solution seems to be the embodiment of a roller with convex segments of the crown, the central section of the crown being cylindrical, see the left part of Fig. 2. The characteristic dimensions of such a roller are its diameter $D$, the width $l_e$, the radius of the convex part of the crown $R$, the width of the convex part $w$ and the curvature radius $r$. The optimization parameters of such a crown profile are then the radius of the convex section $R$ and its width $w$. The optimization of the crown shape itself is carried out in order to achieve the most uniform distribution of the contact stress with the greatest possible load induced by the effects of force $N$.

In order to optimize the profile shape of the crown with convex segments, the Finite Element Method – FEM can be used with some advantage, although it is well-known that in the case of contact tasks, this method is time-consuming as far as calculations are concerned. To achieve the relevant results, a dense network of elements in the contact area is required. This requirement leads to a large number of solved linear algebraic equations. The computational body contact algorithm is based on a numerical iteration of finding the elements in contact, so the numerical solution of the equations of the assignment takes place in several steps. The intent of generating a computational model of a general kinematic pair for FEM purposes is to minimize the size of the two-body contact solution task. This means that the necessary needed number of equations based on the model is established, the solution of which will produce satisfactory results.

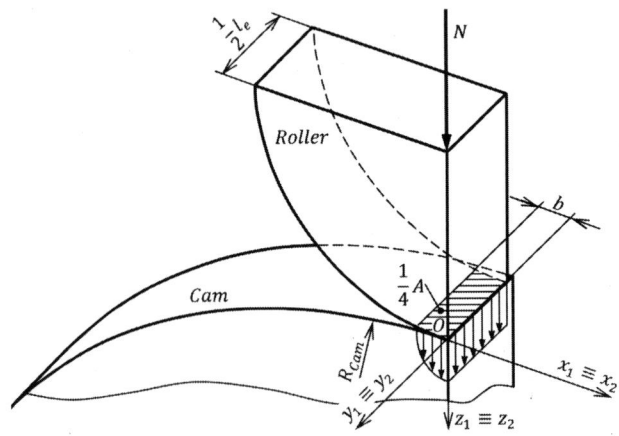

**Fig. 3.**  Schematic drawing of a roller in the contact with a cam

When generating the model of a general kinematic pair for the needs of FEM, it is possible to suppose the assumptions on which Hertz's theory of contact of two elastic bodies is derived, for example, see [1]. Above all, the primary assumption is that the contact area is continuous and much smaller compared to the characteristic dimensions of the bodies in contact. This fact is a basic prerequisite for applying a linear theory of elasticity to analyze the contact of two elastic bodies with sufficiently small deformation. The contact stress is highly concentrated near the contact surface and rapidly decreases with increasing distance from it. The area of stress acting is therefore in the vicinity of the contact of the bodies. As the contact surfaces are dimensionally small compared to the rest of the bodies, the stresses around the contact area are not dependent too much on the shape of the bodies in contact, nor on the attachment of those bodies. Through these assumptions, the definition of boundary conditions is simplified and the application of the theory of elasticity of large bodies is facilitated.

In practical application, the contact of the roller with the cam can be then replaced by the contact of two rotary body segments. One rotating body is a roller with the desired shaping profile and the other rotating body of cylindrical shape replaces the cam. We can afford to do this substitution due to the validity of the Hertz contact theory assumptions. The radius of the cylinder $R_{Cam}$ is the same as the curvature radius of the cam in the point of its contact with the roller. Assuming the parallelism of the axes of the two replacement bodies, one eighth of each of them can be used for the purposes of the computational analyses, see Fig. 3. In the $Oxz$ and $Oyz$ planes shown in Fig. 3, the boundary conditions there are then defined in the symmetry of the solved task accordingly. The lower surface of the body that has a cam meaning is fixed. The upper surface of the roller part is allowed to displace in the direction of the applied force $N$, i.e. in the direction of the $z$ – axis of the $Oxyz$ coordinate system, by the boundary conditions. In this area, force $N$ acting is also evenly distributed as follows

$$p_N = \frac{N}{Dl_e} \ [\text{Nm}^{-2}] \tag{1}$$

When creating an own network of elements, it is important that the area of contact and its close surroundings of both bodies be discretized by a uniform and dense network of elements. With increasing distance from this area, the size of the elements can be gradually increased. For example, the size of this area can be estimated computationally based on Hertz contact theory applied to the contact of cylindrical bodies with parallel axes or to the contact of the general body with the cylindrical body. This issue is published in detail in [1]. This theory makes it possible to calculate the components of deformations and stresses in both bodies in the contact area and its surroundings. Furthermore, the shape and size of the contact surface is determined depending on the load size. Then, on the basis of the data thus determined, the space in the analyzed bodies can be defined to produce an acceptable density of the element network in order to achieve the corresponding results compared to the real state.

The application of the above procedure is demonstrated by analyzing the contact stress distribution in the cam contact area in a nominal width of $l_{Cam} = 20$ mm and the radius of curvature $R_{Cam} = 50$ mm in the vicinity of the contact site with the respective roller. The cam contact stress test is designed to determine the maximum possible reduced stress $\sigma_{red}$ in the part of the cam in question, depending on the greatest possible load of the rollers by force $N$. Knowledge of the stress value is crucial in terms of fatigue damage of the active surfaces of the cam and the follower. Three types of cam crown shapes are considered here: *cylindrical, crowned, part-crown*. The nominal dimensions of each roller type are in accordance with Figs. 1 and 2 listed in Table 1.

**Table 1.** Nominal dimensions of rollers

|  |  | Straight | Crowned | Part-crown |
|---|---|---|---|---|
| Nominal diameter | $D$ [mm] | 35.0 | 35.0 | 35.0 |
| Effective width | $l_e$ [mm] | 18.0 | 18.0 | 18.0 |
| Fillet radius | $r$ [mm] | 0.6 | 0.6 | 0.6 |
| Crown radius | $R$ [mm] | – | 500 | 200 |
| Crown width | $w$ [mm] | – | – | 6.0 |

The characteristic material parameters of the cam and the rollers are shown in Table 2. It is further assumed that the active surfaces of the cam and rollers are heat-treated to an approximate depth of 2 mm to a hardness of 58HRC. On the cam and the rollers, there will thus be created a layer which has different mechanical properties compared to the properties of the base material of the component core. From the experimental strength and elasticity, it is well-known that the tensile strength value $R_m$ is linearly related to the hardness of the material measured by the respective intrusion tests, see [8]. The strength limit $R_m$ of the surface layers can be estimated by calculating on the basis of the knowledge of the predicted hardness to the value: $R_m \approx 2200$ MPa for 58HRC.

**Table 2.** Material characteristics of steels

|  |  | Roller: 100Cr6 | Cam: 16MnCr2 |
|---|---|---|---|
| Mass density | $\rho$ [kg m$^{-3}$] | 7850 | 7850 |
| Young's modulus of elasticity | $E$ [GPa] | 210 | 206 |
| Shear modulus | $G$ [GPa] | 81 | 79 |
| Poisson's ratio | $v$ [–] | 0.3 | 0.3038 |

Strength conditions under a variable load are given by the stress limit $\sigma_h$ which is in agreement with the disturbance caused by the transitory stress. The reduced stresses $\sigma_{red}(\psi, z)$ are limited by the actual strength condition, written as [2]:

$$\max(\sigma_{red}(\psi, z)) < \sigma_h, \quad \psi \in \langle 0, 2\pi \rangle, \quad z \geq 0 \tag{2}$$

For steel the usual values are $\sigma_h \approx 2\sigma_C \approx 0.66R_m$ [2]. However, since the transitory stress limit is the previous relation (2) and may be replaced by the condition:

$$\max(\sigma_{red}(\psi, z)) < 0.66R_m \approx 1500 MPa, \quad \psi \in \langle 0, 2\pi \rangle, \quad z \geq 0 \qquad (3)$$

The magnitude of the applied force $N$ was chosen in such a way so that the theoretical value of the highest possible reduced stress $\max(\sigma_{red}(z))$ is approximately equal to the estimated tensile strength value $R_m$ of the heat-treated cam layer. Significant results are summarized in Table 3. Figure 4 shows the course of the maximum reduced stress in a depth of $z_e$ depending on the half width of the cam for all types of rollers in contact with the cam. Depth $z_e$ expresses the depth under the surface of the cam where just the maximum value of the reduced stress depending on the distance from the contact surface is reached.

**Table 3.** Summary of results

|  |  | Straight | Crowned | Part-crown |
|---|---|---|---|---|
| Load | $N$ [N] | 40000 | 15000 | 35000 |
| Hertzian pressure | $p_H$ [MPa] | 2500 | 2400 | 2400 |
| Maximum reduced stress – Hertz theory | $\sigma_{redMax}$ [MPa] | 1440 | 1370 | 1480 |
| Maximum reduced stress – FEM | $\sigma_{redMax}$ [MPa] | 1970 | 1360 | – |
| Major contact radii | $a$ [mm] | – | 5.555 | – |
| Half width of contact/Minor contact radii | $b$ [mm] | 0.586 | 0.538 | – |
| Depth of maximum reduced stress | $z_e$ [mm] | 0.42 | 0.38 | 0.38 |

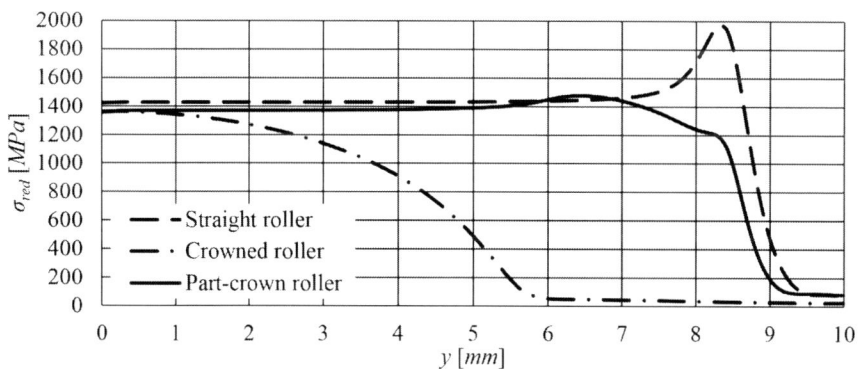

**Fig. 4.** The course of the maximum reduced stress in the depth $z_e$ of a cam

## 4   Conclusions

The geometric shape of the roller crown itself has a significant influence on the stress distribution due to the load and inertia effects in the contact areas of the general kinematic pair. In the case of the cylindrical profile of the roller crown, near the shape discontinuities, and intersections of the cylindrical profile with the cam profile, there is a sharp increase in the size of the reduced stresses. One possible way of ensuring a more even distribution of contact stress is the structural change in the shape of the roller crown. This is one of the reasons for the practical application of crowned rollers. The disadvantage of this solution is that the greatest stress values will be reached in the middle of the contact area. With the logarithmic profile of the roller crown, a uniform stress distribution can be achieved for different load levels of the general kinematic pair, see Fig. 2. However, the production of such a general roller crown profile is a technologically demanding issue in the required precision and quality. A certain compromise solution is the design of a part-crown roller, see Fig. 1. In this case, the shape of the crown itself can be optimized to achieve a most uniform distribution of the contact stress, achieving a greater dynamic load, and a longer general kinematic pair life. The optimization parameters of such a crown profile are, for example, the radius of the convex section and its width. In order to optimize the profile shape of the crown with convex segments, the Finite Element Method can be used with some advantage.

**Acknowledgments.** This paper was created within the work on the project NPU-L01213 – Project supported by the Ministry of Education, Youth and Sports of the Czech Republic.

# References

1. Johnson, K.L.: Contact Mechanics. Cambridge University Press, Cambridge (1985). ISBN 0 521 34796 3
2. Koloc, Z., Václavík, M.: Cam Mechanisms. Elsevier, Amsterdam (1993). ISBN 0-444-98664-2
3. Norton, R.L.: Cam Design and Manufacturing Handbook. Industrial Press Inc., New York (2009). ISBN 978-0-8311-3367-2
4. Hoeprich, M.: Numerical procedure for designing rolling element contact geometry as a function of load cycle, SAE Technical Paper, Series 850764 (1985)
5. de Mul, J.M., Kalker, J.J., Fredriksson, B.: The contact between arbitrarily curved bodies of finite dimensions. ASME J. Tribol. **108**, 140–148 (1986)
6. Reusner, H.: The logarithmic roller profile the key to superior performance of cylindrical and taper roller bearings. Ball Bear. J. **230**, 2–10 (1987)
7. Fujiwara, H., Kawase, T.: Logarithmic Profiles of Rollers in Roller Bearings and Optimization of the Profiles, NTN Technical Review, no. 75, pp. 140–148 (2007)
8. Pluhař, J.: kolektiv: Nauka o materiálech. SNTL, Praha (1989). ISBN 04-205-89

# Forward and Inverse Kinematics in 2-DOF Planar Parallel Continuum Manipulators

Oscar Altuzarra, Diego Caballero$^{(\boxtimes)}$, Francisco J. Campa, and Charles Pinto

University of the Basque Country UPV/EHU, Leioa, Spain
{oscar.altuzarra,diego.caballero,fran.campa,charles.pinto}@ehu.es

**Abstract.** Parallel Continuum Manipulators are parallel mechanisms which movement is due to the deformation of some of their elements, usually flexible rods. Provided that rods are straight in their stress-free reference state and that the movement of the mechanism is within a plane, elliptic integral solution can be used in order to get the deformation of each rod. Since it is a parameterized method, it allows to control all possible solutions related to a certain buckling mode. We will show an unified method to solve both Forward and Inverse Kinematic Position problem of 2-DOF planar Parallel Continuum Manipulators in which two flexible rods are hinged at one extreme and are clamped to the other. The main characteristic of the method proposed is that static equilibrium equations and geometric equations are decoupled.

**Keywords:** Parallel Continuum Manipulators
Compliant mechanisms · Kinematics

## 1 Introduction

Continuum mechanisms are those formed by a set of slender rods linked in a parallel way, and constitute a type of mechanisms itself [1] that are in current research. Some examples can be seen in Fig. 1, in which two rods are hinged together at the end effector. In the right figure in Fig. 1, a prototype for pick and place operations have been mounted.

Some methods have been developed to model rod deformation in the plane [2–5], being the classical Kirchhoff model more appropriate because of its accuracy. In the case in which rods are straight in their stress-free reference state, the movement of the mechanism is within a plane and loads are applied only at the extreme of the rod, elliptic integrals can be used [6]. The main advantage of the elliptic method is that, for a known buckling mode, two parameters defined within a range are used to describe the entire position analysis of a rod. This allows to get all possible solutions when solving the position problem of a mechanisms as a whole.

© Springer Nature Switzerland AG 2019
B. Corves et al. (Eds.): EuCoMeS 2018, MMS 59, pp. 231–238, 2019.
https://doi.org/10.1007/978-3-319-98020-1_27

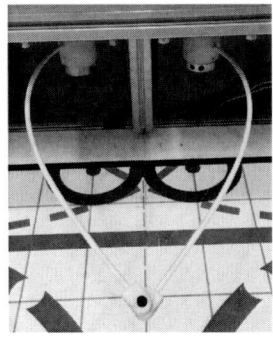  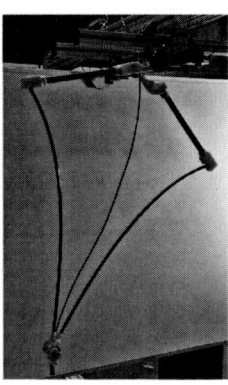

**Fig. 1.** Some planar 2-DOF planar Parallel Continuum Robots

## 2    Position Equations

The position analysis to be done on these mechanisms with 2 flexible links consists in finding the deformed shapes of these rods under the following conditions: the clamped-ends are placed in a pose defined by the inputs, the hinged-ends are joined together at the output coordinates, there is an static-equilibrium of the end-effector subject to the load.

If a local frame is used for the flexible links at the clamped-end, their deformed shapes $\mathbf{p}(s)$ will be ruled by the same equations. The classical Kirchhoff model is used to describe the evolution of the internal force $\mathbf{n}(s)$ and moment $\mathbf{m}(s)$ along the arc-length $s$ of the rod depending on applied loads $\mathbf{f}(s)$ and $\mathbf{l}(s)$:

$$\frac{d\mathbf{n}(s)}{ds} + \mathbf{f}(s) = \mathbf{0}$$

$$\frac{d\mathbf{m}(s)}{ds} + \frac{d\mathbf{p}(s)}{ds} \times \mathbf{n}(s) + \mathbf{l}(s) = \mathbf{0} \tag{1}$$

For planar deformations of rods under no distributed force and moment, $\mathbf{f}$ and $\mathbf{l}$, we get

$$\left\{ \begin{matrix} \frac{dn_x}{ds} \\ \frac{dn_y}{ds} \\ \frac{dm_z}{ds} \end{matrix} \right\} = \left\{ \begin{matrix} 0 \\ 0 \\ n_x \sin\theta - n_y \cos\theta \end{matrix} \right\} \tag{2}$$

Hence, internal force $\mathbf{n}$ is constant.

Bernoulli-Euler law establishes that the bending moment $m_z$ at a point is proportional to the curvature $\kappa$, under the assumption of a linear material-constitutive law:

$$\frac{m_z}{EI} = \frac{d\theta}{ds} = \frac{1}{\rho} = \kappa \tag{3}$$

where $E$ is elastic module, $I$ second moment of area, and $\rho$ is the radius of curvature.

If only end-point load is applied (expressed in terms of magnitude $R$ and orientation $\psi$), for $E$ and $I$ constant, considering Eq. (3), and upon substitution into Eq. (2), we get:

$$\frac{d^2\theta}{d^2 s} = \frac{1}{EI}\left[R\cos\psi\sin\theta - R\sin\psi\cos\theta\right] = \frac{R}{EI}\sin(\theta - \psi) \tag{4}$$

Introducing the geometrical relations of slope $\theta$ on the deformed curve, we get the following system of differential equations:

$$\left\{\begin{array}{c} \frac{dx}{ds} \\ \frac{dy}{ds} \\ \frac{d^2\theta}{d^2 s} \end{array}\right\} = \left\{\begin{array}{c} \cos\theta \\ \sin\theta \\ \frac{R}{EI}\sin(\theta - \psi) \end{array}\right\} \tag{5}$$

These equations can be integrated in terms of elliptic integrals to get:

$$x(\phi_i) = -\sqrt{\frac{EI}{R}}\left[c\psi\left[2E_i - 2E_1 - F_i + F_1\right] + 2k\ s\psi\left[c\phi_i - c\phi_1\right]\right]$$

$$y(\phi_i) = -\sqrt{\frac{EI}{R}}\left[s\psi\left[2E_i - 2E_1 - F_i + F_1\right] - 2k\ c\psi\left[c\phi_i - c\phi_1\right]\right] \tag{6}$$

$$\sqrt{R} = \frac{\sqrt{EI}}{L}\left[F_2 - F_1\right]$$

where $F_i = F(k, \phi_i)$ and $E_i = E(k, \phi_i)$ are the incomplete elliptic integrals of the first and second kind. These equations can be used in the following way, for given values of $k$ and $\psi$, $\phi_i$ varies continuously from $\phi_1$ to $\phi_2$ along the length of the deformed shape defining $x$ and $y$, i.e., coordinates of points on the curve.

Angles $\phi_1$ and $\phi_2$ are defined by the boundary conditions of the rod at extremes. For the rod clamped at one extreme and hinged at the other, we have:

$$\phi_1 = \arcsin\left(\frac{1}{k}\cos\left(\frac{\psi}{2}\right)\right) \tag{7}$$

and

$$\phi_2 = q\frac{\pi}{2}\ \text{ for } q = 1, 3, \ldots \tag{8}$$

where $q$ identifies the so called buckling modes. Each buckling mode has a corresponding number of inflection points, i.e., a mode 1 ($q = 1$), has 1 inflection point, mode 2 ($q = 3$) has 2 inflection points, and so on.

Hence, every set of values in the parameter-space $k$ vs. $\psi$ corresponds to a value of force applied at tip-end $R$ and a deformed shape of the rod, for the buckling mode defined in Eq. (8) as seen in Fig. 2.

The local frame attached to every flexible link $i$ of the mechanism (see Fig. 3), is placed with respect to a fixed global frame as a function of the inputs chosen for

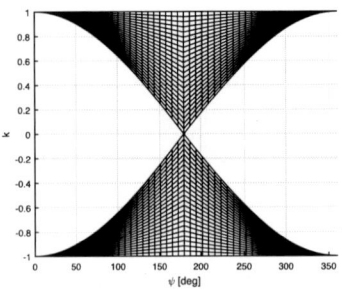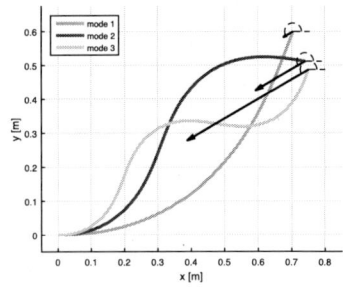

**Fig. 2.** Parameter space for flexible rod clamped-hinged and deformed shapes for a set $k, \psi$

the mechanism. Hence in the general expression for the transformation matrices of each link:

$$\mathbf{T}_i^F = \begin{bmatrix} \mathbf{R}_i^F & \mathbf{p}_{O_i} \\ \mathbf{0}^T & 1 \end{bmatrix} = \begin{bmatrix} \cos\theta_i & -\sin\theta_i & X_{O_i} \\ \sin\theta_i & \cos\theta_i & Y_{O_i} \\ 0 & 0 & 1 \end{bmatrix} \tag{9}$$

$\theta_i$ and $\mathbf{p}_{O_i}$ will depend on inputs chosen.

An static equilibrium of forces at the end-effector $\mathscr{P}$, including not only the end-forces due to the deformation of rods $R^i$ but those from the load, i.e., $\mathbf{f}_{\mathscr{P}}$, must be stated in the global fixed frame using the corresponding rotation matrices for each rod $\mathbf{R}_i^F$:

$$\sum_{i=1}^{2} \mathbf{R}_i^F \cdot \begin{Bmatrix} R_i \cos\psi_i \\ R_i \sin\psi_i \end{Bmatrix} + \mathbf{f}_{\mathscr{P}} = \mathbf{0} \tag{10}$$

The assembly of rods at the end effector $\mathscr{P}$ implies that rod-end coordinates are equal, and the local frames for each rod allow us to use directly expressions for end-coordinates given by Eq. (6). Then,

$$\mathbf{T}_1^F \begin{Bmatrix} x_1 \\ y_1 \\ 1 \end{Bmatrix} - \mathbf{T}_2^F \begin{Bmatrix} x_2 \\ y_2 \\ 1 \end{Bmatrix} = \mathbf{0} \tag{11}$$

In the case of known output coordinates for the end-effector $\mathscr{P}$, we use instead:

$$\mathbf{T}_1^F \begin{Bmatrix} x_1 \\ y_1 \\ 1 \end{Bmatrix} = \mathbf{T}_2^F \begin{Bmatrix} x_2 \\ y_2 \\ 1 \end{Bmatrix} = \mathbf{p}_{\mathscr{P}} \tag{12}$$

Forward and Inverse Kinematics problems are approached in the same way. The goal is to find the parameters of the deformation shapes of the two links, namely $(k_1, \psi_1)$ and $(k_2, \psi_2)$, that fulfill the static equilibrium given by Eq. (10)

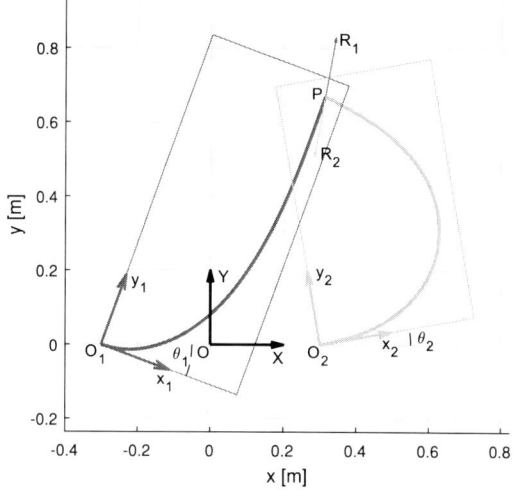

**Fig. 3.** Two rods coupled forming an 2-DOF planar Parallel Continuum Mechanism under no load

and the assembly conditions in Eq. (11) for the Forward Kinematics (FK) or Eq. (12) for the Inverse Kinematics (IK).

As observed, the FK is easier because transformation matrices are defined by given inputs, and only 2 scalar conditions apply (Eq. (11)). A numerical procedure is applied where for each set $(k_1, \psi_1)$ in its range (see Fig. 2), we get $R_1$ from Eq. (6), apply the static equilibrium in Eq. (10) to get $R_2$ and $\psi_2$, obtain $k_2$ from Eq. (6), and check the 2 residues in the assembly conditions in Eq. (11).

Multiple solutions are possible. On the one hand, there are several buckling modes for each rod that can be combined in different ways and specified in Eq. (8) when applied for each rod. On the other hand, finding $k_2$ in Eq. (6) may produce multiple solutions. Finally, the available sets $(k_1, \psi_1)$ may also produce multiple solutions for given buckling modes and single $k_2$ values.

IK needs further discussion depending on the type of inputs of the flexible mechanism. For linear inputs, local frames have a constant orientation, hence constant and known rotation matrices $\mathbf{R}_i^F$. Then, static equilibrium condition can be used as in the FK to determine $(k_2, \psi_2)$ for every prospective $(k_1, \psi_1)$. Also, when applying that the output coordinates are known in Eq. (12), it is quite simple to decouple two scalar conditions independent of the inputs $\lambda_i$ from the two conditions that produce the input values $\lambda_i$. Upon checking residues in the first two scalar conditions we can get the solution sets $(k_1, \psi_1)$, and then find the inputs $\lambda_i$ in the remaining equations. For rotational inputs, we must resort to a simultaneous finding of six unknowns, i.e., $(k_1, \psi_1), (k_2, \psi_2)$ and $\theta_i$, that produce null residues in the six scalar conditions given by Eqs. (10) and (12).

## 3    Case Study

A 2-DOF flexible mechanism in which two flexible rods are clamped to actuated prismatic joints along vertical guides and hinged together at end-effector $\mathscr{P}$ is shown in the center figure in Fig. 1. Transformation matrices to be applied to local frames are defined by constant angles $\theta_i = -\frac{\pi}{2}$, constant coordinates $X_{O_1}$ and $X_{O_2}$, and actuated input-lengths $Y_{O_1} = -\lambda_1$ and $Y_{O_2} = -\lambda_2$, see Fig. 3.

For given inputs, the FK produces multiple solutions as mentioned above. If buckling modes specified for both flexible rods are mode 1, there are inputs for which two solutions exist, as in Fig. 4 on the left, or three solutions are possible, as in Fig. 4 on the right. For a combination of modes 1 and 2 we get two solutions as in Fig. 5.

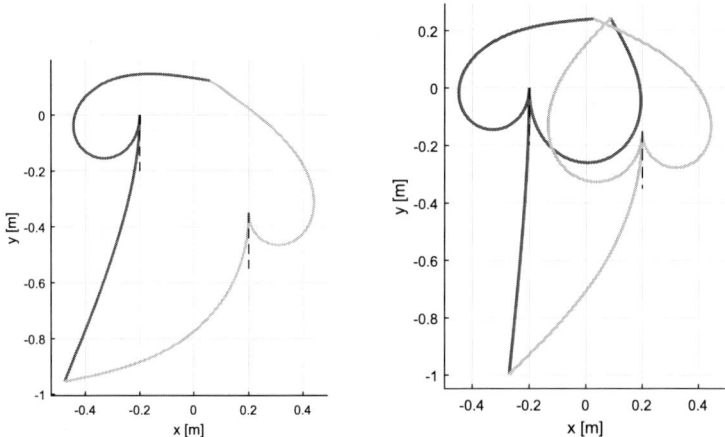

**Fig. 4.** Multiple FK solutions for buckling mode 1

In order to solve the IK, applying the expression for rotation and transformation matrices we get the conditions for static equilibrium as:

$$\begin{bmatrix} 0 & 1 \\ -1 & 0 \end{bmatrix} \begin{Bmatrix} R_1 \cos \psi_1 \\ R_1 \sin \psi_1 \end{Bmatrix} + \begin{bmatrix} 0 & 1 \\ -1 & 0 \end{bmatrix} \begin{Bmatrix} R_2 \cos \psi_2 \\ R_2 \sin \psi_2 \end{Bmatrix} + \mathbf{f}_{\mathscr{P}} = \mathbf{0} \tag{13}$$

and the output given conditions as:

$$\begin{bmatrix} 0 & 1 & -\frac{H}{2} \\ -1 & 0 & -\lambda_1 \\ 0 & 0 & 1 \end{bmatrix} \begin{Bmatrix} x_1 \\ y_1 \\ 1 \end{Bmatrix} = \begin{Bmatrix} X_{\mathscr{P}} \\ Y_{\mathscr{P}} \\ 1 \end{Bmatrix}$$

$$\begin{bmatrix} 0 & 1 & \frac{H}{2} \\ -1 & 0 & -\lambda_2 \\ 0 & 0 & 1 \end{bmatrix} \begin{Bmatrix} x_2 \\ y_2 \\ 1 \end{Bmatrix} = \begin{Bmatrix} X_{\mathscr{P}} \\ Y_{\mathscr{P}} \\ 1 \end{Bmatrix} \tag{14}$$

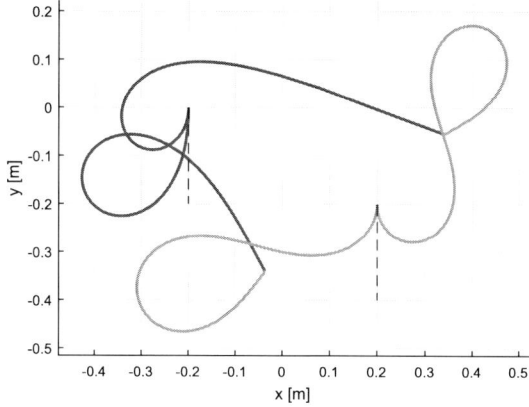

**Fig. 5.** Multiple FK solutions for buckling modes 1 and 2

The procedure to solve is: for each set $(k_1, \psi_1)$ in its range (see Fig. 2), we get $R_1$ from Eq. (6), apply the static equilibrium in Eq. (13) to get $R_2$ and $\psi_2$, obtain $k_2$ from Eq. (6), and check the 2 residues in the assembly conditions from Eq. (14), namely

$$
\begin{aligned}
y_1 - \frac{H}{2} &= X_{\mathscr{P}} \\
y_2 + \frac{H}{2} &= X_{\mathscr{P}}
\end{aligned}
\tag{15}
$$

Once the multiple solutions for the sets $(k_1, \psi_1, k_2, \psi_2)$ are obtained, we apply the other two conditions in Eq. (14) to obtain the unknown inputs $\lambda_1$ and $\lambda_2$, i.e. (Fig. 6)

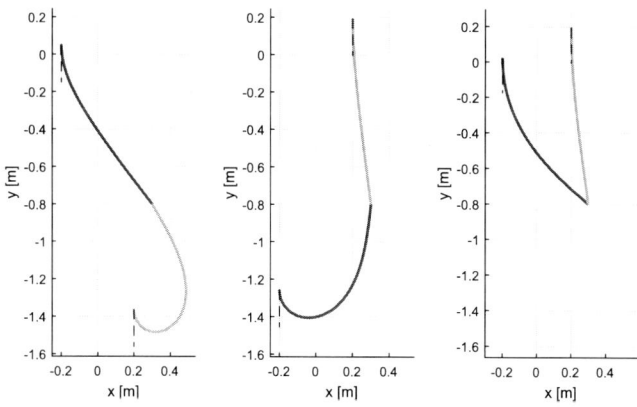

**Fig. 6.** Multiple solutions for IK position problem and for buckling mode 1

$$-x_1 - \lambda_1 = Y_{\mathscr{P}}$$
$$-x_2 - \lambda_2 = Y_{\mathscr{P}}$$

(16)

## 4    Conclusions

Forward and Inverse Kinematic position problem of 2-DOF Planar Continuum Manipulators can be solved decoupling static equilibrium forces and geometric conditions. As in the case of rigid-link mechanism, the concept of multiple solutions has also been analyzed, showing that its number is quite large.

**Acknowledgements.** Authors wish to acknowledge financial support from Spanish Government (DPI2015-64450-R), MEC PhD. grant (FPU15/04422) and Regional Government of the Basque Country (Project IT949-16).

## References

1. Bryson, C.E., Rucker, D.C.: Toward parallel continuum manipulators. In: 2014 IEEE International Conference on Robotics and Automation (ICRA), Hong Kong, pp. 778–785 (2014)
2. Howell, L.L.: Compliat Mechanisms. Wiley, New York (2001)
3. Campanile, L.F., Hasse, A.: A simple and effective solution of the elastica problem. Proc. Inst. Mech. Eng. Part C **222**(12), 2513–2516 (2008)
4. Tolou, N., Herder, J.L.: A seminalytical approach to large deflections in compliant beams under point load. Math. Probl. Eng. (2009)
5. Midha, A., Her, I., Salamon, B.A.: Methodology for Compliant Mechanisms Design: Part I - Introduction and Large-Deflection Analysis. American Society of Mechanical Design, Design Engineering Division, American Society of Mechanical Engineers (ASME) (1992)
6. Zhang, A., Chen, G.: A comprehensive elliptic integral solution to the large deflection problems of thin beams in compliant mechanisms. ASME. J. Mech. Robot. **5**(2) (2013)

# Kinematic Design of Five-Bar Parallel Robot by Kinematically Defined Performance Index for Energy Consumption

Nguyen Duc Sang[✉], Daisuke Matsuura, Yusuke Sugahara, and Yukio Takeda

Tokyo Institute of Technology, Tokyo, Japan
{nguyen.s.aa,takeda.y.aa}@m.titech.ac.jp,
{matsuura,sugahara}@mech.titech.ac.jp

**Abstract.** Evaluating parallel robots' performance is a key to achieve a better design and better understanding of their behavior in operation. In this study, a novel idea of local performance evaluation of robots taking into consideration the directional variation of performance was proposed by introducing a circular trajectory around a local output point. Total displacement of joints was also introduced as a representative kinematically-defined performance index for energy consumption of robots. The proposed index was applied to performance evaluation of five-bar parallel robots and its effectiveness was investigated based on the relationship between its value and the consumed energy obtained by dynamic simulation. Illustrative example of kinematic design of five-bar parallel robot based on the proposed index was presented and its effectiveness was performed by evaluating the energy consumption of the designed robot to achieve pick-and-place motions.

**Keywords:** Robotics · Parallel robot · Performance index · Kinematics Design · Dynamics · Energy consumption · Prescribed workspace Optimization

## 1 Introduction

In recent decades, parallel robots have been widely used in light-weight industry due to their obvious advantages over their counterparts of serial robots. The robots continuously work and their life expectance is supposed of 12–15 years [1]. Therefore, design methodology to reduce energy expenditure of such robots should be thoroughly investigated and several approaches have been proposed so far.

Barreto [1] used a concept of natural motion by attaching springs to active joints in order to support mechanism's movement without need of energy during pick-and-place operations. Pellicciari [2] carried out optimization with respect to energy consumption for trajectories by means of constant time scaling. Brinker performed an integrated design of Delta Parallel Robots based on the kinematic and dynamic performance indices [17]. Some other researchers studied motion/force transmissibility [3–6].

Although the above-mentioned methods are considerably useful, some of them require adding device or only investigate mechanisms from point of view of geometry.

© Springer Nature Switzerland AG 2019
B. Corves et al. (Eds.): EuCoMeS 2018, MMS 59, pp. 239–247, 2019.
https://doi.org/10.1007/978-3-319-98020-1_28

In the definition of local performance index of robots, we usually have to consider their directional variation of the performance. Based on such a local performance index, global performance in the prescribed workspace is evaluated in the kinematic design of robots. Though execution of dynamic performance evaluation of robots such as power and energy consumption as well as acceleration capability and dexterity in their early design stage is desirable, its problem setting and solution procedure are quite complicated. Then, globally optimal solution is not easy to be obtained. To overcome these problems, in this paper, we propose a concept of performance evaluation at a local output point taking into consideration the directional variation of robots' performance. In our idea, a circular trajectory with a small radius around a specified output point is considered as a target trajectory for evaluation. By calculating and evaluating performances such as power consumption along the circular trajectory, local performance evaluation taking into consideration the directional variation can be performed. On the other hand, as a representative kinematically defined measure for energy consumption of robots at a local output point, total displacement of joints of robots carrying out the circular trajectory is introduced. The effectiveness of this index is discussed based on the relationship between the power consumption and the total displacement of joints. Utilizing this idea, kinematic synthesis of five-bar parallel robot is demonstrated for a prescribed rectangular workspace and the obtained design is promise of better performance with less energy consumption. Throughout the paper, we deal with a planar five-bar parallel robot shown in Fig. 1.

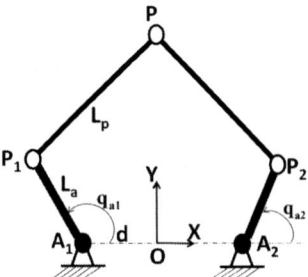

**Fig. 1.** Five-bar parallel robot

## 2 Total Displacements of Joints Along a Circular Trajectory

### 2.1 Inverse Kinematics and Dynamics

In order to calculate power consumption for a given trajectory, we need to first solve inverse kinematics and dynamics to find velocity and actuation torque, respectively. We first introduce kinematics and dynamics briefly about five-bar parallel robot. More details can be found in [7, 8].

We use O-*xy* as the global coordinate. The position of point P and $P_i$ ($i$ = 1, 2) can be written as:

$$\mathbf{p} = [x\,y]^{T}; \; \mathbf{p}_1 = [L_a\cos(q_{a1}) - d \quad L_a\sin(q_{a1})]^{T};$$
$$\mathbf{p}_2 = [L_a\cos(q_{a2}) + d \quad L_a\sin(q_{a2})]^{T}; \tag{1}$$

Then the inverse kinematic problem can be solved by using the following constraint equation: $|PP_i| = L_p$. Expanding this constraint, we could get:

$$\begin{cases} (x - L_a\cos(q_{a1}) + d)^2 + (y - L_a\sin(q_{a1}))^2 = L_p^2 \\ (x - L_a\cos(q_{a2}) - d)^2 + (y - L_a\sin(q_{a2}))^2 = L_p^2 \end{cases} \tag{2}$$

Now, if the position of output point P is known, the input angle can be obtained by

$$q_{ai} = 2\tan^{-1}(d_i) \tag{3}$$

where

$$d_i = \frac{-b_i \pm \sqrt{b_i^2 - 4a_i c_i}}{2a_i}; \quad \begin{aligned} a_i &= L_a^2 + y^2 + (x \pm d)^2 - L_p^2 \pm 2(x \pm d)L_a \\ b_i &= -4yL_a; \; c_i = L_a^2 + y^2 + (x \pm d)^2 - L_p^2 \pm 2(x \pm d)L_a \end{aligned} \tag{4}$$

The dynamic equation for a parallel robot can be expressed as

$$\mathbf{M}(\mathbf{q})\ddot{\mathbf{q}} + \mathbf{C}(\mathbf{q}, \dot{\mathbf{q}})\dot{\mathbf{q}} + \mathbf{G}(\mathbf{q}) = \tau \tag{5}$$

Where $\mathbf{M}(\mathbf{q}), \mathbf{C}(\mathbf{q}, \dot{\mathbf{q}}), \mathbf{G}(\mathbf{q})$ are the mass matrix, the coriolis and centrifugal force matrices and gravity vector, respectively. $\tau$ is the actuation torque vector. $\mathbf{q}, \dot{\mathbf{q}}, \ddot{\mathbf{q}}$ are the position, velocity and acceleration vectors of joints.

## 2.2 Average Power Along a Circular Trajectory

As a manner to quantify the energy consumption of a robot, the average power of the robot during completing a circular trajectory is considered and it is approximated using the expression below.

$$\overline{P} = \frac{1}{N+1}\sum_{i=0}^{N} |\tau_i \dot{q}_i| \tag{6}$$

In Eq. (6), $N$ is the number of all sampled points along a trajectory, $\tau_i, \dot{q}_i$ are the torque and angular velocity of an actuated joint at each instant. For each point $P(x_0, y_0)$ in the workspace, we can generate a circular trajectory as follows (see Fig. 2):

$$\begin{cases} x = x_0 + a\cos(\omega t) \\ y = y_0 + a\sin(\omega t) \end{cases}$$

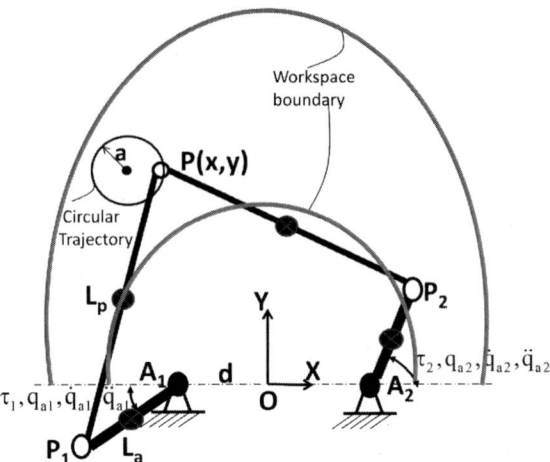

**Fig. 2.** Circular trajectory at a local point in workspace

We assume the average velocity to complete a trajectory to be constant (in this paper, we applied 1 (m/s), $\bar{v} = \frac{2\pi a}{T} = \omega a = 1$ (m/s). When the point P traces the whole circle of the trajectory mentioned above, the output point moves in all directions. Therefore, if we evaluate the performance of the robot at each point along the circular trajectory and summarize the result by setting the radius $a$ a sufficiently small value compared to robot's size, performance evaluation of a robot at a specified point $(x_0, y_0)$ can be enabled to include the directional variation of its performance.

## 2.3  Total Displacement of Joints (TDJ)

When the end-effector of a robot completes a circular trajectory mentioned above, the magnitude of the input displacement differs according to the position of end-effector in the workspace. And, if this total displacement of joints becomes big, the power consumption might become large also. This is understandable because at the same time, if the links have to travel longer, the input velocity and acceleration will be increased, that leads to the growth of power of the actuators. The total displacement of each input joint for a circular trajectory at a point is calculated as shown in Fig. 3.

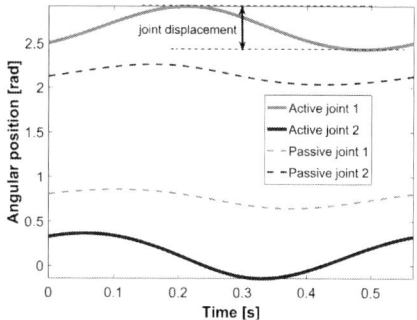

**Fig. 3.** Joint displacements for a circular trajectory

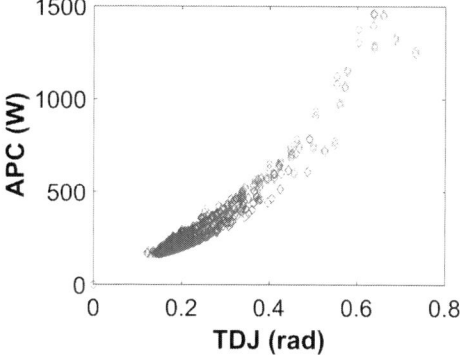

**Fig. 4.** APC vs. TDJ for $a = 0.01$ m

The relationship between power consumption calculated from inverse dynamics and total displacement of joints calculated from inverse kinematics is investigated. Figure 4 shows the relationship between them. In this figure, the vertical axis APC is the average power consumption along the circular trajectory and the horizontal axis TDJ is calculated as the sum of the total displacement of all input joints shown in Fig. 3. From Fig. 4, we can see the linear trend between these indices that shows the effectiveness of the kinematically defined index TDJ as the index used for evaluating the power consumption.

## 3 Kinematic Design of Five-Bar Parallel Robot

### 3.1 Prescribed Workspace

The robot often works within a prescribed workspace of rectangular shape [9], and we intend to synthesize a five-bar parallel robot applying TDJ to clarify the kinematic parameters resulting in small energy consumption when the robot operates there. We assume the size of prescribed workspace is $b \times h = 1 \times 0.25$ (m). We assume as the total length of proximal and distal links $L_a + L_p = 1.5$ (m), the mass of End-effector $m_p = 1$ (kg) and, the total mass of 2 proximal and distal links $m_a + m_p = 2$ (kg). It is reasonable that the mass density (for per length unit) of proximal link is often larger than of distal link, therefore, we assume: $\delta_a = 3\delta_p$.

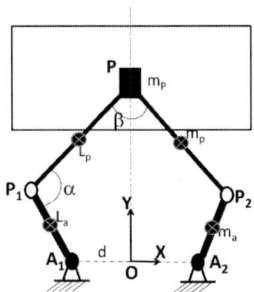

**Fig. 5.** The prescribed workspace

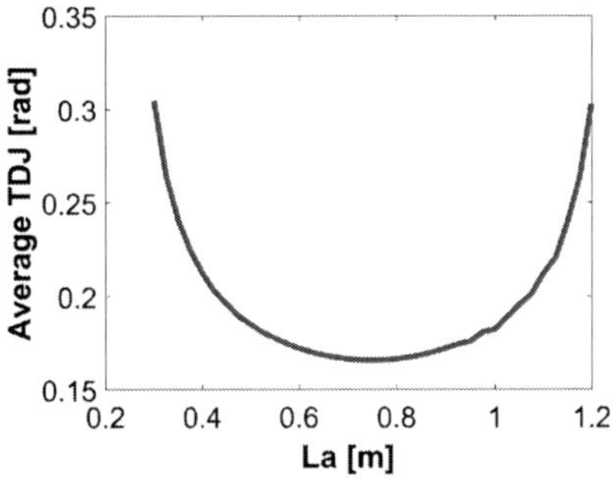

**Fig. 6.** Average TDJ vs. $L_a$

In the next step, we determine the center point of the rectangular workspace. Due to the symmetric configuration of robot structure and workspace, the center was set on the $y$ axis as shown in Fig. 5. In addition, to avoid the singularity and its neighborhood, the center should be far away from them as much as possible. For five-bar robot, two transmission angles $\alpha, \beta$ in Fig. 5 can be used to evaluate the closeness to singularity [7]. If $\alpha, \beta = 0°$ or $180°$, the mechanism is in either input or output singularity. The center of the rectangular workspace therefore was determined to satisfy: $\sin \alpha \sin \beta \to \max$.

The dimensional synthesis problem here can be defined as:

*Minimize: Total displacement of joints for circular trajectories within the prescribed rectangular workspace $b \times h = 1 \times 0.25$ (m).*

*Subject to*: $m_a + m_p = 2$ kg, $L_a + L_p = 1.5$ m ($\delta_a = m_a/L_a = 3\delta_p = 3m_p/L_p$).

Figure 6 shows the relationship between $L_a$ and TDJ, from which we can see that $L_a = L_p = 0.75$ m is the optimal solution (Fig. 7).

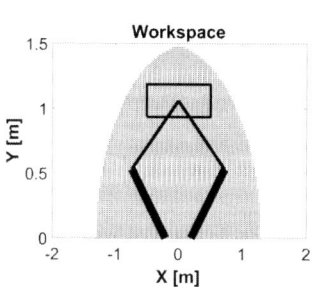

**Fig. 7.** The workspace with $L_a = 0.75$ m

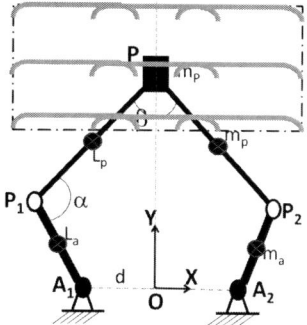

**Fig. 8.** 9 Pick-and-place trajectories

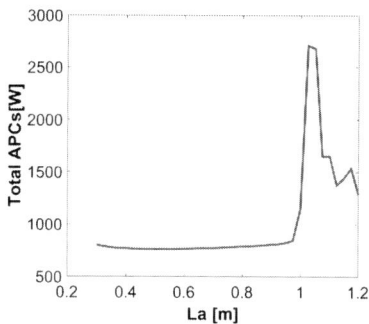

**Fig. 9.** Total average power consumption for 9 PPOs with relation to $L_a$

### 3.2 Evaluation of Energy Consumption for Pick-and-Place Trajectory

In order to verify the optimal design above, we calculate the power consumption for several pick-and-place trajectories (PPO) [10] within the prescribed workspace. Without loss of generality, 9 PPOs equally distributed in the prescribed workspace as shown in Fig. 8 were considered. Figure 9 shows the relationship between $L_a$ and the total average power consumption for 9 PPOs, from which, we can see that the optimal solution $L_a = 0.75$ m obtained by using TDJ as the kinematically defined evaluation index for energy consumption leads to a small energy consumption. Let us note that the power consumption when $L_a \geq 1$ m becomes particularly high because the robot configuration is close to singularity.

## 4 Conclusion

In this paper, in order to enable dimensional synthesis of parallel robots with less energy consumption in the prescribed workspace taking into consideration the directional variation of performance, a novel idea to introduce a circular trajectory around a target point was proposed. Total displacement of joints was also introduced as a representative kinematically-defined performance index for energy consumption of robots. The effectiveness of the proposed idea and index was validated through simulations of a five-bar parallel robot taking into consideration the dynamics of robots. However, in this paper, the design of robot was mainly focused on pick and place operation application and the dynamic model used is still quite simple compared to practical situation. In addition, the present index has similar feature with kinematic isotropy index in [18] and is necessary to make comparison. These problems are left to be addressed in our future work.

# References

1. Barreto, J.P., Scholer, F.J.-F., Corves, B.: The concept of natural motion for pick and place operations. In: New Advances in Mechanisms, Mechanical Transmissions and Robotics, Mechanisms and Machine Science, vol. 46, pp. 89–98. Springer (2017)
2. Pellicciari, M., Berselli, G., Leali, F., Vergnano, A.: A method for reducing the energy consumption of pick-and-place industrial robots. Mechatronics **23**(3), 326–334 (2013)
3. Takeda, Y., Funabashi, H.: Motion transmissibility of in-parallel actuated manipulators. JSME Int. J. Ser. C **38**(4), 749–755 (1995)
4. Takeda, Y., Funabashi, H., Ichimaru, H.: Development of spatial in-parallel actuated manipulators with six degrees of freedom with high motion transmissibility. JSME Int. J. **40** (2), 299–308 (1997)
5. Wang, J., Wu, C., Liu, X.J.: Performance evaluation of parallel manipulators: motion/force transmissibility and its index. Mech. Mach. Theory **45**(10), 1462–1476 (2010)
6. Chen, Y., Liu, X.J., Chen, X.: Dimension optimization of a planar 3-RRR parallel manipulator considering motion and force transmissibility. In: International Conference on Mechatronics and Automation, pp. 670–675 (2013)
7. Liu, X.J., Wang, J., Pritschow, G.: Kinematics, singularity and workspace of planar 5R symmetrical parallel mechanisms. Mech. Mach. Theory **41**(2), 145–169 (2006)
8. Huang, T., Mei, J., Li, Z., Zhao, X., Chetwynd, D.G.: A method for estimating servomotor parameters of a parallel robot for rapid pick-and-place operations. ASME J. Mech. Des. **127** (4), 596–601 (2005)
9. Huang, T., Liu, S., Mei, J., Chetwynd, D.G.: Optimal design of a 2-DOF pick-and-place parallel robot using dynamic performance indices and angular constraints. Mech. Mach. Theory **70**, 246–253 (2013)
10. Xie, Z., Wu, P., Ren, P.: A comparative study on the pick-and-place trajectories for a delta robot. In: ASME Proceedings 40th Mechanisms and Robotics Conference, vol. 5A, pp. V05AT07A040 (2016)
11. Tsai, L.-W.: Robot Analysis: The Mechanics of Serial and Parallel Manipulators. A Wiley-Interscience Publication, Hoboken (1999)
12. Merlet, J.-P.: Parallel Robots. Springer, Cham (2006)
13. Prempraneerach, P.: Workspace and dynamic trajectory tracking of delta parallel robot. In: International Computer Science and Engineering Conference, pp. 469–474 (2014)
14. Khorasani, A., Gholami, S., Taghirad, H.D.: Optimization of KNTU delta robot for pick and place application. In: International Conference on Robotics and Mechatronics, pp. 127–132 (2015)

15. Wu, G., Bai, S., Hjornet, P.: Architecture optimization of a parallel Schonfiles-motion robot for pick-and-place applications in a predefined workspace. Mech. Mach. Theory **106**, 148–165 (2016)
16. Arai, T., Funabashi, H., Nakamura, Y., Takeda, Y., Koseki, Y.: High speed and high precision parallel mechanism. In: IEEE, Proceedings IROS, pp. 1624–1629 (1997)
17. Brinker, J., Corves, B., Takeda, Y.: Kinematic and dynamic dimensional synthesis of extended delta parallel robots. In: Proceedings ISRM 2017, International Symposium on Robotics and Mechatronics, Sydney, pp. 1–8, December 2017
18. Gosselin, C., Angeles, J.: The optimum kinematic design of a planar three-degree-of-freedom parallel manipulator. J. Mech. Transm. Autom. Des. ASME **110**, 35–41 (1988)

# Free Computer Algebra Software and Its Application on Calculative and Graphic Tasks in TMM Course of Bauman University

Kirill Kuprianoff[1]([✉]), Christina Shutova[1], and Andrei Vukolov[2]([✉])

[1] Energetic Engineering Faculty, Bauman Moscow State Technical University,
Moscow, Russian Federation
bdf-1@mail.ru
[2] Institute of Modern Educational Technologies, Bauman Moscow State Technical
University, Moscow, Russian Federation
andrei.vukolov@gmail.com

**Abstract.** Completing the calculation-graphical task at the Theory of machines and mechanisms (TMM) course is one of the most labor-consuming parts of any engineering or scientific project. In Bauman Moscow State Technical University (BMSTU) course projects, including on TMM, are accompanied by a large amount of calculations, which are performed by students using various calculation programs. This paper provides examples of the use of free software (in particular, the Maxima system) and its advantages over widespread proprietary solutions (Mathcad).

**Keywords:** Free software · Open source · TMM
Technical education · Text design documentation · Maxima
Mathcad · Software licensing

## 1 Introduction

Among the mathematical software for symbolic calculations, the most widely known are commercial ones (Maple, Mathematica, Mathcad); this is a very powerful tool for the student, allowing him to automate the part of the work that requires special attention, operating at the same time with an analytical record of the data, i.e., in fact mathematical formulas. Such program can be called a programming environment, with the difference that the usual mathematical notations are used as elements of the programming language. These proprietary solutions have a number of drawbacks [4, 13]: the difficulty of using these software with a variable hardware environment; if several software of a similar type are used, there is often a compatibility problem due to the need for regular updates of operating systems, licenses and hardware, the lack of the ability to deploy and use documentation and/or support software developed by students on their

© Springer Nature Switzerland AG 2019
B. Corves et al. (Eds.): EuCoMeS 2018, MMS 59, pp. 248–255, 2019.
https://doi.org/10.1007/978-3-319-98020-1_29

own or third-party university machines [3,5,9,10,15]. Free software is actively developed all over the world, including the BMSTU [2,12].

The program, which became the topic of the article, works on the same principles and provides similar functionality; the most radical difference is that it is neither commercial nor closed. The main direction where such programs are in demand is higher education. The use of free software for educational needs is a real opportunity for both the university and students and teachers to have legal copies of such software at their disposal without any monetary costs. This solution has a free license and therefore also allows the user to develop and distribute their own programs, documentation and other solutions and products based on them, as well as to place texts, program code and/or executable files in the public domain with strictly defined and limited requirements [11].

## 2  Opportunities of the Computer Algebra System Maxima

In addition to all the basic computing and computing solutions that proprietary software (MathCad) has, maxima has additional capabilities that increase the efficiency of the work.

1. Parametrization of solutions, which makes them universal, makes it possible to identify analytical dependencies between the initial parameters, is a direct way to developing libraries for large engineering tasks.
2. Maxima is able to import/export text into LaTeX and XML/HTML. In fact, the formulas are displayed in the usual mathematical notation, but they can be edited and copied to other documents like ordinary text, which further simplifies the creation of text documentation for engineering calculations.

A distinctive feature of Mathcad from most other modern mathematical applications is its construction on the principle of WYSIWYG ("What You See Is What You Get"). Therefore, it is very easy to use, in particular, because there is no need to first write a program that implements certain mathematical calculations, and then run it for execution. Instead, simply enter mathematical expressions using the built-in formula editor, and in the form closest to the standard, and immediately get the result. For effective work with Mathcad's editor is enough basic user skills. Mathcad's listing has the ability to present the results in a finished form.

Despite this, the graphical representation is limited to a set of formulas, the built-in Greek language and support for text characters, which is often not enough for the standards of textual engineer documents, such as an explanatory note and etc. In this case, Mathcad does not have the ability to output data to a file and in this sense is an isolated system, which makes the user's capabilities largely limited. Maxima is a programming language in which the entire calculation part is written first and then the user receives a graphic output. Data of the executed calculation can always be written to a file in a convenient form for further integration into any of the editors, whose capabilities are much more extensive than the built-in Mathcad editor. The downside is that new users will have to learn the programming language which Maxima supports, but this disadvantage is rather small in comparison with possibilities that Maxima is giving for calculations and data presentation.

## 3  Usable Features: Cam Mechanism Synthesis

Maxima input language is a superset based on Common Lisp development environment. It is a functional programming language, which represents the program as linear lists of symbols. The simplest program in both Maxima and Common Lisp looks like:

```
1       a:1 ;
2       b:2 ; - defining global variables
3       a+b ; - defining the function
```

In the course of work, a program was created to synthesize the profile of a cam mechanism with a flat translational follower according to user input. In the first stage, the calculation of follower motion law is performed, an array $A$ consisting of pieces of the graph of initial motion law and an array of coordinates of the stitching points of these pieces of $X$. (Fig. 1).

```
A:[a,
  −a,
  0,
  −b·a,
  b·a
];

X: makelist(concat(x,i),i,length(A));
x1:fu/2;
x2:fu/2;
x3: 0;
x4:fs/2;
x5:fs/2;
```

**Fig. 1.** Listing of creating of the simple Maxima arrays with built-in functions

To create universal solution, we use parametrization (parameters $a$ and $b$ within the original array $A$), which allows our program to work on the output not only with numerical values, but also operate with symbols that can be visually used to detect analytic dependencies between parameters, $b$ is responsible for determining the ratio of the approach and removal angles of the follower (an additional parameter $c$ is introduced with respect to which an analytical solution is obtained, Fig. 2).

Prior to working with the main program, functions were written with help of which the source arrays are processed. The function $sh2(el1, el2, xsh)$ does the stitching of pieces of the graph $(el1, el2)$ at the given points $xsh$.

```
1      sh2(el1,el2,xsh):=
2      block([f1: el1,f2: el2,xsh: xsh],
3      C: ev(f1,x=xsh)-ev(f2,x=xsh),
4      f2: f2+C
5      );
```

The integration function $ishfun(A, X)$ and differentiation $dshfun(A, X)$ calculate the necessary initial parameters $aq$, $vq$, $s$.

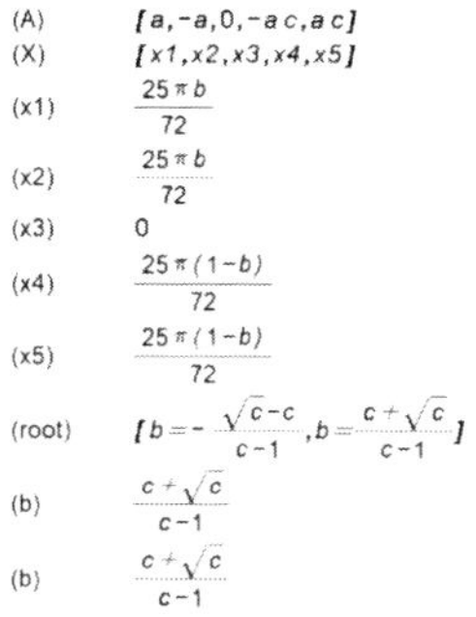

| (A) | $[a,-a,0,-ac,ac]$ |
|---|---|
| (X) | $[x1,x2,x3,x4,x5]$ |
| (x1) | $\dfrac{25\pi b}{72}$ |
| (x2) | $\dfrac{25\pi b}{72}$ |
| (x3) | $0$ |
| (x4) | $\dfrac{25\pi(1-b)}{72}$ |
| (x5) | $\dfrac{25\pi(1-b)}{72}$ |
| (root) | $[b=-\dfrac{\sqrt{c-c}}{c-1}, b=\dfrac{c+\sqrt{c}}{c-1}]$ |
| (b) | $\dfrac{c+\sqrt{c}}{c-1}$ |
| (b) | $\dfrac{c+\sqrt{c}}{c-1}$ |

**Fig. 2.** Obtaining of parametrized solution in Maxima

```
1      ishfun(A,X):=
2      block([A: A,X: X],
3      xk: 0,
4      A[1]: integrate(A[1],x),
5      for i:1 thru length(A)-1 do
6      [ A[i+1]: integrate(A[i+1],x),
7      xk: xk+X[i],
8      A[i+1]: sh2(A[i],A[i+1],xk) ]
9      );
```

Working with functions is a big advantage of Maxima:

- The function can be called from various places in the program, thus avoiding the repetition of code snippets.
- You can use the same function in different programs.
- Functions increase modularity of the program and facilitate its design.
- The use of functions facilitates the reading and understanding of the program and speeds up the search and correction of errors.
- Ability to create libraries from sets of functions for solving specific engineering tasks.

For example, Mathcad (Fig. 3) does not provide the possibility of creating functions at core level. It is limited with the built-in library of standard functions. Also the possibility of parametrization is extremely inconvenient and the free output of symbolic expressions is limited with locked integration of Maple V core functions.

One of the main stages of designing the cam mechanism [6] is to determine the coordinates of cam profile which must being represented as drawing in the course project. Main advantage of Maxima here is the ability to create a file in which you can put an array of output data (in our case, coordinates of the points as shown on figures). This will allow us to work with the obtained values regardless of the program and simplify the development of engineering documentation (for example, it is possible to plot this graph in LaTeX environment using PGFplots [7,13]).

In the MathCad program, the output of data to a separate file is not possible, which makes it difficult to use the data received from program for presentation in documentation (Fig. 4).

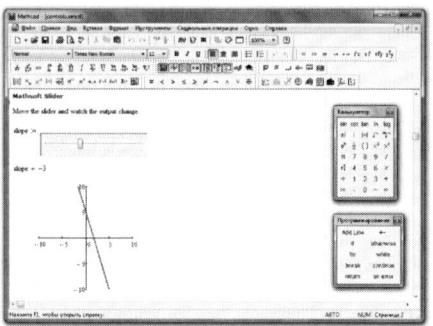

**Fig. 3.** Mathcad interface

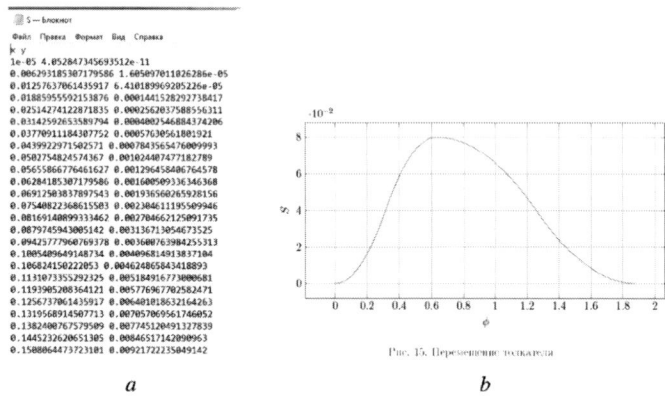

a                                          b

**Fig. 4.** Output of point coordinates from Maxima and their representation in LaTeX

## 4    Working with Maxima for Students

At first glance it may seem that it is a problem for a student to learn the Maxima programming language - CommonLisp, but this is not right. The language uses the most simple constructions of the assignment type, and any mathematical

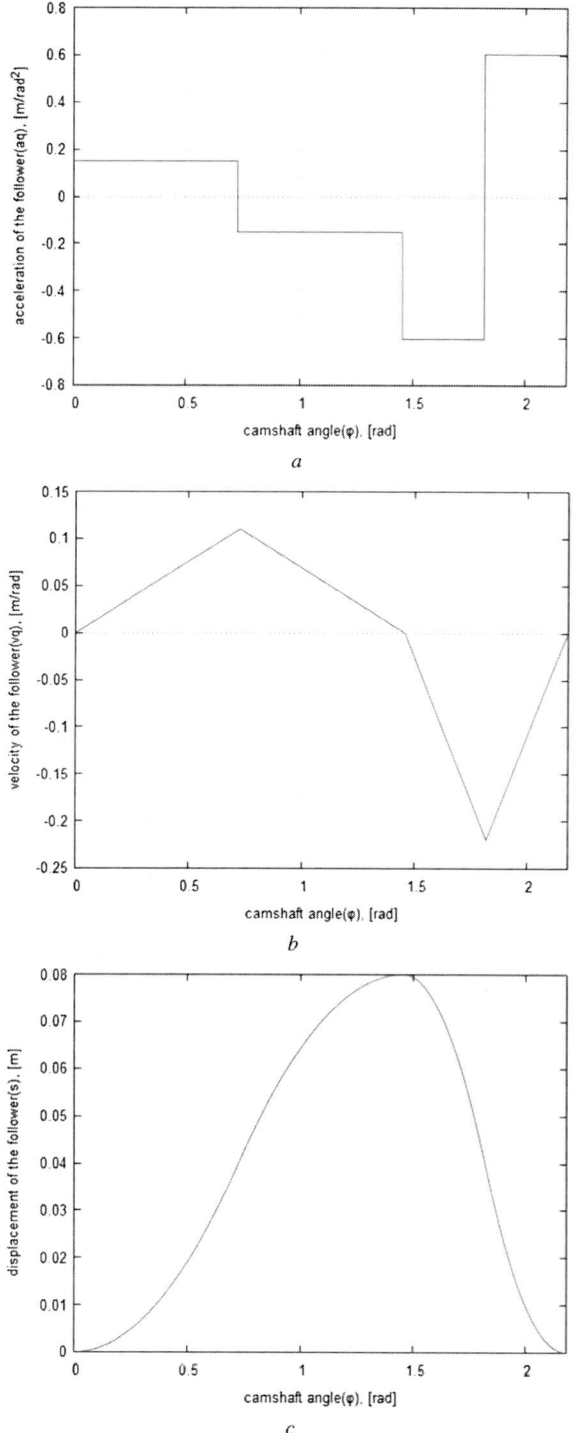

**Fig. 5.** Cam mechanism motion law calculated with Maxima

action is performed by an intuitive-understand built-in function, i.e. all learning process comes down to finding the required mathematical functions in the manual. The absence of the usual graphic symbols is easily leveled out by the fact that when you learn all the necessary functions, the productivity of the work is increased due to the simplicity of the set of these functions (including those that can be written by the user himself).

For example the set of functions was developed to transform standard cam motion law into Maxima array of floats. The pictures drawn by GNUplot within Maxima graphical interface wxMaxima are presented on Fig. 5.

It can be seen from the stages described above that in case of several students working on the same project, the ability to create exportable functions and exchange them with co-workers can speed up the development process. Moreover the students can create libraries of functions for specific tasks, develop them independently and use them for studies [14]. As an addition this work requires creation of readable documentation because the developed solution would be used by people who did not being involved into development.

## 5   Conclusion

The described features make Maxima almost universal tool for solving calculation-graphic tasks. Using the example of a typical TMM course project, Maxima can:

- Accelerate the process of solving;
- Optimize the documenting process;
- Statically analyze the obtained solution, including finding errors.

Studying Maxima makes it possible to expand the range of tools used by students to solve mechanical tasks, and also improves programming skills, that is necessary in the future. Thus, Maxima can be noted as one of the most effective programs for solving computational and graphic problems, especially in connection with LaTeX. There are also many analogs of free programs of the maxima type created for solving technical tasks (Scilab, GNU Octave), working similarly in numeric mode [1,8] and having a similar set of capabilities, and, as a result, same advantages over proprietary solutions.

**Acknowledgements.** Authors want to acknowledge Ph.D., doc. Lyudmilla Chernaya for her advice about computer algebra systems.

## References

1. Alexeev, E.R., Chesnokova, O.V., Rudchenko, E.A.: SCILAB: Solving of Engineering and Mathematical Tasks (in Russian). ALTLinux Library. ALTLinux; BINOM. Laboratory of Knowledge Management (2008)
2. Belonozhko, P.P., Belous, V.V., Karpenko, A.P., Khramov, D.A.: Instruments for automated numeric estimation of trainees' meta competences (in Russian). Sci. Educ.: Electron. Sci. J. (10) (2015). http://technomag.bmstu.ru/doc/821623.html

3. Corral Abad, E., Gómez-García, M.J., Ruiz Blázquez, R., Bustos-Caballero, A., García-Prada, J.C.: Using an android app for teaching mechanism and machine theory. In: Castejón, C., García-Prada, J.C., et al. (eds.) Proceedings of ISEMMS-2017 the 2nd International Symposium on the Education in Mechanism and Machine Science. Universidad Carlos III de Madrid, Madrid. Springer (2018). Preprint edn

4. Ghosh, R., Glott, R., Krieger, B., Robles, G.: Free/libre and open source software: survey and study. Technical report, International Institute of Infonomics, University of Maastricht (2002)

5. Kovalenko, V.A.: State of deployment analysis of free software in institutions of the educational system in russia (in Russian). Pedagogical Educ. Russia **6**, 188–192 (2013)

6. Leonov, I.V., Baryshnikova, O.O., Kuzenkov, V.V., Sinitsin, V.V., Tarabarin, V.B.: Usage of Mathcad system while developing of course projects and hometasks of theory of mechanisms and machines (in Russian). Bauman Moscow State Technical University, Moscow (2004)

7. Norserium: Plotting graphs in LaTeX—PGFPlots (in Russian) (2015). https://habrahabr.ru/post/250997/

8. O'Hara, K., Kay, J.: Open source software and computer science education. Technical report, Rowan University, Department of Computer Science (2003)

9. Pires, P.S.M., Rogers, D.A.: Free/open source software: an alternative for engineering students. In: Proceedings of 32th ASEE/IEEE Frontiers in Education Conference (2002)

10. Prokurat, G., Vukolov, A., Strukova, A., Egorova, O.: Generated graphics and game development software in engineering education: perspectives and experience of usage. In: Castejón, C., García-Prada, J.C., et al. (eds.) Proceedings of ISEMMS-2017 the 2nd International Symposium on the Education in Mechanism and Machine Science. Universidad Carlos III de Madrid, Madrid. Springer (2018). Preprint edn

11. Roberts, J.A., Hann, I.H., Slaughter, S.A.: Understanding the motivations, participation, and performance of open source software developers: a longitudinal study of the Apache projects. Manag. Sci. **52**(7), 984–999 (2006). https://doi.org/10.1287/mnsc.1060.0554

12. Sorokin, M.O., Khoteev, S.D.: Free instruments for organization of information systems development in building (in Russian). Youth Her. Tech. Sci. (1) (2013). http://sntbul.bmstu.ru/doc/532852.html

13. Titov, A., Vukolov, A.: Free and open source software for technical texts editing, its advantages and experience of usage on TMM training in Bauman University. In: Castejón, C., García-Prada, J.C., et al. (eds.) Proceedings of ISEMMS-2017 the 2nd International Symposium on the Education in Mechanism and Machine Science. Universidad Carlos III de Madrid, Madrid. Springer (2018). Preprint edn

14. Vorotnikov, S., Pushkin, A., Saschenko, D.: Experience of creation of distance training system for robotics in Bauman Moscow state technical university (in Russian). In: Proceedings of VIII International Conference "Vibration Machines and Technologies", Kursk, pp. 267–274 (2008)

15. Vukolov, A.: Free and Open Source Software Applications for Education of TMM Discipline in Bauman University, pp. 253–260. Springer, Cham (2017). https://doi.org/10.1007/978-3-319-44156-6_26

# Similitude of Scaled and Full Scale Linkages

Simon Laudahn[1(✉)], Magnus Sviberg[2], Lukas Wiesenfeld[2], Franz Haberl[2],
Johannes Haidl[2], Kassim Abdul-Sater[1], and Franz Irlinger[1]

[1] Technical University of Munich, Munich, Germany
{simon.laudahn,kassim.abdul-sater,irlinger}@tum.de
[2] Webasto Convertibles GmbH, Hengersberg, Germany

**Abstract.** This paper studies approaches for determining an optimal scaling factor for the building of scaled linkage prototypes that take similar deformation behavior to the original linkage into account. Beam theory is used for determining scaling laws with neglecting transversal contraction. A more general approach is shown, using a three-dimensional linear-elastic case. Both approaches take weight and external loads into account. It is shown that an optimal scaling factor depends on the material properties of model and original. FEM is used for validation. According to a FEM validation the scaling factor obtained by the beam theory approach was better suited for our example.

**Keywords:** Model linkages · Similitude · Scaling factors
Mechanism prototyping

## 1 Introduction

Linkages or mechanisms are used in different applications like for example pick and place, medical systems, furnitures, or automotive. For the design of linkages synthesis and analysis procedures are used. Those methods are supported by CAD, FEM and Multibody Simulation. However before building a prototype of a new kinematic device it is often desired to build a simpler functional model first. Building a model can be faster and at a lower cost, compared to building a real prototype of original size. These models are often scaled and made out of a different material than their original counterpart.

### 1.1 Kinematic Models

Model building of mechanisms dates back at least as far as about 1830 when models (Schubert-Models) build out of cedar wood were used for the teaching of engineers [10]. Similarly the Reuleaux-Models built out of metal and dating back to about 1890 were used and can still be found at some universities. In modern days kinematic models are often produced with additive manufacturing.

B. Corves et al. (Eds.): EuCoMeS 2018, MMS 59, pp. 256–264, 2019.
https://doi.org/10.1007/978-3-319-98020-1_30

Examples are Gosselins and Hamels Agile Eye (cf. [5]) or a 1:1-model of a convertible car roof linkage (cf. [7]). Both were made with selective laser sintering of polyamide PA 12. At the Chair of Micro Technology and Medical Device Technology at TUM selective laser sintering of polyamide as well as laser cutting of wood plates are also used for building scaled kinematic models. Examples are a 1:4 scaled polyamide model of a chair with integrated lifting mechanism [4] and a model of a spatial car door hinge that was carried out as a 1:5 model [1] as well as a full scale hybrid prototype made of a combination of steel parts and laser sintered polyamide parts [11]. All these models are used for the evaluation of functionality and concepts. According to [7] kinematic functional models give the user a haptic impression of a linkage's motion.

## 1.2 Similitude Theory

There are methods of dimensional analysis like Buckingham's *pi theorem* [3] that are suited to find dimensionless numbers that can be used to determine the quality of a model system for specific predictions about the corresponding original system. For example the Reynolds number or the Prandtl number are commonly used in the fields of fluid dynamics and thermodynamics to decide on the applicability of a model design. From these numbers laws are derived about how different physical dimensions have to be scaled in order for the model to work [13].

There are examples for the application of similitude in the field of structural mechanics as well. In [15] similitude observations of laminated plates and shells are made for aerospace applications, and also a historical overview about structural similitude is provided. In recent years [6] used FEM and analysis of frequency response functions to study the similitude of static and dynamic responses between scaled and full sized models. As an example they used an aircraft wing. [14] used equations obtained by dimensional analysis to predict the stress and displacement for scaled models of large structural parts of derricks. They used the original material properties for the model. Using parts of wind turbines as an application [2] worked with different material properties to build models for composite I-beams.

## 1.3 Drawbacks on Kinematic Models

Although the above mentioned kinds of kinematic models make the motion of a mechanism tangible, other factors like stiffness or weight of the mechanism's links are usually neglected. The display of the real deformation behavior is not subject of the observation. That means for example that it can not be expected from a polyamide model with an arbitrary scaling factor $n$ of a full scale prototype made of steel to show a similar scaled deformation behavior when exposed to a (scaled) external load. This is of course due to different material properties of steel and polyamide (e.g. elastic modulus, density, ...) and due to different scaling of geometric properties such as length ($\propto n$), cross-section area ($\propto n^2$, affecting e.g. section modulus) and volume ($\propto n^3$, affecting weight).

## 1.4  Task Description and Expected Advantages

In our project we want to increase the informative value of kinematic models, by taking deformational aspects into account. Deformations occur in consequence to internal stresses within the part material, which originate from dynamic and static loads on the mechanism. During the movement of a mechanism the direction and magnitude of these loads vary over time. Therefore the possible stress conditions within mechanism parts are manifold. By analyzing the relations of original mechanisms and their scaled models the rules of scaling of geometric dimensions and loads is to be determined in order to obtain a similar deformational behavior for the model and the original. External loads, weight and material properties of the model and the original shall be taken into account. Since for examination every mechanical system (like a mechanism) can be divided into several subsystems (mechanism links) with appropriate boundary conditions on the interfaces to other subsystems, in this approach we want to focus on the scaling behavior of a single mechanism link. Forces and their moments from neighboring (moving) links can be interpreted as (variable) external loads on the observed link, which have to be scaled accordingly. The results shall than later be transferred to entire mechanisms.

## 2  Derivation of Scaling Laws

By using the principles and laws of linear elasticity, we want to derive the rules that have to be applied to scaling in order to obtain a model with a similar deformational behavior as it can be expected from its original. For now we will focus on the quasistatic condition, i.e. that dynamic effects will be neglected in this paper.

### 2.1  Formula Symbols and Scaling Factors

For physical properties the notations of Table 1 are used. They will be used with the index $S$ for scaled model and index $O$ for original. E.g. $E_S$ and $E_O$ are the elastic moduli of the materials of scaled and original part.

Scaling factors are denoted with $n$ and are defined as the original property divided by the model property. An index is used to indicate the scaled property. For example $n_E = \frac{E_O}{E_S}$ is the scaling factor of the elastic modulus. If used without an index, $n$ denotes the scaling factor of the geometric length: $n := n_l$.

### 2.2  Beam Theory

As a first assessment of applicable scaling factors beam theory is used which neglects transversal contraction. When considering a simple straight horizontal cantilever beam with constant cross section that is fixed on one end and applied with a vertical force and a torque on its other end we receive the maximum deflection of its loose end with Eq. (1).

$$w_{max,O} = w_{max,G,O} + w_{max,F,O} + w_{max,M,O} \tag{1}$$

**Table 1.** Used formula symbols

| Symbol | Unit | Description | Symbol | Unit | Description |
|--------|------|-------------|--------|------|-------------|
| $M$ | Nm | External moment of force | $\sigma$ | Pa | Stress |
| $F$ | N | External force | $\epsilon$ | | Strain |
| $g$ | $\frac{m}{s^2}$ | Gravitational acceleration | $E$ | Pa | Elastic modulus |
| $w_{max}$ | m | Max. displacement | $G$ | Pa | Shear modulus |
| $l, b, h$ | m | Length, width, height | $\nu$ | | Poisson's ratio |
| $I$ | $m^4$ | Second moment of inertia | $\rho$ | $\frac{kg}{m^3}$ | Density |
| $A$ | $m^2$ | Area of beam cross section | $n_X$ | | Scaling factor of property $X$ |

Hereby $w_{max,G,O}$, $w_{max,F,O}$ and $w_{max,M,O}$ are the superposed maximum deflections from gravity, external force and external torque, respectively. Their formulas are:

$$w_{max,G,O} = \frac{\rho_O \cdot g \cdot A_O \cdot l_O^4}{8 \cdot E_O \cdot I_O} \tag{2}$$

$$w_{max,F,O} = \frac{F_O \cdot l_O^3}{3 \cdot E_O \cdot I_O} \tag{3}$$

$$w_{max,M,O} = \frac{M_O \cdot l_O^2}{2 \cdot E_O \cdot I_O} \tag{4}$$

(cf. [9]).

For a scaled model the relations can be written analogously:

$$w_{max,G,S} = \frac{\rho_S \cdot g \cdot A_S \cdot l_S^4}{8 \cdot E_S \cdot I_S} \tag{5}$$

$$w_{max,F,S} = \frac{F_S \cdot l_S^3}{3 \cdot E_S \cdot I_S} \tag{6}$$

$$w_{max,M,S} = \frac{M_S \cdot l_S^2}{2 \cdot E_S \cdot I_S} \tag{7}$$

It is assumed that the gravitational acceleration $g$ will not be scaled, since this is usually not an option in most applications.

When using the above defined scaling factors $n_X = \frac{X_O}{X_S}$ to express all model properties in Eqs. (5)–(7) by their original counterparts and pulling the scaling factors to the front we obtain:

$$\frac{1}{n_w} \cdot w_{max,G,O} = \frac{n_E \cdot n_I}{n_\rho \cdot n_A \cdot n_l^4} \cdot \frac{\rho_O \cdot g \cdot A_O \cdot l_O^4}{8 \cdot E_O \cdot I_O} \tag{8}$$

$$\frac{1}{n_w} \cdot w_{max,F,O} = \frac{n_E \cdot n_I}{n_F \cdot n_l^3} \cdot \frac{F_O \cdot l_O^3}{3 \cdot E_O \cdot I_O} \tag{9}$$

$$\frac{1}{n_w} \cdot w_{max,M,O} = \frac{n_E \cdot n_I}{n_M \cdot n_l^2} \cdot \frac{M_O \cdot l_O^2}{2 \cdot E_O \cdot I_O} \tag{10}$$

Comparing Eqs. (8)–(10) with Eqs. (2)–(4) yields the following relations between the different scaling factors:

$$\frac{1}{n_w} = \frac{n_E \cdot n_I}{n_\rho \cdot n_A \cdot n_l^4} \tag{11}$$

$$\frac{1}{n_w} = \frac{n_E \cdot n_I}{n_F \cdot n_l^3} \tag{12}$$

$$\frac{1}{n_w} = \frac{n_E \cdot n_I}{n_M \cdot n_l^2} \tag{13}$$

For further considerations some specifications are made. All geometric lengths shall be scaled similarly, i.e.: $n_l = n_b = n_h = n_w =: n$. This leads to the following relations for the other two used geometric properties: $n_A = n^2$ and $n_I = n^4$. Hereby Eqs. (11)–(13) can be simplified to:

$$n = \frac{n_E}{n_\rho} \tag{14}$$

$$n_F = n_E \cdot n^2 = \frac{n_E^3}{n_\rho^2} \tag{15}$$

$$n_M = n_E \cdot n^3 = \frac{n_E^4}{n_\rho^3} \tag{16}$$

Equation (14) indicates that the scaling factor $n$ for geometric lengths can not be chosen freely, but is instead dependent on the selected materials of original and scaled model. Equations (15) and (16) are the scaling laws for external loads.

Note that by using different scaling factors for length ($n_l$) and cross section dimensions ($n_b$ and $n_h$) the option to choose $n_l$ freely can be gained, when scaling a model. Under condition $n_w = n_l$ the scaling laws then are: $n_b = n_h = \sqrt{\frac{n_l^3 \cdot n_\rho}{n_E}}$, $n_F = \frac{n_l^4 \cdot n_\rho^2}{n_E}$ and $n_M = \frac{n_l^5 \cdot n_\rho^2}{n_E}$. However this might only be applicable to beam-like structures, which have one major length dimension and smaller cross section dimensions.

In Table 2 the geometric scaling factors for combinations of the materials steel ($E_{st} = 210\,\mathrm{GPa}$, $\rho_{st} = 7860\,\frac{\mathrm{kg}}{\mathrm{m}^3}$), aluminum ($E_{al} = 70.0\,\mathrm{GPa}$, $\rho_{al} = 2699\,\frac{\mathrm{kg}}{\mathrm{m}^3}$), brass ($E_{br} = 90.0\,\mathrm{GPa}$, $\rho_{br} = 8410\,\frac{\mathrm{kg}}{\mathrm{m}^3}$) and polyamide PA 12 ($E_{pa} = 1.70\,\mathrm{GPa}$, $\rho_{pa} = 900\,\frac{\mathrm{kg}}{\mathrm{m}^3}$) are shown.

A few remarks on the scaling factors from Table 2:

A scaling factor $n > 1$ indicates that the scaled model is smaller than the original. It is not possible to build a scaled model out of the same material as the original, when weight and external loads shall be taken into account for the deformation ($n = 1.00$). An aluminum model of a steel original is almost not scaled in size ($n = 1.03$). It could however have the advantage of being manufactured more quickly since aluminum is easier to machine than steel. Building a PA 12 model of an steel original as it has been often done for model mechanisms

**Table 2.** Examples for the geometric scaling factor $n$

| $n = \dfrac{l_O}{l_M}$ | | Original | | | |
|---|---|---|---|---|---|
| | | Steel | Aluminum | Brass | PA 12 |
| Model | Steel | 1.00 | 0.971 | 0.401 | 0.0707 |
| | Aluminum | 1.03 | 1.00 | 0.413 | 0.0728 |
| | Brass | 2.50 | 2.42 | 1.00 | 0.177 |
| | PA 12 | 14.1 | 13.7 | 5.67 | 1.00 |

with selective laser sintering is not sufficient, since the model would have to be relatively small ($n = 14.1$). Brass on the contrary could be a feasible material to substitute a steel original since the material combination shows a practical scaling factor of $n = 2.50$.

## 2.3   Finite Element Method Approach

A FEM simulation with the material combination steel original – brass model (Poisson's ratios: $\nu_{st} = 0.324$, $\nu_{br} = 0.372$) of the simple, beam-like mechanism link from Fig. 1a is used to validate the results from the beam theory approach. In the simulation the part is fixed on the two mounting holes on the right side and charged with a external force with $x$-, $y$- and $z$-component on the mounting hole on the other side. For this study the scaling factor $n$ is varied and the resulting model is calculated in FEM with accordingly scaled external load. The results are shown in Table 3. The best results were obtained for a scaled model of $n = 2.5$, which matches the results from the beam bending approach, although the latter neglects transversal contraction. This might be due to the beam-like shape of the example part or due to the used loading, e.g. the ratio of external to internal forces and might therefore not hold in general. Finding the best scaling factor for a complete mechanism can be achieved by a larger FEM study that calculates the structure with different scaling factors $n$ and for all load cases (scaled according to Eqs. (15) and (16)) that occur during its motion. The errors between the expected deflections ($w_S = \frac{w_O}{n}$) and the ones obtained from FEM can then be used to pick the scaling factor $n$ that induces the most similar deformational behavior to the original. However this method can be computational expensive and time consuming. This is why we consider the following discussion based on the linear elastic theory.

## 2.4   General Linear-Elastic Approach

The beam theory approach does not take transversal contraction into account. A more general approach can be done with Hooke's Law (cf. [12]) in three dimensions. For an isotropic body the relation between the principle strains $\varepsilon = (\varepsilon_1, \varepsilon_2, \varepsilon_3)^T$ and the principal stresses $\sigma = (\sigma_1, \sigma_2, \sigma_3)^T$ can be formulated as (cf. [8]):

**Table 3.** Results of FEM validation

| $n$ | 2.35 | 2.4 | 2.45 | 2.5 | 2.55 | 2.6 | 2.65 |
|---|---|---|---|---|---|---|---|
| Error max. displacement | 0.0654 | 0.0426 | 0.0208 | −0.0001 | −0.0202 | −0.0394 | −0.0580 |
| Error max. strain | −0.0091 | −0.0106 | −0.0112 | −0.0107 | −0.0110 | −0.0121 | −0.0118 |

$$\varepsilon = \mathrm{K} \cdot \boldsymbol{\sigma}, \text{ where } \mathrm{K} = \begin{bmatrix} c_1 & c_2 & c_2 \\ c_2 & c_1 & c_2 \\ c_2 & c_2 & c_1 \end{bmatrix} \tag{17}$$

The values $c_1$ and $c_2$ in Eq. (17) depend only on material properties: $c_1 = \frac{1}{E}$, $c_2 = \frac{2 \cdot G - E}{2 \cdot E \cdot G}$. The relation between $E$, $\nu$ and $G$ is given as $G = \frac{E}{2 \cdot (1 + \nu)}$. Scaling factors for $c_1$ and $c_2$ can be calculated as:

$$n_{c_1} = \frac{E_S}{E_O} \tag{18}$$

$$n_{c_2} = \frac{E_S \cdot G_S \cdot (2 \cdot G_O - E_O)}{E_O \cdot G_O \cdot (2 \cdot G_S - E_S)} \tag{19}$$

It is still assumed that all geometric lengths shall be scaled similarly with the same scaling factor $n$. A comparison of physical dimensions provides then the following relations for the scaling factors of stresses and strains, where $i = 1, 2, 3$:

$$n_\sigma = \frac{\sigma_{i,O}}{\sigma_{i,S}} = n_\rho \cdot n \tag{20}$$

$$n_\varepsilon = \frac{\varepsilon_{i,O}}{\varepsilon_{i,S}} = 1 \tag{21}$$

Writing Eq. (17) for both the scaled model and the original and using the identity of the strains from Eq. (21) we obtain

$$\mathrm{K}_S \cdot \boldsymbol{\sigma}_S = \mathrm{K}_O \cdot \boldsymbol{\sigma}_O. \tag{22}$$

Using the scaling factor definitions to substitute the model properties in Eq. (22) and rearranging we receive an equation of the form $\boldsymbol{a} \cdot n = \boldsymbol{b}$:

$$\begin{bmatrix} c_{1,O} \cdot \sigma_{1,O} + c_{2,O} \cdot (\sigma_{2,O} + \sigma_{3,O}) \\ c_{1,O} \cdot \sigma_{2,O} + c_{2,O} \cdot (\sigma_{1,O} + \sigma_{3,O}) \\ c_{1,O} \cdot \sigma_{3,O} + c_{2,O} \cdot (\sigma_{1,O} + \sigma_{2,O}) \end{bmatrix} \cdot n = \begin{bmatrix} \frac{c_{1,O}}{n_{c_1} \cdot n_\rho} \cdot \sigma_{1,O} + \frac{c_{2,O}}{n_{c_2} \cdot n_\rho} \cdot (\sigma_{2,O} + \sigma_{3,O}) \\ \frac{c_{1,O}}{n_{c_1} \cdot n_\rho} \cdot \sigma_{2,O} + \frac{c_{2,O}}{n_{c_2} \cdot n_\rho} \cdot (\sigma_{1,O} + \sigma_{3,O}) \\ \frac{c_{1,O}}{n_{c_1} \cdot n_\rho} \cdot \sigma_{3,O} + \frac{c_{2,O}}{n_{c_2} \cdot n_\rho} \cdot (\sigma_{1,O} + \sigma_{2,O}) \end{bmatrix}. \tag{23}$$

This is an overdetermined system of equations with the unknown $n$. Therefore an exact solution for $n$ does not exist in general. Furthermore an optimal $n$ depends on the stress condition (=values of $\boldsymbol{\sigma}_O$). That means at every material point within a part the optimal scaling factor $n$ is different in general. For a given principal stress $\sigma_O$ we obtained an optimal $n$ with Gauß's least square method.

The solution we propose to find an optimal $n$ for a part is to calculate the best $n$ for a multitude of stress conditions and then use the mean value of all these $n$ for scaling. The stress conditions can for example be obtained by an FEM simulation of the original part. For the scaled model external loads have to be scaled according to $n_F = n_\rho \cdot n^3$ and $n_M = n_\rho \cdot n^4$.

As example the material combination steel original – brass model is used, because this pair showed a practical scaling factor of $n = 2.50$ in the beam theory approach. Doing so with the FEM results of the mechanism part depicted in Fig. 1a, with 1455 element stress conditions the distribution for $n$ of Fig. 1b is acquired. The mean value of this distribution is 2.573 with a small standard deviation of 0.01107. This result is close to the one obtained by the beam theory approach.

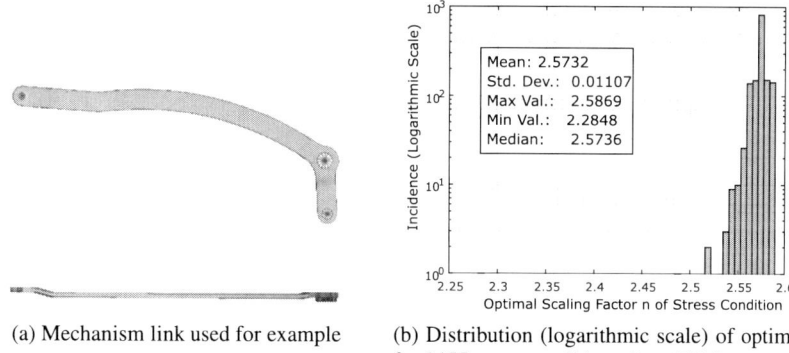

(a) Mechanism link used for example

(b) Distribution (logarithmic scale) of optimal $n$ for 1455 stress conditions from FEM

**Fig. 1.** Example of mechanism link

## 3    Conclusion and Outlook

Two approaches are shown to determine an optimal scaling factor for a model part that take deformation into account. All considerations are quasi static and with respect to weight and external loads. The results of both approaches are similar, however for now it appears in our example of a curved mechanism link that the beam theory approach performs better than the more general approach. To examine this issue further FEM validations with different part geometries and external loads have to be carried out. However for real applications the beam theory approach might already be sufficient since taken simplifications in this approach might be negligible, due to more weighty factors one is faced when building real models, like varying material properties, joint tolerances and manufacturing precision. For further considerations the change of focus from individual links to building whole model linkages has still to be done. For this among other things joints with defined transmission torques have to be constructed to give the model a scaled haptic movement behavior.

**Acknowledgements.** We would like to thank the Bavarian Research Foundation (BFS) for funding this research project under contract number AZ-1181-15. Thanks also goes to the project partners of Webasto Convertibles GmbH and BMW AG.

# References

1. Abdul-Sater, K., Winkler, M.M., Irlinger, F., Lueth, T.C.: Three-position synthesis of origami-evolved, spherically constrained spatial revolute–revolute chains. J. Mech. Robot. **8**(1) (2016)
2. Asl, M.E., Niezrecki, C., Sherwood, J., Avitabile, P.: Vibration prediction of thin-walled composite I-beams using scaled models. Thin-Walled Struct. **113**, 151–161 (2017)
3. Buckingham, E.: On physically similar systems: illustrations of the use of dimensional equations. Phys. Rev. E **4**, 345 (1914)
4. D'Angelo, L.T., Abdul-Sater, K., Pfluegl, F., Lueth, T.C.: Wheelchair models with integrated transfer support mechanisms and passive actuation. J. Med. Devices **9**(1) (2015)
5. Ebert-Uphoff, I., Gosselin, C.M., Rosen, D.W., Laliberte, T.: Rapid prototyping for robotics. In: Cutting Edge Robotics (2005)
6. Gang, X., Wang, D., Su, X.: A new similitude analysis method for a scale model test. Key Eng. Mater. **439–440**, 704–709 (2010)
7. Gebhardt, A.: Additive Fertigungsverfahren, 5th edn. Hanser, Munich (2016)
8. Gross, D., Hauger, W., Schröder, J.: Technische Mechanik 4, 8th edn. Springer, Cham (2011)
9. Gross, D., Hauger, W., Schröder, J., Wall, W.A.: Technische Mechanik 2, 10th edn. Springer, Cham (2009)
10. Hermann von Helmholtz-Zentrum für Kulturtechnik: Getriebesammlung Technische Universität Dresden. Universitätssammlungen in Deutschland - Das Informationssystem zu Sammlungen und Museen an deutschen Universitäten. http://www.universitaetssammlungen.de/sammlung/83
11. Laudahn, S., Irlinger, F., Lueth, T.C., Abdul-Sater, K.: Auslegung und Rapid Prototyping einer räumlichen Fahrzeugtürkinematik unter Berücksichtigung von Singularitätsbetrachtungen sphärischer 4–Gelenkgetriebe. In: 18. VDI Bewegungstechnik, pp. 37–48. VDI Verlag (2016)
12. Mase, G.T., Smelser, R.E., Mase, G.E.: Continuum Mechanics for Engineers, 3rd edn. Taylor & Francis Group, Abingdon (2010)
13. Polifke, W., Kopitz, K.: Wärmeübertragung Grundlagen, analytische und numerische Methoden, 2nd edn. Pearson-Verlag, London (2009)
14. Shehadeh, M., Shennawy, Y., El-Gamal, H.: Similitude and scaling of large structural elements: case study. Alex. Eng. J. **54**, 147–154 (2015)
15. Simitses, G., Starnes, J., Rezaeepazhand, J.: Structural similitude and scaling laws for plates and shells: a review. In: 41st Structures, Structural Dynamics, and Materials Conference and Exhibit, Atlanta, USA (2000)

# Mechanical Transmissions and Gears

# Load Distribution in Four-Point Contact Slewing Bearings Considering Manufacturing Errors and Ring Flexibility

Iker Heras[✉], Josu Aguirrebeitia, Mikel Abasolo, and Ibai Coria

Faculty of Engineering, University of the Basque Country (UPV/EHU),
Bilbao, Spain
{iker.heras,josu.aguirrebeitia,mikel.abasolo,
ibai.coria}@ehu.eus

**Abstract.** This work proposes an approach to calculate the load distribution among balls in a four-point contact slewing bearing considering manufacturing errors and ring stiffness. The model is established upon the formulation and minimization of the potential energy of the system. Calculations are done for a particular bearing considering different assumptions and the results are presented. It is concluded that manufacturing errors affect significantly the load distribution in idling conditions, but not under external loads. Moreover, ring flexibility is proved to have a great effect in the load distribution in idling conditions.

**Keywords:** Slewing-bearing · Four-point contact · Manufacturing errors
Ring flexibility

## 1 Introduction

Slewing bearings are large sized rolling bearings employed for orientation purposes. They are used in machines like tower cranes, radio-telescopes or solar trackers, for example. Furthermore, slewing bearings are used for yaw and pitch rotations in wind turbine generators. Wind Energy has been experiencing a constant growth in importance over recent years [1, 2], and nowadays is the second largest form of power generation capacity in Europe after Natural Gas [2]. Therefore, the Wind Energy sector demands a deeper knowledge of these components to gain expertise in the design process and to obtain more reliable and cost effective Wind Turbine Generators.

The load distribution problem seeks to find how the applied loads are distributed among the different rolling elements. Thus, it is required to be solved in order to ensure that a particular bearing fulfils load requirements [3]. Moreover, these loads can be used to further calculate the friction torque, which is required in order to dimension the actuation system. The first approach to solve the load distribution in four-point contact slewing bearings was proposed by Zupan and Prebil [4], but many other manuscripts dealt with this issue since then [5–8]. The models proposed in these works are established upon the contact geometrical interference, and Finite Element (FE) analyses are always required to simulate rings flexibility. More recent studies carried out by Starvin and Manisekar [9] and Aithal et al. [10] demonstrated via FE calculations that

© Springer Nature Switzerland AG 2019
B. Corves et al. (Eds.): EuCoMeS 2018, MMS 59, pp. 267–274, 2019.
https://doi.org/10.1007/978-3-319-98020-1_31

manufacturing errors can significantly affect the load distribution in large diameter angular contact ball bearings.

In the EUCOMES 2016, the authors presented a new analytical approach to calculate ball-raceway interferences due to manufacturing errors assuming rigid rings [11]. This model, in combination with FE calculations, was later used to evaluate the effect of these errors in the friction torque [12]. The current work extends the analytical approach in [11, 12] to implement ring flexibility and external loads.

## 2  Interferences Calculation Model

In this section, the original model in [11, 12] for rigid rings is briefly introduced and then a novel procedure is presented to implement ring flexibility and external loads.

### 2.1  Model for Rigid Rings

The basis of the approach lies on the simulation of ball-raceway contacts through the mechanism in Fig. 1. Depending on the manufacturing errors, a gap (diagonal 1 in Fig. 1, i.e. points $P^1$ and $P^3$) or a interference (diagonal 2 in Fig. 1, i.e. points $P^2$ and $P^4$) can exist in a contact. The location of the centres of the raceways ($O^i$) is determined from experimental measurements, as explained in [11]. These centres are then linked by traction only springs, which natural lengths are also a function of the real shape of the raceways. To determine the interferences in the contacts after the assembly of the bearing (after inserting the balls and before applying any external load), the potential energy of the system is formulated, which is a function of the relative position of one ring respect to the other:

$$U_{contact} = \frac{2}{3} \sum_{b=1}^{B} \left[ K_{Tot}^{1b} \left( \delta_{Tot}^{1b} \right)^{5/2} + K_{Tot}^{2b} \left( \delta_{Tot}^{2b} \right)^{5/2} \right] \tag{1}$$

Where $B$ is the number of balls, $K_{Tot}^{ib}$ is the total stiffness of the spring $i$ of the ball $b$ and $\delta_{Tot}^{ib}$ is its total elongation (the summation of the interferences in each contact pair). The final relative position will be the one with the minimum associated energy. Therefore, the final spatial configuration, and thus contact interferences, can be determined by minimizing (1).

### 2.2  Implementation of Ring Flexibility

In (1) rigid rings are assumed, so only the potential energy associated to ball-raceway contacts is considered. Nevertheless, ring deformations have a great influence in large sized slender bearings [2–4], so it must be considered in slewing bearings to obtain accurate results. For this purpose, the FE static condensation method is used. In order to obtain the stiffness matrix of any ring, a fully parametric FE model was built in ANSYS® (see Fig. 2). From this model, the stiffness matrix of each ring condensed to the centres of the raceways ($O^i$ in Fig. 1, master nodes in Fig. 2a) can be obtained. Since no load is transmitted between rings in the circumferential direction, only axial and radial degrees of freedom are considered. Note that the span angle corresponding to each ball will vary depending on the number of balls (see Fig. 2b and c).

Once the stiffness matrices of the rings are obtained, they can be implemented in the analytical model. Now, the final position of the centres of the raceways will not be only a function of the rigid body displacements of one ring respect to the other, but also of the elastic deformations of the rings. Moreover, these deformations will also accumulate elastic energy. Consequently, the total potential energy of the system in idling conditions will be given by:

$$U_{idling} = U_{contact} + U_{rings} \tag{2}$$

Where the potential energy due to the elastic deformations of the rings can be calculated as follows:

$$U_{rings} = \frac{1}{2} \left[ \{D_{out}\}^T [K_{out}] \{D_{out}\} + \{D_{in}\}^T [K_{in}] \{D_{in}\} \right] \tag{3}$$

Where $[K_{out}]$ and $[K_{in}]$ are the stiffness matrices of outer and inner rings respectively and $\{D_{out}\}$ and $\{D_{in}\}$ are the displacements of the centres of the raceways due to the elastic deformation of the rings. As it was done with rigid rings, the contact interferences can be obtained by minimizing (2).

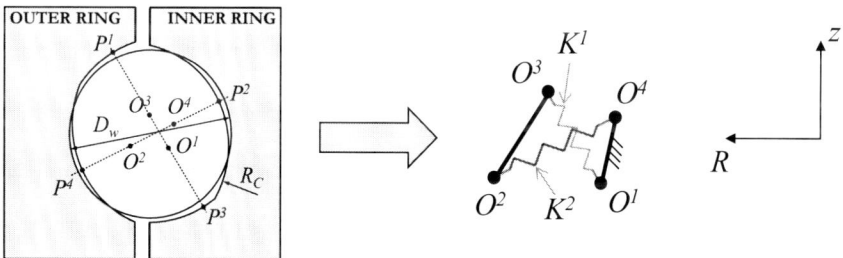

**Fig. 1.** Graphical representation of the mechanism of the analytical model.

## 2.3   External Loads Application

The potential energy of the system will now be reformulated for cases with external loads. It is known that the change in the potential energy of a system due to an applied conservative load is equal to the negative of the work done by it. For a load $F$ applied along a displacement $\delta$:

$$F = -\frac{dU}{d\delta} \quad \rightarrow \quad dU = -F\,d\delta \quad \rightarrow \quad U = -F \cdot \delta = -W \tag{4}$$

Thus, the total potential energy of our system when external loads are applied can be calculated by deducting the work done by these loads to expression (2):

$$U_{total} = U_{contact} + U_{rings} + U_{loads} = U_{contact} + U_{rings} - W_{loads} \qquad (5)$$

Where $W_{loads}$ is:

$$W_{loads} = F_a \delta_a + F_r \delta_r + M_t \theta_t \qquad (6)$$

Subscripts $a$, $r$ and $t$ refer to axial and radial loads and tilting moment respectively. Once more, minimizing (5) the final position for given external loads (or the reaction forces for certain imposed displacements) could be found.

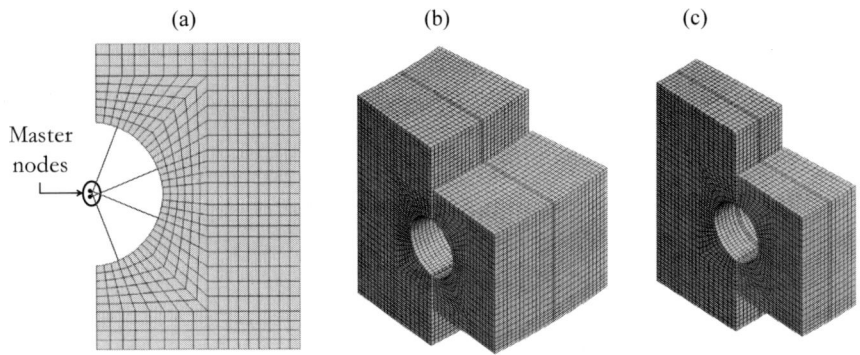

**Fig. 2.** FE model: (a) master nodes; (b) sector for 32 balls; (c) sector for 67 balls.

## 3  Results and Discussion

Measurements of a particular bearing were taken [11] and the proposed methodology was applied to calculate ball-raceway interferences in idling conditions assuming rigid rings (Fig. 3) and considering ring flexibility (Fig. 4). From the comparison, a large effect of ring flexibility can be seen. For the nominal ball, the average interference decreases from 5 μm for rigid rings (Fig. 3a) to 3 μm for deformable rings (Fig. 4a). Moreover, the difference between the maximum and the minimum interferences also decreases, from 14 μm for rigid rings to 12 μm for deformable rings. Thus, by considering ring flexibility lower interferences are obtained, and the distribution is also smoother. This happens because the rings are deformed due to ball-raceway contact loads. When the ball preload increases (defined the preload as an increase of a few microns of its diameter), the effect is more noticeable. Thus, for the preload of +20 μm, the average interference decreases from 25 μm for rigid rings (Fig. 3b) to 14 μm for deformable rings (Fig. 4b). With deformable rings, an increment in the preload does not mean an equal increment in the interferences, as happens with rigid rings, because the higher the preload is, the higher the contact loads are, and therefore the more the rings are deformed. Moreover, with higher preloads the interferences distribution becomes even smoother.

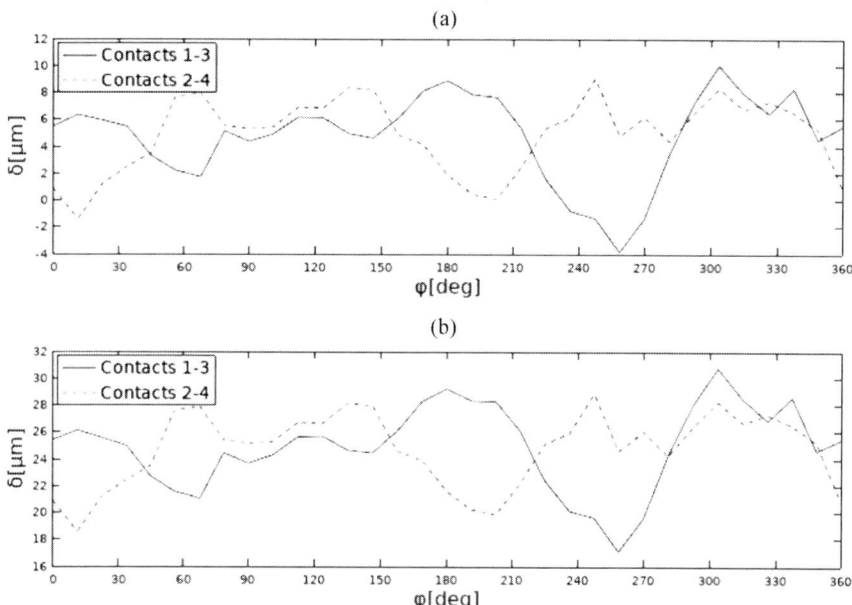

**Fig. 3.** Interferences for rigid rings with 32 balls: (a) nominal ball; (b) +20 μm.

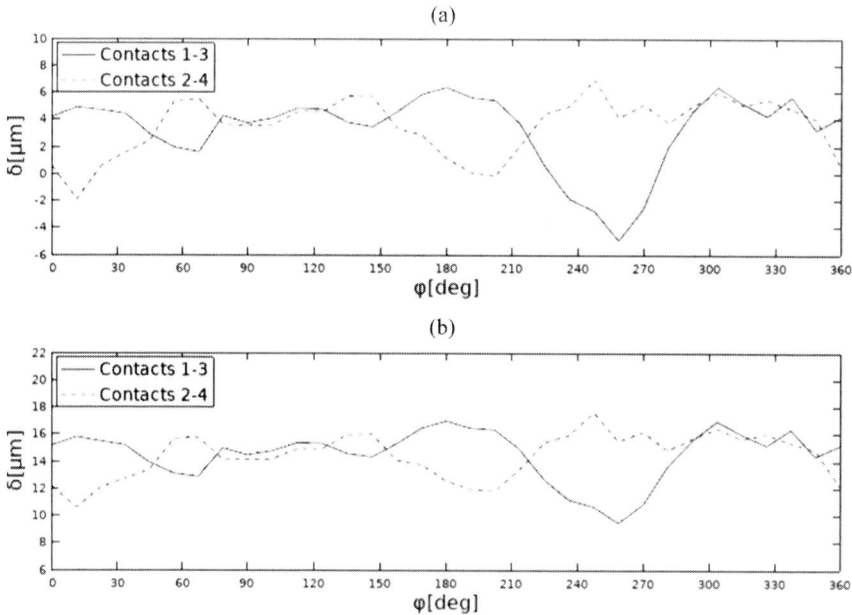

**Fig. 4.** Interferences for deformable rings with 32 balls: (a) nominal ball; (b) $\delta_P = +20$ μm.

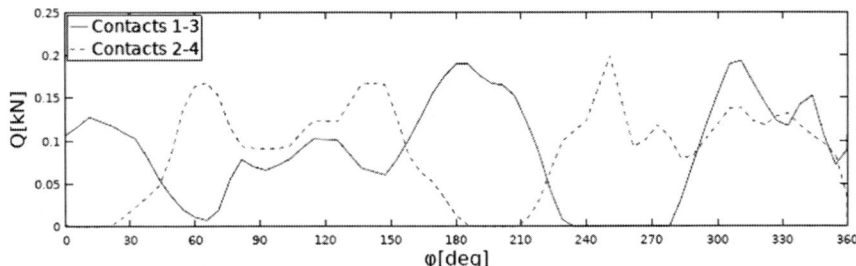

**Fig. 5.** Load distribution without preload and no applied loads for 67 balls.

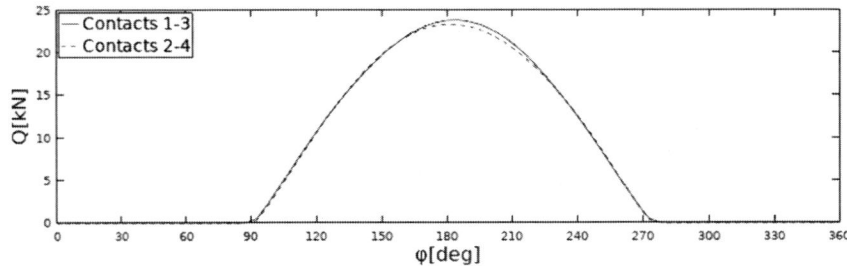

**Fig. 6.** Load distribution without preload and for a radial load for 67 balls.

Figures 5 and 6 show the load distribution considering ring flexibility for the idling case and for a radial load respectively. The applied load is one half the static load capacity of the bearing. From these plots, it can be concluded that under external loads, the effect of manufacturing errors on the load distribution is residual.

Apart from the fact that the model considers manufacturing errors and ring flexibility, it offers another important advantage. When the stiffness of the ball-raceway contact is simplified by means of a beam-spring mechanism in FE calculations, as done by Smolnicki [13] or Daidié [14], and a radial load is applied, the mechanism leaves this plane, as represented in Fig. 7. When this happens, an unreal radial stiffness appears due to the misalignment of the springs. Since the circumferential degree of freedom is not considered in the proposed semi-analytical model, this problem is avoided, thus offering more reliable results for load cases which involve radial displacements (Fig. 6). Moreover, once the stiffness matrices have been calculated, the proposed approach is much faster than a FE model with any simplified mechanism. The described model only requires FE analysis for the calculation of the stiffness matrices, and then any load case can be solved quickly, while regular FE models require one calculation (or load step) for each load case.

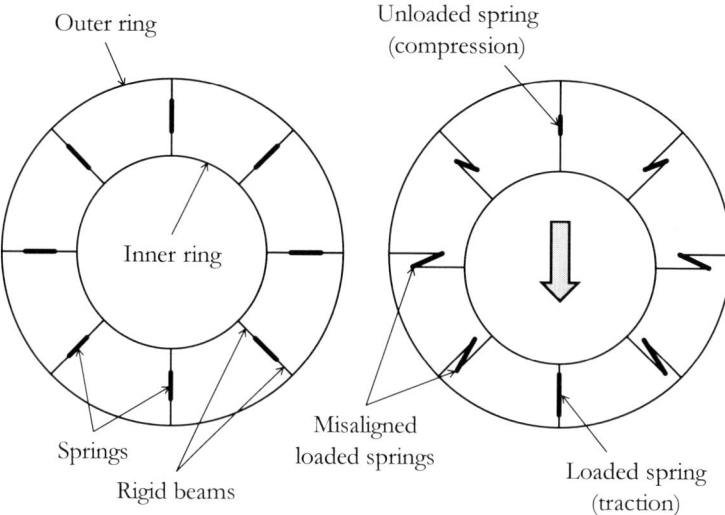

**Fig. 7.** Schematic representation of the misalignment of Daidié's mechanism when a radial load is applied.

## 4  Conclusions

In this work, a semi-analytical model to calculate the load distribution in four-point contact slewing bearing considering manufacturing errors and ring flexibility is proposed. From the application of the model to a particular case, it is concluded that ring deformations have a significant influence on the ball-raceway interferences due to manufacturing errors or preload. Moreover, the load distribution is demonstrated to be affected by manufacturing errors in idling conditions, but this effect becomes residual when external loads are applied.

**Acknowledgments.** This research work is a result of the close collaboration that the authors maintain with the company Iraundi S.A. The authors wish also to acknowledge the financial support of the Ministry of Economy and Competitiveness through project number DPI2017-85487-R, and the Basque Government through project number IT947-16.

## References

1. The European Wind Energy Association: 2015 European Statistics (2016)
2. Wind Europe: 2016 European Statistics (2017)
3. ISO 76:2006. Rolling bearings - Static load ratings (2006)
4. Zupan, S., Prebil, I.: Carrying angle and carrying capacity of a large single row ball bearing as a function of geometry parameters of the rolling contact and the supporting structure stiffness. Mech. Mach. Theor. **36**, 1087–1103 (2001). https://doi.org/10.1016/S0094-114X(01)00044-1

5. Amasorrain, J.I., et al.: Load distribution in a four contact-point slewing bearing. Mech. Mach. Theor. **38**, 479–496 (2003). https://doi.org/10.1016/S0094-114X(03)00003-X

6. Olave, M., et al.: Design of four contact-point slewing bearing with a new load distribution procedure to account for structural stiffness. J. Mech. Des. **132**, 21006 (2010). https://doi.org/10.1115/1.4000834

7. Aguirrebeitia, J., et al.: Effect of the preload in the general static load-carrying capacity of four-contact-point slewing bearings for wind turbine generators: theoretical model and finite element calculations. Wind Energy **17**, 1605–1621 (2014). https://doi.org/10.1002/we.1656

8. Plaza, J., et al.: A new finite element approach for the analysis of slewing bearings in wind turbine generators using superelement techniques. Meccanica **50**, 1623–1633 (2015). https://doi.org/10.1007/s11012-015-0110-7

9. Starvin, M.S., Manisekar, K.: The effect of manufacturing tolerances on the load carrying capacity of large diameter bearings. Sadhana **40**, 1899–1911 (2015)

10. Aithal, S., et al.: Effect of manufacturing errors on load distribution in large diameter slewing bearings of fast breeder reactor rotatable plugs. Proc. Inst. Mech. Eng. Part C J. Mech. Eng. Sci. 1–12 (2015). https://doi.org/10.1177/0954406215579947

11. Heras, I., et al.: Calculation of the ball raceway interferences due to manufacturing errors and their influence on the friction moment in four-contact-point slewing bearings. In: 6th European Conference on Mechanism Science, Nantes, France (2016) https://doi.org/10.1007/978-3-319-44156-6_1

12. Heras, I., et al.: Friction torque in four contact point slewing bearings: effect of manufacturing errors and ring stiffness. Mech. Mach. Theor. **112**, 145–154 (2017). https://doi.org/10.1016/j.mechmachtheory.2017.02.009

13. Smolnicki, T., Rusiński, E.: Superelement-based modeling of load distribution in large-size slewing bearings. J. Mech. Des. **129**, 459 (2007). https://doi.org/10.1115/1.2437784

14. Daidié, A., et al.: 3D simplified finite elements analysis of load and contact angle in a slewing ball bearing. J. Mech. Des. **130**, 82601 (2008). https://doi.org/10.1115/1.2918915

# The Rotary into Helical Transmission by the "Friction Cam-Helical Follower" Kinematic Pair

Jesús Meneses[✉], Eduardo Corral[✉], Cristina Castejón[✉],
Higinio Rubio[✉], and Juan Carlos García-Prada[✉]

MaqLab Research Group, Universidad Carlos III de Madrid, Getafe, Spain
{meneses,ecorral,castejon,hrubio,
jcgprada}@ing.uc3m.es

**Abstract.** A cam-follower type kinematic pair, which converts continuum rotary motion into helical back and forth motion is presented. The rotary motion of the cam transmits, on one hand, the usual linear back and forth motion to the follower; on the other hand, the cam also transmits, by friction, a rotational movement to the follower, so that it is animated by a combined reciprocating and helical motions. The rotation of the cam fully determines the translation motion of the follower, whereas the rotation motion of the follower is not defined directly from the cam rotation motion. The reason is that slippage between the cam and the follower is kinematically unavoidable and it also depends on dynamic parameters like the cam-follower normal force and coefficient of friction. Hence, the dynamics must be solved in order to obtain the full law of motion of the follower from the law of motion of the cam. For that purpose, a dynamical model of the kinematic pair and the corresponding equations of motion are presented in this work. These equations have been numerically solved with a MatLab code, and the model has been validated comparing these results with those provided by the multibody simulation program CREO (mechanism module).

**Keywords:** Helical gear · Transmission · Kinematic pair · Cam Friction

## 1 Introduction

The rotation-helical transmission is hardly mentioned in the literature. Since the nineteenth century, the helical movement has been achieved through the combination of a rotary motor with a linear actuator. Such is the case of Tomas A. Edison's first phonograph [1]. The patented "Adjustable Angle Helix Generator for Edge and Radial Relief Sharpening" [2] performs continuous helical movement, it has a very specific field of application and no published work on the related kinematics and dynamics has been found by the authors. The "spiral motor" [3–6] generates helical movement, however it has been mainly designed (and operates) as a linear actuator [4, 5]. The authors have found a single article in which it is suggested to take advantage of a 2-DoF induction helical motor for some "industrial applications as grinders, augers,

© Springer Nature Switzerland AG 2019
B. Corves et al. (Eds.): EuCoMeS 2018, MMS 59, pp. 275–283, 2019.
https://doi.org/10.1007/978-3-319-98020-1_32

drilling an milling spindles, robotic arms and drives for medical tools and prostheses" [6]. The motivation of this article is to develop/improve the patented "Dispositivo Automático para Biopsias Cutáneas (Automatic Device for Skin Biopsy)" [7] for producing back and forth helical movement with a single kinematic pair. The kinematics and dynamics of this device would be developed and published as a continuation of this work.

## 2 Kinematics of the "Friction Cam-Helical Follower" Pair. Sliding Velocity at the Contact Point

A novel kinematic pair called "friction cam-helical follower" is proposed. For the sake of simplicity, an eccentric circumference cam profile and a flat follower are considered. In Fig. 1 front and top views of the pair are represented. The position of the cam, described by the angular coordinate $\varphi_1$, determines the follower longitudinal position, described by the linear coordinate $z_2$ (see Fig. 1), but not its angular position, since no pure rolling occur, but sliding. Actually, as it can be seen later, pure rolling only can take place when $\varphi_1 = 0, \pi, 2\pi, \ldots$. In other words, $\varphi_1$ and $z_2$ are dependent coordinates, whereas the angular coordinates $\varphi_1$ and $\varphi_2$ (and the corresponding angular velocities and accelerations) are independent a priori.

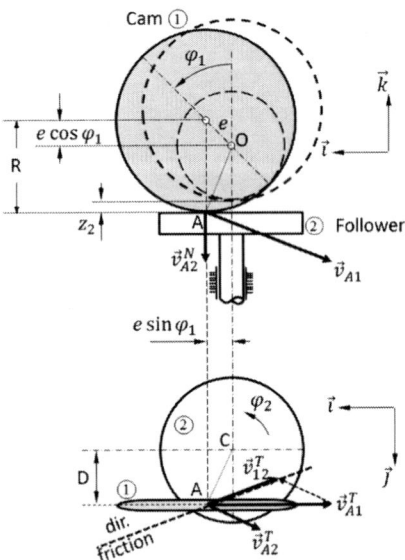

**Fig. 1.** Front and top views of the "friction cam (1) - helical follower (2)" kinematic pair. The front view of the cam in its initial position is shown in dashed line. The velocity components of point A as belonging to (1) and (2), as well as the corresponding sliding velocity are shown.

$$z_2 = -e(1 - \cos \varphi_1) \tag{1}$$

$$\dot{z}_2 = -e\dot{\varphi}_1 \sin \varphi_1 \tag{2}$$

$$\ddot{z}_2 = -e\ddot{\varphi}_1 \sin \varphi_1 - e\dot{\varphi}_1^2 \cos \varphi_1, \tag{3}$$

The sliding velocity can be expressed as a function of the angular velocities, but these are only obtained by solving the system dynamics. In fact, the linear velocity of the contact point A on the cam, $\vec{v}_{A1}$, and that of the contact point on the follower, $\vec{v}_{A2}$, are obtained from the angular velocities of both bodies in the following way (see Fig. 1):

$$\vec{v}_{A1} = \dot{\varphi}_1 \vec{J} \times \overrightarrow{OA} = \underbrace{-(R - e \cos \varphi_1)\dot{\varphi}_1 \vec{i}}_{\vec{v}_{A1}^T} \underbrace{- e\dot{\varphi}_1 \sin \varphi_1 \vec{k}}_{\vec{v}_{A1}^N = \vec{v}_{A2}^N} \tag{4}$$

$$\vec{v}_{A2} = \vec{v}_{A1}^N + \dot{\varphi}_2 \vec{k} \times \overrightarrow{CA} = \underbrace{-D\dot{\varphi}_2 \vec{i} + e\dot{\varphi}_2 \sin \varphi_1 \vec{J}}_{\vec{v}_{A2}^T} \underbrace{- e\dot{\varphi}_1 \sin \varphi_1 \vec{k}}_{\vec{v}_{A2}^N = \vec{v}_{A1}^N} \tag{5}$$

The sliding velocity (for example, the velocity of point A on the cam as seen from the follower) results:

$$\vec{v}_{12}^T = \vec{v}_{A1}^T - \vec{v}_{A2}^T = [D\dot{\varphi}_2 - (R - e \cos \varphi_1)\dot{\varphi}_1]\vec{i} - e\dot{\varphi}_2 \sin \varphi_1 \vec{J} \tag{6}$$

Sliding velocity can vanish only at the highest and lowest positions of the follower, in which $\varphi_1 = n\pi$ (see Fig. 2); indeed at these two positions, the velocities of point A on the cam and point A on the follower, have the same direction. Moreover, setting the

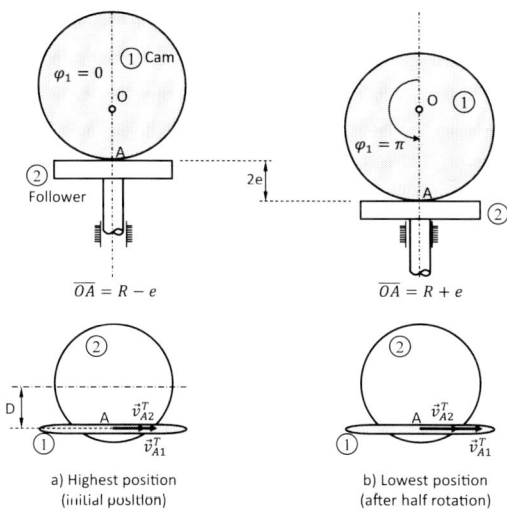

a) Highest position
(initial position)

b) Lowest position
(after half rotation)

**Fig. 2.** Front and top views of the "friction cam (1) - helical follower (2)" kinematic pair at: (a) highest (initial) and (b) lowest positions of the follower.

sliding velocity to zero at these positions, Eq. (6) becomes the cam-follower rolling (without sliding) condition: $D\dot{\varphi}_2 = (R \pm e)\dot{\varphi}_1$. Actually, rolling will take place at those positions, only if there is enough friction. But what is clear from Eq. (6) and from Fig. 1, is that out of those positions, slippage will always occur for $e \neq 0$. The sliding velocity expressed in Eq. (6) is crucial in the dynamic analysis, as the friction force has the same direction.

## 3   Dynamic Analysis. Equations of Motion

In Fig. 3, the forces and torques involved in the proposed kinematic pair are shown (constraint forces have been excluded). Let us describe them, by putting together those acting on each element:

- Acting on the cam, there is an applied torque, $M_1$, as well as the contact force exerted by the follower, which is decomposed into tangential (friction force), $\vec{F}_{21}^T$, and normal force, $\vec{F}_{21}^N$, components.
- The follower is under the opposite contact forces exerted by the cam, $\vec{F}_{12}^T = -\vec{F}_{21}^T$, and $\vec{F}_{12}^N = -\vec{F}_{21}^N$. An axial force, $F_2$, and moment, $M_2$, should also be considered acting on the follower: the former is necessary to maintain the follower in contact with the cam; the latter is any (optional) resistive torque the mechanism works against. The motion of the follower is constrained by a cylindrical kinematic pair, but this will be considered as ideal: both axial friction force and torque will be neglected (they could be added to $F_2$ and $M_2$ respectively).

As stated above, the friction force exerted by the cam (part 1 in Figs. 1 and 3) on the follower (part 2 in Figs. 1 and 3), $\vec{F}_{12}^T$, has precisely the same direction as the

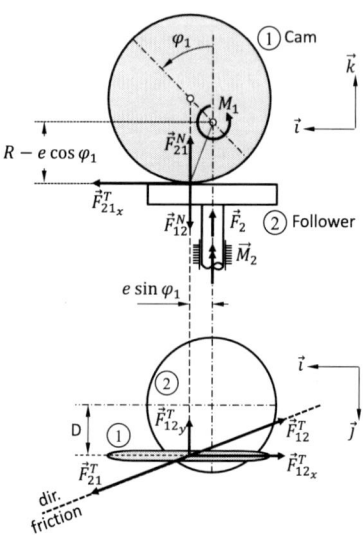

**Fig. 3.** Forces and torques involved in the "friction cam (1) - helical follower (2)" pair.

sliding velocity vector $\vec{v}_{12}^T$, as expressed in (6), and indicated in Fig. 1. Furthermore, this friction force is proportional to the cam-follower normal contact force [8] (except for the singular points, if pure rolling occurs), so we can write:

$$\vec{F}_{12}^T = \mu |\vec{F}_{12}^N| \frac{\vec{v}_{12}^T}{|\vec{v}_{12}^T|}, \tag{7}$$

where $\mu$ is the cam-follower coefficient of friction.

According to (6), the friction force in (7) is decomposed into the following $x$ and $y$ components:

$$F_{12x}^T = \frac{\mu |\vec{F}_{12}^N| [D\dot{\varphi}_2 - (R - e \cos \varphi_1)\dot{\varphi}_1]}{\sqrt{[D\dot{\varphi}_2 - (R - e \cos \varphi_1)\dot{\varphi}_1]^2 + e^2 \dot{\varphi}_2^2 \sin^2 \varphi_1}} \tag{8a}$$

$$F_{12y}^T = -\frac{\mu |\vec{F}_{12}^N| e\dot{\varphi}_2 \sin \varphi_1}{\sqrt{[D\dot{\varphi}_2 - (R - e \cos \varphi_1)\dot{\varphi}_1]^2 + e^2 \dot{\varphi}_2^2 \sin^2 \varphi_1}} \tag{8b}$$

According to Fig. 3, the rotational $y$-component equation of motion for the cam is:

$$M_1 - F_{21}^N e \sin \varphi_1 - F_{21x}^T (R - e \cos \varphi_1) = I_1 \ddot{\varphi}_1, \tag{9}$$

where $I_1$ is the cam moment of inertia with respect to its fixed point O. Taking into account the action-reaction law, $F_{21x}^T = -F_{12x}^T$, and using Eqs. (8a), (9) yields:

$$M_1 - F_{21}^N e \sin \varphi_1 + \frac{\mu F_{21}^N (R - e \cos \varphi_1)[D\dot{\varphi}_2 - (R - e \cos \varphi_1)\dot{\varphi}_1]}{\sqrt{[D\dot{\varphi}_2 - (R - e \cos \varphi_1)\dot{\varphi}_1]^2 + e^2 \dot{\varphi}_2^2 \sin^2 \varphi_1}} - I_1 \ddot{\varphi}_1 = 0, \tag{10}$$

where it has been also assumed $|\vec{F}_{12}^N| = F_{21}^N$, since the normal force on the cam must always be at the $z$-positive direction. The $z$-translational equation of motion for the follower is (weight is neglected or considered as part of $F_2$):

$$F_2 - F_{21}^N = m_2 \ddot{z}_2 \tag{11}$$

Or, in terms of angular variables, using Eq. (3) we have:

$$F_2 - F_{21}^N + m_2 e \ddot{\varphi}_1 \sin \varphi_1 + m_2 e \dot{\varphi}_1^2 \cos \varphi_1 = 0 \tag{12}$$

Meanwhile, the rotational $z$-component equation of motion for the follower results:

$$M_2 - F_{12x}^T D + F_{12y}^T e \sin \varphi_1 = I_2 \ddot{\varphi}_2 \tag{13}$$

And introducing Eqs. (8a) and (8b) in (13), we get:

$$M_2 + \frac{\mu F_{21}^N \left[ D(R - e \cos \varphi_1)\dot{\varphi}_1 - (D^2 + e^2 \sin^2 \varphi_1)\dot{\varphi}_2 \right]}{\sqrt{[D\dot{\varphi}_2 - (R - e \cos \varphi_1)\dot{\varphi}_1]^2 + e^2 \dot{\varphi}_2^2 \sin^2 \varphi_1}} - I_2\ddot{\varphi}_2 = 0 \qquad (14)$$

In summary, the equations of motion for the proposed pair, is a system of three second-order nonlinear differential Eqs. (10), (12) and (14):

$$\begin{cases} M_1 - F_{21}^N e \sin \varphi_1 + \dfrac{\mu F_{21}^N (R - e \cos \varphi_1)[D\dot{\varphi}_2 - (R - e \cos \varphi_1)\dot{\varphi}_1]}{\sqrt{[D\dot{\varphi}_2 - (R - e \cos \varphi_1)\dot{\varphi}_1]^2 + e^2 \dot{\varphi}_2^2 \sin^2 \varphi_1}} - I_1\ddot{\varphi}_1 = 0 \\[2mm] F_2 - F_{21}^N + m_2 e \ddot{\varphi}_1 \sin \varphi_1 + m_2 e \dot{\varphi}_1^2 \cos \varphi_1 = 0 \\[2mm] M_2 + \dfrac{\mu F_{21}^N \left[ D(R - e \cos \varphi_1)\dot{\varphi}_1 - (D^2 + e^2 \sin^2 \varphi_1)\dot{\varphi}_2 \right]}{\sqrt{[D\dot{\varphi}_2 - (R - e \cos \varphi_1)\dot{\varphi}_1]^2 + e^2 \dot{\varphi}_2^2 \sin^2 \varphi_1}} - I_2\ddot{\varphi}_2 = 0 \end{cases} \qquad (15)$$

The system of Eq. (15) will be solved it this article via the inverse dynamics approach, where the motion of the cam is prescribed by the function $\varphi_1(t)$, then the motion of the follower and the torque required on the cam, are both calculated. Introducing this function and its derivatives into Eq. (15) leads to a system of two differential equations in $M_1(t)$ and $\varphi_2(t)$. For instance, if the cam rotates at a constant angular velocity, $\omega$:

$$\varphi_1 = \omega t; \ \dot{\varphi}_1 = \omega; \ \ddot{\varphi}_1 = 0 \qquad (16)$$

Thus, the second of Eq. (15) gives:

$$F_{21}^N = F_2 + m_2 e \omega^2 \cos \omega t, \qquad (17)$$

and the equations of motion are reduced to a system of two differential equations, where the unknowns are the motor torque on the cam, $M_1(t)$, and the follower rotation angle $\varphi_2(t)$:

$$\begin{cases} M_1 - (F_2 + m_2 e \omega^2 \cos \omega t)e \sin \omega t + \dfrac{\mu (F_2 + m_2 e \omega^2 \cos \omega t)(R - e \cos \omega t)[D\dot{\varphi}_2 - \omega(R - e \cos \omega t)]}{\sqrt{[D\dot{\varphi}_2 - \omega(R - e \cos \omega t)]^2 + e^2 \dot{\varphi}_2^2 \sin^2 \omega t}} = 0 \\[2mm] M_2 + \dfrac{\mu (F_2 + m_2 e \omega^2 \cos \omega t)\left[ D(R - e \cos \omega t)\omega - (D^2 + e^2 \sin^2 \omega t)\dot{\varphi}_2 \right]}{\sqrt{[D\dot{\varphi}_2 - \omega(R - e \cos \omega t)]^2 + e^2 \dot{\varphi}_2^2 \sin^2 \omega t}} - I_2\ddot{\varphi}_2 = 0 \end{cases} \qquad (18)$$

The second of Eq. (18) has been numerically solved for the follower rotation, $\varphi_2(t)$ (using "ode45" from MATLAB ®, [9] with $\varphi_2(0) = \dot{\varphi}_2(0) = 0$), and then $\dot{\varphi}_2(t)$ replaced into the first of Eq. (18) to obtain the required motor torque $M_1(t)$. Figure 4 these three functions, $\varphi_2(t)$, $\dot{\varphi}_2(t)$ and $M_1(t)$, have been represented, for (in S.I. of units): $\omega = 20\pi$ (10 rps), $R = D = 2e = 0.03$, $m_1 = 0.015; m_2 = 0.045$, $I_1 = 10^{-5}$; $I_2 = 5 \cdot 10^{-5}$, $F_2 = 5$; $M_2 = 0$, and several values of the cam-follower coefficient of friction, $\mu = 0.2, 0.6$ and 1.

Note that, insofar as $R = D$, the angular position and velocities of the follower can be compared with those of the cam (as done in Fig. 4): actually, if in addition $e = 0$, and there was enough friction so that the cam and the follower could roll without sliding, their angular position and velocity would be equal to each other (also, no reciprocating motion would occur).

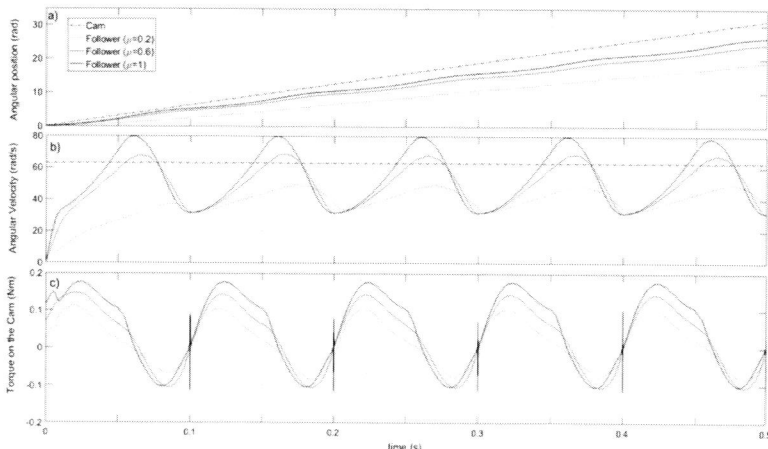

**Fig. 4.** For a prescribed uniform cam rotation $\omega = 20\pi$ rad/s and for different coefficients of friction: (a) follower rotation angle, and (b) angular velocity functions; (c) required torque on the cam; when a constant axial force, $F_2 = 5$ N is applied on the follower to ensure the cam-follower contact. Cam rotation angle and angular velocity have also been represented in chain line in (a) and (b).

As it can be seen in graphs (b) and (c) of Fig. 4, the higher the coefficient of friction, the lower the number of cycles to reach the steady state, and the higher the average angular velocity reached by the follower, but also the higher the energy dissipated in friction. The last statement can be checked by integrating the torque on a steady state cycle.

Once the equations of motions have been solved, Eq. (6) is reclaimed to get any component of the sliding velocity, or even its modulus:

$$v_{12}^T = \sqrt{[D\dot{\varphi}_2 - (R - e\cos\varphi_1)\dot{\varphi}_1]^2 + (e\dot{\varphi}_2\sin\varphi_1)^2} \tag{19}$$

In Fig. 5 the sliding velocity is plotted against time for the same parameters used in Fig. 4, showing that when the coefficient of friction is large enough, no sliding occurs at the highest positions of the follower, when the cam just completes a whole number of revolutions (see Fig. 2a). Some decrease in the sliding velocity, but not so drastic, can also be appreciated at lowest positions of the follower, when the cam makes half integer number of revolutions (see Fig. 2b). This result is logical since in the latter case, the lever arm, $\overline{OA} = R - e$ is smaller than in the former, $\overline{OA} = R + e$, and so it is the velocity of point A on the cam; therefore, less friction is required for pure rolling at highest positions than it is at lowest positions of the follower. Note that the value of the sliding velocity at t = 0 comes from the fact that the cam has the prescribed rotation velocity (constant) whereas the initial angular velocity of the follower is zero (In fact, the initial sliding velocity is equal to $\omega(R - e)$, regardless of the coefficient of friction).

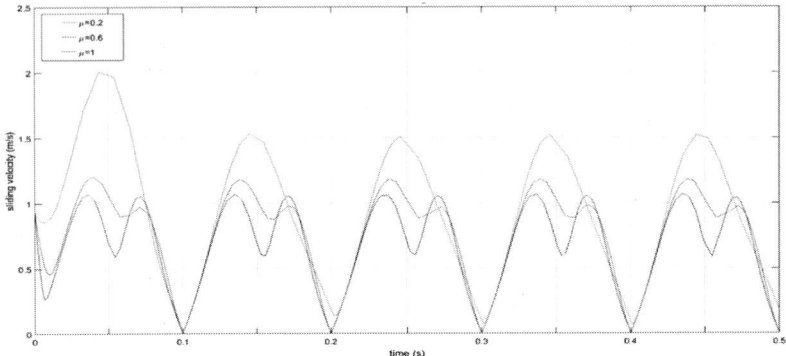

**Fig. 5.** Cam-follower sliding velocity against time, for the same parameters as in Fig. 4

In order to validate the model presented in this article, the results it provides have been compared with those obtained with the simulation program Creo 2.0 "Mechanism", for the "cam prescribed rotation at constant velocity" case. The Creo model is depicted in Fig. 6. The simulation was done for $\mu = 0.6$, and the same parameters. As shown in Fig. 6, the coincidence between the graphics for the follower angular position, velocity, and the cam-follower sliding velocity, as calculated from the presented model (Eqs. 18 and 19), and the corresponding ones as simulated by Creo, is remarkable.

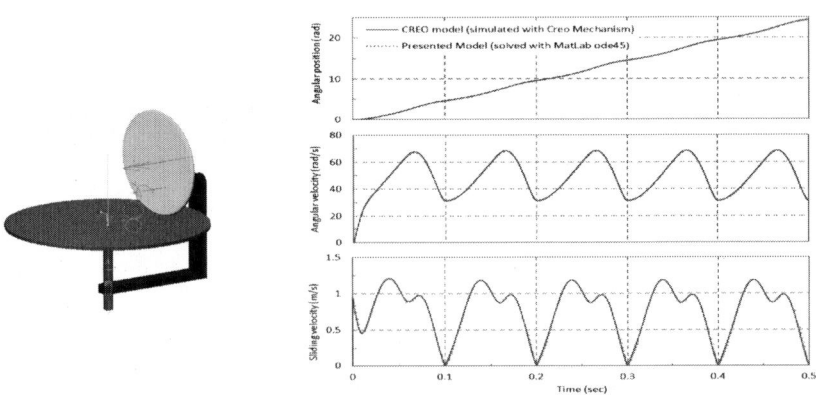

**Fig. 6.** Left: the "friction cam - follower" as modelled with Creo. Right: comparison between the graphics for the follower angular position, angular velocity and cam-follower sliding velocity as calculated from the presented model and the Creo simulations.

## 4    Conclusions and Future Work

In this paper the kinematics and dynamics of the "friction cam-helical follower" kinematic pair have been studied. The corresponding equations of motion have been presented and implemented in a MATLAB ® code that provides the full law of motion of the follower from the law of motion of the cam. The model has been validated comparing these results with those provided by the multibody simulation program CREO (mechanism module). The developed MATLAB ® code computing time is considerably shorter than that of CREO for this model.

Among the advantages of the kinematic pair is the ease of varying the parameters of the transmission, by adjusting the distance D or the eccentricity of the cam, e; the disadvantage is the existence of sliding friction.

As a future work, a model of the automatic device for skin biopsy [7], based on the "friction cam-helical follower" presented in this article, will be developed. The mechanical parameters of this mechanism can be optimized with the aid of the corresponding extension of the above-mentioned numerical code.

**Acknowledgements.** The authors acknowledge the Spanish Found through the project "MAQSTATUS", DPI 2015-69325-C2-1-R, for financial support.

## References

1. Edison, T.A.: Phonograph or Speaking Machine. US Patent Office, No. 200,521 (1878)
2. Harding, W.: Adjustable Angle Helix Generator for Edge and Radial Relief Sharpening. United States Patent, No. 3,600,860 (1971)
3. Kwon, H., Fujimoto, Y.: FEM analysis of high thrust spiral motor. In: The 8th IEEE International Workshop on Advanced Motion Control. AMC 2004 (2004). https://doi.org/10.1109/amc.2004.1297943
4. Smadi, I.A., Omori, H., Fujimoto, Y.: On direct-drive motion of a spiral motor. In: IECON 2010 - 36th Conference on IEEE Industrial Electronics Society (2010). https://doi.org/10.1109/iecon.2010.5675163
5. Kominami, T., Fujimoto, Y.: Studies on thrust characteristics of high-thrust spiral motor. Electr. Eng. Jpn. **177**(2) (2011). https://doi.org/10.1002/eej.21168
6. Caruso, M., Cecconi, V., Di Dio, V., Di Tomasso, A.O., Genduso, F., La Cascia, D., Liga, R., Miceli, R.: Speed control of a two-degrees of freedom motor with rotor helical motion for industrial applications. In: AEIT Conference, pp. 1–6 (2014)
7. Grillo, E., Vañó, S., Jaén, P., Castejón, C., Meneses, J., García-Prada, J.C., Rubio, H.: Dispositivo Automático para Biopsias Cutáneas. Oficina Española de Patentes y Marcas, No. ES2537831 B1 (12.06.2015). Int. Class.: A61B10/02 (2006.01) (2015)
8. Marques, F., Flores, P., Claro, J.C.P., et al.: A survey and comparison of several friction force models for dynamic analysis of multibody mechanical systems. Nonlinear Dyn. **86**, 1407–1443 (2016)
9. Corral, E., Meneses, J., García-Prada, J.C.: Forward and Inverse Dynamics and Quasi-Static Analysis of Mechanizes with MATLAB®. InTech (2016). https://doi.org/10.5772/63372

# Design and Experiences of a Planetary Gear Box for Adaptive Drives

Konstantin Ivanov[1]([⊠]), Claudia Aide Gonzalez-Cruz[2],
Marco Ceccarelli[3], Assylbek K. Ozhiken[4], and Daniele Cafolla[3]

[1] Almaty University of Power Engineering and Telecommunications,
Almaty 050013, Kazakhstan
ivanovgreek@mail.ru
[2] Univesidad Autónoma de Querétaro,
76010 Santiago de Querétaro, Querétaro, Mexico
claudia.aide.gonzalez@gmail.com
[3] LARM, University of Cassino and South Latium, 03043 Cassino, Italy
{ceccarelli, cafolla}@unicas.it
[4] Al-Farabi Kazakh National University, Almaty 050040, Kazakhstan
ozhikenll@gmail.com

**Abstract.** This paper presents a design of an adaptive gear variator together with lab experiences for its operation characterization. The proposed design is based on a solution with a differential planetary gear system in which the second degree of freedom is activated when the input torque exceeds the possibility for the first degree of freedom. Both kinematical and mechanical designs are presented with the peculiarities of the design. Experimental validation is worked out with a proper test-bed and test results are discussed for a characterization of the presented variator with its peculiar functioning.

**Keywords:** Gear system · Planetary gears · Gear variators · Design
Performance analysis

## 1 Introduction

The concept "variator" means the frictional mechanism with the controllable transfer ratio [1]. The following frictional mechanisms are used as the variators: crown variator and conic variator with intermediate rollers and belt transmission with compound wedge pulleys. The control of the crown variator and conic variator is carried out by change of position of intermediate roller. While the control of the belt transmission is carried out by change of diameter of wedge pulley.

The main deficiency of a frictional variator is the low reliability and the complexity of its control. In this way, the more reliable mechanism is the hydro-mechanical transmission (CVT) uniting the fluid converter with a stepped gearing. In this transmission the fluid converter carries out the smooth change of the transfer ratio in narrow limits of each transfer step. However, the deficiencies that the hydro-mechanical CVT presents are: (a) a complex design, (b) a complex and not quite adequate control system of the gear change, and (c) ruptures of a transferred power stream leading to blows, to mention a few.

© Springer Nature Switzerland AG 2019
B. Corves et al. (Eds.): EuCoMeS 2018, MMS 59, pp. 284–291, 2019.
https://doi.org/10.1007/978-3-319-98020-1_33

Attempts of the use of double coupling step transfers to soften the forward motion at the switching steps lead to the complication of the design [2].

The gear variator is a new wheelwork branch with constant engagement of toothed wheels and with variable transfer ratio. Theoretical preconditions of gear variators have been developed by Ivanov [3–5]. The construction of experimental prototypes of adaptive gear variator has been created on the basis of the execution of the necessary and the sufficient adaptation conditions (a condition of existence of a toothed variator) [8–12].

The objective of the present work is the description of a pilot model of an adaptive gear variator at Almaty University of Power Engineering and Telecommunications and the experimental confirmation of its properties on a test-bed.

## 2   Description of Adaptive Gear Variator

The adaptive gear variator is intended for use in the module of lifting of loads. The working principle behind the design is that a constant power of the gear variator, the lift velocity should be adapted to the variable weight of the freight. Thus, the engine capacity is defined according to the necessary work to reach the minimum lift velocity for the freight at its maximum weight and viceversa, the maximum lift velocity that can be reached at the minimum freight weight. In this way, it will be possible to profit all the engine capacity.

The design of a gear variator is developed on the basis of the theory of an adaptive gear variator [3–5]. The proposal design contains: (a) a necessary adaptation condition, (b) a sufficient adaptation condition, and (c) an independent start condition. The necessary adaptation condition is reduced to the equation of the force adaption effect.

$$\omega_{H2} = M_{H1}\omega_{H1}/M_{H2}, \tag{1}$$

where $\omega_{H2}$ is the output angular velocity, $\omega_{H1}$ is the input angular velocity, $M_{H1}$ is the input motive moment and $M_{H2}$ is the output moment. This condition allows to define output angular velocity on the set of constant parameters for the input power and the resistance moment.

The sufficient adaptation condition is reduced to the relation among the teeth number, $z_i$ $(i = 1..6)$, of the mechanism wheels

$$-\frac{z_8}{z_7} = \frac{z_3z_4z_5 - z_1z_5z_6}{z_3z_4z_6 - z_1z_4z_5} \tag{2}$$

From this condition it is possible to select a known teeth number for the toothed wheels 7 and 8 in the parallel gearing of the planetary kinematic chain.

The independent start condition is defined by the equality of the lengths of the input and output carriers, which is expressed by means of the teeth number of the wheels.

$$z_1 + z_2 = z_4 + z_5 \tag{3}$$

This condition allows to select the teeth number of the toothed wheels of a planetary kinematic chain.

Considering the three conditions after mentioned, by means of Eqs. (1–3), an adaptive gear variator is developed. Figure 1 shows the kinematic design of the proposed gear variator. It contains the frame 0, the carrier $H_1$, the closed contour with toothed wheels 1-2-3-6-5-4 and the carrier $H_2$. The closed contour contains the satellite 2, the block of solar wheels 1-4, the block of ring wheels 3-6 and the satellite 5. The mechanism has additional doubling constraint in the form of a tooth gearing 8, 7, connecting the input carrier $H_1$ with the output satellite 5.

The mechanism presented on Fig. 1 contains the mobile closed contour with additional constraint and the doubling transfer 8-7 with a speed coincidence center,

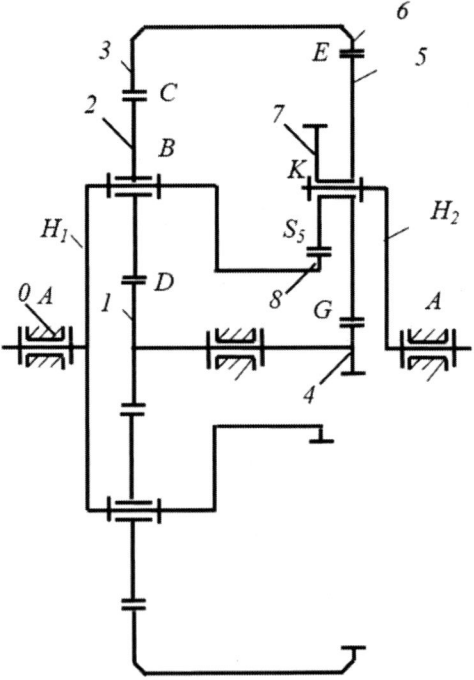

**Fig. 1.** Kinematic design of the proposed adaptive gear variator.

creating a shifting bearing (carrying out function of the moment lever). This mechanism provides force adaptation to a variable load as it satisfies necessary and sufficient conditions of force adaptation [14, 15].

## 3  Design and Work of Adaptive Gear Variator

Design documentation under the developed circuit design is executed and an adaptive gear variator is designed. The knots design of the adaptive gear variator at the Almaty University of Power Engineering and Telecommunications (AUPET) is presented on

Fig. 2. The CAD model in Fig. 2(b) shows the main components of the variator: the covers, the input and output carriers and the closed gear chain formed by the sun gear (1), the four satellites (2), the internal epicyclic gears (3-6), the two satellite gears (5) and the sun gear (4). It also shows the additional doubling constraint formed by the gears (8-7), which connect the input carrier with the output satellite (5).

The built prototype of the adaptive gear variator is shown in Fig. 2(b). It shows: (1) input half-coupling, (2) input cover with the input bearing of the case, (3) mobile case with gear rings and an output cover, (4) input carrier as an assembly, (5) output carrier as an assembly, (6) output bearing of the case, (7) output half-coupling.

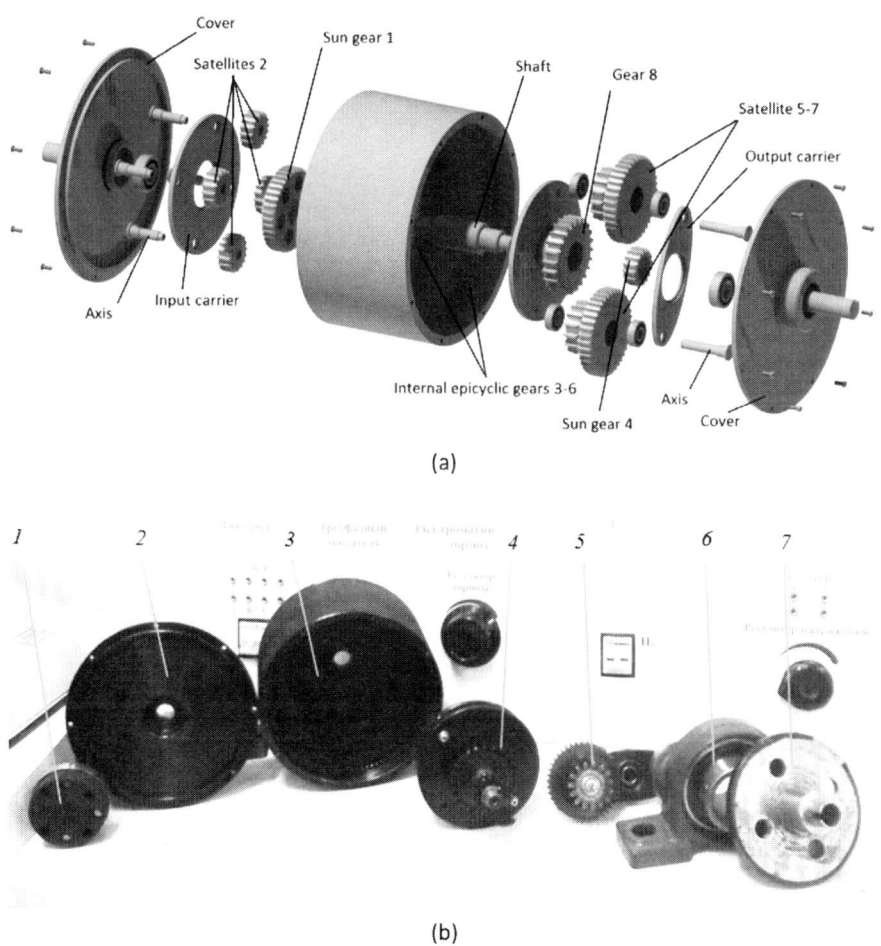

Fig. 2. Design solution for the AUPET gear adaptive variator: (a) the CAD design with components; (b) the built prototype: (1) input half-coupling, (2) input cover, (3) mobile case with gear rings, (4) input carrier as an assembly, (5) output carrier as an assembly, (6) output case bearing, (7) output half-coupling.

The adaptive gear train operates as follow: When the friction forces between gears is high, satellite gears (5) are locked to the sun gear (4) and the epicyclic internal gear (6), then the movement is transmitted from the input carrier to the output carrier. However, when the friction forces between gears are small, the satellite gears (5) run like output link holding the output carrier in a fixed position and activating the second degree of freedom of the adaptive gear variator. The design parameters of the variator gears are listed in Table 1.

**Table 1.** Parameters of the adaptive variator gears at AUPET.

| Parameters | Number of teeth, z | Modulus [mm] | Radio [mm] |
|---|---|---|---|
| Input sun gear | 40 | 2 | 400 |
| Input satellite gear | 6 | 2 | 160 |
| Input epicyclic internal gear | 72 | 2 | 720 |
| Output sun gear | 16 | 2 | 160 |
| Output satellite gear | 40 | 2 | 400 |
| Output epicyclic internal gear | 96 | 2 | 960 |
| Output satellite | 15 | 3 | 225 |
| Additional transmission gear | 24 | 3 | 360 |
| Input and output carriers | – | – | 675 |

## 4  Test Bed Description

Tests for the adaptive gear variator were performed in the AUPET test-bed, see Fig. 3. It consists of a basis (tractive) engine (1) that drive the adaptive gear variator (2), which leads the auxiliary (brake) engine (3), which produces a resistant loading torque. The instrumented panel (4) measures different physical variables, such as resistant torque $M_R$, angular velocity $n_R$, current strength $I_R$, voltage $U_R$ and power $P_R$. While, the monitoring system (5) lets the system performance supervision.

In order to analyze the performance of the adaptive gear variator, a set of tests were conducted following two main objectives: (1) prove the achievement of the force adaptation effect under operating conditions and (2) evaluate the possibility of the motion starts in the presence of an initial resistance torque. Thus, by means of tests, the resistance torque $M_R$ and the angular velocity $\omega_R$ of the output variator shaft can be experimentally defined.

The tests are accomplished at constant traction power and with a smooth increase of the resistance torque from null to maximum. While, the shaft variator parameters were measured at the different loading conditions.

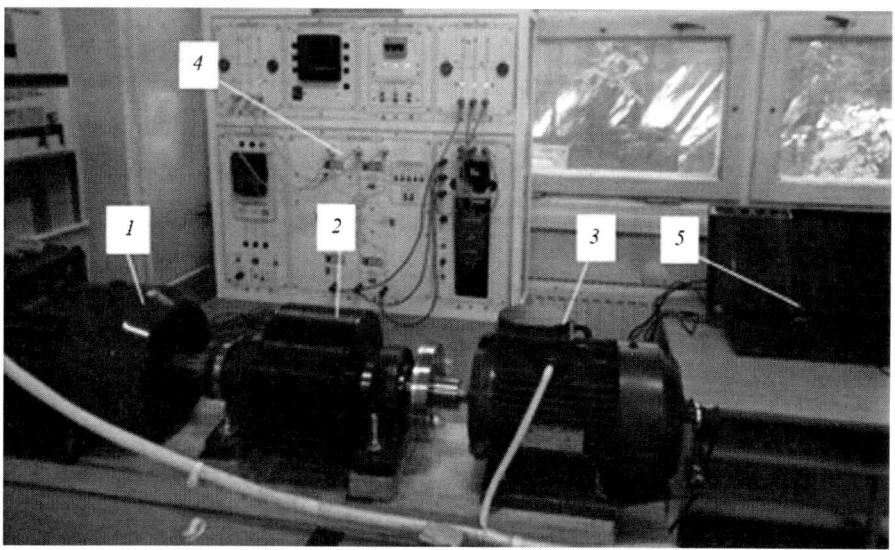

**Fig. 3.** AUPET test bed: 1-basic (tractive) engine, 2-gear adaptive variator, 3-auxiliary (brake) engine, 4-instrument panel, 5-monitoring system.

## 5 Experimental Analysis

The received parameters of the brake engine and the variator at different loading conditions are listed in Table 2. It can be seen that at bigger resistance torque the angular velocity of the variator output shaft and its power decrease. Furthermore, when the output torque is minimum, the variator gear ratio is $u = 1$, which means that the first degree of freedom operates, and the movement is directly transmitted from the input carrier to the output carrier. However, as the resistance torque increases, the output shaft velocity decreases, and the gear ratio increases by the action of the second degree of freedom.

**Table 2.** Brake engine and variator parameters

| Point | Resistance torque $M_R$ [Nm] | Number of revolution $n_R$ [rpm] | Angular velocity $\omega_R$ [s$^{-1}$] | Current strength $I_R$ [A] | Voltage $U_R$ [V] | Power $P_R$ [W] | Gear ratio $u$ |
|-------|------|------|------|------|------|------|------|
| A | 155.8 | 54 | 5.6 | 4.02 | 24.5 | 295.4 | 17.04 |
| B | 77.2 | 109 | 11.4 | 3.41 | 34.3 | 350.8 | 8.48 |
| C | 72.1 | 117 | 12.2 | 3.28 | 36.8 | 362.1 | 7.92 |
| D | 28.8 | 292 | 30.5 | 2.78 | 64.3 | 536.2 | 3.17 |
| E | 17.9 | 468 | 49.0 | 2.57 | 75.3 | 580.5 | 1.96 |
| F | 9.1 | 920 | 96.3 | – | – | – | 1.00 |

The effect of force adaptation or self-regulation characteristic of the gear variator can be seen in Fig. 4. A varying loading force demands a change in the angular velocity of the variator output shaft. The maximum freight weight 155.8 Nm is driven at the minimum operational velocity 5.6 cps (54 rpm), while the minimum freight weight 9.1 Nm is driven at 96.3 cps (920 rpm). It can be seen that as the freight weight decreases, the driven speed of the variator in the output shaft increases.

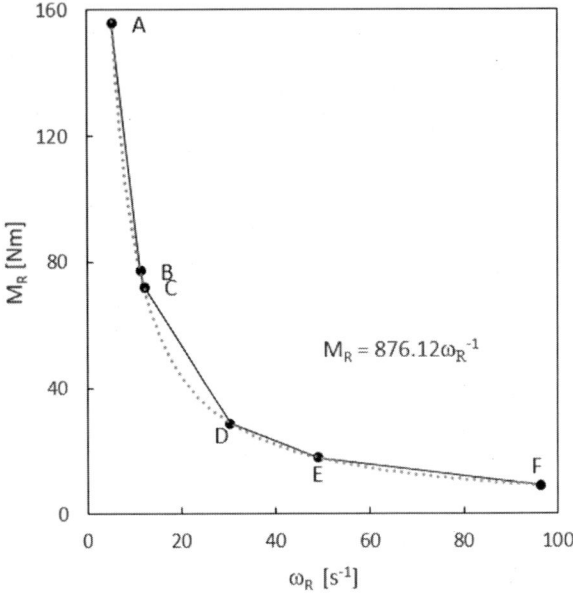

**Fig. 4.** Experimental tractive characteristic of gear variator.

The results in Fig. 4 allow to characterize the traction of the gear variator as a function of the angular velocity of the resistance load $M_R = f(\alpha, \omega_R)$. From a numeric regression the gear variator traction can be represented as

$$M_R = \alpha \cdot \omega_R^{-1} \tag{4}$$

which means that the load $M_R$ that can be driven is inversely proportional to the angular operational speed with a gain $\alpha = 876.12$. It is assumed that the value of the constant gain $\alpha$ is given by the engine parameters and the driven load.

## 6   Conclusions

The paper presents an experimental characterization of a new adaptive planetary gear variator by using a specific test-bed. The peculiarities of the variator are presented by test results to prove the feasibility of the proposed gear variator in the design structure and operation performance. The proposed adaptive gear variator is the highly effective self-controlled transmission gear which can be used in machines with variable loads for a wide range of applications.

**Acknowledgments.** The second author C. A. González-Cruz acknowledges the Mexican Government Foundation CONACYT for the fellowship of postdoctoral studies at LARM in the academic year 2017–2018. The author A. K. Ozhiken wishes to gratefully acknowledge Kazakh National University through grant No. 4-2667 for permitting an internship at LARM of Cassino University from September to December of 2018.

# References

1. Lang, K.R.: Continuously Variable Transmissions. An Overview of CVT Research Past, Present, and Future (2000). http://www.cs.montana.edu/∼halla/egen310/cvt/cvt.pdf
2. Kim, S., Oh, J., Choi, S.: Gear shift control of a dual-clutch transmission using optimal control allocation. Mech. Mach. Theory **113**, 109–125 (2017)
3. Ivanov, K.S.: Theory of continuously variable transmission with two degrees of freedom: paradox of mechanics. In: ASME 2012 International Mechanical Engineering Congress and Exposition, Volume 3: Design, Materials and Manufacturing, Parts A, B, and C, Houston, Texas, USA, pp. 933–942 (2012). https://doi.org/10.1115/IMECE2012-85762
4. Ivanov, K.S., Jilisbaeva, K.: Paradox in the mechanism science. In: New Trends in Educational Activity in the Field of Mechanism and Machine Theory, pp. 132–138. Springer, Cham (2014)
5. Ivanov, K.S., Ceccarelli, M., Tultaev, B.: Force adaptation in robot transmissions. Int. J. Adv. Robot Autom. **2**(2), 1–8 (2017). Symbiosis
6. Ceccarelli, M., Balbaev, G., Ivanov, K.: Experimental test relation of a new planetary transmission. Int. J. Mech. Control **15**(2), 3–7 (2014)
7. Balbayev, G., Ceccarelli, M.: Design and characterization of a new planetary gear box. In: New Advances in Mechanisms, Transmissions and Applications, pp. 91–98. Springer, Dordrecht (2014)
8. Ivanov, K.S.: Discovery of the force adaptation effect. In: Proceedings of 11th World Congress in Mechanism and Machine Science, Tianjin, China, pp. 581–585 (2004)
9. Ivanov, K.S.: Gear automatic adaptive variator with constant engagement of gears. In: Proceedings of the 12th World Congress in Mechanism and Machine Science, Besancon, France, pp. 182–188 (2007)
10. Ivanov, K.S.: The simplest automatic transfer box. In: World Congress on Engineering (ICME), London, UK, pp. 1179–1184 (2010)
11. Ivanov, K.S., Yaroslavceva, E.K.: Device of automatic and continuous change of a twisting moment-and changes of a corrected speed of output shaft depending on a tractive resistance, Germany. Patent No. 20 2012 101 273.1 (2012)
12. Ivanov, K.S., Dinasilov, A.D., Yaroslavceva, E.K.: Gear variator - scientific reality. In: Mechanisms, Transmissions and Applications, pp. 169–176. Springer, Cham (2015)

# On Meshing Limit Line of ZC1 Worm Pair

Yaping Zhao[✉] and Xiaodong Sun

Northeastern University, Shenyang, China
zhyp_neu@163.com, 784433792@qq.com

**Abstract.** The theory of the meshing limit line of a ZC1 worm pair is fully established on the basis of meshing theory for gear drives. A method to judge the existence of the meshing limit line is proposed. Meanwhile such a method can provide the reasonable initial value for iteratively solving the nonlinear equation encountered during determining the meshing limit line. The numerical example is provided for validation and verification and the result shows, usually, the working length of a ZC1 worm cannot be in excess of the half of its thread length.

**Keywords:** $ZC_1$ worm · Meshing limit line · Nonlinear equations
Meshing function · Gear geometry

## 1 Introduction

As reported in the literature [1, 2], about in the 1930s, Gustave Niemann originally invented a concave-convex worm-gear drive. The worm helicoid is ground by a disk-shaped grinding wheel with a toroidal generating surface. During machining the worm blank, the crossing angle between the axial lines of the blank and the grinding wheel is the lead angle of the worm on its pitch cylinder [3]. Afterwards a worm obtained in such a way is known as the Niemann worm [4]. In China national standards, the symbol ZC1 is utilized to represent the Niemann worm.

In the next few decades, the meshing theory of the Niemann worm drive was continuously studied by Litvin [5], Wu [6], Wang [7] and so on. However, the meshing limit line was not paid enough attention to. The meshing limit line is usually defined as an enveloping line of the instantaneous contact line on the helicoid of the worm [8, 9]. In the existing literature, the meshing limit function of a ZC1 worm drive was obtained in accordance with the vector multiplication. Some researchers drew the meshing limit line on the helicoidal surface of the ZC1 worm by virtue of conjecture but the specific solution method was not given at all [6].

In this work, the theory of solving meshing limit line for a ZC1 worm is fully established. The numerical example investigation is also implemented.

© Springer Nature Switzerland AG 2019
B. Corves et al. (Eds.): EuCoMeS 2018, MMS 59, pp. 292–298, 2019.
https://doi.org/10.1007/978-3-319-98020-1_34

## 2 Computing Principle of Meshing Limit Line for $ZC_1$ Worm Pair

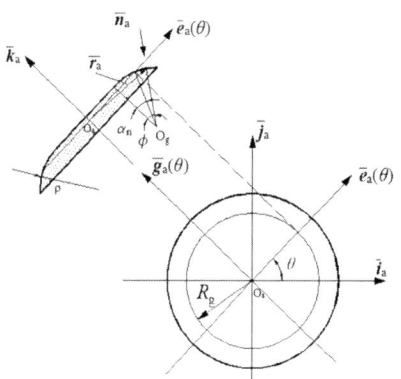

**Fig. 1.** Generating surface of grinding wheel in $\sigma_a$.

A coordinate system $\sigma_a\{O_a; \vec{i}_a, \vec{j}_a, \vec{k}_a\}$ is fixed with the grinding wheel and the unit vector $\vec{k}_a$ is along its axial line as shown in Fig. 1. The intersecting line of an axial section of the grinding wheel and its generating torus is a segment of circular arc with the radius of $\rho$. By means of the circular and sphere vector functions [10], the equation of the toroidal generating surface, $\Sigma_g$, of the grinding wheel and the unit normal vector of $\Sigma_g$ can be represented as

$$(\vec{r}_a)_a = x_a\vec{i}_a + y_a\vec{j}_a + z_a\vec{k}_a, \ (\vec{n}_a)_a = -\vec{m}_a(\theta, \phi) \tag{1}$$

where $x_a = \cos\theta(\rho\sin\phi + R_g - \rho\sin\alpha_n)$, $y_a = \sin\theta(\rho\sin\phi + R_g - \rho\sin\alpha_n)$ and $z_a = \rho(\cos\phi - \cos\alpha_n)$. Here $\theta$ and $\phi$ are the curvilinear coordinates, $R_g$ is the nominal radius of the grinding wheel, and $\alpha_n$ is its shape angle.

As shown in Fig. 2, the unit vector $\vec{k}_{o1}$ of a static coordinate system $\sigma_{o1}\{O_1; \vec{i}_{o1}, \vec{j}_{o1}, \vec{k}_{o1}\}$ is along the axial line of the worm blank. The crossing angle between the axial line, $\vec{k}_a$, of the grinding wheel and the axial line, $\vec{k}_{o1}$, of the worm is $\gamma$, which is the lead angle of the worm on its pitch cylinder. The unit vector $\vec{i}_{o1}$ is along the common perpendicular $\overrightarrow{O_1O_a}$ to the preceding axial lines. The original point $O_1$ is the middle point of the thread length of the worm. During grinding the worm blank, it performs rotating motion around $\vec{k}_{o1}$ and the grinding wheel makes translational motion along $\vec{k}_{o1}$. The blue position of $\sigma_a$ indicates the initial position of the grinding wheel whereas the black position is the current position.

In $\sigma_{o1}$, the equation of the surface of action for machining the worm can be attained as

$$(\vec{r}_1)_{o1} = (x_a + a_d)\vec{i}_{o1} + y_{o1}\vec{j}_{o1} + z_{o1}\vec{k}_{o1}, \ \phi(0) = \arctan(B_d/A_d) \tag{2}$$

where $y_{o1} = y_a\cos\gamma - z_a\sin\gamma$, $z_{o1} = y_a\sin\gamma + z_a\cos\gamma - p\varphi$, $A_d = \sin\theta(a_d\cos\gamma + p\sin\gamma) - \rho\cos\alpha_n\sin\gamma\cos\theta$, and $B_d = \sin\gamma\cos\theta(R_g - \rho\sin\alpha_n) + a_d\sin\gamma - p\cos\gamma$. Here $\varphi$ is the rotary angle of the worm work-blank, $a_d$ is the processing center distance for grinding the worm, and $p$ is its spiral parameter.

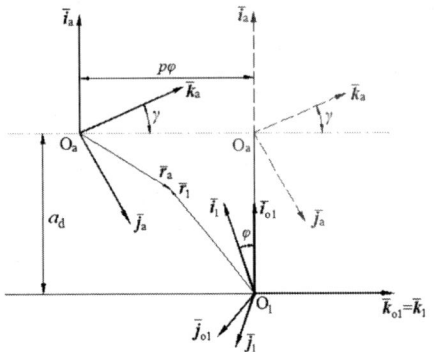

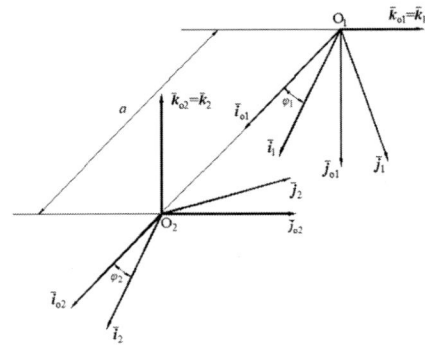

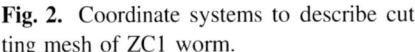

**Fig. 2.** Coordinate systems to describe cutting mesh of ZC1 worm.

**Fig. 3.** Coordinate systems to describe mesh of worm drive.

As depicted in Fig. 3, the unit vector $\vec{k}_{o2}$ of a static coordinate system $\sigma_{o2}\{O_2; \vec{i}_{o2}\vec{j}_{o2}, \vec{k}_{o2}\}$ is along the axial line of the worm gear. The forgoing axial line $\vec{k}_{o2}$ is perpendicular to the axial line $\vec{k}_{o1}$ of the worm. Both the unit vectors $\vec{i}_{o1}$ and $\vec{i}_{o2}$ are along the common perpendicular of the two axial lines $\vec{k}_{o1}$ and $\vec{k}_{o2}$.

The equation of the surface of action during machining the worm gear can be procured in $\sigma_{o2}$ as

$$(\vec{r}_2)_{o2} = (x_{o1}^* - a)\vec{i}_{o2} + z_o\vec{j}_{o2} - y_{o1}^*\vec{k}_{o2}, \quad \phi(\theta) = \arctan(B_d/A_d), \quad \Phi = 0 \quad (3)$$

where $x_{o1}^* = x_{o1}\cos(\varphi_1 - \varphi) - y_{o1}\sin(\varphi_1 - \varphi)$, $y_{o1}^* = x_{o1}\sin(\varphi_1 - \varphi) + y_{o1}\cos(\varphi_1 - \varphi)$. Herein $\varphi_1$ is the rotation angle of the worm in the process of the mesh with the worm gear, and $a$ and $\Phi$ are the center distance and the meshing function of the worm pair, respectively.

In order to work out the meshing function $\Phi$, the expression of the unit normal vector of $\Sigma_g$ in $\sigma_{o1}$ should be acquired in advance. Actually, such an expression can be acquired from the second expression in Eq. (1) as

$$(\vec{n}_a)_{o1} = n_x\vec{i}_{o1} + n_y\vec{j}_{o1} + n_z\vec{k}_{o1} \quad (4)$$

where $n_x = -\cos\theta \sin\phi$, $n_y = -\cos\gamma \sin\theta \sin\phi + \sin\gamma \cos\phi$, and $n_z = -\sin\gamma \sin\theta \sin\phi - \cos\gamma \cos\phi$.

The meshing function of the worm pair can thus be ciphered out as

$$\Phi = \frac{1}{i_{12}}\left[\left(n_y p\varphi + \frac{b_A}{\cos\theta}\right)\sin(\varphi_1 - \varphi) + (b_B - n_x p\varphi)\cos(\varphi_1 - \varphi) + n_z(a - i_{12}p)\right] \quad (5)$$

where $b_A = \sin\theta(a_d n_y + p n_z)/\sin\gamma$, and $b_B = -p n_z \cot\gamma + a_d \sin\theta \sin\phi/\sin\gamma$. Here $i_{12}$ is the transmission ratio of the worm pair.

The value of the angle $\varphi$ on the meshing limit line should satisfy the following quadratic equation as

$$a_\varphi \varphi^2 + b_\varphi \varphi + c_\varphi = 0 \tag{6}$$

where $a_\varphi = p^2 \cos^2\theta\left(n_x^2 + n_y^2\right)$, $b_\varphi = -p\cos\theta\left(n_x b_B \cos\theta - n_y b_A\right)$, and $c_\varphi = b_A^2 + \cos^2\theta\left(b_B^2 - C^2\right)$. The two solutions of Eq. (6) can be figured out as $\varphi^{(i)} = A_\varphi^{(i)}/a_\varphi$, $i = 1, 2$, in which $A_\varphi^{(i)} = -b_\varphi \pm \sqrt{b_\varphi^2 - a_\varphi c_\varphi}$.

Then value of $\varphi_1$ on the meshing limit line can be determined from the following expressions

$$\sin(\varphi_1 - \varphi) = -\frac{n_y p\, \varphi\, \cos\theta + b_A}{n_z \cos\theta(a - i_{12}p)}, \quad \cos(\varphi_1 - \varphi) = \frac{n_x p\varphi - b_B}{n_z(a - i_{12}p)} \tag{7}$$

From the above, after a value of $\theta$ is given, the related values of $\phi$, $\varphi$ and $\varphi_1$ on the meshing limit line can be worked out, respectively. As a result, a meshing limit point and its conjugate point can be determined on the tooth surfaces of the worm pair.

## 3   Solving Method of Meshing Limit Line

Letting the symbol $A_1$ denote the intersecting point between the meshing limit line and the worm addendum, in terms of the first expression in Eq. (1), the equation to determine the value of $\theta$ at the point $A_1$ can be represented as

$$f_{A_1}(\theta) = \sqrt{(x_a + a_d)^2 + (y_a \cos\gamma - za \sin\gamma)^2} - r_{a1} = 0 \tag{8}$$

where $r_{a1}$ is the radius of the addendum circle of the worm.

Letting the symbol $A_5$ denote the intersecting point between the meshing limit line and the arc at the worm gear addendum as shown in Fig. 4, by means of the geometric relation, the equation to determine the value of $\theta$ at the point $A_5$ can be built up as $(i = 1, 2)$

$$f_{A_5}^{(i)}(\theta) = \sqrt{\left[\left(n_z p^2 A_\varphi^{(i)} - a_\varphi b_B y_{o1}\right)\cos\theta - a_\varphi b_A(x_a + a_d)\right]^2 + t^2} - r_{g2} Ca_\varphi \cos\theta = 0, \tag{9}$$

where $t = \sqrt{\left[W_t - aCa_\varphi\cos\theta\right]^2 + C^2\cos^2\theta\left[a_\varphi(z_{o1} + p\varphi) - pA_\varphi^{(i)}\right]} - aCa_\varphi\cos\theta$, $C = n_z(a - i_{12}p)$, $W_t = pA_\varphi^{(i)}\cos\theta\left[n_x(x_a + a_d) + n_y y_{o1}\right] - a_\varphi\left[(x_a + a_d)b_B\cos\theta - y_{o1}b_A\right]$.

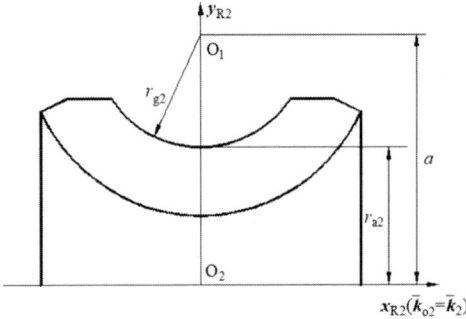

**Fig. 4.** Arc at the worm gear addendum.

Generally speaking, the existence of the meshing limit line can be boiled down to the existence of the solution of the nonlinear equations $f_{A_1}(\theta) = 0$ and $f_{A_5}^{(i)}(\theta) = 0$ $(i = 1, 2)$ in Eqs. (8) and (9). One way to detect the solution existence of them is to draw the images of the functions $f_{A_1}(\theta)$ and $f_{A_5}^{(i)}(\theta)$. If the image has an intersecting point with the abscissa axis in the solution domain, this shows that a solution of the nonlinear equation is of existence.

According to geometric relation depicted in Fig. 5, the maximum and minimum values of $\phi$ can be figured down as

$$\phi_{max} = \arcsin\left(\frac{\rho \sin \alpha_n + h_f + \Delta h}{\rho}\right), \quad \phi_{min} = \arcsin\left(\frac{\rho \sin \alpha_n - h_a - \Delta h}{\rho}\right). \quad (10)$$

On the basis of Eq. (10), the relevant value range of $\theta$ can be ascertained from the second expression in Eq. (2) as $U_1 \cup U_2 = \left[\theta_{min}^{(1)}, \theta_{max}^{(1)}\right] \cup \left[\theta_{min}^{(2)}, \theta_{max}^{(2)}\right]$.

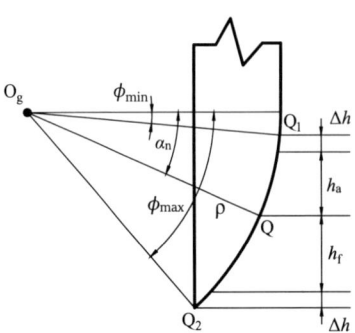

**Fig. 5.** Schematic diagram to determine value range of $\phi$.

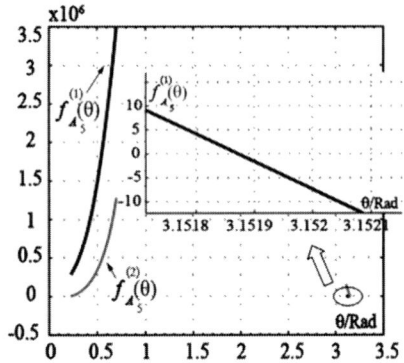

**Fig. 6.** Function image of $f_{A_5}^{(i)}(\theta)$ $(i = 1, 2)$.

## 4 Numerical Example Study

The leading processing parameters are $R_g = 120$ mm, $\alpha_n = 22.9985°$, and $\rho = 49.5$ mm. The main parameters of the worm pair are $a = 250$ mm, $i_{12} = 19$, the modulus of worm drive $m = 9$ mm, the number of worm thread $Z_1 = 2$, and the modification coefficient $\xi_2 = 1$. In this study, letting $\Delta h = 1$ mm leads to the value range of $\theta$ as $U_1 \cup U_2 = [13°, 40°] \cup [178°, 181°]$.

Due to the space limitation, the computation of the meshing limit point $A_5$ is taken to be an example to show the method proposed in the current work. The image of the function $f_{A_5}^{(i)}(\theta)$, $i = 1, 2$, is portrayed over $U_1 \cup U_2$ in Fig. 6. It reflects that the related nonlinear equation $f_{A_5}^{(i)}(\theta) = 0$ exists a solution in the interval $U_2$. The value of $\theta$ at the zero point can be used as the initial value to solve the aforesaid nonlinear equation iteratively. The value of $\varphi$ needs to be figured out from the expression $\varphi^{(1)}(\theta)$ mentioned in Sect. 2. The obtained numerical results regarding the point $A_5$ are $\theta = 180.5906°$, $\phi = 32.5956°$, $\varphi = -186.0279°$, and $\varphi_1 = -187.9187°$.

After obtaining the points $A_1$ and $A_5$, the accurate value range of $\theta$ on the meshing limit line can be determined. Ergo, some points on the meshing limit line can be ascertained by setting the value of $\theta$ within the afore-mentioned value range. In the current study, three such points, i.e. the points $A_2$ to $A_4$ in Figs. 7 and 8 are adopted. Finally, the meshing limit line can be drawn by virtue of the interpolation method based on the five points $A_1$ and $A_5$.

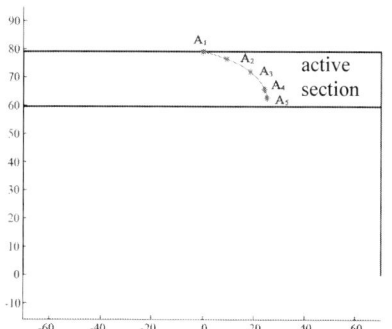

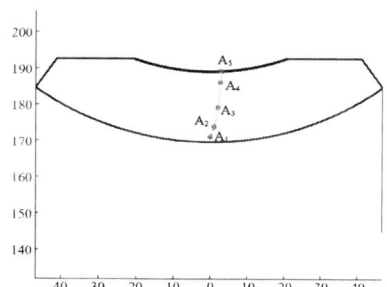

**Fig. 7.** Meshing limit line in axial section of worm.

**Fig. 8.** Conjugate line of meshing limit line on tooth surface of worm gear.

In fact, the meshing limit line is a spiral line in 3D space and its real image is not beneficial to show the axial working length of the worm. In the current work, the projection drawings of the meshing limit line and its conjugate line are drawn in the axial sections of the worm gearing in Figs. 7 and 8. According to the computing outcome, the points on the right side of the meshing limit line satisfy the relationship

$$\left(n_y p\varphi + \frac{b_A}{\cos\theta}\right)^2 + (b_B - n_x p\varphi)^2 - n_z^2(a - i_{12}p)^2 \geq 0$$

so that the right side of the meshing limit line on the worm helicoid is the active section of the worm.

## 5  Conclusions

The theory of meshing limit line is fully established for the ZC1 worm drive and some basic and important results are obtained, for instance the meshing function. A method of judging the existence of the meshing limit line is proposed. Meanwhile, by means of geometric drawing, whether a nonlinear equation has a solution over the given solving domain can be detected. A nice initial value can be achieved to solve a nonlinear equation iteratively.

The numerical results show that the meshing limit line always exists on the spiral surface of a ZC1 worm and the conjugate line of the meshing limit line is approximately located at the middle part of the tooth surface of the mating worm gear. The profile shift cannot clear off the meshing limit line. Owing to the existence of the meshing limit line, the working length of the ZC1 worm is less than half of the worm thread length. Thus it can be seen, the existence of the meshing limit line goes against improving the contact ratio of the worm pair and will affect the bearing capacity.

**Acknowledgments.** This study was funded by National Natural Science Foundation of China (51475083, U1708254), and the Fundamental Research Funds for the Central Universities (N160304012).

## References

1. Litvin, F.L.: Development of Gear Technology and Theory of Gearing. NASA Reference Publication, Cleveland (1997)
2. Dudas, I.: The Theory and Practice of Worm Gear Drives. Penton Press, London (2000)
3. Litvin, F.L., et al.: Gear Geometry and Applied Theory, 2nd edn. Cambridge University Press, Cambridge (2004)
4. Crosher, W.: Design and Application of the Worm Gear. ASME Press, New York (2002)
5. Litvin, F.L.: Meshing Principle for Gear Drives (齿轮啮合理论) (trans. D Chun). Shanghai Science and Technology Press, Shanghai (1964). (in Chinese)
6. Wu, H., et al.: Design of Worm Drive (蜗杆传动设计), vol. 1. Mechanical Industry Press, Beijing (1986). (in Chinese)
7. Wang, S.: Circular Arc Cylindrical Worm Drive (圆弧圆柱蜗杆传动). Tianjin University Press, Tianjin (1991). (in Chinese)
8. Dong, X.: Foundation of Meshing Theory for Gear Drives (齿轮啮合理论基础). China Machine Press, Beijing (1989). (in Chinese)
9. Wu, D., Luo, J.: Meshing Theory for Gear Drives (齿轮啮合理论). Science Press, Beijing (1985). (in Chinese)
10. Zhao, Y., Zhang, Y.: Determination of the Most Dangerous Meshing Point for Modified Hourglass Worm Drives. ASME J. Mech. Des. **135**(3), 034503/1–034503/5 (2013)

# Load Transfer Among Spur Gear Teeth with Tip Relief Under Non-nominal Loading Conditions

Miguel Pleguezuelos, Miryam B. Sánchez, and José I. Pedrero[✉]

UNED, Madrid, Spain
{mpleguezuelos,msanchez,jpedrero}@ind.uned.es

**Abstract.** Profile modifications are often used on gear teeth to ensure a smooth start of contact, avoiding noise, vibrations and dynamic loads induced by a sooner contact outside the pressure line which occurs due to unavoidable tooth deflections under load. A suitable tip relief on the driven gear teeth delays the start of contact, locating it at the appropriate point of the pressure line. However, the amount of relief depends on the deflection, and consequently on the load, which means that it should be calculated for a specific load, obviously coincident with the nominal load, which may be non-suitable for non-nominal, overloaded or underloaded operating conditions. In this paper, the behaviour of gear teeth with profile modifications under non-nominal loading conditions is studied. The load transfer between couples of teeth is simulated and described by analytic, approximate equations. The influence of the non-nominal load on the effective contact ratio is also presented.

**Keywords:** Spur gears · Profile modifications · Load sharing
Meshing stiffness

## 1 Introduction

The tooth deflections under load produce a relative rotation between both meshing gears, delaying the driven gear respect to the driving gear. Due to this relative rotation, the root of the driving tooth hits with the tip of the driven tooth before reaching the theoretical inner point of contact, at a point outside the pressure line. This sooner, non-conjugate contact results in a shock between both meshing teeth, inducing noise, vibrations and dynamic load, which reduce the load capacity and the gears life.

These problems can be avoided by means of profile modifications. A suitable tip relief on the driven gear teeth delays the start of contact and locates it at the appropriate point of the pressure line, ensuring a smooth contact. The small amount of material eliminated at the tooth tip (few microns) has no significant influence on the tooth stiffness; however, the gear meshing stiffness may be affected due to the different contact conditions, and consequently some important transmission parameters, as the load sharing among couples of teeth in simultaneous contact or the transmission error, may be influenced too.

© Springer Nature Switzerland AG 2019
B. Corves et al. (Eds.): EuCoMeS 2018, MMS 59, pp. 299–306, 2019.
https://doi.org/10.1007/978-3-319-98020-1_35

The above means that these transmission parameters can be controlled with the amount, length and shape of tip relief, and some studies have been published on the influence of tip relief on the gear pair behavior. Velex et al. [1] presented some theoretical developments showing that transmission error fluctuations heavily depend on profile relief definition. Bruyere and Velex [2] used the perturbation method to obtain approximate, closed-form expressions for profile relief that minimize the fluctuations of quasi-static transmission errors under load. Bruyere et al. [3] provided analytical formulations for the depths and extents of symmetric profile modifications minimizing the fluctuations of quasi-static transmission error in narrow-faced spur and helical gears. Li [4] investigated the effect of misalignments and profile modifications on tooth engagements of spur gear pairs. Ghosh and Chakraborty [5] studied the influence of tooth profile modification of spur gear pair on reducing the level of vibrations. Recently, the authors presented a study on the influence of profile modifications on the load sharing ratio [6] and the quasi-static transmission error [7] both for standard and high contact ratio spur gears. The load transfer between couples of unmodified teeth has been also studied and the actual contact ratio has been obtained.

The amount of profile modification required to shift the start of contact to its theoretical location depends on the teeth deflection, which depends on the transmitted load. Consequently, the appropriate modification for the nominal loading conditions may be not so appropriate for other operating conditions. This paper presents a study on the behavior of gear teeth with profile modifications under non-nominal loading conditions. The curve of load sharing ratio and the effective contact ratio have been also obtained.

## 2    Contact Interval of Unmodified Teeth Under Load

The load $F_i$ and the tooth pair deflection $\delta_i$ of the pair $i$ verify:

$$F_i = K_i \delta_i$$
$$F_i = \frac{K_i}{\sum_j K_j} F_T \quad \Rightarrow \quad \delta_i = \frac{F_T}{\sum_j K_j} = \delta \tag{1}$$

where $F_T$ is the total load, $K_i$ the mesh stiffness of the couple of teeth at the considered contact position and the sum is extended to all the couples in simultaneous contact. From Eq. (1), all the couples in simultaneous contact have equal deflection, and consequently the delay angle of the driven gear is $\varphi_2 = \delta/r_{b2}$.

Due to this delay the contact starts at a point of the wheel output circle –point $I$ in Fig. 1– outside the pressure line, in which the pinion profile hits the output point of the delayed wheel profile. From Fig. 1:

$$\theta_{inn} - \frac{\Delta - \delta}{r_{b1}} - v(r_{1I}) + \phi_1 - \alpha_t' = 0 \tag{2}$$

where $\delta$ is the deflection of tooth pairs in contact at this moment, $\alpha_t'$ the operating pressure angle, $r_{1I}$ and $v(r_{1I})$ the radius and the polar angle on the involute at point $I$,

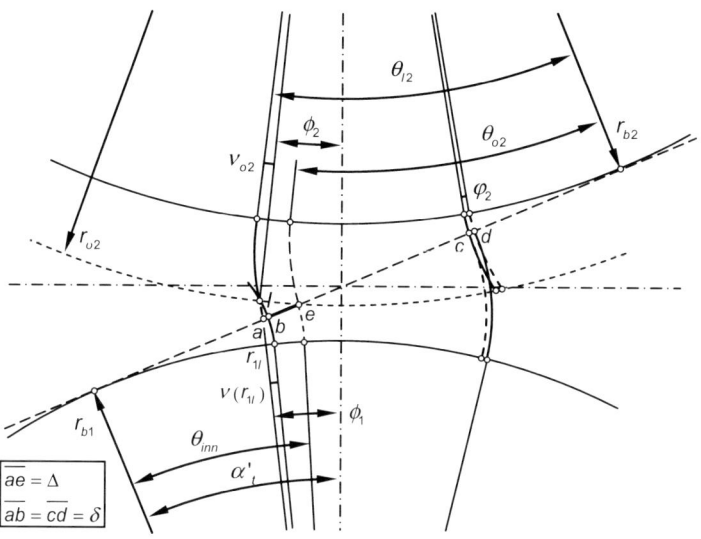

**Fig. 1.** Start of contact of loaded teeth

and $r_{b1}$ the pinion base radius. $r_{1I}$ and $\phi_1$ can be obtained by solving the triangle $(O_1, O_2, I)$ –being $O_1$ and $O_2$, the rotation centers of pinion and wheel, respectively–, and angle $\phi_2$ is:

$$\phi_2 = \theta_{I2} - v_{o2} - \alpha'_t = \frac{\Delta}{r_{b2}} + \operatorname{atan}\sqrt{\left(\frac{r_{o2}}{r_{b2}}\right)^2 - 1} - \alpha'_t \tag{3}$$

$r_{o2}$ being the wheel output radius. From Fig. 1, the additional contact angle due to the tooth deflection $\delta$ is described by $(\Delta - \delta)/r_{b1}$. Distance $\Delta$ can be computed from Eq. (2) as a function of $\delta$, however a strongly non-linear equation should be solved. An approximate, accurate value can be obtained from [6]:

$$\left(\frac{\Delta - \delta}{r_{b1}}\right) \approx \sqrt{\frac{1}{C_{p-i}}\left(\frac{\delta}{r_{b1}}\right)} \tag{4}$$

Also from Fig. 1, the increment of the contact ratio due to sooner contact is:

$$\Delta\varepsilon_i = \frac{Z_1}{2\pi}\left(\frac{\Delta - \delta}{r_{b1}}\right) \approx \frac{Z_1}{2\pi}\sqrt{\frac{1}{C_{p-i}}\left(\frac{\delta}{r_{b1}}\right)} \tag{5}$$

where $Z_1$ is the number of teeth on pinion. In Eq. (5), $\delta$ is the teeth deflection just before the contact of the new couple of teeth ($i = 0$), which from Eq. (1) is:

$$\delta = \frac{F_T}{\sum\limits_{j>0} K_j} \tag{6}$$

For non-nominal load:

$$F'_T = C_F F_T \quad \Rightarrow \quad \delta' = C_F \delta \quad \Rightarrow \quad \Delta \varepsilon'_t = \sqrt{C_F} \Delta \varepsilon_i \tag{7}$$

## 3   Load Transfer Among Couples of Unmodified Teeth

If $\phi$ is the angle rotated by the pinion from the effective start of contact, the extra interval of contact is described by $0 \leq \phi \leq (\Delta - \delta)/r_{b1}$. As seen in Fig. 1, for a pinion rotation $\phi$ inside this interval, it is verified:

$$(\Delta - \delta)_{\phi=0} - (\Delta - \delta)_\phi = r_{b1}\phi \tag{8}$$

The load at the new couple of teeth ($i = 0$) inside this extra interval and the load at the rest of couples are given by:

$$\begin{aligned} F_0(\phi) &= K_0(\phi)(\delta(\phi) - \delta_G(\phi)) \\ F_{i>0}(\phi) &= K_i(\phi)\delta(\phi) \end{aligned} \tag{9}$$

and, consequently,

$$F_T = \sum_{i \geq 0} F_i(\phi) = \sum_{i \geq 0} K_i(\phi)\delta(\phi) - K_0(\phi)\delta_G(\phi)$$
$$\delta(\phi) = \frac{F_T - K_0(\phi)\delta_G(\phi)}{\sum\limits_{i \geq 0} K_i(\phi)} \tag{10}$$

where $\delta_G$ corresponds to the rotation of the pinion until hitting the wheel tooth, which can be computed from Eqs. (8) and (4):

$$\left(\frac{\delta_G(\phi)}{r_{b1}}\right) = C_{p-i}\left(\frac{(\Delta - \delta)_\phi}{r_{b1}}\right)^2 = C_{p-i}\left(\frac{(\Delta - \delta)_{\phi=0} - r_{b1}\phi}{r_{b1}}\right)^2 \tag{11}$$

Substituting Eqs. (10) and (11) in Eq. (9), as well as the total load obtained from Eq. (6), the load sharing ratio along the extra interval is described by:

$$R_0(\phi) = \frac{F_0(\phi)}{F_T} = \frac{K_0(\phi)}{\sum\limits_{j \geq 0} K_j(\phi)}\left(1 - \frac{\delta_G(\phi)}{\delta(0)}\right)$$
$$= \frac{K_0(\phi)}{\sum\limits_{j \geq 0} K_j(\phi)}\left(1 - \left(\frac{(\Delta - \delta)_{\phi=0} - r_{b1}\phi}{(\Delta - \delta)_{\phi=0}}\right)^2\right) \tag{12}$$

corresponding to a parabola with vertex at $\phi = (\Delta - \delta)_{\phi=0}/r_{b1}$, i.e., at the theoretical inner point of contact. The load sharing ratio for the rest of tooth pairs is:

$$R_{i>0}(\phi) = \frac{F_i(\phi)}{F_T} = \frac{K_i(\phi)}{\sum\limits_{j\geq 0} K_j(\phi)} \left(1 + \frac{K_0(\phi)}{\sum\limits_{j>0} K_j(\phi)} \left(\frac{(\Delta - \delta)_{\phi=0} - r_{b1}\phi}{(\Delta - \delta)_{\phi=0}}\right)^2\right) \quad (13)$$

Figure 2(left) represents the load sharing ratio along the extra contact interval for a spur gear pair with $1 \leq \varepsilon \leq 2$.

## 4  Modified Teeth

Non-conjugate contact outside the pressure line produces edge contact at the wheel tip, which increases the contact stress levels and the wear, and reduces the efficiency and the life. These problems can be solved by a suitable profile modification, which locates the start of contact at the theoretical inner point of contact.

A tip relief is described by $\delta_R = \delta_R(\theta)$ for $\theta_{r2} \leq \theta_2 \leq \theta_{o2}$, where $\theta$ is the angular parameter of the pinion involute which meshes with point $\theta_2$ of the wheel, and subscripts $r$ and $o$ denote the inner point of modification and the outer point of the profile. Obviously, to shift the start of contact to the theoretical inner point of contact, the amount of modification at the outer point of the profile, $\delta_{Ro} = \delta_R(\theta_{inn})$, should be equal to the teeth deflection at this position, given by Eq. (6). As the delay angle is equal for all the teeth, the deflection of the couple $i$ will be:

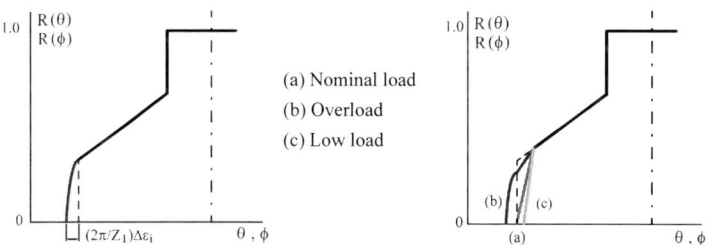

**Fig. 2.**  Load sharing ratio: (left) unmodified profile, (right) tip relieved profile

$$\delta^i(\theta) = \delta(\theta) - \delta_R^i(\theta) \quad (14)$$

consequently, the load will be:

$$F_{i \geq 0}(\theta) = K_i(\theta)\left(\delta(\theta) - \delta_R^i(\theta)\right)$$
$$F_T = \sum_{i \geq 0} F_i(\theta) = \sum_{i \geq 0} K_i(\theta)\delta(\theta) - \sum_{i \geq 0} K_i(\theta)\delta_R^i(\theta)$$
$$\delta(\theta) = \frac{F_T + \sum_{i \geq 0} K_i(\theta)\delta_R^i(\theta)}{\sum_{i \geq 0} K_i(\theta)} \tag{15}$$
$$R_i(\theta) = \frac{K_i(\theta)}{\sum_{j \geq 0} K_j(\theta)}\left(1 - \frac{1}{F_T}\sum_{j \geq 0} K_j(\theta)\left(\delta_R^i(\theta) - \delta_R^j(\theta)\right)\right)$$

The last two equations provide the relation between the tip relief and the delay angle and the load sharing ratio, allowing to control the quasi-static transmission error [7] and the load transfer curve [6], respectively.

If assumed the profile modification to affect exclusively the inner interval of maximum tooth pair contact, then $\delta_R^j(\theta) = 0$ for $j \neq 0$. Consequently:

$$R_0(\theta) = \frac{K_0(\theta)}{\sum_{j \geq 0} K_j(\theta)}\left(1 - \frac{1}{F_T}\sum_{j > 0} K_j(\theta)\delta_R^0(\theta)\right)$$
$$R_{i > 0}(\theta) = \frac{K_i(\theta)}{\sum_{j \geq 0} K_j(\theta)}\left(1 + \frac{1}{F_T}K_0(\theta)\delta_R^0(\theta)\right) \tag{16}$$

### 4.1   Overloaded Teeth

Overloaded teeth start the contact before the theoretical inner point of contact even for modified profiles. In this case, the deflections verify:

$$\delta^0(\phi) = \delta(\phi) - \delta_G(\phi) - \delta_R^0(\phi)$$
$$\delta^{i > 0}(\phi) = \delta(\phi) - \delta_R^i(\phi) \tag{17}$$

where $\delta_G(\phi)$ is the rotation of the pinion required to contact the wheel tooth tip at a position described by $(\Delta - \delta)_G$ for delay angle described by $\delta(\theta)$. Obviously, $\delta_G$ and $(\Delta - \delta)_G$ verify Eq. (4). Computing $F_i(\phi)$ and $F_T$ as above, it is obtained:

$$\delta(\phi) = \frac{F_T + \sum_{i \geq 0} K_i(\phi)\delta_R^i(\phi) + K_0(\phi)\delta_G(\phi)}{\sum_{i \geq 0} K_i(\phi)}$$
$$R_0(\phi) = \frac{K_0(\phi)}{\sum_{j \geq 0} K_j(\phi)}\left(1 - \frac{r_{b1}\delta_R^0(\phi) + C_{p-i}\left((\Delta-\delta)_{\phi=0} - r_{b1}\phi\right)^2}{r_{b1}\delta_R^0(\phi) + C_{p-i}(\Delta-\delta)_{\phi=0}^2}\right) \tag{18}$$
$$R_{i > 0}(\phi) = \frac{K_i(\phi)}{\sum_{j \geq 0} K_j(\phi)}\left(1 + \frac{K_0(\phi)}{\sum_{j > 0} K_j(\phi)}\frac{r_{b1}\delta_R^0(\phi) + C_{p-i}\left((\Delta-\delta)_{\phi=0} - r_{b1}\phi\right)^2}{r_{b1}\delta_R^0(\phi) + C_{p-i}(\Delta-\delta)_{\phi=0}^2}\right)$$

where the load sharing ratio has been calculated regarding the condition $\delta_R(\theta) = 0$ for $j \neq 0$. For the contact beyond the theoretical inner point of contact the load sharing ratio remains described with Eq. (16), but in this case the amount of modification at the outer point of the profile, $\delta_{Ro} = \delta_R(\theta_{inn})$, is not equal to the teeth deflection at this position $\delta(\theta_{inn})$, given by Eq. (6), but $C_F \delta(\theta_{inn})$. It has been represented in Fig. 2(right), line (b). In this case, the increment of the contact ratio is:

$$\Delta\varepsilon_i = \frac{Z_1}{2\pi} \left( \frac{(\Delta - \delta)_{\phi=0} - \delta_R^0(0)}{r_{b1}} \right) \tag{19}$$

## 4.2 Low Loaded Teeth

For low load, under the nominal value, the contact starts beyond the theoretical inner point of contact. Once again, the load sharing ratio is described with Eq. (16). The effective inner point of contact can be computed by doing $R_0(\theta) = 0$ in Eq. (16); however, a strongly non-linear equation should be solved. Figure 2(right) line (c) shows an example.

# 5 Conclusions

The start of contact and the load transfer among couples of teeth in simultaneous contact have been studied for standard and high contact ratio spur gears, considering both nominal and non-nominal loading conditions. Approximate equations for the offset angle and the additional contact ratio have been presented. By means of them, analytical expressions for the load sharing ratio along the effective contact interval, including the additional interval with contact outside the pressure line, have been obtained.

Gear teeth with profile modifications have been also considered. The effective start of contact, the load sharing ratio, the effective contact ratio and the offset angle have been all calculated for nominal loading, overloading and low loading conditions. In all the cases, the load transfer is described by a parabolic function along the additional contact interval (with contact outside the pressure line), while the shape of profile modification governs the load sharing along the contact interval inside the pressure line.

Of course, the downloading process of the teeth at the other limit of the contact interval, around the outer point of contact, can be studied in the same way and the same results are obtained.

**Acknowledgements.** Thanks are expressed to the Spanish Council for Scientific and Technological Research for the support of the project DPI2015-69201-C2-1-R, "Load Distribution and Strength Calculation of Gears with Modified Geometry", as well as the School of Engineering of UNED for the support of the action 2017-MEC27, "Calculation Models of Cylindrical Gears".

# References

1. Velex, P., Bruyere, J., Houser, D.R.: Some analytical results on transmission errors in narrow-faced spur and helical gears: influence of profile modifications. J. Mech. Des. **133**(3), 11 p. (2011). https://doi.org/10.1115/1.4003578. Article no. 031010
2. Bruyere, J., Velex, P.: Derivation of optimum profile modifications in narrow-faced spur and helical gears using a perturbation method. J. Mech. Des. **135**(7), 8 p. (2013). https://doi.org/10.1115/1.4024374. Article no. 071009
3. Bruyere, J., Gu, X., Velex, P.: On the analytical definition of profile modifications minimizing transmission error variations in narrow-faced spur and helical gears. Mech. Mach. Theory **92**, 257–272 (2015)
4. Li, S.: Effect of misalignment error, tooth modifications and transmitted torque on tooth engagements of a pair of spur gears. Mech. Mach. Theory **83**, 125–136 (2015)
5. Ghosh, S.S., Chakraborty, G.: On optimal tooth profile modification for reduction of vibration and noise in spur gear pairs. Mech. Mach. Theory **105**, 145–163 (2016)
6. Pedrero, J.I., Pleguezuelos, M., Sánchez, M.B.: Load sharing model for spur gears with tip relief. In: Proceedings of the International Conference on Gears 2017, Munich (2017)
7. Pedrero, J.I., Pleguezuelos, M., Sánchez, M.B.: Control del error de transmisión cuasiestático mediante rebaje de punta en engranajes rectos de perfil de evolvente. Anales de Ingeniería Mecánica 21 (2018)

# Mechanics of Robots and Manipulators

# Deflection Modeling of a Manipulator for Mechanical Design

Tobias Weiser[1]([✉]) and Burkhard Corves[2]

[1] KUKA Roboter GmbH, Augsburg, Germany
tobias.weiser@kuka.com
[2] Department of Mechanism Theory, Dynamics of Machines and Robotics,
RWTH Aachen University, Aachen, Germany
corves@igmr.rwth-aachen.de

**Abstract.** A manipulator model for use in a design process of manipulator mechanics is shown in this paper. The target is to minimize deformation and vibrations in application. The focus is to model manipulator components and merge them in a manipulator model. The proposed model uses the Floating Frame of Reference approach for modeling the structural parts of the manipulator. The elastic modeling of bearings is introduced. An approach to consider stiffness and damping of bolted joint is presented. The cable stiffness is calculated. A simple drivetrain model is derived.

**Keywords:** Manipulator design · Elastic multibody system
Bearing stiffness · Cable stiffness · Joint stiffness

## 1 Introduction

This paper presents a model of a manipulator with rotational joints for use in a mechanical design process. The scope of this model is to support reducing deformation and vibration through appropriate design of the manipulator as well as dimensioning and selection of the manipulator components. Since the model is used in a design process there is no requirement on real-time calculation.

The manipulator is modeled component oriented as an elastic multibody system. The contribution of manipulator components to the vibration and deformation behaviour is presented by a model approach for each component.

Besides modeling for mechanical design, manipulator dynamics were used for controller algorithms [11] or trajectory planning [19]. Additionally, models of dynamics of flexible manipulators have been developed in parallel to the design of controller algorithms, mechanics and trajectory planning [21].

The link elements are modeled with a distributed link flexibility using the floating frame of reference approach. [22] shows an overview of elastic multibody system dynamics publications.

An approach to consider the bearings contribution to the manipulator dynamics is introduced. In rotordynamics the bearing stiffness is required to

© Springer Nature Switzerland AG 2019
B. Corves et al. (Eds.): EuCoMeS 2018, MMS 59, pp. 309–316, 2019.
https://doi.org/10.1007/978-3-319-98020-1_36

model elastic suspended systems [4]. The bearing stiffness is relevant in noise effects [18] and proper design of rolling elements due to noise reduction [16]. Furthermore, the stiffness is used for predicting and analyzing bearing damage [24].

The stiffness of cables in the manipulator influences manipulator deflections and vibrations. The origins of the cable modeling are investigations on dynamic effects for wire ropes [9], strands for power transmission lines [13] and suspension bridges [1]. Also stress and fatigue calculations for insulations [17] and wires were performed [8].

The joints connect all components, therefore their stiffness and damping contribute to the manipulator dynamics. The joint damping and stiffness is shown. Damping modeling and identification is described in [12].

## 2   Structural Parts

The structural parts represent the kinematic dimensions of the manipulator and connect the joints $i$ and $i+1$ (see Fig. 1). In this model, the structural parts are modeled with the floating frame of reference approach [23]. The motion $\mathbf{r}$ of the link body $B$ is split into the reference motion $\mathbf{r}_B$, the motion of the material points $\mathbf{u}_0$ and the displacement field $\mathbf{u}_F$ (1).

$$\mathbf{r} = \mathbf{r}_B + \mathbf{u}_0 + \mathbf{u}_F \tag{1}$$

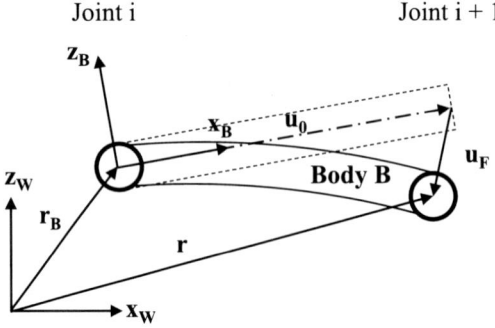

**Fig. 1.** Floating frame of reference

We denote the generalized Newton-Euler equations of motion with the beam mass $m$, $[3 \times 3]$-Identity Matrix $\mathbf{I}_3$, the translational acceleration $\mathbf{a}_{B,t}$ and the angular acceleration $\alpha_{B,r}$ of the beam, the second derivative with respect to time of the elastic coordinates $\ddot{\mathbf{q}}_{B,f}$, the inertia tensor $\mathbf{J}$, the center of mass position $\mathbf{r}_{cm}$, the inertia coupling matrices $\mathbf{C}_t$ and $\mathbf{C}_r$, the structural mass matrix $\mathbf{M}_r$, the external forces $\mathbf{f}_{e,t}, \mathbf{f}_{e,r}, \mathbf{f}_{e,f}$, the gyroscopic and centripetal terms $\mathbf{h}_{\omega,t}, \mathbf{h}_{\omega,r}, \mathbf{h}_{\omega,f}$, the structural stiffness $\mathbf{K}_e$ and the damping matrix $\mathbf{D}_e$ [23]:

$$\begin{pmatrix} m\mathbf{I}_3 & m\mathbf{r}_{cm} & \mathbf{C}_t \\ m\mathbf{r}_{cm} & \mathbf{J} & \mathbf{C}_r \\ \mathbf{C}_t & \mathbf{C}_r & \mathbf{M}_e \end{pmatrix} \begin{pmatrix} \mathbf{a}_{B,t} \\ \alpha_{B,r} \\ \ddot{\mathbf{q}}_{B,f} \end{pmatrix} = \begin{pmatrix} \mathbf{h}_{\omega,t} \\ \mathbf{h}_{\omega,r} \\ \mathbf{h}_{\omega,f} \end{pmatrix} - \begin{pmatrix} \mathbf{0} \\ \mathbf{0} \\ \mathbf{K}_e\mathbf{q}_{B,f} + \mathbf{D}_e\dot{\mathbf{q}}_{B,f} \end{pmatrix} + \begin{pmatrix} \mathbf{f}_{e,t} \\ \mathbf{f}_{e,r} \\ \mathbf{f}_{e,f} \end{pmatrix} \tag{2}$$

To calculate the beam deformation, the boundary conditions have to be selected [7]. The Table 1 shows an overview for a serial manipulator with rotating axes.

**Table 1.** Link: Boundary conditions of beam: C - clamped, S - supported, F - free

| Link type | Bending y $i$ | Bending y $i+1$ | Bending z $i$ | Bending z $i+1$ | Torsion Elongation $i$ | Torsion Elongation $i+1$ |
|-----------|-----------|-------------|-----------|-------------|--------------------|----------------------|
| Base | C | F | C | F | C | F |
| Twist | C | F | S | F | C | F |
| Rotating | F | F | F | F | F | F |

## 3    Bearings

In a rotational joint bearings enable the rotation and pass loads of the whole manipulator to its base. This chapter presents the bearing stiffness of single- and double-row roller bearings. Ball bearings, tapered bearings and cross-roller bearings can be simulated. The loads $F_{x,y,zbm}$, $M_{x,ybm}$ and the displacements $\delta_{x,y,zbm}$, $\beta_{x,ybm}$ acting on a bearing (Fig. 2(a)) are transformed on each rolling element $i$ in row $j$ (Fig. 2(b)) [25]. The force and torque equilibrium $\mathbf{F}_{EB}(\mathbf{q}_B)$ is calculated with the displacement vector $\mathbf{q}_B$, the contact angle under load $\alpha_{ij}$, the distance $e$ between both rows, the row coefficient $c_1 = [-1,1]$ , the force acting on each rolling element $Q_j^i$ and the position angle $\psi_j^i$ of the rolling element:

$$\mathbf{q}_B = \begin{bmatrix} \delta_{xm}\delta_{ym}\delta_{zm}\beta_{xm}\beta_{ym} \end{bmatrix}^T \tag{3}$$

$$\mathbf{F}_{EB}(\mathbf{q}_B) = \begin{bmatrix} F_{xbm} \\ F_{ybm} \\ F_{zbm} \\ M_{xbm} \\ M_{ybm} \\ M_{zbm} \end{bmatrix} = \sum_{i=1}^{2}\sum_{j=1}^{Z} Q_j^i(\mathbf{q}_B) \begin{bmatrix} cos(\alpha_j^i(\mathbf{q}_B))cos(\psi_j^i) \\ cos(\alpha_j^i(\mathbf{q}_B))sin(\psi_j^i) \\ sin(\alpha_j^i(\mathbf{q}_B)) \\ R^*(\mathbf{q}_B)sin(\psi_j^i) \\ -R^*(\mathbf{q}_B)cos(\psi_j^i) \\ 0 \end{bmatrix} \tag{4}$$

$$R^*(\mathbf{q}_B) = R\ sin(\alpha_j^i(\mathbf{q}_B)) - c_1\ e\ cos(\alpha_j^i(\mathbf{q}_B)) \tag{5}$$

The load $Q_j^i(\mathbf{q}_B)$ on each rolling element $j$ in row $i$ depends on the type of roller bearing $W$ and is calculated with the Hertz'ian exponent $n$. For elliptical

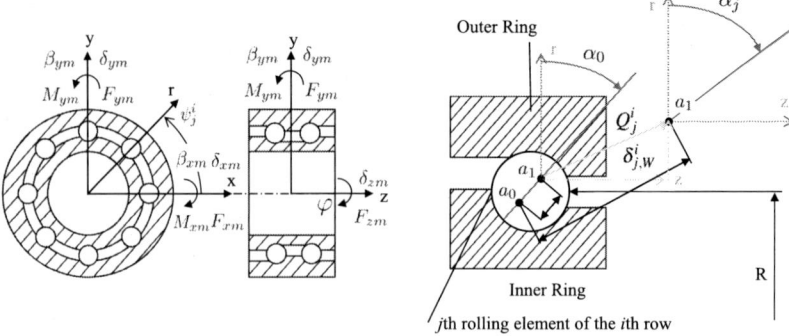

(a) Bearing loads and displacements          (b) Loaded bearing

**Fig. 2.** Bearing stiffness

Hertz'ian contact (ball - elastic half-space) we assume $n = 3/2$. For a rectangular contact (cylinder - elastic half-space) we assume $n = 10/9$ [10]. Depending on the geometry and the material properties of the roller we denote the load $Q_j^i(\mathbf{q}_B)$ on each rolling element $j$ in row $i$ with the Hertz'ian stiffness constant $K_n$ and the displacement of the centers of the raceway groove curvature $\delta_{j,W}^i(\mathbf{q}_B, F_{x,y,zbm}, M_{x,ybm})$:

$$Q_j^i(\mathbf{q}_B) = K_n \cdot \delta_{j,W}^i(\mathbf{q}_B, F_{x,y,zbm}, M_{x,ybm}) \, (\psi_j^i)^n \qquad (6)$$

The $[5 \times 5]$, full occupied stiffness matrix $\mathbf{K}(\mathbf{q}_B)$ can be calculated by means of the partial differentiation of the force and torque vectors $\mathbf{F}_{EB}(\mathbf{q}_B)$ according to the generalized coordinates of the bearing deformation $\mathbf{q}_B$ in Eq. (3) [6].

$$\mathbf{K}(\mathbf{q}_B) = \frac{\partial \mathbf{F}_{EB}(\mathbf{q}_B)}{\partial \mathbf{q}_B} \qquad (7)$$

For the deflection analysis damping and friction is neglected. For future modal analysis or positioning analysis the damping and the friction of bearings have to be considered.

## 4   Bolted Joints

Manipulator components are connected by joints. Generally, most joints are bolted or crimped. The normal contact stiffness $k_{nrm}$ and the transverse contact stiffness of a joint $k_{trn}$ is calculated with the radius of the contact area $a$, Young's modulus $E_i$, Poisson's ratio $\nu_i$ and the shear modulus $G_i$ for each material $i = 1, 2$ in the contact [15]:

$$\frac{1}{E^*} = \frac{1-\nu_1^2}{E_1} + \frac{1-\nu_2^2}{E_2}$$
$$k_{nrm} = 2aE^* \tag{8}$$
$$\frac{1}{G^*} = \frac{1-\nu_1^2}{G_1} + \frac{1-\nu_2^2}{G_2}$$
$$k_{trn} = 2aG^* \tag{9}$$

The damping behaviour of the structural parts through material damping is global, whereas the joint damping is local and dependent on the eigenmodes [2]. The joint damping is up to three times higher than the material damping [20].

The damping torques $M_{JD}$ are calculated with the relative manipulator joint speed $\dot{\varphi}$ and the damping coefficient $d_J$ [5].

$$M_{JD} = d_J\dot{\varphi} \tag{10}$$

Both stiffness and damping of the joint is dependent on e.g. the joint load, the surface roughness and the corrosion of the contact.

## 5   Cable

Cables in a manipulator transport power, electric signals and other media like hydraulic fluids from the manipulator base to all structural parts. A cable consists of multi-level helices. The first level helix consists of conductors and in the second level helix wires are wrapped around the center line of the conductor (Fig. 3(a)) with the lay angle $\alpha_{cn,wm}$. If we simplify the cable of a hollow shaft joint to a solid copper wire the stiffness is up to 10% of the gearbox torsional stiffness. This ratio is not negligible and therefore detailed modeling is required. This chapter shows how to model, derive and consider cable stiffness in a manipulator model.

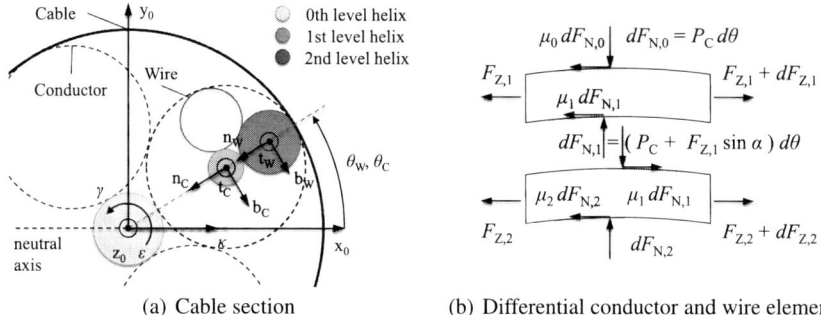

(a) Cable section                    (b) Differential conductor and wire element

**Fig. 3.** Cable stiffness

The cable stiffness consists of two parts of bending forces. The first one is pure bending due to the cable strains of torsion $\gamma$, bending $\kappa$ and elongation $\varepsilon$

(Fig. 3(a)). These strains are iteratively transformed from the lowest helix level $\{\mathbf{x}_0, \mathbf{y}_0, \mathbf{z}_0\}$ to each higher helix level of the conductor $cn$ $\{\mathbf{t}_C, \mathbf{n}_C, \mathbf{b}_C\}$ and the wire $wm$ $\{\mathbf{t}_W, \mathbf{n}_W, \mathbf{b}_W\}$ (Fig. 3(a)).

The second part of tensile and bending loads in each wire is the interlayer slip and friction. The differential equations for the tensile force $F_{Z,1}, F_{Z,2}$ of a differential conductor $cn$ and wire element $wm$ in each layer with the friction coefficient $\mu_0$ and the force $P$ by the pressure of the jacket, the friction coefficient $\mu_1, \mu_2$ between the conductors of different layers, the numbers of conductors $cm$ and wires $wm$ in the ultimate $K_1$ and penultimate layer $K_2$ and the lay angle $\alpha_1, \alpha_2$ must be solved [8].

$$dF_{Z,1} = \mu_1 \left( sin(\alpha_1) * F_{Z,1} \right) d\theta + \mu_0 P d\theta \tag{11}$$

$$dF_{Z,2} = \mu_2 \left( (sin(\alpha_1) F_{Z,1} + P) \frac{K_1}{K_2} + sin(\alpha_2) F_{Z,2} \right) d\theta +$$

$$\mu_1 \left( sin(\alpha_1) F_{Z,1} + P \right) \frac{K_1}{K_2} d\theta \tag{12}$$

Finally with the conductor and wire forces $F_{Z_m,n}$ we obtain the sum of wire and conductor torques $M_{m,n}$ and the overall wire bending $\kappa_w^*$ and bending stiffness $EI$ [14]. Depending on the loads and strains, the conductors and wire are either in stick $EI_{stick}$ or in slip $EI_{slip}$ state. The torsional stiffness $GJ$ is derived similarly.

$$EI = \frac{M_{m,n}}{\kappa_w^*} = \begin{cases} EI_{slip} \\ EI_{stick} \end{cases} \tag{13}$$

## 6   Drivetrain

In this approach, the drivetrain consists of at least the drive as well as a transmission element like a gearbox, belt drive or spur gears. The total stiffness of the drivetrain is concentrated in a linear spring with the stiffness $c_{AS}$ [3].

## 7   Results

The dynamics of the components described in the previous chapters are assembled in a manipulator model. The model is implemented within Modelica, which is an object oriented model language.

For validation of the manipulator model, static experiments with a KR 6 R900 sixx have been performed at the IGMR, RWTH Aachen. For multiple robot positions a force of $F_{x,y,z} = 99.871N$ is applied on the flange of the robot in three directions of the world coordinate system $\{X, Y, Z\}$.

The experiment results show slightly higher deflections. The stiffness of the manipulator model is higher.

For one robot position the Table 2 shows typical results for loading the manipulator flange with the force $F_{X,S}, F_{Y,S}, F_{Z,S}$ of the model and $F_{X,E}, F_{Y,E}, F_{Z,E}$ of the experiment.

**Table 2.** Static displacements

| Deflection | $F_{X,S}$ | $F_{X,E}$ | $F_{Y,S}$ | $F_{Y,E}$ | $F_{Z,S}$ | $F_{Z,E}$ |
|---|---|---|---|---|---|---|
| X in $mm$ | 0.531 | 0.577 | 0 | 0.002 | −0.483 | −0.475 |
| Y in $mm$ | 0 | −0.075 | 0.813 | 1.037 | 0 | 0.011 |
| Z in $mm$ | −0.484 | −0.498 | 0 | −0.037 | −1.044 | −1.181 |

# 8    Conclusions and Outlook

In the previous chapters, different approaches of components which contribute to the manipulator system dynamics, deflection and vibration characteristics were shown. The subsystems of the structural parts as a flexible multibody system, the bearings, the cable and joints are merged in a manipulator model. This model can be used in an iterative design process or in an optimization loop. Dynamic validation measurements are currently in progress.

The drivetrain model is simple and consists of a drive and a reducing element like a gearbox. Experiments show that the current manipulator model is stiffer than the experimental specimen. Further investigations in drivetrain components like belt drives, shafts or gears are required.

The cable model will be extended by merging the multi-level helix approach with friction [8] to the thin wire theory.

Experiments for more detailed information about the joint stiffness and damping for an a priori use in the design process will be performed. One is to create a database of different joint sizes and perform parameter identification. Another step is to identify the joint model parameters by step by step identification during assembly [12].

Finally, an optimization algorithm with uncertainties will be set up for use in the design process.

# References

1. Betti, R., Yanev, B.: Conditions of suspension bridge cables: New York city case study. Transp. Res. Rec.: J. Transp. Res. Board **1654**, 105–112 (1999)
2. Bograd, S., Schmidt, A., Gaul, L.: Joint damping prediction by thin layer elements. In: Proceedings of the IMAC 26th Society of Experimental Mechanics Inc. Bethel, CT (2008)
3. De Luca, A., Tomei, P.: Theory of robot control. Elastic Joints, pp. 179–218 (1996)
4. Gasch, R., Nordmann, R., Pfützner, H.: Rotordynamik. Springer, Heidelberg (2006)
5. Großmann, K., Rudolph, H.: Dämpfungsbeschreibung für die modellgestützte dynamische Strukturanalyse: Modellierungsansätze zur Beschreibung strukturrelevanter Dämpfung an Werkzeugmaschinen. ZWF Zeitschrift für wirtschaftlichen Fabrikbetrieb **103**(11), 767–773 (2008)
6. Gunduz, A.: Multi-dimensional stiffness characteristics of double row angular contact ball bearings and their role in influencing vibration modes. Ph.D. thesis, The Ohio State University (2012)

7. Heckmann, A.: On the choice of boundary conditions for mode shapes in flexible multibody systems. Multibody Syst. Dyn. **23**(2), 141–163 (2010)
8. Inagaki, K., Ekh, J., Zahrai, S.: Mechanical analysis of second order helical structure in electrical cable. Int. J. Solids Struct. **44**(5), 1657–1679 (2007)
9. Le Marrec, L., Zhang, D., Ostoja-Starzewski, M.: Three-dimensional vibrations of a helically wound cable modeled as a timoshenko rod. Acta Mechanica **229**, 1–19 (2017)
10. Lim, T., Singh, R.: Vibration transmission through rolling element bearings, part i: bearing stiffness formulation. J. Sound Vibr. **139**(2), 179–199 (1990)
11. Moberg, S.: Modeling and control of flexible manipulators. Ph.D. thesis, Linköping University Electronic Press (2010)
12. Niehues, K.K.: Identifikation linearer Dämpfungsmodelle fuer Werkzeugmaschinenstrukturen, vol. 318. Herbert Utz Verlag (2016)
13. Papailiou, K.: On the bending stiffness of transmission line conductors. IEEE Trans. Power Deliv. **12**(4), 1576–1588 (1997)
14. Papailiou, K.O.: Die Seilbiegung mit einer durch die innere Reibung, die Zugkraft und die Seilkrümmung veränderlichen Biegesteifigkeit. Ph.D. thesis, ETH Zürich (1995)
15. Popov, V.: Kontaktmechanik und Reibung: von der Nanotribologie bis zur Erdbebendynamik. Springer, Heidelberg (2016)
16. Qian, W., Jacobs, G.: Dynamic simulation of cylindrical roller bearings. Technical report, Lehrstuhl und Institut für Maschinenelemente und Maschinengestaltung (2014)
17. Raghava, B.V., Reddy, S.K.: Deformation and durability studies of insulation polymers. Ph.D. thesis, University of Akron (2008)
18. Razpotnik, M., Bischof, T., Boltežar, M.: The influence of bearing stiffness on the vibration properties of statically overdetermined gearboxes. J. Sound Vibr. **351**, 221–235 (2015)
19. Reiner, M.J.: Modellierung und Steuerung von strukturelastischen Robotern. Dissertation, Technical University Munich, Germany (2010)
20. Rudolph, H., Ihlenfeldt, S.: Dämpfung in verspannten Fugen - Teil1. ZWF Zeitschrift für wirtschaftlichen Fabrikbetrieb **111**(7–8), 439–444 (2016)
21. Seifried, R.: Dynamics of Underactuated Multibody Systems: Modeling, Control and Optimal Design, vol. 205. Springer, Heidelberg (2013)
22. Shabana, A.A.: Flexible multibody dynamics: review of past and recent developments. Multibody Syst. Dyn. **1**(2), 189–222 (1997)
23. Shabana, A.A.: Dynamics of Multibody Systems. Cambridge University Press, Cambridge (2013)
24. Tadina, M., Boltežar, M.: Improved model of a ball bearing for the simulation of vibration signals due to faults during run-up. J. Sound Vibr. **330**(17), 4287–4301 (2011)
25. Weiser, T., Corves, B.: Modelling of roller bearings. In: Proceedings of the 12th International Modelica Conference. Linköping University Electronic Press (2017)

# Self-Motion of the 3-PPPS Parallel Robot with Delta-Shaped Base

Damien Chablat[1]([⊠]), Erika Ottaviano[2], and Swaminath Venkateswaran[3]

[1] CNRS, Laboratoire des Sciences du Numérique de Nantes,
UMR CNRS 6004, 44321 Nantes, France
Damien.Chablat@cnrs.fr
[2] DICeM, University of Cassino and Southern Lazio,
via G. Di Biasio 43, 03043 Cassino, FR, Italy
ottaviano@unicas.it
[3] Ecole Centrale de Nantes, Laboratoire des Sciences du Numérique de Nantes,
UMR CNRS 6004, 44321 Nantes, France
swaminath.venkateswaran@ls2n.fr

**Abstract.** This paper presents the kinematic analysis of the 3-PPPS parallel robot with an equilateral mobile platform and an equilateral-shaped base. Like the other 3-PPPS robots studied in the literature, it is proved that the parallel singularities depend only on the orientation of the end-effector. The quaternion parameters are used to represent the singularity surfaces. The study of the direct kinematic model shows that this robot admits a self-motion of the Cardanic type. This explains why the direct kinematic model admits an infinite number of solutions in the center of the workspace at the "home" position but has never been studied until now.

**Keywords:** Parallel robots · 3-PPPS · Singularity analysis
Kinematics · Self-motion

## 1 Introduction

It has been shown that by applying simplifications in parallel robot design parameters, self-motions may appear [1–3]. For example, Bonev et al. demonstrated that all singular orientations of the popular 3-RRR spherical parallel robot design (known as the Agile Eye) correspond to self motions [4], but arguably this design has the "best" spherical wrist. The Cardanic motion can be found as self-motion for two robots in literature, the 3-RPR parallel robot [5] and the PamInsa robot [6].

Most of the examples for a fully parallel 6-DOF manipulator can be categorized by the type of their six identical serial chains namely UPS [7,9–12], RUS [13] and PUS [14]. However for all these robots, the orientation of the workspace is rather limited due to the interferences between the legs.

To solve this problem, new parallel robot designs with six degrees of freedom appeared recently having only three legs with two actuators per leg. The Monash

© Springer Nature Switzerland AG 2019
B. Corves et al. (Eds.): EuCoMeS 2018, MMS 59, pp. 317–324, 2019.
https://doi.org/10.1007/978-3-319-98020-1_37

Epicyclic-Parallel Manipulator (MEPaM), called 3-PPPS is a six DOF parallel manipulator with all actuators mounted on the base [15]. Several variants of this robot have been studied were the three legs are made with three orthogonal prismatic joints and one spherical joint in series. The first two prismatic joints of each legs are actuated. In the first design, the three legs are orthogonal [16]. For this design, the robot can have up to six solutions to the Direct Kinematic Problem (DKP) and is capable of making non-singular assembly mode change trajectories [16]. In [15], the robot was made with an equilateral mobile platform and an equilateral-shape base. In [17,18], the robot was designed with an equilateral mobile platform and a U-shaped base. For this design, the Direct Kinematic Model (DKM) is simple and can be solved with only quadratic equations.

The common point of these three variants is that the parallel singularity is independent of the position of the end-effector. This paper presents the self-motion of the 3-PPPS parallel robot derived from [15] with an equilateral mobile platform and an equilateral-shaped base.

The outline of this article is as follows. In the next section, the manipulator architecture as well as its associated constraint equations are explained. Followed by that, calculations for parallel singularities as well as self-motions are presented. The article then ends with conclusions.

## 2   Mechanism Architecture

The robot under study is a simplified version of the MEPaM that has been developed at the Monash University [15,19]. This architecture is derived from the 3-PPSP that was introduced earlier [20].

### 2.1   Geometric Parameters

For this parallel robot, the three legs are identical and it consists of two actuated prismatic joints, a passive prismatic joint and a spherical joint (Fig. 1). The axes of first three joints of each leg form an orthogonal reference frame.

The coordinates of the point $C_1$ are $\rho_{1x}$, $\rho_{1y}$ and $\rho_{1z}$, wherein the last two are actuated. We assume an origin $A_i$ for each legs and the equations are given by:

$$\mathbf{A}_1 = [2, 0, 0]^T \tag{1}$$
$$\mathbf{A}_2 = [-1, \sqrt{3}, 0]^T \tag{2}$$
$$\mathbf{A}_3 = [-1, -\sqrt{3}, 0]^T \tag{3}$$

The coordinates of $C_2$ and $C_3$ are obtained by a rotation around the $z$ axis by $2\pi/3$ and $-2\pi/3$ respectively and the equations are as follows:

$$\mathbf{C}_1 = [\rho_{1x}, \rho_{1y}, \rho_{1z}]^T \tag{4}$$
$$\mathbf{C}_2 = [-\rho_{2x}/2 - \sqrt{3}\rho_{2y}/2, \sqrt{3}\rho_{2x}/2 - \rho_{2y}/2, \rho_{2z}]^T \tag{5}$$
$$\mathbf{C}_3 = [-\rho_{3x}/2 + \sqrt{3}\rho_{3y}/2, -\sqrt{3}\rho_{3x}/2 - \rho_{3y}/2, \rho_{3z}]^T \tag{6}$$

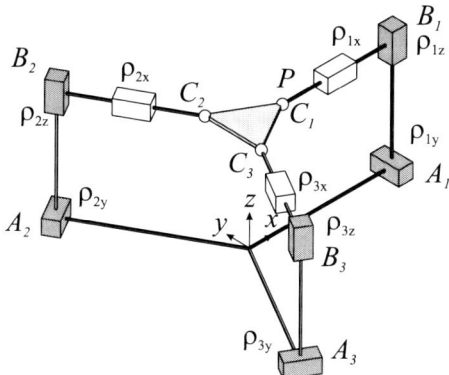

**Fig. 1.** A scheme for the 3-PPPS parallel robot and its parameters in its "home" position with the actuated prismatic joints in blue, the passive joints in white and the mobile platform drawn in green with $x = 1/\sqrt{3}, y = 0, z = 0, q_1 = 1, q_2 = 0, q_3 = 0, q_4 = 0$

Three locations are now written to describe the mobile platform in the moving frame for an equilateral triangle whose edge lengths are set to one.

$$\mathbf{V}_1 = [0, 0, 0]^T \tag{7}$$
$$\mathbf{V}_2 = [-\sqrt{3}/2, 1/2, 0]^T \tag{8}$$
$$\mathbf{V}_3 = [-\sqrt{3}/2, -1/2, 0]^T \tag{9}$$

Generally in the robotics community, Euler or Tilt-and-Torsion angles are used to represent the orientation of the mobile platform [16]. These methods have a physical meaning, but there are singularities to represent certain orientations. The unit quaternions give a redundant representation to define the orientation but at the same time it gives a unique definition for all orientations. The rotation matrix $\mathbf{R}$ is described by:

$$\mathbf{R} = \begin{bmatrix} 2q_1^2 + 2q_2^2 - 1 & -2q_1q_4 + 2q_2q_3 & 2q_1q_3 + 2q_2q_4 \\ 2q_1q_4 + 2q_2q_3 & 2q_1^2 + 2q_3^2 - 1 & -2q_1q_2 + 2q_3q_4 \\ -2q_1q_3 + 2q_2q_4 & 2q_1q_2 + 2q_3q_4 & 2q_1^2 + 2q_4^2 - 1 \end{bmatrix} \tag{10}$$

Here $q_1 \geq 0$ and $q_1^2 + q_2^2 + q_3^2 + q_4^2 = 1$. We can write the coordinates of the mobile platform using the previous rotation matrix as:

$$\mathbf{C}_i = \mathbf{R}\mathbf{V}_i + \mathbf{P} \quad \text{where} \quad \mathbf{P} = [x, y, z]^T \tag{11}$$

Thus, we can write the set of constraint equations with the position of $C_i$ in the both reference frames by the relations:

$$\rho_{1y} = y \tag{12}$$

$$\rho_{1z} = z \tag{13}$$

$$(2q_1q_4 - x)\sqrt{3} + 2q_1{}^2 + 3q_2{}^2 - q_3{}^2 - y - 2\rho_{2y} = 1 \tag{14}$$

$$-\sqrt{3}q_1q_3 + \sqrt{3}q_2q_4 - q_1q_2 - q_3q_4 + \rho_{2z} = z \tag{15}$$

$$(2q_1q_4 + x)\sqrt{3} - 2q_1{}^2 - 3q_2{}^2 + q_3{}^2 - y - 2\rho_{3y} = -1 \tag{16}$$

$$-\sqrt{3}q_1q_3 + \sqrt{3}q_2q_4 + q_1q_2 + q_3q_4 + \rho_{3z} = z \tag{17}$$

## 2.2   Constraint Equations

The main problem is to find the location of the mobile platform by looking for the values of the passive prismatic joints $[\rho_{1x}, \rho_{1y}, \rho_{1z}]$ as proposed in [21]. The distances between any couple of points $C_i$ are given by:

$$||\mathbf{C}_1 - \mathbf{C}_2|| = ||\mathbf{C}_1 - \mathbf{C}_3|| = ||\mathbf{C}_2 - \mathbf{C}_3|| = 1 \tag{18}$$

This method is also used by [15] for the 3-PPPS by using the dialytic elimination [8]. A fourth degree polynomial equation with complicate coefficients is then obtained. However, no self-motion was detected.

Unfortunately, when we want to solve the DKM for the "home" position, i.e. $\rho_{iy} = 0$ and $\rho_{iz} = 0$ and $i = 1, 2, 3$, we have an infinite number of solutions, which correspond to the self-motion. This result remains the same if we set $\rho_{1z} = \rho_{2z} = \rho_{3z}$. Assuming that this motion is in a plane parallel to the plane (0xy), we write the system coefficients for $\rho_{iz} = 0$ with $i = 1, 2, 3$. The fourth degree polynomial then degenerates and a quadratic polynomial equation is obtained which is of the form:

$$9(\rho_{1y} + \rho_{2y} + \rho_{3y})^2\rho_{1x}^2 +$$
$$6\sqrt{3}(\rho_{1y} + \rho_{2y} + \rho_{3y})^2(\rho_{2y} - \rho_{3y})\rho_{1x} +$$
$$(\rho_{1y} + \rho_{2y} + \rho_{3y})^2(\rho_{1y}^2 + 2\rho_{1y}\rho_{2y} + 2\rho_{1y}\rho_{3y} + 4\rho_{2y}^2 - 4\rho_{2y}\rho_{3y} + 4\rho_{3y}^2 - 3) = 0 \tag{19}$$

Here all the terms of the equation cancel out each other when $(\rho_{1y} + \rho_{2y} + \rho_{3y} = 0)$. With these conditions, the robot becomes similar to the 3-RPR parallel robot for which its self-motion produces a Cardanic motion. Since the self-motion corresponds to singular configurations, the objective now will be to locate these singularities with respect to other singularities. Geometrically, the self-motion has been described when the motion axes of passive joints intersect at a single point and an angle of $2\pi/3$ is formed between each axis. An example is shown in Fig. 2 where the green mobile platform is one assembly mode obtained in the "home" position. This phenomenon has already been characterized for the 3-RPR parallel robot [5] and the PamInsa robot in [6]. To validate this result, we use the Siropa library programmed under Maple and its "InfiniteEquations" function [22]. Three condition are obtained which are $(\rho_{1y} + \rho_{2y} + \rho_{3y} = 0)$, $\rho_{1z} = \rho_{2z}$ and $\rho_{2z} = \rho_{3z}$.

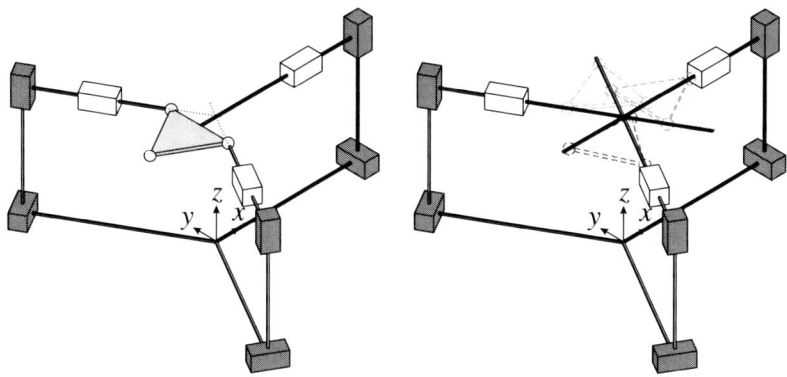

**Fig. 2.** Set of postures to describe the Cardanic motion starting from the "home" position

## 3  Singularity Analysis

The singular configurations of the 3-PPPS robot have been studied in several articles with either a parametrization of orientations using Euler angles or Quaternions [17]. Serial and parallel Jacobian matrices can be computed by differentiating the constraint equations with respect to time [23–25]. These Jacobian serial and parallel matrices must satisfy the following relationship

$$\mathbf{At} + \mathbf{B}\dot{\rho} = 0 \tag{20}$$

where $\mathbf{t}$ is the twist of the moving platform and $\dot{\rho}$ is the vector of the active joint velocities.

According to the leg topology of the 3-PPPS robot, there is no serial singularity because the determinant of the $\mathbf{B}$ matrix does not vanish. Using the same approach as in [19], we can determine the matrix $\mathbf{A}$ and its determinant can be factorized as follows:

$$\left(q_1{}^2 - q_2{}^2 - q_3{}^2 + q_4{}^2\right)\left(q_1 - q_4\right)\left(q_1 + q_4\right) = 0 \tag{21}$$

To calculate this result, we use the "ParallelSingularties" function from the Siropa library [22]. To represent this surface, we eliminate $q_1$ thanks to the relation on the quaternions

$$\left(q_2{}^2 + q_3{}^2 + 2q_4{}^2 - 1\right)\left(2q_2{}^2 + 2q_3{}^2 - 1\right) = 0 \tag{22}$$

One of the surface represents a cylinder and the other an ellipsoid. Figure 3 depicts these surfaces bounded by the unit sphere.

By writing the conditions $(\rho_{1y} + \rho_{2y} + \rho_{3y} = 0)$ and $\rho_{1z} = \rho_{2z} = \rho_{3z}$ with the constraint equations, the Groebner basis elimination method makes it possible to obtain a set of equations that depends on $q_2$, $q_3$ and $q_4$.

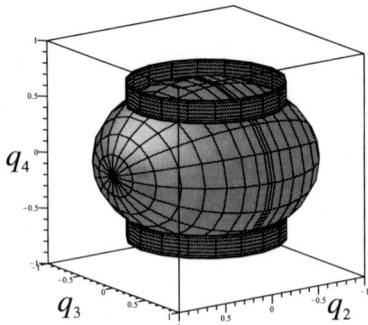

**Fig. 3.** Parallel singularity of the 3-PPPS robot with quaternion representation

$$q_2 q_4 = 0$$
$$q_3 q_4 = 0$$
$$q_2 \left( q_2{}^2 + q_3{}^2 - 1 \right) = 0$$
$$q_3 \left( q_2{}^2 + q_3{}^2 - 1 \right) = 0 \qquad (23)$$
$$q_4{}^3 - q_4 = 0$$
$$q_2{}^4 + \left( q_3{}^2 + q_4{}^2 - 1 \right) q_2{}^2 + q_3{}^2 q_4{}^2 = 0$$

The intersection of these six equations and the Eq. 22 is a unit radius circle centered at the origin of the plane $(0q_2 q_3)$ as shown in Fig. 4. This circle is tangential to the parallel singularity shown in Fig. 3.

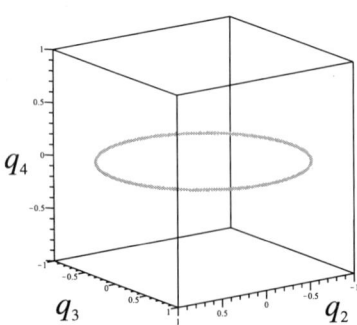

**Fig. 4.** Location of self-motions in the workspace.

As the singularity does not depend on the position, the self-motion exists for an infinite number of positions of the mobile platform. This phenomenon may not be found if numerical methods are used to solve the DKM because the relations $(\rho_{1y} + \rho_{2y} + \rho_{3y} = 0)$ and $\rho_{1z} = \rho_{2z} = \rho_{3z}$ are never completely satisfied.

## 4   Conclusions and Perspectives

In this article, we studied the parallel robot 3PPPS to explain the presence of a self-motion at the "home" position. This movement is a Cardanic movement that has already been studied on the 3-RPR parallel robot or the PamInsa robot. Self-motions can often explain the problems of solving the DKM when using algebraic methods. The calculation of a Groebner base makes it possible to detect this problem but the characterization of this movement for robots with six degrees of freedom is difficult because of the size of the equations. Other robots, such as the CaPaMan at the University of Cassino have the same type of singularity despite the phenomenon never being studied. The objective of future work will be to identify architectures with passive prismatic articulations connected to the mobile platform which is capable of having the same singularity conditions.

## References

1. Husty, M., Zsombor-Murray, P.: A special type of singular Stewart-Gough platform. In: Lenarcic, J., Ravani, B. (eds.) Advances in Robot Kinematics and Computational Geometry, pp. 449–458. Kluwer Academic Publishers, The Netherlands (1994)
2. Karger, A.: Singularities and self motions of a special type of platforms. In: Lenarcic, J., Thomas, F. (eds.) Advances in Robot Kinematics, pp. 449–458. Kluwer Academic Publishers, The Netherlands (2002)
3. Wohlhart, K.: Synthesis of architecturally mobile double-planar platforms. In: Lenarcic, J., Thomas, F. (eds.) Advances in Robot Kinematics, pp. 473–482. Kluwer Academic Publishers, The Netherlands (2002)
4. Bonev, I.A., Chablat, D., Wenger, P.: Working and assembly modes of the Agile Eye. In: Proceedings of the 2006 IEEE International Conference on Robotics and Automation, May 15–19, Orlando, Florida (2006)
5. Chablat, D., Wenger, P., Bonev, I.: Self motions of a special 3-RPR planar parallel robot. In: 10th International Symposium on Advances in Robot Kinematics, Ljubljana, Slovenia, 25–29 June, pp. 221–228. Kluwer Academic Publishers (2006)
6. Briot, S., Bonev, I., Chablat, D., Wenger, P., Arakelian, V.: Self-motions of general 3-RPR planar parallel robots. Int. J. Robot. Res. **27**(7), 855–866 (2008)
7. Merlet, J.P.: Parallel robots, vol. 128. Springer, The Netherlands (2006)
8. Angeles, J.: Fundamentals of Robotic Mechanical Systems: Theory, Methods, and Algorithms. Springer, New York (2007)
9. Pierrot, F., Shibukawa, T.: From hexa to hexam. In: Internationale Parallel kinematic-Kolloquium (IPK 1998), Zurich, pp. 75–84 (1998)
10. Corbel, D., Company, O., Pierrot, F.: Optimal design of a 6-dof parallel measurement mechanism integrated in a 3-dof parallel machine-tool. In: IEEE International Conference on Intelligent Robots and Systems, Nice, p. 7, September 2008
11. Stoughton, R., Arai, T.A.: Modified Stewart platform manipulator with improved dexterity. IEEE Trans. Robot. Autom. **9**(2), 166–173 (1993)
12. Ji, Z., Li, Z.: Identification of placement parameters for modular platform manipulators. J. Robot. Syst. **16**(4), 227–236 (1999)

13. Honegger, M., Codourey A., Burdet, E.: Adaptive control of the Hexaglide, a 6 dof parallel manipulator. In: IEEE International Conference on Robotics and Automation, Albuquerque, April, 21–28, pp. 543–548 (1997)
14. Hunt, K.H.: Structural kinematics of in parallel actuated robot arms. J. Mech. Trans. Autom. Des. **105**(4), 705–712 (1983)
15. Chen, C., Gayral, T., Caro, S., Chablat, D., Moroz, G.: A six-dof epicyclic-parallel manipulator. J. Mech. Robot. Am. Soc. Mech. Eng. **4**(4), 041011-1-8 (2012)
16. Caro, S., Wenger, P., Chablat, D.: Non-singular assembly mode changing trajectories of a 6-DOF parallel robot. In: ASME Design Engineering Technical Conferences & Computers and Information in Engineering Conference IDETC/CIE, Chicago, USA, 12–15 August (2012)
17. Chablat, D., Baron, L., Jha, R.: Kinematics and workspace analysis of a 3-PPPS parallel robot with U-shaped base. In: ASME 2017 International Design Engineering Technical Conferences and Computers and Information in Engineering Conference, Cleveland, Ohio, USA, 6–9 August (2017)
18. Chablat, D., Baron, L., Jha, R., Rolland, L.: The 3-PPPS parallel robot with U-shape Base, a 6-DOF parallel robot with simple kinematics. In: 16th International Symposium on Advances in Robot Kinematics, Bologna, 1–5 July (2018)
19. Caro, S., Moroz, G., Gayral, T., Chablat, D., Chen, C.: Singularity analysis of a six-DOF parallel manipulator using Grassmann-Cayley algebra and Gröbner bases. In: Proceedings of an International Symposium on the Occasion of the 25th Anniversary of the McGill University Centre for Intelligent Machines, Montréal, Canada, November 2010, pp. 341–352 (2010)
20. Byun, Y.K., Cho, H.-S.: Analysis of a novel 6-DOF, 3-PPSP parallel manipulator. Int. J. Robot. Res. **16**(6), 859–872 (1997)
21. Parenti-Castelli, V., Innocenti, C.: Direct displacement analysis for some classes of spatial parallel mechanisms. In: Proceedings of the 8th CISM-IFTOMM Symposium on Theory and Practice of Robots and Manipulators, pp. 126–130 (1990)
22. Jha, R., Chablat, D., Barin, L., Rouillier, F., Moroz, G.: Workspace, joint space and singularities of a family of delta-like robot. Mech. Mach. Theory Workspace **127**, 73–95 (2018)
23. Gosselin, C., Angeles, J.: Singularity analysis of closed-loop kinematic chains. IEEE Trans. Robot. Autom. **6**(3), 281–290 (1990)
24. Sefrioui, J., Gosselin, C.: Singularity analysis and representation of planar parallel manipulators. Robots Auton. Syst. **10**, 209–224 (1992)
25. Chablat, D., Wenger, Ph.: Working modes and aspects in fully-parallel manipulator. In: Proceedings IEEE International Conference on Robotics and Automation, pp. 1964–1969, May 1998

# Workspace and Singularity Analysis of a 3-RUU Parallel Manipulator

Thomas Stigger$^{(\boxtimes)}$, Martin Pfurner, and Manfred Husty

Unit Geometry and CAD, University of Innsbruck, 6020 Innsbruck, Austria
{thomas.stigger,martin.pfurner,manfred.husty}@uibk.ac.at

**Abstract.** The aim of this paper is to give a detailed examination of the workspace and the input and output singularities of the 3-RUU parallel manipulator in the translational operation mode. The examination is achieved by using algebraic constraint equations derived with the help of the Linear Implicitization Algorithm. The computation of the workspace is not limited to the boundaries of the translational operation mode but tailored with the conditions for this special case.

The investigation of the singularities include the input singularities mapped into a Study subspace and into the joint space. Also the output singularities are computed and mapped into the Study subspace and the joint space, therewith a complete singularity investigation of the translational operation mode of the 3-RUU parallel manipulator is provided.

**Keywords:** 3-RUU · Singularity analysis · Singularities · Workspace
Algebraic geometry

## 1 Introduction

There are many different approaches to perform the kinematic analysis of parallel manipulators (PM). More recently dual quaternions to describe the displacement group of the Euclidean space $SE(3)$ and the use of Study's kinematic mapping have proven to be very successful to obtain information about the global behavior of parallel manipulators. One aim in this kinematic analysis is to derive algebraic constraint equations to solve the direct and inverse kinematics, to describe the complete workspace, operation modes and all singularities (see e.g. [6]).

There are numerous papers investigating workspace and singularities of parallel manipulators especially of the Stewart-Gough platform e.g. by Dasgupta and Mruthyunjaya [2], Borràs et al. [3], Nawratil [7,8] but also closely related manipulators were already investigated like the 4-RUU in Amine et al. [1].

There are also numerous papers on lower dof parallel manipulators, most of them use for the kinematic analysis vector loop equations or screw theory. The global kinematics of lower dof parallel manipulators using algebraic constraint equations on the other hand can explain the overall kinematic behavior, e.g. determining different operation modes, transition between the operation modes and all singularities of the manipulator (see as an example [9]).

© Springer Nature Switzerland AG 2019
B. Corves et al. (Eds.): EuCoMeS 2018, MMS 59, pp. 325–332, 2019.
https://doi.org/10.1007/978-3-319-98020-1_38

In this paper, using results of [10], a complete singulartiy analysis of the translational operation mode of a 3-RUU parallel manipulator is given. Furthermore, an algebraic representation of the workspace is derived, which would be very useful in task planning. Section 2 recalls the manipulator architecture and recalls the algebraic constraint equations that describe the motion capabilities of the manipulator, Sect. 3 gives the algorithm to derive the workspace equations, which unfortunately due to their length cannot be displayed in this paper. In Sect. 4 a complete singularity analysis is given comprising a complete description of input and output singularities in the kinematic image space (Study space) as well as in the joint space. The last section concludes the results.

## 2   Manipulator Architecture and Constraint Equations

As there is a detailed explanation of the 3-RUU parallel manipulator's architecture in Stigger et al. [10] its architecture is only briefly explained. The manipulator consists of three identical RUU-limbs. Each of the limbs comprises three joints, an actuated revolute (R) joint and two passive universal (U) joints. Each vertex of the equilateral base triangle is connected via a RUU limb with a vertex of the equilateral moving frame triangle. The first and the last axis in each limb are tangent to the circum-circles of base and moving triangular platform (Fig. 1). The geometric design of each limb is described with Denavit-Hartenberg (DH) parameters. [4]

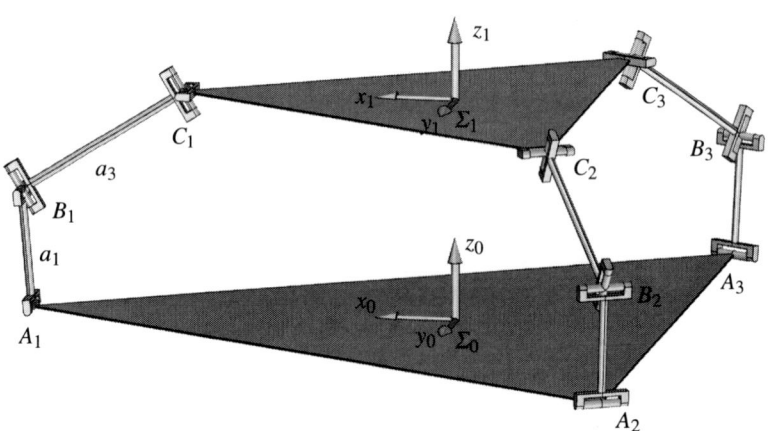

**Fig. 1.** 3-RUU parallel manipulator

### 2.1   Constraint Equations

For the complete workspace and singularity analysis algebraic constraint equations of the 3-RUU chains are necessary. These equations have been derived with different methods in [10]. Therefore, due to lack of space, the lengthy derivation of the constraint equations is omitted here and the reader is referred to this paper. The canonical constraint equations in the best adapted coordinate systems for an RUU limb (Fig. 2) are

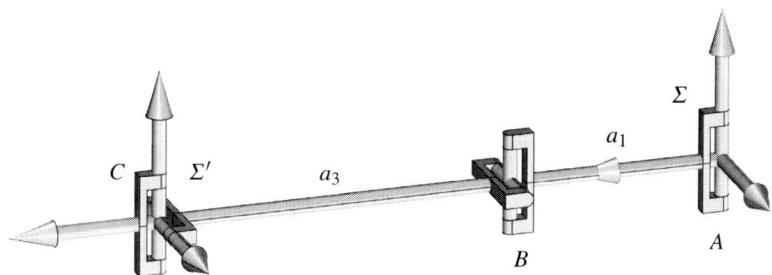

**Fig. 2.** A RUU limb

$$f_1 := \left((x_0x_1 - x_2x_3)\left(v_1{}^2 - 1\right) - (2x_0x_2 + 2x_1x_3)\,v_1\right)a_1 + 2\left(v_1{}^2 + 1\right)(x_0y_0 + x_3y_3) = 0$$

$$f_2 := -\left(x_0{}^2 + x_1{}^2 + x_2{}^2 + x_3{}^2\right)\left(v_1{}^2 + 1\right)a_1{}^2 + \left(4\left(y_1x_0 - y_0x_1 + y_3x_2 - y_2x_3\right)v_1{}^2\right.$$
$$+ 8\left(-x_0y_2 + x_1y_3 + x_2y_0 - x_3y_1\right)v_1 + 4\left(y_2x_3 - y_3x_2 - y_1x_0 + y_0x_1\right)\right)a_1$$
$$+ \left(\left(x_0{}^2 + x_1{}^2 + x_2{}^2 + x_3{}^2\right)a_3{}^2 - 4\left(y_2{}^2 + y_3{}^2 + y_0{}^2 + y_1{}^2\right)\right)\left(v_1{}^2 + 1\right) = 0, \qquad (1)$$

where $x_i, y_i$ are the (image space) coordinates that determine the possible poses of the $\Sigma'$ coordinate system with respect to the $\Sigma$ base system and $v_1$ is the algebraic value of the input angle of the first revolute joint. From the canonical constraint equations the general constraint equations are found by applying transformations that move the legs from the origin of the base system to their appropriate positions on the manipulator. The general constraint equations are obtained by performing transformations of the $x_i, y_i$ coordinates in the seven-dimensional projective (Study) space $\mathbb{P}^7$ that are applied to the set of canonical constraint equations. It is important to note that these transformations do not change the degree of the constraint equations. For each limb the appropriate transformations are applied to $f_1$ and $f_2$. In total this results in six general constraint equations $g_1, \ldots, g_6$. Together with the equation of the Study quadric $g_7 : x_0y_0 + x_1y_1 + x_2y_2 + x_3y_3 = 0$ and the normalization condition $g_8 : x_0^2 + x_1^2 + x_2^2 + x_3^2 = 0$ these equations yield a complete kinematic description of the manipulator. The three input parameters in the general constraint equation are denoted by $t_i, i = 1, 2, 3$. Therefore the set $\mathcal{W} = \{g_1, \ldots, g_8\}$ forms a three parameter system of constraint equations.

## 3   Workspace Description

The system $\mathcal{W}$ determines a three dimensional variety on the Study quadric that is the kinematic image of the workspace of the manipulator. But unfortunately it consists of six parametric constraint equations. For path planning it would be more convenient to have a system of equations without the input parameters describing the workspace only in image space coordinates. Inspection of the constraint equations shows that they are only pairwise dependent on the same input parameter: $g_1$ and $g_2$ for example depend only on $t_1$. Eliminating $t_1$ via computing the resultant of $g_1$ and $g_2$ yields an equation only depending on the Study parameters. Using the same procedure on $g_3, g_4$ thereby eliminating $t_2$ as well as on $g_5, g_6$ eliminating $t_3$, three equations are obtained depending on the Study parameters. The three resulting equations are independent of the input parameters and yield a suitable description of the workspace for path planning. As the computation is straightforward from the set of Eq. 1 and because of lack of space the three equations are not shown. But it can be stated explicitly that the whole translational three-space

$$x_0 = 1, \; x_1 = x_2 = x_3 = y_0 = 0 \tag{2}$$

lies in the workspace, a fact that confirms once more that the manipulator has a translational operation mode, although it is not the only mode.

## 4   Singularity Analysis

Singularities in parallel mechanisms can occur either because one limb or the platform itself is in a singular position. Furthermore, both can take place at the same time meaning that a limb and the platform are in singular positions. In [10] two operation modes of the 3-RUU have been identified. In the sequel a complete singularity analysis of the translational operation mode will be given. Therefore, it is assumed that condition 2 holds from now on. Referring to Gosselin and Angeles [5] and using the system $\mathscr{W}$ yields

$$\mathbf{J_o \dot{y}} + \mathbf{J_i \dot{t}} = 0, \tag{3}$$

where $\mathbf{J_o} = \frac{\partial g}{\partial y_i}$, $\mathbf{J_i} = \frac{\partial g}{\partial t_i}$ and $\dot{\mathbf{y}} = [\dot{y}_1, \dot{y}_2, \dot{y}_3]^T$, $\dot{\mathbf{t}} = [\dot{t}_1, \dot{t}_2, \dot{t}_3]^T$, the subscript $o$ refers to *output*, $i$ refers to *input*. In the following input and output singularities will be discussed separately in joint space as well as in the kinematic image space.

### 4.1   Input Singularities

If one limb reaches the boundary of its workspace then it is in a singular position. Following Gosselin, Angeles [5] this occurs when $\mathbf{J_i}$ is rank deficient. As there are two general constraints per leg, the Study quadric $g_7$ and the normalization condition $g_8$, the system $\mathscr{W}$ consists of eight equations. The input vector on the other hand $\dot{\mathbf{t}}$ only has 3 variables. The resulting $8 \times 3$ Jacobian matrix is obviously not square. The algebraic condition for rank deficiency is that all twenty $3 \times 3$ sub-determinants have to vanish. When the condition for the translational mode (2) is substituted in the system all equations but one vanish, the only remaining is

$$32 \left( y_1{}^2 t_1 + y_2{}^2 t_1 + y_3{}^2 t_1 + 4\, y_1\, t_1 - 3\, y_3\, t_1 + 3\, y_1 + 6 \right)$$

$$\left( 4\, \sqrt{3} t_2 y_2 + 2\, t_2 y_1{}^2 + 2\, t_2 y_2{}^2 + 2\, t_2 y_3{}^2 + 3\, \sqrt{3} y_2 - 4\, t_2 y_1 - 6\, t_2 y_3 - 3\, y_1 + 12 \right)$$

$$\left( 4\, \sqrt{3} t_3 y_2 - 2\, t_3 y_1{}^2 - 2\, t_3 y_2{}^2 - 2\, t_3 y_3{}^2 + 3\, \sqrt{3} y_2 + 4\, t_3 y_1 + 6\, t_3 y_3 + 3\, y_1 - 12 \right) = 0. \quad (4)$$

This equation factors into three different parts. Every factor is related to one leg of the mechanism. Since only one of the three factors has to be fulfilled in order to satisfy the singularity condition only the first of the factors is discussed. The other two yield similar results. Adding this factor to the system $\mathscr{W}$ and computing a Groebner base eliminating $t_1, t_2$ and $t_3$ yields

$$y_1{}^4 + y_2{}^4 + y_3{}^4 + 2( y_1{}^2 y_2{}^2 + y_1{}^2 y_3{}^2 + y_2{}^2 y_3{}^2) + 8( y_1{}^3 + y_2{}^2 y_1 + y_3{}^2 y_1)$$

$$+ 7\, y_1{}^2 - 9\, y_3{}^2 - 36( y_1 + 1) = 0 \qquad (5)$$

This is the singularity surface in the kinematic image space related to the first leg and has the following interpretation: when a pose of the manipulator in the translational operation mode is chosen such that Eq. 5 is fulfilled, then the first leg is either stretched out or folded. The surface can be visualized in the three dimensional $y_1, y_2, y_3$ Study space representing all translations (Fig. 3) and an easy computation shows that it is a torus. The other two factors in Eq. 4 yield also tori which are rotated by 120° and 240° about the $y_3$-axis. All three surfaces have two singular points in common. These points correspond to poses where all three legs are singular. Intersection curves of a pair of tori correspond to poses where two legs are singular. Note that the compound of the three surfaces determines the boundary of the workspace of the 3-RUU manipulator in the translational operation mode.

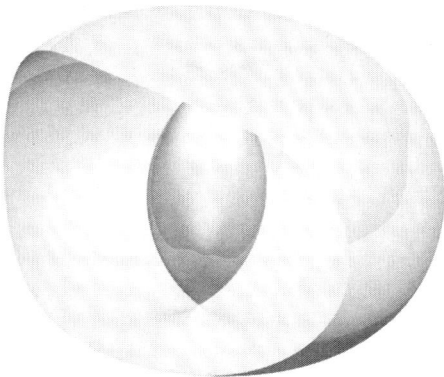

**Fig. 3.** Spindle torus as singularity surface

For practical control purposes it would be much better to have a representation of the input singularities in joint space, as to avoid these inputs during operation. Using system Eq. (4) one can map the singularities also into the joint space by eliminating the image space coordinates instead of the joint coordinates from the system of equations consisting of $\mathscr{W}$ and one of the factors of Eq. 4. Computing a Groebner base eliminating $y_1, y_2$ and $y_3$ yields a polynomial of degree 12, depending on $t_1, t_2$ and $t_3$ which is the same for each of the three factors of. Because of lack of space the polynomial is not shown, but the singularity surface in the joint space is displayed in Fig. 4.

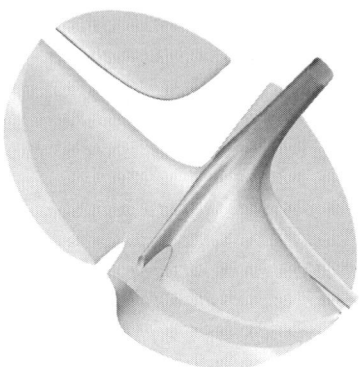

**Fig. 4.** singularity surface in the joint space

## 4.2 Output Singularities

Computing the output singularities via Eq. (3), the condition $\det(\mathbf{J_o}) = 0$ has to be fulfilled. For the 3-RUU PM there are 8 constraint equations and also 8 Study parameters, yielding an $8 \times 8$ Jacobian matrix. As the aim of this paper is to investigate the translational operation mode the conditions (2) are substituted into the Jacobian and only a $3 \times 3$ sub-matrix remains. Computing the determinant of this sub-matrix one obtains

$$
\begin{aligned}
&12 + 8\,t_2{}^2 y_3 - 4\,t_3{}^2 y_1 + 8\,t_3{}^2 y_3 + 9\,t_1 t_2 + 8\,t_1{}^2 y_1 + 8\,t_1{}^2 y_3 + 9\,t_1 t_3 + 8\,t_1 y_3 \\
&-4\,t_2{}^2 t_3{}^2 + 9\,t_2 t_3 + 8\,t_2 y_3 + 8\,t_3 y_3 - 4\,t_1{}^2 t_2{}^2 - 4\,t_1{}^2 t_3{}^2 + 8\,t_1{}^2 t_3 y_3 + 8\,t_1 t_3{}^2 y_3 \\
&+6\,t_1 t_3 y_3 + 8\,t_1{}^2 t_3{}^2 y_3 + 8\,t_2{}^2 t_3{}^2 y_3 + 8\,t_2{}^2 t_3 y_3 + 8\,t_2 t_3{}^2 y_3 + 6\,t_2 t_3 y_3 - 8\,t_2{}^2 t_3{}^2 y_1 \\
&-9\,t_1 t_2{}^2 t_3 - 6\,t_2{}^2 t_3 y_1 + 4\,t_1{}^2 t_2{}^2 y_1 - 9\,t_1{}^2 t_2 t_3 + 4\,t_1{}^2 t_3{}^2 y_1 + 4\,\sqrt{3} t_2{}^2 y_2 - 12\,t_1{}^2 t_2 t_3{}^2 \\
&+6\,t_1{}^2 t_3 y_1 - 4\,\sqrt{3} t_3{}^2 y_2 - 6\,t_2 t_3{}^2 y_1 - 9\,t_1 t_2 t_3{}^2 - 12\,t_1{}^2 t_2{}^2 t_3 - 12\,t_1 t_2{}^2 t_3{}^2 - 12\,t_1{}^2 t_2{}^2 t_3{}^2 \\
&+6\,t_1{}^2 t_2 y_1 + 8\,t_1{}^2 t_2 y_3 + 8\,t_1 t_2{}^2 y_3 + 6\,t_1 t_2 y_3 + 8\,t_1{}^2 t_2{}^2 y_3 + 12\,t_1 + 8\,y_3 + 12\,t_2 + 12\,t_3 \\
&+4\,t_1{}^2 + 4\,t_2{}^2 + 4\,t_3{}^2 - 4\,t_2{}^2 y_1 + 6\,t_1 t_2{}^2 t_3 y_3 + 6\,t_1 t_2 t_3{}^2 y_3 + 8\,t_1{}^2 t_2{}^2 t_3{}^2 y_3 \\
&-2\,\sqrt{3} t_1{}^2 t_3 y_2 + 4\,\sqrt{3} t_1 t_2{}^2 y_2 - 4\,\sqrt{3} t_1 t_3{}^2 y_2 + 2\,\sqrt{3} t_2{}^2 t_3 y_2 - 2\,\sqrt{3} t_2 t_3{}^2 y_2 + 4\,\sqrt{3} t_1{}^2 t_2{}^2 y_2 \\
&-4\,\sqrt{3} t_1{}^2 t_3{}^2 y_2 + 2\,\sqrt{3} t_1{}^2 t_2 y_2 + 8\,t_1{}^2 t_2{}^2 t_3 y_3 + 8\,t_1{}^2 t_2 t_3{}^2 y_3 + 8\,t_1 t_2{}^2 t_3{}^2 y_3 + 6\,t_1{}^2 t_2 t_3 y_3 = 0\ (6)
\end{aligned}
$$

In the translational operation mode only three constraint equations of the system $\mathscr{W}$ remain. These constraint equations together with the determinant Eq. (6) yield a system $\mathscr{L}$ of four equations in six parameters, $t_1, t_2, t_3, x_1, x_2, x_3$.

Computing resultants one obtains by eliminating successively $t_1, t_2$ and $t_3$ one equation with two significant factors and a third numerical one. One of the two factors is a torus equation to the power of three, shown in Eq. (7).

$$x^4 + 2x^2y^2 + 2x^2z^2 + y^4 + 2y^2z^2 + z^4 - 16x^2 - 16y^2 - 9z^2 + 64 \qquad (7)$$

The second factor yields a surface of degree 28 which was not possible to be plotted at once, but two planar intersections are shown in Figs. 5 and 6.

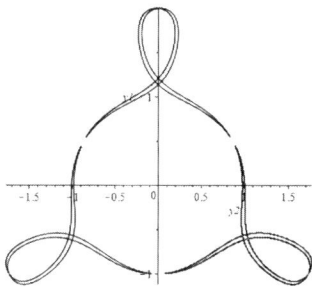

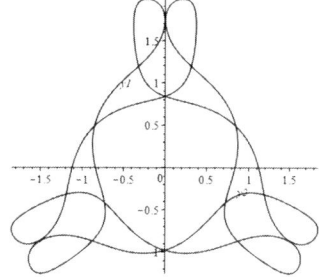

**Fig. 5.** Intersection at $y_3 = 0.01$          **Fig. 6.** Intersection at $y_3 = 0.1$

In the same way as the input singularities are computed it is possible to compute the output singularities in joint space via successive eliminating the Study parameters $y_1, y_2, y_3$ form the system $\mathscr{L}$. After this process a surface of degree 30 (Fig. 7) remains representing the output singularities in joint space . Now all input and output singularities of the translational operation mode of the 3-RUU PM are determined in the space of Study parameters and in joint space. By intersecting these varieties one can easily find poses where the manipulator is simultaneously input and output singular. An admittedly impractical pose being both input and output singular is shown in Fig. 8.

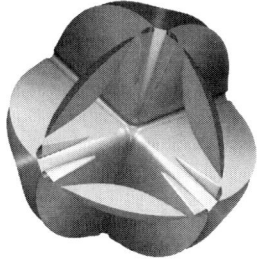

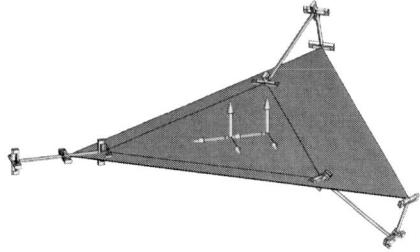

**Fig. 7.** Output singularities in joint space

**Fig. 8.** Simultaneous input and output singularity

## 5    Conclusion

A complete singularity analysis of the translational operation mode of the 3-RUU parallel manipulator was given. Although a special design in the base and platform was used herein the describing equation are huge. Other designs would be possible but computations may run out of memory. Using a system of algebraic constraint equations all input and output singularities could be computed in the Study space and in the joint space. Furthermore a characterization of the overall workspace of the manipulator was found. Future research should focus on the singularities of the general operation mode and on those poses of the end effector where the manipulator can switch between the modes.

**Acknowledgements.** This work was conducted with the support of the University of Innsbruck and the support of the FWF project KAPAMAT (I 1750-N26).

## References

1. Amine, S., Tale Masouleh, M., Caro, S., Wenger, P., Gosselin, C.: Singularity conditions of 3T1R parallel manipulators with identical limb structures. J. Mech. Robot. **4**(1), 1851–1863 (2012). https://doi.org/10.1115/1.4005336
2. Bhaskar: The stewart platform manipulator: a review. Mech. Mach. Theory **35**(1), 15–40 (2000). https://doi.org/10.1016/S0094-114X(99)00006-3. http://www.sciencedirect.com/science/article/pii/S0094114X99000063
3. Borràs, J., Thomas, F., Torras, C.: Singularity-invariant leg rearrangements in stewart-gough platforms. In: Lenarcic, J., Stanisic, M.M. (eds.) Advances in Robot Kinematics: Motion in Man and Machine, pp. 421–428. Springer, Netherlands (2010)
4. Denavit, J., Hartenberg, R.S.: A kinematic notation for lower-pair mechanisms based on matrices. Trans. ASME E J. Appl. Mech. **22**, 215–221 (1955)
5. Gosselin, C., Angeles, J.: Singularity analysis of closed-loop kinematic chains. IEEE Trans. Robot. Autom. **6**(3), 281–290 (1990). https://doi.org/10.1109/70.56660
6. Husty, M.L., Pfurner, M., Schröcker, H.P., Brunnthaler, K.: Algebraic methods in mechanism analysis and synthesis. Robotica **25**, 661–675 (2007)
7. Nawratil, G.: Stewart Gough platforms with linear singularity surface. In: 19th International Workshop on Robotics in Alpe-Adria-Danube Region (RAAD 2010), pp. 231–235 (2010). https://doi.org/10.1109/RAAD.2010.5524579
8. Nawratil, G.: Stewart Gough platforms with non-cubic singularity surface. Mech. Mach. Theory **45**(12), 1851–1863 (2010). https://doi.org/10.1016/j.mechmachtheory.2010.08.005. http://www.sciencedirect.com/science/article/pii/S0094114X10001394
9. Schadlbauer, J., Walter, D.R., Husty, M.L.: The 3-RPS parallel manipulator from an algebraic viewpoint. Mech. Mach. Theory **75**, 161–176 (2014)
10. Stigger, T., Nayak, A., Caro, S., Wenger, P., Pfurner, M., Husty, M.: Algebraic analysis of a 3-RUU parallel manipulator. In: 16th International Symposium on Advances in Robot Kinematics (2018, to submitted)

# Design and Direct Position Analysis of a New 3T1R Parallel Manipulator with Low Coupling Degree

Huiping Shen[1]($\boxtimes$), Zhengxiao Xu[1], Ke Xu[1], Shaoping Bai[2],
Jiaming Deng[1], Guanglei Wu[3], and Ting-li Yang[1]

[1] School of Mechanical Engineering, Changzhou University,
Changzhou 213016, People's Republic of China
shp65@126.com
[2] Department of Mechanical and Manufacturing Engineering,
Aalborg University, Aalborg, Denmark
shb@make.aau.dk
[3] School of Mechanical Engineering, Dalian University of Technology,
Dalian 116024, People's Republic of China
gwu@dlut.edu.cn

**Abstract.** In this paper, a new 3T1R partially motion-decoupled parallel manipulator (PM) with low coupling degree of $\kappa = 1$ and one redundant actuation chain is proposed. The topological design of the PM is introduced. Then the direct position analysis is developed by the molding method based on ordered single-opened-chain (SOC) units. The direct kinematics modeling method based on the SOC units is simple but powerful, which can be applicable to any PMs.

**Keywords:** Parallel mechanism · Direct position analysis · Motion decoupling
Coupling degree · Redundant actuation

## 1 Introduction

The topological design and performance investigation of four DOFs 3T1R parallel mechanisms (PMs) have being attracted more attentions.

Reboulet et al. designed a wrist with three translations and one rotation [1]. Pierrot et al. proposed a new type of four DOF parallel robot H4 [2]. Luc Rolland suggested a 3T1R PM for industrial grasp [3]. Briot et al. developed a decoupled 3T1R parallel robot in the application of pick-and-place [4]. Huang et al. invented a 3T1R Cross-type IV high-speed robot [5]. Liu et al. developed X4 robot prototype with one moving platform [6]. A parallel Schönflies-motion robot admitting a rectangular workspace was recently proposed by Wu and Bai, et al. [7]

It is noted that the most of above stated 3T1R PMs have higher coupling degree $\kappa$ of 2, and are not input-output motion decoupled, which lead to their forward kinematics, inverse dynamics analysis and motion control are more complex.

© Springer Nature Switzerland AG 2019
B. Corves et al. (Eds.): EuCoMeS 2018, MMS 59, pp. 333–339, 2019.
https://doi.org/10.1007/978-3-319-98020-1_39

Coupling degree ($\kappa$) is a topological constant for any *Assur Kinematic Chain(AKC)* with zero DOF. It stands for complexity of its topological structure and also reflects complexity of solutions of direct kinematic and inverse dynamic analysis. We have proved that the greater the coupling degree, the complex the solutions of direct kinematic and inverse dynamic analysis will be [8, 12]. Therefore the PMs with low coupling degree ($\kappa = 0$ or 1) are favorable in practical applications.

The loop method are widely adopted to build the input-output position equations. However, these position equations contain a number of unknown variables. The solutions of these position equations could be calculated by numerical methods [9] or algebraic methods [10] that need very complex mathematical derivation and deduction.

In this paper, a new four DOFs 3T1R PM with one redundant chain is proposed. The new 3T1R PM features low coupling degree ($\kappa = 1$) and input-output partial motion decoupling. Then a simple and general direct position analysis method based on ordered single-opened-chain (SOC) units is developed for the direct kinematics of this PM. So-called SOC unit is serial connections of some joints and links,

## 2   Design of 3T1R Parallel Mechanism

The topological structure of the 3T1R PM is proposed inspired from Ref. [11], shown in Fig. 1. Between the moving platform 1 and base platform 0, there is an hybrid chain, containing a sub parallel mechanism (sub-PM) consisted of limbs A, B and C, as well as two simple chains $R_{31}$-$S_{32}$-$S_{33}$(links 10 and 11)and $R_{21}$-$S_{22}$-$S_{23}$(links 12 and 13). The PM is partially symmetrical about the plane $o$-$xz$ comparing with the structure of existing fully symmetrical H4, I4 or Par4 PMs that have four identical legs etc.

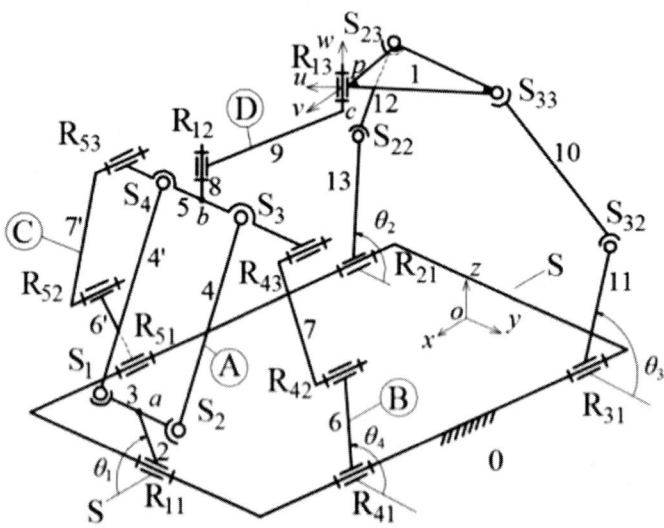

**Fig. 1.** A new 3T1R parallel mechanism

The arrangement of axes of revolute joints on the base platform 0 is $R_{11} \perp R_{41}$, $R_{11} \perp R_{51}$, and the axes of $R_{41}$ and $R_{31}$, $R_{51}$ and $R_{21}$ coincide respectively. Hence the whole 3T1R manipulator is denoted as 2-(RPa-3R)$\perp$2R + 2-RSS. Here, symbol "$\perp$" stands for perpendicular. The function of the redundant actuation limb C of the PM is to raise the stiffness and eliminate the singularity when the PM is at its singular positions, which does not affect the result of the kinematic analysis stated in the text.

According to the composition principle of mechanism based on single-open-chain (SOC) units [11, 12], this PM includes three independent loops. The limbs A and B constitute the first loop which is denoted as $\{R_{11}$-P*-P*-R*-R*-$R_{43} \| R_{42} \| R_{41}\}$, also be regarded the first sub-PM. The second loop consists of the former first sub-PM, adding limb D and limb $R_{31}$-$S_{32}$-$S_{33}$ (i.e., SOC$_2$), which is denoted as $\{R_{12}$-$R_{13}$-$S_{33}$-$S_{32}$-$R_{31}\}$, also be regarded the second sub-PM. The third loop is composed of the second sub-PM and adding limb SOC$_3$, which is denoted as $\{S_{23}$-$S_{22}$-$R_{21}\}$. We have calculated the degree of freedom of this PM is 4 and the output motion of the moving platform is three-translation and one-rotation.

That is, the first loop itself constitutes the first AKC, i.e., denoted as $AKC_1$. Its coupling degree is $\kappa_1 = 0$ [11], which means the direct kinematics solution of $AKC_1$ can be easily and directly calculated. It is known that other two loops, i.e., the second and third loops, form the second AKC, denoted as $AKC_2$. The coupling degree of the $AKC_2$ is calculated as $\kappa_2 = 1$ [11], which means that direct kinematics solutions of the $AKC_2$ must be solved simultaneously by both the second and third loop instead of by only one of these two loops, but only one virtual input variable needs to be assigned since its coupling degree is only one.

Thus, this PM contains two AKCs, i.e., $AKC_1$ and $AKC_2$, their coupling degrees are respectively $\kappa_1 = 0$ and $\kappa_2 = 1$. The direct position solution of the PM can be transferred into direct position kinematics solution of AKCs. While position kinematic solutions of an AKC can be further transferred into kinematic solutions of several ordered SOC units with three different constraint degree value, i.e., $\Delta > 0$, $\Delta = 0$ and $\Delta < 0$. Constraint degree reflects the constraint property of a SOC on AKCs [11, 12]. This is the basic idea of direct kinematics molding method based on the SOC units that will be described in the following section.

## 3   Direct Position Analysis

### 3.1   The Direct Kinematics Based on SOC Units

Let $\theta_1$, $\theta_2$, $\theta_3$ and $\theta_4$ be input angles of the joints $R_{11}$, $R_{21}$, $R_{31}$ and $R_{41}$, respectively, their definitions are shown in Fig. 2. The attitude angle $\alpha$ of the moving platform is the angle between axis $u$ and $x$, and the virtual input $\delta*$ assigned is defined as an angle between EG side and $x$ axis.

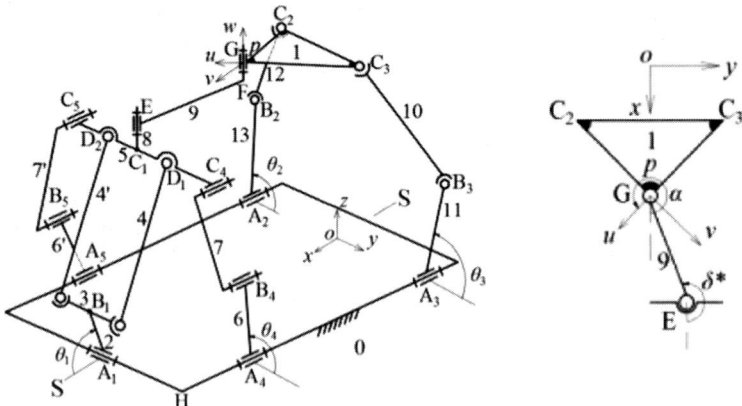

**Fig. 2.** Kinematics modeling

The structure parameters of the PM are denoted in following:

$A_3A_4 = A_4H = l_1$, $A_1H = l_2$, $GC_2 = GC_3 = l_3$, $A_iB_i = l_4(i = 1 \sim 4)$, $B_iC_i = l_5(i = 1 \sim 4)$;
$C_1D_1 = D_1C_4 = l_6$, $EF = l_7$, $C_1E = FG = l_8$.

### The solution of AKC₁

The position of the output link 5, $(x_{c1}, y_{c1}, z_{c1})$, is easily calculated due to zero coupling degree as follows:

From Fig. 2 we know the output motion of link 5 is only two-translation, so link $D_2C_4$ is always parallel to x axis, i.e. $x_{C_1} = l_1$ and the coordinate of point $C_4$ is always $(l_1, y_{C_1} + 2l_6, z_{C_1})$.

According to length constraints $B_iC_i = l_5, (i = 1, 4)$, we get

$$Ay_{C_1} + Bz_{C_1} = C \tag{1}$$

where

$$A = 2(y_{B_4} - 2l_6) \ , \quad B = 2(z_{B_4} - z_{B_1}) \ , \quad C = (y_{B_4} - 2l_6)^2 + z_{B_4}^2 - z_{B_1}^2 - (x_{B_1} - l_1)^2$$

If $A = B = 0$, then $C = -(x_{B_1} - l_1)^2 = 0$. Hence A and B are not zero at the same time due to $l_4 < l_1$ and $x_{B_1} > l_1$. Thus, there are two cases below

(a) when $A = 0$,

$$\begin{cases} z_{C_1} = \dfrac{C}{B} \\ y_{C_1} = \pm\sqrt{l_5^2 - (x_{B_1} - l_1)^2 - (z_{B_1} - z_{C_1})^2} \end{cases} \tag{2}$$

(b)  when A $\neq$ 0,

$$
\begin{cases}
z_{C_1} = \frac{K \pm \sqrt{K^2 - 4MN}}{2M} \\
y_{C_1} = \frac{C - Bz_{C_1}}{A}
\end{cases}
\tag{3}
$$

where

$$M = A^2 + B^2 , \quad K = 2(BC + z_{B_1}A^2) , \quad N = A^2[(x_{B_1} - l_1)^2 + z_{B_1}^2 - l_5^2] + C^2$$

### The solution of $AKC_2$

Because $AKC_2$ contains two loops and its coupling degree is $\kappa_2 = 1$ its position must be solved simultaneously by two following loops as follows:

(1)  In the $\{R_{12}\text{-}R_{13}\text{-}S_{23}\text{-}S_{22}\text{-}R_{21}\}$ whose constrain degree value $\Delta_1$ is one (i.e., $\Delta_1 = 1$), one virtual input variable should be assigned.

From Fig. 2, the coordinates of point $p$ on the moving platform are

$$
\begin{cases}
x^* = l_1 + l_7 \cos \delta^* \\
y^* = y_{C_1} + l_7 \sin \delta^* \\
z^* = z_{C_1} + 2l_8
\end{cases}
\tag{4}
$$

In the moving coordinate system $p\text{-}uvw$, coordinates of points $C_2$ and $C_3$ are $(0, -l_3, 0)$, $(-l_3, 0, 0)$ respectively.

Let $\mathbf{R}(z, \alpha)$ be the rotation matrix of the moving coordinate system relative to the frame coordinate system. Then we have

$$C_2 = \mathbf{R}(z, \alpha) \cdot {}^P C_2 + p = [l_3 \sin \alpha + x*, -l_3 \cos \alpha + y*, z*]^T \tag{5}$$

$$C_3 = \mathbf{R}(z, \alpha) \cdot {}^P C_3 + p = [-l_3 \cos \alpha + x*, -l_3 \sin \alpha + y*, z*]^T \tag{6}$$

According to length constraint $B_2 C_2 = l_5$, we can obtain

$$J \sin \alpha + P \cos \alpha + Q = 0 \tag{7}$$

We have

$$\alpha = 2\arctan \frac{J \pm \sqrt{J^2 + P^2 - Q^2}}{P - Q} \quad P \neq Q \tag{8}$$

where

$$J = 2l_3 x^* , \quad P = 2l_3(y_{B_2} - y^*) , \quad Q = l_3^2 + (y_{B_2} - y^*)^2 + (z_{B_2} - z^*)^2 + (x^*)^2 - l_5^2$$

(2) In SOC$_3$ {$S_{33}$-$S_{32}$-$R_{31}$}whose constrain degree is negative one, i.e., $\Delta_2 = -1 < 0$, one constraint equation must be established

Due to length constraint $B_3C_3 = l_5$ in {$S_{33}$-$S_{32}$-$R_{31}$}, we can obtain

$$F(\delta^*) = (x_{B_3} - x_{C_3})^2 + (y_{B_3} - y_{C_3})^2 + (z_{B_3} - z_{C_3})^2 - l_5^2 \qquad (9)$$

By constantly changing the value of virtual input $\delta^*$, we can make $F(\delta^*) \to 0$ until the real angle $\delta^*$ can be obtained [13]. Then replace the real value angle $\delta^*$ into Eqs. (4) and (8) to obtain the coordinates of point $p$ and the attitude angle $\alpha$ of the moving platform 1.

From Eq. (4), we find $x = \phi(\theta_1, \theta_2, \theta_3, \theta_4)$, $y = f(\theta_1, \theta_2, \theta_3, \theta_4)$, $z = \varphi(\theta_1, \theta_4)$ and $\alpha = \eta(\theta_1, \theta_2, \theta_3, \theta_4)$. Since $z$ coordinates of the moving platform 1 are determined by only inputs $\theta_1$ and $\theta_4$, we say the PM has input-output (I-O) partial motion decoupling. The inverse position formula are also derived but omitted because of the space limitation.

Let structural parameters of the PM are respectively

$l_1 = 300$, $l_2 = 300$, $l_3 = 150$, $l_4 = 250$, $l_5 = 800$, $l_6 = 100$, $l_7 = 200$, $l_8 = 25$ (*units : mm*) And let a group of four actuated input angles be respectively: $\theta_1 = 37.23°$, $\theta_2 = 156.22°$, $\theta_3 = 57.18°$ and $\theta_4 = 21.43°$

The range of the virtual input angle $\delta*$ is changed from 0 to $2\pi$. Using one-dimensional search method, two sets of real direct solutions are obtained as below. i.e., $x_1 = 135.1471$, $y_1 = -204.3738$, $z_1 = 819.8335$, $\alpha_1 = -100.02°$, and $x_1 = 103.3202$, $y_1 = -54.8413$, $z_1 = 819.8335$, $\alpha_1 = 9.84°$.

Substituting the first set of data into the inverse solutions, the four input angles are obtained as below.

$$\theta_1' = 37.2308°, \; \theta_2' = 156.2227°, \; \theta_3' = 57.1812° \text{ and } \theta_4' = 21.4286°$$

The results are consistent with the four input angles given.

## 4   Conclusions

In this paper, a 4-DOF low coupling degree ($\kappa = 1$) and partially motion decoupled 3T1R PM manipulator with one redundant actuation limb is designed, which make this PM have potential applications.

The direct kinematics modeling based on the SOC units is a simple but powerful method, which can be applicable to any PMs. The other work relating workspace and rotation ability, singularity analysis, prototype development and dynamic analysis for the PM will be discussed in the another paper.

**Acknowledgments.** This research is sponsored by the NSFC (No. 51475050 and No. 51375062), Jiangsu Key Development Project (No.BE2015043) and Jiangsu Scientific and Technology Transformation Fund Project (No.BA2015098).

# References

1. Reboulet, C., et al.: Rapport d'avancement projet VAP, thème 7, phase 3. Rapport de Recherche 7743, CNES/DERA, January 1991
2. Pierrot, F., Company, O.: H4: a new family of 4-DOF parallel robots. In: 1999 IEEE/ASME International Conference on Advanced Intelligent Mechatronics, Atlanta, GA, USA, pp. 508–513 (1999)
3. Rolland, L.: The Manta and the Kanuk: novel 4 dof parallel mechanism for industrial handling. In: Proceedings of ASME Dynamic Systems and Control Division IMECE 1999 Conference, 14–19 November, Nashville, USA, vol. 67, pp. 831–844 (1999)
4. Briot, S., Bonev, I.A.: Pantopteron-4: a new 3T1R decoupled parallel manipulator for pick-and-place applications. Mech. Mach. Theory **45**(5), 707–721 (2010)
5. Huang, T., Zhao, X., Mei, J.: A parallel mechanism with three-translation and one-rotation: China, 202528189U[P], 14 November 2012
6. Liu, X., Xie, F., Wang, L.A.: Four degree of freedom parallel mechanism with single platform for realizing SCARA motion: China, 201210435375.1[P], 13 February 2013
7. Wu, G., Bai, S., Hjørnet, P.: Architecture optimization of a parallel Schönflies-motion robot for pick-and-place applications in a predefined workspace. Mech. Mach. Theory **09**(5), 148–165 (2016)
8. Yang, T.-L.: Theory of Robot Mechanism Topology. China Machine Press, Beijing (2004)
9. Husain, M., Waldron, K.J.: Direct position kinematics of the 3-1-1-1 Stewart platforms. Trans. ASME J. Mech. Des. **116**(4), 1102–1107 (1994)
10. Rojas, N., Thomas, F.: On closed-form solutions to the position analysis of Baranov trusses. Mech. Mach. Theory **50**, 179–196 (2012)
11. Yang, T.-L., Liu, A., Shen, H., et al.: Topological structure synthesis of 3T1R parallel mechanism based on POC equations. In: Proceedings of 9th International Conference on Intelligent Robotics and Applications, ICIRA 2016. LNCS, Japan, vol. 9834, pp. 147–161 (2016)
12. Yang, T., Liu, A.X., Shen, H.P., et al.: Topology Design of Robot Mechanisms. Springer, Heidelberg (2018)
13. Shen, H.P., Ding, K.L., Yang, T.-L.: Configuration analysis of complex planar linkages and manipulators. Mech. Mach. Theory **35**(3), 353–362 (2000)

# Parameterized Inverse Kinematics of Parallel Mechanism Based on CGA

X. M. Huo[1(✉)], B. B. Lian[1,2(✉)], Tao Sun[1], and Y. M. Song[1]

[1] Tianjin University, Tianjin, China
{xmhuo, stao, ymsong}@tju.edu.cn
[2] KTH Royal Institute of Technology, Stockholm, Sweden
binbin2@kth.se

**Abstract.** A parameterized inverse kinematic model is the theoretical basis for performance analysis, design and control of parallel mechanism (PM). Current methods are either computationally expensive or difficult to get analytical form. To deal with this problem, this paper proposes a parameterized method by conformal geometric algebra (CGA). Based on the description and computation of screw motions in CGA, closure equations about successive screw displacements of any PM can be formulated. Joint displacements of each limb and screw parameters of end-effector are then solved in an analytical manner. The proposed method is exemplified by a 3 degree-of-freedom (DoF) PM, which shows high efficiency in deriving the analytical inverse kinematic model.

**Keywords:** Inverse kinematics · Parallel mechanism
Conformal geometric algebra · Joint displacement

## 1 Introduction

Parallel mechanisms (PM) consist of base platform connected with end-effector by several kinematic limbs. Due to the parallel topology, PMs have potential advantages in stiffness, dynamic and accuracy. The analysis, design and control of PMs have been the hot spots in mechanism research community in the past few decades. Among them, inverse kinematics involves obtaining joint displacements of limbs under given position and orientation of end-effector. It is the analytical basis for the development of PMs, for example, workspace analysis and precise performance modelling [1].

Up to present, the most common way to deal with the inverse kinematics of PMs is geometric method [11]. Vector algebra is usually applied to form the vector-loop equations of each limb relating centers of base platform and end-effector. By taking advantage of specific geometric characteristics, joint displacements are solved in sequence. However, the inverse kinematic problem of PMs is tackled case by case in the geometric method. There is little attention in obtaining a general inverse kinematic model for any PM.

Another way to formulate inverse kinematic model of PMs is algebraic method [9]. Initially, Denavit-Hartenberg (D-H) parameters were introduced to the analysis of PMs. The closure equations were established with the coordinate transformation from base platform to end-effector. Motivated by this work, screw motion description was

© Springer Nature Switzerland AG 2019
B. Corves et al. (Eds.): EuCoMeS 2018, MMS 59, pp. 340–346, 2019.
https://doi.org/10.1007/978-3-319-98020-1_40

applied. Since the pose of end-effector is the resultant screw motion of all the joints in each limb, closure equations about successive screw displacements can be easily expressed [6, 7]. For a long period, transformation matrix or its exponential form are applied in motion description. Hence, closure equation is performed by matrix multiplication or B-C-H formula [8]. Although current algebraic methods have the potential to obtain analytical inverse kinematic model, they involve complex computation and approximate result with higher order items.

The key to derive joint displacements analytically and efficiently lies in selecting a proper mathematic tool. It should be able to describe and compute screw motions concisely. To this end, CGA [4] has been introduced as a promising solution. CGA is a unified framework integrating different algebraic systems, thus can covers the whole process of invention, analysis and optimal design of PMs avoiding mutual transformations of different algebraic languages. Nevertheless, CGA is well known for direct geometry calculation and effective rigid body motion description. The former feature enables efficient computation of actuator inputs concerning special geometric characteristics of joints, referred to the work in [2, 5]. The latter contributes a general method for formulating closure equations of mechanisms in concise manner [3]. It is noted that only 8 algebraic elements would be applied in a CGA form instead of the 16 factors in transformation matrix, leading to more efficient computation process. However, D-H parameter descriptions were still dominant in these works, which would bring inconvenience to the inverse kinematics of PMs as multiple closed-loops are involved. Based on our previous work about topology synthesis [10], this paper attempts to propose an analytical and efficient method for inverse kinematics of any types of PMs. The motion axes are described as lines directly and the both merits of CGA are made full use.

Having outlined the state-of-art, this paper is organized as follows. Section 2 formulates closure equations about successive screw displacements of PMs in CGA form. Section 3 illustrates the prosed inverse kinematic analysis method of PMs. A typical example is given to demonstrate the effectiveness of the approach in Sect. 4 and conclusions are drawn in Sect. 5.

## 2   Closure Equations Based on CGA

For a PM with $f$- degree-of-freedom (DoF), the screw motion $\boldsymbol{M}_{PM}$ of the end-effector is considered as the composition of $f$ screw motions in sequence. This can be performed by the geometric product [4] in CGA as follows,

$$\boldsymbol{M}_{PM} = \boldsymbol{M}_1\boldsymbol{M}_2\cdots\boldsymbol{M}_i\cdots\boldsymbol{M}_f, \ i = 1, 2, \cdots, f \tag{1}$$

where $\boldsymbol{M}_i$, defined as motor, is the 1-DoF screw motion. It is given by,

$$\boldsymbol{M}_i = \left(\cos\frac{\theta_i}{2} - \sin\frac{\theta_i}{2}l^R_{M,i}\right)\left(1 - \frac{t_i}{2}l^P_{M,i}\right) \tag{2}$$

herein, $\theta_i$ is the angular displacement around rotation axis $l^R_{M,i}$. $t_i$ represents the linear displacement along translation axis $l^P_{M,i}$. $l^R_{M,i}$ and $l^P_{M,i}$ are expressed as,

$$l_{M,i}^R = u_{M,i}I_3 + m_{M,i}e_\infty = \sum_{k=1}^{3} u_{M,k,i}e_k I_3 + \sum_{k=1}^{3} m_{M,k,i}e_k e_\infty \tag{3}$$

$$l_{M,i}^P = u_{M,i}e_\infty = \sum_{k=1}^{3} u_{M,k,i}e_k, e_\infty \tag{4}$$

where $u_{M,i}$ and $m_{M,i}$ are direction and position vectors of the screw axis. Their coordinates are $\begin{pmatrix} u_{M,1,i} & u_{M,2,i} & u_{M,3,i} \end{pmatrix}$ and $\begin{pmatrix} m_{M,1,i} & m_{M,2,i} & m_{M,3,i} \end{pmatrix}$, respectively. $m_{M,i} = r_i \times u_{M,i}$. $r_i$ is the position vector pointing from the origin $O$ to an arbitrary point on the motion axis. $\times$ denotes the cross product of vectors. The orthogonal elements $e_1$, $e_2$, $e_3$ and $e_\infty$ [4] are the basic elements of CGA,

$$e_1 \times e_2 = e_3, \; e_k^2 = 1, \; e_\infty^2 = 0, \; k = 1, 2, 3 \tag{5}$$

$e_2e_3, e_3e_1, e_1e_2, e_1e_\infty, e_2e_\infty, e_3e_\infty$ are elements produced by the geometric product of even base generators. The geometric product of elements is denoted by juxtaposition. For example, the geometric product of the generators $e_1$, $e_2$, $e_3$ is written as $e_1e_2e_3$, which is regarded as unit pseudoscalar $I_3$. Based on the original definition of geometric algebra, the geometric product of generators is anti-commutative, such as $e_+e_- = -e_-e_+$, $e_3e_+e_- = -e_-e_+e_3$. Taking geometric product of a motor $M$ and its conjugate $\tilde{M}$ leads to,

$$M\tilde{M} = 1 \tag{6}$$

herein, $\tilde{M} = \tilde{M}_P\tilde{M}_R$, $\tilde{M}_R = \cos\frac{\theta}{2} + \sin\frac{\theta}{2}l_M^R$, $\tilde{M}_P = 1 + \frac{t}{2}l_M^P$.

Closure equations about successive screw displacements of PMs could be derived,

$$M_{PM} = M_{L,j}, \; j = 1, 2, \cdots, m \tag{7}$$

where $M_{L,j}$ denotes the finite motion of $j$th open-loop limb,

$$M_{L,j} = M_{L,1,j}M_{L,2,j} \cdots M_{L,i_j,j} \cdots M_{L,n_j,j}, \; i_j = 1, 2, \cdots, n_j \tag{8}$$

herein, $M_{L,i_j,j}$ expresses the screw motion of $i_j$th 1-DOF joint of $j$th open-loop limb with joint displacements $\theta_{i_j,j}$ and $t_{i_j,j}$.

## 3   Inverse Kinematic Analysis of PMs

Taking geometric product of $\tilde{M}_{PM}$ on both side of Eq. (7) yields,

$$\tilde{M}_{PM}M_{PM} = M_{A_j} \tag{9}$$

where $M_{A_j} = 1$ is the function of joint displacements and screw parameters of the end-effector,

$$M_{A_j} = \tilde{M}_{PM} M_{L,1,j} M_{L,2,j} \cdots M_{L,i,j} \cdots M_{L,n_j,j} \tag{10}$$

Measuring the right of Eq. (10) as,

$$M_{A_j} = \prod_{g_R=1}^{q_R} \left( \cos \frac{\hat{\theta}_{g_R,j}}{2} - \sin \frac{\hat{\theta}_{g_R,j}}{2} \hat{I}^R_{M,g_R} \right) \prod_{g_P=1}^{q_P} \left( 1 - \frac{\hat{t}_{g_P,j}}{2} \hat{I}^P_{M,g_P} \right) \tag{11}$$

where $\hat{I}^R_{M,g_R}$ $(g_R = 1, \cdots, q_R, q_R \leq 3)$ and $\hat{I}^P_{M,g_P}$ $(g_P = 1, \cdots, q_P, q_P \leq 3)$ are linear-independent rotation and translation axes respectively. Equations about unknown joint displacements of limbs could be formulated as,

$$\begin{cases} \hat{\theta}_{g_R,j} = f_{g_R} \left( \theta_i, t_i, \theta_{i,j}, t_{i,j} \right) = 0 \\ \hat{t}_{g_P,j} = f_{g_P} \left( \theta_i, t_i, \theta_{i,j}, t_{i,j} \right) = 0 \end{cases} \tag{12}$$

Having the description and calculation of screw motions based on CGA at hand, the joint displacements could be obtained in an analytical manner by the following three steps.

**Step 1**: Once an $f$-DoF PM is given, its position and orientation can be readily described in CGA form.

**Step 2**: Formulate closure equations about successive screw displacements and converse them as Eq. (11).

**Step 3**: Construct the equations about joint displacements of each limb and screw parameters of end-effector based on Step 2 and then solve these equations.

## 4   Example

A 3-DoF PM with one translation and two rotations proposed in our previous work [10] is sketched here as an example. It is composed of fixed base, moving platform, two identical PRS (P, R and S denote the prismatic, revolute and spherical joint, respectively) limbs and a PRRU (U is the universe joint) limb, as illustrated in Fig. 1. Frame $O$-$xyz$ is assigned to the center $O$ of the moving platform. The $z$-axis is along the motion axis of the R joint that connects to moving platform in the PRRU limb. The $x$-axis is along line $A_1 A_2$ at the initial configuration.

The position and orientation of the end-effector could be described as,

$$M_{PM,1T2R} = M_1 M_2 M_3$$
$$= \left( 1 - \frac{t_1}{2} e_3 e_\infty \right) \left( c \frac{\theta_2}{2} - s \frac{\theta_2}{2} e_3 e_1 \right) \left( c \frac{\theta_3}{2} - s \frac{\theta_3}{2} e_2 e_3 \right) \tag{13}$$

**Step 1**, write the closure equations about successive screw displacements as,

$$M_{\text{PM,1T2R}} = M_{\text{L}_j} = M_{\text{L},1,j}M_{\text{L},2,j}M_{\text{L},3,j}M_{\text{L},4,j}M_{\text{L},5,j} \tag{14}$$

where

$$M_{\text{L},1,j} = 1 - \frac{t_{1,j}}{2}l^{\text{P}}_{\text{M},1,j}, M_{\text{L},i_R,j} = c\frac{\theta_{i_R,j}}{2} - s\frac{\theta_{i_R,j}}{2}l^{\text{R}}_{\text{M},i_R,j}(i_R = 2,3,4,5), l^{\text{P}}_{\text{M},1,j} = e_3e_\infty,$$
$$l^{\text{R}}_{\text{M},i_R,j} = u_{\text{M},i_R,j}I_3 + \left(r_{i_R,j} \times u_{\text{M},i_R,j}\right)e_\infty, u_{\text{M},2,j_w} = u_{\text{M},3,j_w} = u_{\text{M},4,3} = e_2(j_w = 1,2),$$
$$u_{\text{M},4,j_w} = u_{\text{M},2,3} = u_{\text{M},3,3} = e_1, u_{\text{M},5,j} = e_3, r_{2,j} = r_{3,j} + A_jB_j, r_{3,3} = r_3e_2,$$
$$r_{i_s,j_w} = r_{j_w}(-1)^{j_w+1}e_1(i_s = 3,4,5), A_{j_w}B_{j_w} = l_{j_w}\cos\varphi_{j_w}e_1 + l_{j_w}\sin\varphi_{j_w}e_3,$$
$$A_3B_3 = l_3\cos\varphi_3e_2 + l_3\sin\varphi_3e_3$$

herein, $l_j$ and $r_j$ are the lengths of $A_jB_j$ and $OA_j$, respectively. $\varphi_{j_w}$ ($\varphi_3$) is the angle between $A_{j_w}B_{j_w}$ ($A_3B_3$) and x-axis (y-axis). Equation (14) can also expressed as,

$$\tilde{M}_{\text{PM,1T2R}}M_{\text{PM,1T2R}} = \tilde{M}_{\text{PM,1T2R}}M_{\text{L}_j} = M_{\text{A}_j} \tag{15}$$

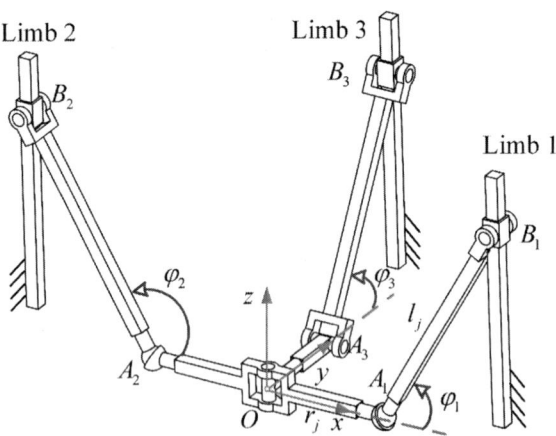

**Fig. 1.** 2PRS-PRRU parallel mechanism

**Step 2**, measure $M_{\text{A}_j}$ as Eq. (11) and formulate equations about the unknown joint displacements.

Here, the first two PRS limbs are taken as an example to show the process. Considering the first two limbs, Eq. (15) could be converted to,

$$M_{\text{A}_{j_w}} = \hat{M}_{\text{L},1,j_w}\hat{M}_{\text{L},2,j_w}\hat{M}_{\text{L},3,j_w}\hat{M}_{\text{L},4,j_w}\hat{M}_{\text{L},5,j_w}, j_w = 1,2$$

where

$$\hat{\theta}_{3,j_w} = \theta_{2,j_w} + \theta_{3,j_w} - \theta_2, \hat{\theta}_{4,j_w} = \theta_{4,j_w} - \theta_3, \hat{\theta}_{5,j_w} = \theta_{5,j_w}, \hat{I}_{\mathrm{M},1,j_w}^{\mathrm{P}} = e_3 e_\infty, i_{\mathrm{P}} = 1,2,$$

$$\hat{M}_{\mathrm{L},i_{\mathrm{P}},j_w} = \tilde{M}_3 \tilde{M}_2 \left( 1 - \frac{\hat{t}_{i_{\mathrm{P}},j_w}}{2} \hat{I}_{\mathrm{M},i_{\mathrm{P}},j_w}^{\mathrm{P}} \right) M_2 M_3, \hat{M}_{\mathrm{L},5,j_w} = \cos \frac{\hat{\theta}_{5,j_w}}{2} - \sin \frac{\hat{\theta}_{5,j_w}}{2} I_{\mathrm{M},5,j_w}^{\mathrm{R}},$$

$$\hat{M}_{\mathrm{L},3,j_w} = \tilde{M}_3 \left( \cos \frac{\hat{\theta}_{3,j_w}}{2} - \sin \frac{\hat{\theta}_{3,j_w}}{2} e_3 e_1 \right) M_3, \hat{M}_{\mathrm{L},4,j_w} = \cos \frac{\hat{\theta}_{4,j_w}}{2} - \sin \frac{\hat{\theta}_{4,j_w}}{2} e_2 e_3,$$

$$\hat{t}_{1,j_w} = t_{1,j_w} - t_1 + \sin \theta_{2,j_w} \left( l_{j_w} \cos \varphi_{j_w} + (-1)^{j_w+1} r_{j_w} \right)$$
$$+ 2 \sin^2 \frac{\theta_{2,j_w}}{2} \sin \varphi_{j_w} l_{j_w} + (-1)^{j_w+1} r_{j_w} \left( -2 \sin \theta_{2,j_w} \sin^2 \frac{\theta_{3,j_w}}{2} + \cos \theta_{2,j_w} \sin \theta_{3,j_w} \right),$$

$$\hat{t}_{2,j_w} = - \sin \theta_{2,j_w} \sin \varphi_{j_w} l_{j_w} + 2 \sin^2 \frac{\theta_{2,j_w}}{2} \left( l_{j_w} \cos \varphi_{j_w} + (-1)^{j_w+1} r_{j_w} \right)$$
$$+ (-1)^{j_w+1} r_{j_w} \left( \sin \theta_{2,j_w} \sin \theta_{3,j_w} + 2 \cos \theta_{2,j_w} \sin^2 \frac{\theta_{2,j_w}}{2} \right), \hat{I}_{\mathrm{M},2,j_w}^{\mathrm{P}} = e_1 e_\infty$$

**Step 3**, formulate unknown joint displacements as,

$$\begin{cases} \hat{t}_{g_{\mathrm{P}},j_w} = 0 \\ \hat{\theta}_{g_{\mathrm{R}},j_w} = 0 \end{cases}, g_{\mathrm{P}} = 1,2, g_{\mathrm{R}} = 3,4,5 \tag{16}$$

Solving Eq. (16) yields,

$$\theta_{5,j_w} = 0, \theta_{4,j_w} = \theta_3, \theta_{3,j_w} = \theta_2 - \theta_{2,j_w},$$
$$\theta_{2,j_w} = \varphi_{j_w} - \arccos\left( -d_{2,j_w} / l_{j_w} + \cos \varphi_{j_w} \right),$$
$$t_{1,j_w} = d_{1,j_w} - 2 \sin^2 \frac{\theta_{2,j_w}}{2} \sin \varphi_{j_w} l_{j_w} - \cos \varphi_{j_w} \sin \theta_{2,j_w} l_{j_w}, d_{1,j_w} = t_1 - (-1)^{j_w+1} r_{j_w} \sin \theta_2,$$
$$d_{2,j_w} = -2 (-1)^{j_w+1} r_{j_w} \sin^2 \frac{\theta_2}{2}$$

Similarity, the joint displacements of the PRRU limb could be solved as,

$$\theta_{5,3} = - \arctan(\sin \theta_3 \tan \theta_2), \theta_{4,3} = \arcsin(\sin \theta_2 \cos \theta_3), d_{1,3} = t_1 + r_3 \sin \alpha,$$
$$\theta_{3,3} = \arctan(\sin \theta_3 / (\cos \theta_3 \cos \theta_2)) - \theta_{2,3}, \theta_{2,3} = -\varphi_3 + \arccos\left( d_{2,3} / l_3 - \cos \varphi_3 \right),$$
$$t_{1,3} = d_{1,3} - 2 l_3 \sin^2 \frac{\theta_{2,3}}{2} \sin \varphi_3 + l_3 \cos \varphi_3 \sin \theta_{2,3}, d_{2,3} = -2 r_3 \sin^2 \frac{\alpha}{2}, \alpha = \theta_{3,3} + \theta_{2,3}$$

## 5   Conclusion

A CGA based method is proposed in this paper for parameterized inverse kinematics of PMs. First of all, successive screw motion of any PM can be concisely formulated by CGA. By concerning geometric features of joints, equations of joint displacements are then reconstructed and solved. A 3-DoF PM is taken as example to demonstrate the proposed method. It shows that the CGA based method deals with the inverse kinematic problem in an analytical and efficient manner. This lays a solid theoretical foundation for the development of PMs.

**Acknowledgments.** This work was supported in part by the National Natural Science Foundation of China under Grant 51475321, Grant 51675366, in part by International Postdoctoral Exchange Fellowship Program No. 32 Document of OCPC, 2017.

# References

1. Araujo-Gómez, P., Mata, V., Díaz-Rodríguez, M., et al.: Design and kinematic analysis of a novel 3UPS/RPU parallel kinematic mechanism with 2T2R motion for knee diagnosis and rehabilitation tasks. J. Mech. Robot. **9**, 061004 (2017)
2. Carbajal-Espinosa, O., Izar-Bonilla, F., Díaz-Rodríguez, M.: Inverse kinematics of a 3 DOF parallel manipulator: a conformal geometric algebra approach. In: Proceedings of 2016 IEEE-RAS 16th International Conference on Humanoid Robots (Humanoids) (2016)
3. Fu, Z.T., Yang, W.Y., Yang, Z.: Solution of inverse kinematics for 6R robot manipulators with offset wrist based on geometric algebra. J. Mech. Robot. **5**(3), 310081–310087 (2013)
4. Hestenes, D.: Old wine in new bottles: a new algebraic framework for computational geometry. In: Bayro-Corrochano, E., Sobczyk, G. (eds.) Geometric Algebra with Applications in Science and Engineering. Birkhauser, Boston (2001)
5. Kim, J.S., Jin, H.J., Park, J.H.: Inverse kinematics and geometric singularity analysis of a 3-SPS/S redundant motion mechanism using conformal geometric algebra. Mech. Mach. Theory **90**, 23–36 (2015)
6. Kohli, D., Soni, A.H.: Kinematic analysis of spatial mechanisms via successive screw displacements. J. Eng. Ind. **97**(2), 739 (1975)
7. Hestenes, D.: New Foundations for Classical Mechanics, 2nd ed. Springer, Netherlands (1999)
8. Milenkovic, P.: Series solution for finite displacement of planar four-bar linkages. ASME J. Mech. Robot. **3**, 014501-1–014501-7 (2011)
9. Raghavan, M., Roth, B.: Inverse kinematics of the general 6R manipulator and related linkages. ASME J. Mech. Des. **115**, 502–508 (1993)
10. Song, Y.M., Han, P.P., Wang, P.F.: Type synthesis of 1T2R and 2R1T parallel mechanisms employing conformal geometric algebra. Mech. Mach. Theory **121**, 475–486 (2018)
11. Tsai, L.-W.: Robot Analysis and Design: The Mechanics of Serial and Parallel Manipulators. Wiley, Hoboken (1999)

# On the Full-Spin Dexterous Orientation Workspace of Spherical Parallel Robots of 3-R̲RR-Type

Khaled A. Arrouk$^{(\boxtimes)}$, B. Chedli Bouzgarrou, and Grigore Gogu

Université Clermont Auvergne, CNRS, SIGMA Clermont, Institut Pascal,
63000 Clermont-Ferrand, France
{khaled.arrouk,belhassen-chedli.bouzgarrou,
grigore.gogu}@sigma-clermont.fr

**Abstract.** This paper highlights the use of graphical techniques for workspace characterisation of spherical parallel robotic manipulators (SPRMs). Kinematic analysis of this type of mechanism is revisited through a 3D representation of the orientation operational space in a spherical coordinate system in which the radial coordinate corresponds to the last rotation of the *Z-Y-X* Euler angles sequence. The full-spin orientation workspace (FSOW) is defined as the region attainable for any spin rotation, third rotation around *X*, of the mobile platform (MPF). This new type of workspace can be considered as a dexterous workspace for SPRMs. A geometric approach, enabling efficient and rapid determination and representation of the FSOW, is introduced. Several examples are given to illustrate the implementation of the proposed method in a CAD environment.

**Keywords:** Spherical parallel manipulators · Full-spin orientation workspace
Geometric approaches · CAD-based graphical techniques

## 1 Introduction

One of the most important design criteria of parallel robotic manipulators (PRMs) is the workspace. In the literature, several types of workspaces have been studied. We can cite for instance, the total workspace, constant orientation workspace, inclusive maximal workspace, dexterous workspace, etc. The dexterous workspace can be defined as the set of locations (points) of the end-effector characteristic point (EECP) for which all orientations are possible [1], or as the volume within which the robot end-effector (EE) has complete manipulative capability. With a reference point in the dexterous workspace, the EE can be completely rotated about any axis through that point [2]. Generally, the dexterous workspace is a very small subset of the total reachable workspace, and in some cases this subset is empty for some robots [3]. Compared to serial architectures, PRMs have generally smaller operational workspace. Dexterous workspace has not been sufficiently deepened in the literature for SPRMs despite some valuable contributions related to this type of manipulators [4–6]. This paper highlights the use of graphical techniques for workspace characterization of SPRMs while considering a new type of workspace: full-spin orientation workspace (FSOW). This work

© Springer Nature Switzerland AG 2019
B. Corves et al. (Eds.): EuCoMeS 2018, MMS 59, pp. 347–354, 2019.
https://doi.org/10.1007/978-3-319-98020-1_41

is in line with our recent developments on orientation workspace analysis through a specific parameterizing and 3D representation of the operational space [7, 8]. The rest of this paper is organized as follows. Section 2 presents the parameterization used for geometric modeling of the SPRM 3-RRR. We also introduce a new graphical representation of the orientation space. This representation is used for the determination of the 3D workspace of the SPRM using CAD environment facilities. In Sect. 3, we propose a new graphical method to determine the FSOW of SPRM of 3-RRR-type.

## 2   Workspace Analysis of 3-RRR-Type SPRM

A SPRM of 3-RRR-type is made up of a fixed base (FB) and a moving platform (MPF), connected by three limbs of RRR-type. Each limb consists of two links connected to each other by a revolute joint: a proximal link attached to the FB, and a distal link attached to the MPF by revolute joints. Figure 1 displays a CAD model of a SPRM of 3-RRR-type and a serial spherical limb (SSL) of RRR-type isolated from the rest of the mechanism. The SPRM is characterized by the fact that all joint axes intersect at a common point, which is called the geometric center of the mechanism and referred to as $O$, see Fig. 1. This architecture enables each point of the MPF to move on a spherical surface centered in point $O$. So a SPRM of 3-RRR-type provides 3-DOFs of pure rotation about the point $O$.

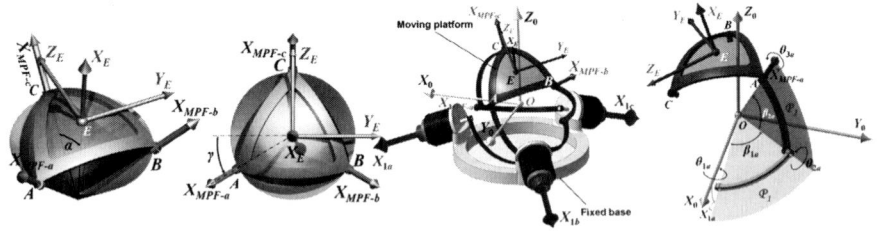

**Fig. 1.** Parametrized CAD model of SPRM 3-RRR-type [8]

### 2.1   Geometric Modeling

We denote by $R_0(O\text{-}X_0Y_0Z_0)$ the fixed frame attached to the FB at the geometric center of the mechanism $O$ and $R_E(E\text{-}X_EY_EZ_E)$ attached to the MPF at the EECP $E$. The sequence of Euler angles we have adopted is such that the third rotation corresponds to the spin angle about $X_E$-axis. In this paragraph, the parameters used in geometric modeling of the spherical 3-RRR are introduced. Readers can refer to our recent works [7, 8] for more details about the parameterization of SPRM 3-RRR-type. These parameters are involved in workspace determination. They are divided into **5 sets** defined as follows: **Set 1:** intrinsic parameters of the FB defined by angles $\xi$ and $v_k$ ($k = a, b, c$) that characterize the design of the FB. $v_k$ designates the first rotation angle around $Z_0$ and $\xi$ is the second rotation angle about $Y_{0k}$. These two successive rotations transform the $X_0$ axis into the axis of the motorized revolute joint of the limb $k$ on the fixed base. For a symmetric design $v_a = 0$ deg, $v_b = 120$ deg and $v_c = 240$ deg. **Set 2:** intrinsic parameters of the MPF defined by the angles $\alpha$ and $\gamma_k$ ($k = a, b, c$), see Fig. 1. The two successive rotations of angle $\gamma_k$ about $X_E$ and $\alpha$ angle about $Z_{MPF\text{-}k}$ or $Y_{MPF\text{-}k}$ transform the $X_E$-axis into $X_{MPF\text{-}k}$, axis of the third passive revolute joint of the limb $k$ on the MPF. It can be highlighted that the frame $R_{Ek}(I\text{-}X_{MPF\text{-}k}Y_{MPF\text{-}k}Z_{MPF\text{-}k})$ is attached to the MPF at its vertex ($I = A, B, C$). **Set 3:** joint and design parameters of the limbs. It contains the parameters used for one serial spherical limb $k$ of the SPRM of 3-RRR-type. Each limb $k$ is defined by two design angular parameters $\beta_{1k}$ and $\beta_{2k}$ ($k = a, b, c$) corresponding respectively to the lengths of its proximal and distal links. The joint space configuration of a limb $k$ is given by $\theta_{1k}$, $\theta_{2k}$ and $\theta_{3k}$ joint variables. **Set 4:** motorized joint space parameters. It regroups the motorized joints parameters. For the SPRM of 3-RRR-type, the joint parameters correspond to the rotation angles of the motorized revolute joints mounted on the fixed base denoted by $\theta_{1a}$, $\theta_{1b}$ and $\theta_{1c}$ (**Set 4** is a subset of **Set 3**). **Set 5:** operational space parameters. It contains the parameters of rotation from $R_0$ to $R_E$ defined by the Z-Y-X sequence of Euler angles denoted by $\phi_z$, $\phi_y$ and $\phi_x$. The configuration of the MPF in the operational space is given by the rotation matrix $\mathbf{R}_{OE}$.

$$\mathbf{R}_{0E} = \mathbf{R}_{Z_0}(\phi_z)\,\mathbf{R}_{Y_1}(\phi_y)\,\mathbf{R}_{X_2}(\phi_x) \tag{1}$$

This configuration can be expressed in terms of geometric parameters of the FB, the MPF and limbs as well as joint variables by using the successive rotations product. The transformation $\mathbf{T}_{0E}^{k}$ associated with a limb $k$ can be identified with the rotation matrix $\mathbf{R}_{OE}$.

$$\mathbf{T}_{0E}^{k} = \mathbf{R}_{Z_0}(v_k)\,\mathbf{R}_{Y_{0k}}(\xi)\,\mathbf{R}_{X_{1k}}(\theta_{1k})\mathbf{R}_{Z'_{1k}}(\beta_{1k})\mathbf{R}_{X_{2k}}(\theta_{2k})\mathbf{R}_{Z_{3k}}(\beta_{2k})\mathbf{R}_{X_{3k}}(\theta_{3k})\mathbf{R}_{Ek}^{T}$$

$$\mathbf{T}_{0E}^{k} = \mathbf{R}_{0E} \tag{2}$$

where $\mathbf{R}_{Ek}$ is the rotation matrix from the MPF coordinate system $R_E$ to the platform local frame $R_{MPF\text{-}k}$ attached to the vertex $I = A, B, C$ and $k = a, b, c$. This matrix is a function of MPF parameters $\alpha$ and $\gamma_k$.

## 2.2 Total Orientation Workspace (TOW) Determination and Representation

In our previous works [7–9] we have shown that the workspace of a parallel manipulator can be obtained by the intersection of the vertex workspaces associated with each limb isolated from the whole mechanism and having the MPF as end-effector (EE). This approach has been implemented for the determination and representation of the total workspace of planar and spherical PRMs by using a graphical resolution in a CAD environment, since the operational space is of dimension less than or equal to 3. So, the fundamental idea in our method for obtaining the 3D TOW representation of SPRM 3-RRR-type resides in finding the three regions attainable by the EECP attached to each serial spherical limb. It is noteworthy that, the boundaries of these regions, embedded in 3D space, are completely, for these types of mechanisms, defined by the so-called serial singularities. Additionally, the vertex space $W_{Ek}$ ($k = a$, $b$, $c$) of each SSL can be obtained by using its direct kinematic model. Since the SPRM 3-$\underline{R}$RR-type is the parallel association of three identical SSLs, the orientation space attainable by the EECP can be obtained by the intersection of all vertex spaces accessible by the MPF considered as EE attached to SSLs. Employing the formulations presented in Sect. 2.1 of this article, one can generate the point cloud corresponding to the 3D vertex volume attainable by the MPF attached to each SSL. It is worth mentioning, that the point clouds are obtained by fixing the joint angle $\theta_{2k}$ at 0 then at $\pi$ while varying angles $\theta_{1k}$ and $\theta_{3k}$ from 0 to $2\pi$ with a given discretization step. After importing the 3D scattered point clouds, the 3D solid volumes associated with the vertex volumes of the SSL are constructed using several Computer-Aided Design functionalities. Then, one can obtain the TOW of the SPRM 3-$\underline{R}$RR-type by applying Boolean intersection operations on the 3D reconstructed volumes. Figure 2 presents 3D Vertex Total Orientation Workspaces (Vertex-TOW) and the Robot-TOW of the SPRM 3-$\underline{R}$RR-type. Readers can refer to [7, 8] for more details about the reconstruction of the TOW of SPRM 3-$\underline{R}$RR-type.

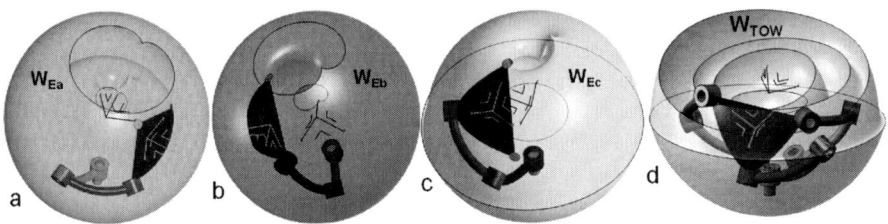

**Fig. 2.** Procedure for obtaining the TOW of SPRM 3-$\underline{R}$RR-type: vertex total orientations workspaces $W_{Ek}$ of SSLs (a, b, c), Robot-TOW of the SPRM 3-$\underline{R}$RR-type (d)

## 2.3 Constant-Spin Orientation and the Full-Spin Orientation Workspaces

After obtaining the 3D Vertex-TOWs and the Robot-TOW of the SPRM, Fig. 2, the Constant-Spin Orientation Workspaces (CSOWs) can be extracted. This type of workspace can be defined as follows: Vertex-CSOW is the spherical region attainable by the MPF for any fixed desired value of the spin angle $\phi_x$ when the MPF is connected to the SSL. Robot-CSOW definition is similar to that of Vertex-CSOW, and it can be defined as the spherical region accessible by the MPF for a fixed value of the spin angle $\phi_x$ when the MPF is connected at the same time to all the SSLs. In other words, the Robot-CSOWs are the spherical zones that can be attained by the MPF for a given value of the $\phi_x$ after closing the kinematic chains. CSOWs are obtained by cutting out the volume of the Vertex-TOW and Robot-TOW by spheres corresponding to different spin angles $\phi_x$. It is notable that the Robot-CSOW for a given value of $\phi_x$ is nothing but the intersection between the Vertex-CSOWs associated to the three SSLs. Figure 3 depicts two *"anatomic"* views of the 3D Vertex-TOW (left), and 3D Robot-TOW (right) after cutting them by some spherical slices corresponding to different CSOWs.

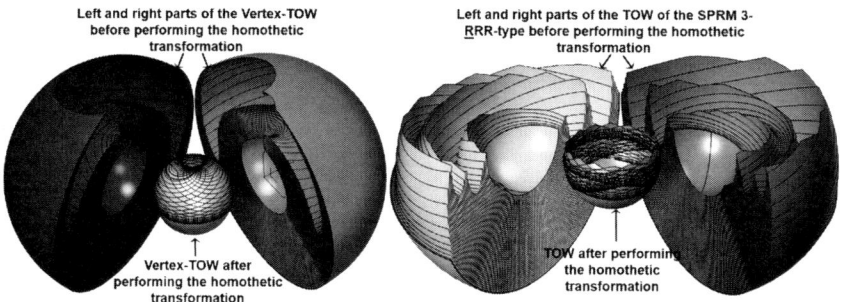

**Fig. 3.** Anatomic views of the Vertex-TOW of one SSL $\underline{R}$RR-type (left) and 3D TOW of SPRM 3-$\underline{R}$RR-type (right)

The other type of workspace that we can introduce concerns the Full-Spin Orientation Workspace (FSOW). Here we also have to distinguish the Vertex-FSOW and Robot-FSOW. The first one can be defined as the region of the space which can be reached, at any spin rotation angle, by the EECP of the MPF attached to the SSL when this latter is considered isolated from the whole mechanism. For the full SPRM 3-$\underline{R}$RR-type, the FSOW is the region attainable by the EECP at any spin rotation angle of the MPF after closing the kinematic chains. It is noteworthy that the Robot-FSOW is nothing but the intersection of all Vertex-FSOWs attainable by the MPF attached to the SSLs. The technique adopted to obtain the FSOW of the previously obtained 3D Vertex-TOW and Robot-TOW can be resumed as follows: based on the use of the CSOWs as an intermediate step, we apply a homothetic transformation on each spherical slice (CSOWs), of radius $\rho$ in $[\rho_{min}, \rho_{max}]$, to bring it back on to the sphere at the scale of the robot. Where $\rho$ corresponds to the radial distance calculated from $O$ replacing the spin rotation angle $\phi_x$. For better comprehension, we have created anatomic views before and after applying the homothetic transformation as illustrated in Fig. 3. By considering all these surfaces on the same supporting sphere, it is easy to recognize a common region corresponding quite simply to the Vertex-FSOW for the SSL and Robot-FSOW (Fig. 3). This procedure can be formulated as following:

$$\text{FSOW} = \bigcap_{\phi_x=0}^{2\pi} \text{CSOW}(\phi_x) \qquad (3)$$

## 3 Geometric Method for the Determination of Full-Spin Orientation Workspace

In this section we present a geometric method that enables the construction of the Vertex-FSOW and the Robot-FSOW without any need to use the numerical procedure presented in Sect. 2. Indeed, it can be observed that the Vertex-FSOW depends only on the geometric parameters $\beta_{1k}$ and $\beta_{2k}$ of the SSL and the semi-angle $\alpha$ of the MPF

(Fig. 1). However, the inclination of the Vertex-FSOW is governed only by the angle $\xi$, see Fig. 1 in Sect. 2.

## 3.1 Principle of the Method

If a point on the unit sphere, corresponding to given operational space parameters $\phi_z$ and $\phi_y$ of the MPF, satisfies full-spin condition, this implies that the MPF can perform $2\pi$ rotation around its axis $X_E$ while all the SSLs remain assembled with the MPF. This means that the axis $X_{MPF-k}$ of the local frame $R_{Ek}$ attached to the vertex ($I = A, B, C$) of the MPF can coincide with the axis $X_{3k}$ of the revolute joint between the distal link of the limb $k$ and the MPF. Given that these axes are normal to the unit sphere, this condition can be interpreted geometrically as follows: the geometric entity swept by the axis $X_{MPF-k}$, when achieving a full spin rotation around $X_E$, must be totally inside the geometric entity swept by the axis $X_{3k}$ for all values of joint variables ($\theta_{1k}, \theta_{2k}$) in $[0,2\pi] \times [0,2\pi]$. In fact, the geometric entity swept by the axis $X_{MPF-k}$ when achieving a full spin rotation around the axis $X_E$ is a conic surface, with half angle $\alpha$ and vertex point $O$. This surface is denoted by $S (X_E, \alpha)$. The geometric entity swept by the axis $X_{3k}$ for all values of joint variables ($\theta_{1k}, \theta_{2k}$) in $[0,2\pi] \times [0,2\pi]$ is a 3D volume denoted by $\mathcal{V}$. It is delimited by conic surfaces depending on limb lengths $\beta_{1k}$ and $\beta_{2k}$. Therefore, the full-spin orientation space is the set of points on the unit sphere for which the conic surface $S$ can be totally inside the volume $\mathcal{V}$, as illustrated in Fig. 4.

## 3.2 Application Cases

In order to illustrate the proposed method, several designs of the SPRM of 3-RRR-type are considered. For a symmetric design, the Vertex-FSOWs associated with all SSLs have the same topology. Therefore, the construction procedure of the Vertex-FSOW is applied for only one SSL. The other Vertex-FSOWs are obtained by applying rotations on one Vertex-FSOW around the vertical axis $Z_0$ of the fixed base by 120 deg and 240 deg. For space limitation reason two architectures have been presented: **Case-1:** $\beta_{1k} = \beta_{2k}$ (<90 deg) and $\beta_{1k} + \beta_{2k} + \alpha < 180$ deg, **Case-2:** $\beta_{1k} = \beta_{2k}$ (<90 deg) and $\beta_{1k} + \beta_{2k} + \alpha > 180$ deg. The geometrical method for obtaining the Vertex-FSOW is illustrated graphically in 2D in Fig. 5(a) for **Case-1** and (b) for **Case-2**. For **Case-1**, the SPRM 3-RRR-type has: $\beta_{1k} = \beta_{2k} = 60$ deg, $\alpha = 45$ deg and $\xi = 75$ deg. The procedure for constructing the Vertex-FSOW and the Robot-FSOW can be resumed as following: from the schema Fig. 5(a), one can conclude that the Vertex-FSOW is confined within 3 surfaces. The 1st and the 3rd ones are spherical caps having as radii $\rho_{min}$ and $\rho_{max}$ respectively. Between the previous surfaces there is a conical surface having as semi-angle the value $\Omega = \beta_{1k} + \beta_{2k} - \alpha$. Where, $\rho_{min, max}$ correspond to the minimal and maximal radial distances (for $\phi_x = 0$ deg, $\phi_x = 360$ deg) respectively. This radial distance $\rho$ is calculated from $O$. It replaces the spin rotation angle $\phi_x$. It must be noted that the axis of the conical surface is nothing but the axis of the first revolute joint of the SSL of RRR-type. For **Case-2**, the SPRM 3-RRR-type has: $\beta_{1k} = \beta_{2k} = 80$ deg, $\alpha = 45$ deg and $\xi = 30$ deg. We have found that each SSL has 2 non-connex FSOWs, Fig. 5(b). The 1st one is obtained as previously illustrated for **Case-1**. The 2nd one is obtained similarly but with $\Omega = |180 - (\beta_{1k} + \beta_{2k} + \alpha)|$.

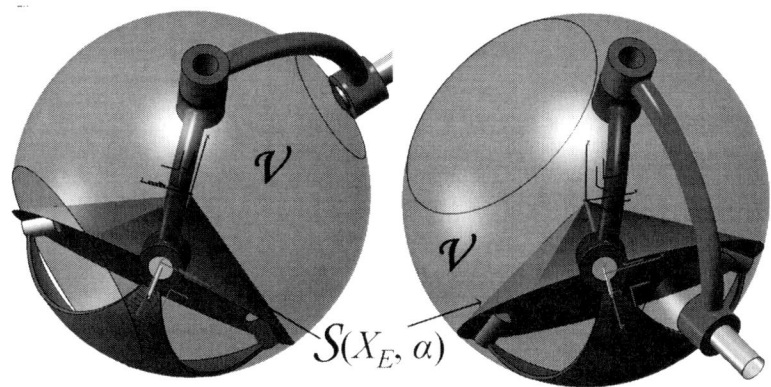

**Fig. 4.** Graphical interpretation of the proposed method for Vertex-FSOW: EECP does not belong (left), belongs (right) to the SSL Vertex-FSOW

After construction the first volume associated with the 1st Vertex-FSOW, one can apply easily 2 rotations on this volume, by 120 deg and 240 deg, to obtain the Vertex-FSOWs attainable by the other SSLs, as depicted in Fig. 6(a, c). Then to obtain the Robot-FSOW, one can apply two Boolean intersection operations to extract the common volume, as illustrated in Fig. 6(b, d).

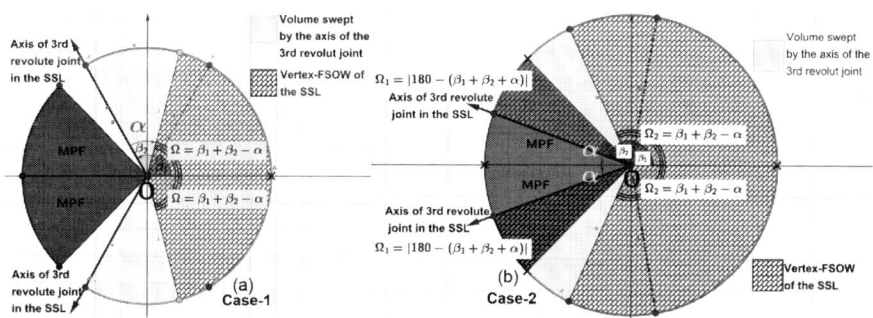

**Fig. 5.** Graphical method for obtaining the Vertex-FSOW of one SSL RRR-type

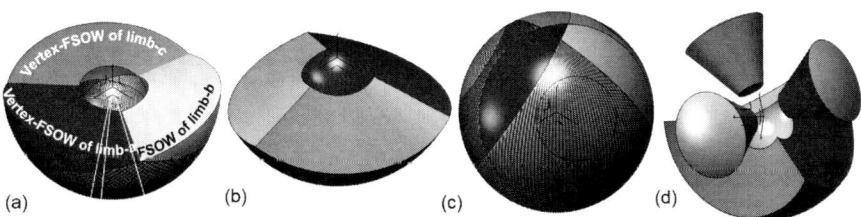

**Fig. 6.** Vertex-FSOWs: Case-1 (a), Case-2 (c), Total Robot-FSOW: Case-1 (b), Case-2 (d)

## 4   Conclusions

The dexterous workspace of SPRMs has been addressed by introducing the full-spin orientation workspace (FSOW). A new geometric approach has been presented to determine the dexterous domain attainable by the EECP of the SPRM 3-R̲RR-type. Based on the design parameters, one can easily determine the vertex FSOW of each limb. Thereafter, the FSOW of the robot is obtained by Boolean intersection of vertex FSOWs of limbs. The different examples of applications demonstrate the efficiency of the method in the preliminary design stages of SPRM. Our future works will focus on finding the various types of singularities of the SPRM of 3-R̲RR-type and determining how they affect its dexterity.

## References

1. Merlet, J.-P.: Parallel Robots, 2nd edn. Springer, Heidelberg (2005). ISBN 978-1-4020-4133-4
2. Vijaykumar, R., Tsai, M.J., Waldron, K.J.: Geometric optimization of serial chain manipulator structures for working volume and dexterity. Int. J. Robot. Res. **5**(2), 91–103 (1986)
3. Zacharias, F.: Knowledge Representations for Planning Manipulation Tasks. Technische Universität München, Diss (2011)
4. Chaker, A., Laribi, M.A., Romdhane, L., Zeghloul, S.: Synthesis of spherical manipulator for dexterous medical task. In: 2nd IFToMM International Symposium on Robotics and Mechatronics, Shangai, China, 3–5 November 2011
5. Bai, S.P.: Optimum design of spherical parallel manipulators for a prescribed workspace. Mech. Mach. Theory **45**(2), 200–211 (2010)
6. Wu, G., Caro, S., Bai, S., Kepler, J.: Dynamic modeling and design optimization of a 3-DOF spherical parallel manipulator. Robot. Auton. Syst. **62**(10), 1377–1386 (2014)
7. Arrouk, K.A., Bouzgarrou, B.C., Gogu, G.: On the workspace representation and determination of spherical parallel robotic manipulators. In: Wenger, P., Flores, P. (eds.) New Trends in Mechanism Science: Theory and Industrial Applications, pp. 131–139. Springer, Cham (2016). ISBN 978-3-319-44156-6
8. Arrouk, K.A., Bouzgarrou, B.C., Gogu, G.: Workspace characterization and kinematic analysis of general spherical parallel manipulators revisited via graphical based approaches. Mech. Mach. Theory **122**, 404–431 (2018)
9. Arrouk, K.A., Bouzgarrou, B.C., Gogu, G.: CAD-based unified graphical methodology for solving the main problems related to geometric and kinematic analysis of planar parallel robotic manipulators. Robot. Comput. Integr. Manuf. **37**, 302–321 (2016)

# Solving Inverse Kinematics of a Planar Dual-Backbone Continuum Robot Using Neural Network

Ebrahim Shahabi and Chin-Hsing Kuo$^{(\boxtimes)}$

Department of Mechanical Engineering, National Taiwan University of Science and Technology, Taipei, Taiwan
{dl0603805, chkuo717}@mail.ntust.edu.tw

**Abstract.** The inverse kinematics of multiple-backbone continuum robots is a highly non-linear problem. Traditional methods for solving such kinds of problems include the inverse transformation, geometric approach, etc., which are relatively complex and have multiple solutions. On the other hand, the pseudo-rigid-body model (PRBM) is one simple approach for solving the forward kinematics of multiple-backbone continuum robots, but currently, it is still not applicable for solving the inverse kinematics problem. In this paper, we present a strategy for solving the inverse kinematics problem of a dual-backbone continuum robot using PRBM and the Artificial Neural Network (ANN). The strategy firstly computes the forward kinematic solutions of the dual-backbone robot via the PRBM approach. The obtained solutions are then used to build up an ANN model for solving the inverse kinematics of the dual-backbone continuum robot. Based on the Bayesian Regularization training algorithm, the accuracy of the ANN results after linear regression is 99.99%. Finally, we compared the errors between the inputs of the forwards kinematics problem and the output (solutions) of the inverse kinematics problem solved by the ANN model.

**Keywords:** Artificial Neural Network · Inverse kinematic · Continuum robot
Dual-backbone · Pseudo-rigid-body model

## 1 Introduction

In last decade continuum robot has been one of the hot subjects in the field of robotics. Unlike traditional robots, continuum robot doesn't have any discrete part and is usually concreted by fluid or cables. One particular feature of the continuum robots is the infinity degree of freedom (DoF). With this excellent ability, continuum robot can avoid obstacles, but it is completely dependent on the kinematic analysis of the robot. Burgner-Kahrs et al. [1] classified continuum robots into three types of structures, i.e., the single backbone, multi-backbone and concentric tube models. Among which, the structures of the multi-backbone type of continuum robots are formed by several in-parallel super-elastic wires. The super-elastic wire has a large deflection, and this feature can make a wide range of workspace for the robots.

© Springer Nature Switzerland AG 2019
B. Corves et al. (Eds.): EuCoMeS 2018, MMS 59, pp. 355–361, 2019.
https://doi.org/10.1007/978-3-319-98020-1_42

Most of the studies have been on the single backbone robot [3–6], whereas researches on kinematics in the multi-backbone are rare. Some researches [6, 7, 9, 10] proposed a theory to solve the kinematics on hyper-redundant multi-backbone robots. On the other hand, Constant-curvature assumption [11] is a famous theory for solving the kinematics of continuum robots. In this theory, each wire is assumed to have a circular shape, and the kinematics of their robots can be solved with considerable simplicity. Xu et al. [8] developed a multi-backbone robot and solved the kinematics with constant-curvature assumption theory. Recently, Bryson and Rucker [2] proposed a type of continuum parallel manipulator that has six flexible in-parallel compliant legs made of steel music wires. They used the Cosserat rod theory for solving the kinematics of this robot.

Pseudo-rigid-body model (PRBM) [9] is a useful method for analyzing the kinematics of non-rigid body mechanisms. The PRBM 1R (one revolute joint) [10] represents a flexural beam by two rigid links. Then this model was extended to PRBM 2R [11] and PRBM 3R [12] for large-deflection beams. Kuo et al. [13] presented the continuum kinematics of a planar dual-backbone robot based on PRBM 3R. Other types of PRBM have also developed for different purposes [14].

ANN is a mathematical algorithm model that imitates the pattern of human being's neural network. Currently, ANN has been widely used in the robotics. The ANN method has several advantages, for example, unique solution, self-adaptive, fast solving speed, etc. Some researchers had used the ANN to find the inverse kinematics of the planar-mechanism problems [15, 16]. Recently, ANN has been employed to find the positions of the end-effector and rotation of the joints [17–19].

In this paper, we used the Back Propagation Neural Network (BPNN) algorithm for solving the inverse kinematics of the dual-backbone planar mechanism that is formulated by PRBM. First of all, we solved the forward kinematics of the planar robot with PRBM and then used the solutions to build up a BPNN for solving the inverse kinematics problem. Lastly, we compared the result to show the error between ANN and PRBM 3R model.

## 2   Geometry of the Dual-Backbone Continuum Robot

As shown in Fig. 1, the dual-backbone continuum robot is formed by two super-elastic NiTi-alloy wires connected to the end-effector and a frame disk. Three specific dimension parameters show us the geometry of the frameworks and the end-effector: $d_b$, $d_p$, and $d_e$. Changing the lengths of the two NiTi wires ($L_1$ and $L_2$) affects the position of the end-effector. The position of the end-effector explained with two referencing coordinate systems $O(x, y)$ and $E(u, v)$ are attached to the frame disk and end-effector, respectively. The origin of $E(e_x, e_y)$ and the angle $\Theta$ represent the position and the orientation of the end-effector, respectively.

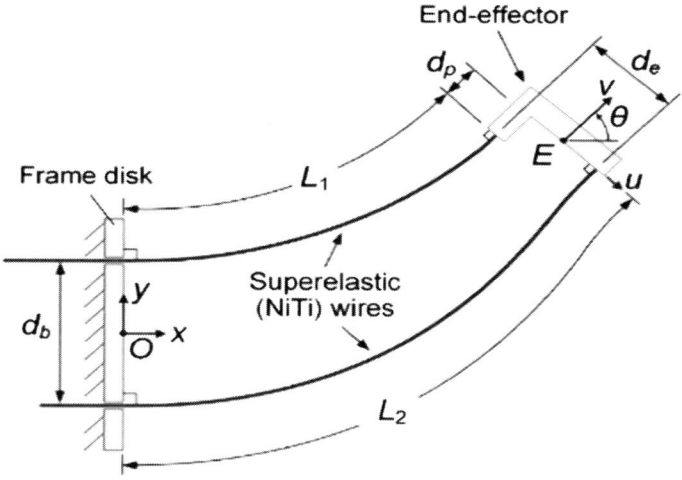

**Fig. 1.** Geometry of the mechanism [13]

# 3 Kinematics of the Robot

## 3.1 Pseudo-Rigid-Body Model

Kuo et al. [13] solved the forward kinematics and inverse kinematics of a planar dual-backbone continuum robot based on PRBM 3R. Accordingly, they formulated a two-wire robot as a system of nonlinear equations. They further studied some experimental models, also compared it with finite element model. In their PRB 3R modeling, the maximum percentage error was 0.47% and 1.96% in the $x$- and $y$-directions, respectively. Notable of their research is solved displacement-displacement kinematics. In fact, they solved the relationship between the wire lengths and end-effector's location. They showed how the load-displacement PRBM could be extended to apply for the displacement-displacement problem.

## 3.2 Forward and Inverse Kinematics by PRBM

Forward kinematics is a way to find the location and orientation of the end-effector, i.e., ($\Theta$, $e_x$, $e_y$), with the given lengths of $L_1$ and $L_2$. In forward kinematics of the two-wire continuum robots, there are eight scalar equations with eight unknowns in the PRBM 3R formulation [17]. The end-effector's orientation and position can be solved from the set of equations. However, on the other hand, it was shown in [17] that the inverse kinematics of the two-wire continuum robot is not solvable since the system of equations has an insufficient number of equations.

## 3.3 Inverse Kinematics by Neural Networks

Neural networks (NN) has lots of advantages compared to equation method, capability, and flexibility to learn, simple structure, etc. Also, NN has some disadvantages, for example, this method requires a series of basic information for training, the answers are approximate, and there are some errors in it. Some works focus on solving the inverse kinematics of the continuum robots [22], rigid body robots [23, 24], cable robots [18], etc., with different models of NN.

As noted earlier, it is impossible to solve the inverse kinematics of planar mechanism via the PRBM 3R method. For the first time, this research focus on solving the inverse kinematics of the dual-backbone continuum robot by BPNN [20]. The Bayesian method was chosen to train the NN since it does not require validation data, which is very useful when the problem has limited data [21]. With this respected mean squared error (MSE) and regression (R) are $7.52792 \text{ e}^{-4}$ and $9.99 \text{ e}^{-1}$, respectively.

For this purpose, BPNN is employed here to calculate the inverse kinematics from the forward kinematics of the dual-backbone super-elastic continuum robot. Figure 2 illustrates the relationship between the kinematics of our model. The forward Kinematics is comparable to a function that takes the inputs $L_1$ and $L_2$, in our case, to the outputs $\Theta$, $e_x$, and $e_y$, while the inverse function brings the latter back to the former. Therefore, by using the inputs and outputs of the forward kinematics, we can train a neural network for inverse kinematics.

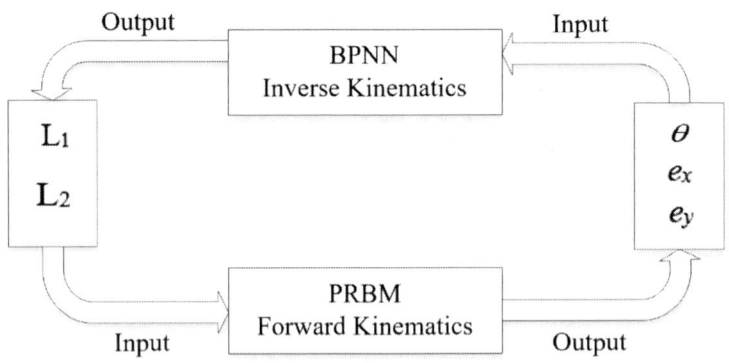

**Fig. 2.** The relationship between forward and inverse kinematics

In this research, the BPNN was constructed via the neural network toolbox in Matlab® version 2016b with a hidden layer with eight neurons. The NN was trained and tested with 80 and five sets of data, respectively. The results of our BPNN are presented in Figs. 3 and 4. The output of the regression is the result of our networks, so Fig. 3 shows that the accuracy of the network is around 99.99%. Figure 4 illustrates the output performance for training and test. Furthermore, Table 1 illustrates the comparison between the output of BPNN and the input of forward kinematics. As it can be seen, the maximum error for $L_1$ and $L_2$ are 3.8% and 0.79%, respectively. Besides, as $L_1$ and $L_2$ get closer to zero, the error expands.

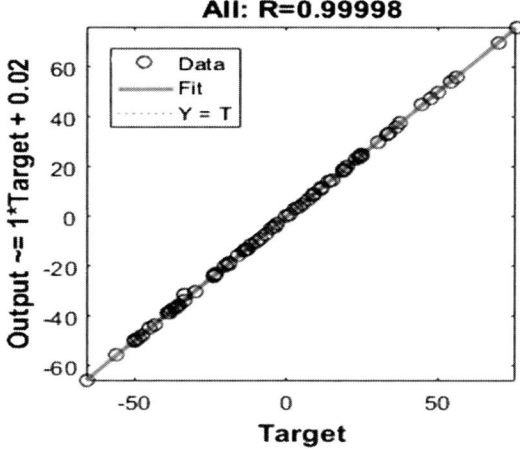

**Fig. 3.** Output accuracy of BPNN

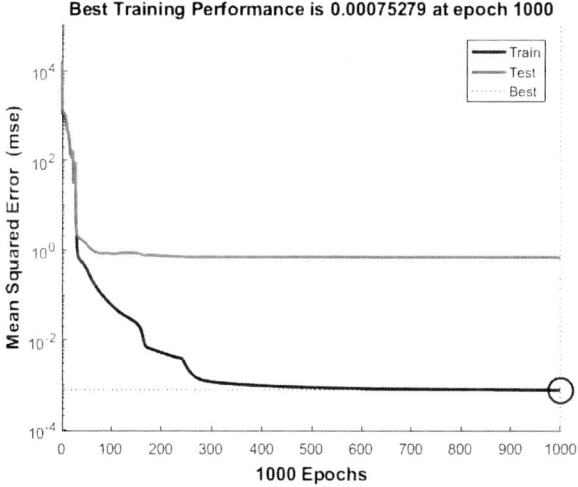

**Fig. 4.** Output performance

**Table 1.** Comparison between BPNN and the forward kinematics solutions

| Test no. | Input of forward kinematics (mm) | Output of BPNN (mm) | Error of BPNN % |
|----------|----------------------------------|---------------------|-----------------|
| 1 | $L_1 = -50$ | $L_1 = -50.1371$ | $L_1 = 0.27$ |
|   | $L_2 = 24.7$ | $L_2 = 24.8945$ | $L_2 = 0.79$ |
| 2 | $L_1 = 36$ | $L_1 = 35.9891$ | $L_1 = 0.03$ |
|   | $L_2 = -23$ | $L_2 = -23.0338$ | $L_2 = 0.14$ |
| 3 | $L_1 = 9$ | $L_1 = 9.1645$ | $L_1 = 1.8$ |
|   | $L_2 = -13$ | $L_2 = -13.0523$ | $L_2 = 0.4$ |
| 4 | $L_1 = -23.8$ | $L_1 = -23.8106$ | $L_1 = 0.04$ |
|   | $L_2 = 24.9$ | $L_2 = 24.9027$ | $L_2 = 0.01$ |
| 5 | $L_1 = 3.8$ | $L_1 = 3.6525$ | $L_1 = 3.8$ |
|   | $L_2 = -44.9$ | $L_2 = -44.8740$ | $L_2 = 0.05$ |

# 4 Conclusion

This paper presented a novel solution to solve the inverse kinematics of a planar dual-backbone continuum robot. Via using Back Propagation Neural Network created, we solved the aforementioned system that cannot be solved by traditional PRBM methods. Through constructing an ANN in Matlab, we found that this model has a simple structure and works fast to solve the inverse kinematics of the dual-backbone continuum robot with acceptable accuracy.

In the future, we will use other types of Artificial Neural Networks, such as, support vector machine, $k$-nearest neighbor, etc. to compare with this network.

# References

1. Burgner-Kahrs, J., Rucker, D.-C., Choset, H.: Continuum robots for medical applications: a survey. IEEE Trans. Rob. **31**, 1261–1280 (2015)
2. Bryson, C.E., Rucker, D.C.: Toward parallel continuum manipulators. In: IEEE International Conference on Robotics & Automation, Hong Kong, China (2014)
3. Chirikjian, G.-S., Burdick, J.-W.: The kinematics of hyper-redundant robot locomotion. IEEE Trans. Robot. Autom. **11**, 781–793 (1995)
4. Li, C., Rahn, C.-D.: Design of continuous backbone, cable-driven robots. ASME J. Mech. Des. **124**, 265–271 (2002)
5. Gravagne, I.-A., Walker, I.-D.: Manipulability, force, and compliance analysis for planar continuum manipulators. IEEE Trans. Robot. Autom. **18**, 263–273 (2002)
6. Jones, B.-A., Walker, I.-D.: Kinematics for multi-section continuum robots. IEEE Trans. Rob. **22**, 43–55 (2006)
7. Webster III, R.-J., Jones, B.-A.: Design and kinematic modeling of constant curvature continuum robots: a review. Int. J. Robot. Res. **29**, 1661–1683 (2010)
8. Xu, K., Simaan, N.: Analytic formulation for kinematics, statics, and shape restoration of multi-backbone continuum robots via elliptic integrals. ASME J. Mech. Robot. **2**, 011006 (2010)
9. Howell, L.-L.: Compliant Mechanisms. Wiley, Hoboken (2001)
10. Howell, L.-L., Midha, A.: Parametric deflection approximations for end-loaded, large-deflection beams in compliant mechanisms. ASME J. Mech. Des. **117**, 156–165 (1995)
11. Kimball, C., Tsai, L.-W.: Modeling of flexural beams subjected to arbitrary end loads. ASME J. Mech. Des. **124**, 223–235 (2002)
12. Su, H.-J.: A pseudo-rigid-body 3R model for determining large deflection of cantilever beams subject to tip loads. ASME J. Mech. Robot. **1**, 021008 (2009)
13. Kuo, C.-H., Chen, Y.-C., Pan, T.-Y.: Continuum kinematics of a planar dual-backbone robot based on pseudo-rigid-body model: formulation, accuracy, and efficiency. In: ASME International Design Engineering Technical Conferences and Computers and Information in Engineering Conference, p. V05AT08A015 (2017)
14. Chen, G., Xiong, B., Huang, X.: Finding the optimal characteristic parameter for 3R pseudo-rigid-body model using an improved particle swarm optimizer. Precis. Eng. **35**, 505–511 (2011)
15. Wang, L.-C.T., Chen, C.-C.: A combined optimization method for solving the inverse kinematics problems of mechanical manipulators. IEEE Trans. Robot. Autom. **7**, 489–499 (1991)

16. Aristidou, A., Lasenby, J.: FABRIK: a fast, iterative solver for the inverse kinematics problem. Graph. Models **73**, 243–260 (2011)
17. Duka, A.-V.: Neural network based inverse kinematics solution for trajectory tracking of a robotic arm. Procedia Technol. **12**, 20–27 (2014)
18. Shahabi, E., Hosseini, M.A.: Kinematic synthesis of a novel parallel cable robot as the artificial leg. Int. J. Adv. Des. Manuf. Technol. **9**, 1–10 (2016)
19. Assal, S.-F.M., Watanabe, K., Izumi, K.: Neural network learning from hint for the inverse kinematics problem of redundant arm subject to joint limits. In: IEEE/RSJ International Conference on Intelligent Robots and Systems, Canada, pp. 1477–1482 (2005)
20. Hecht-Nielsen, R.: Theory of the back propagation neural network. In: Neural Networks for Perception, pp. 65–93 (1992)
21. Dutta, S., Gupta, J.-P.: PVT correlations for Indian crude using artificial neural networks. J. Petrol. Sci. Eng. **72**, 93–109 (2010)
22. Wu, G., Shi, G., Shi, Y.: Modeling and analysis of a parallel continuum robot using artificial neural network. In: IEEE International Conference on Mechatronics (ICM), pp. 153–158, February 2017
23. Pannawit, S., Sento, A., Yuttana, K.: Inverse kinematics solution using neural networks from forward kinematics equations. In: 9th International Conference on Knowledge and Smart Technology, pp. 61–65 (2017)
24. Chao, M., Zhang, Y., Cheng, J., Wang, B., Zhao, Q.: Inverse kinematics solution for 6R serial manipulator based on RBF neural network. In: International Conference on Advanced Mechatronic Systems, pp. 350–355 (2016)

# Theoretical Kinematics

# Optimal Design of Tensegrity Mechanisms Used in a Bird Neck Model

Matthieu Furet[1], Anders van Riesen[1,2], Christine Chevallereau[1], and Philippe Wenger[1(✉)]

[1] Laboratoire des Sciences du Numérique de Nantes (LS2N), CNRS, Ecole centrale de Nantes, 44321 Nantes, France
Philippe.Wenger@ls2n.fr
[2] University of Twente, 7500 AE Enschede, The Netherlands

**Abstract.** This paper deals with the optimal design of an antagonistically actuated X-shape Snelson tensegrity mechanism to be used in a preliminary bird neck model made of a series of cascaded such mechanisms. The mechanism, subject to its own weight and to the weight of the subsequent mechanisms, is designed to maximize its wrench feasible workspace under given maximal actuation forces. Moreover, the mechanism is constrained to stand in a prescribed rest configuration. The optimized parameters are the link lengths and the springs stiffness.

**Keywords:** Tensegrity · Wrench feasible workspace · Design · Stiffness

## 1 Introduction

This work falls within the frame of the AVINECK project, a collaboration project with biologists, in which a robotic model of a bird neck shall be designed and built. Birds use their neck as an arm. It exhibits very interesting properties such as a high dexterity (e.g. the vulture can tear meat inside a carcass), a high dynamics (e.g. the woodpecker makes holes with high-frequency motions) or a high payload-to-weight ratio (e.g. the parrot can hang from a cage bar using its beak and thus carry its own weight). Contrary to muscular hydrostats (elephant trunks or cephalopod tentacles), bird necks have a skeletal spine like snakes, but contrary to the latter, bird necks do not lie on the ground. The concept of tensegrity has been chosen in this project as a general paradigm able to link the interests of biologists and roboticists. A tensegrity structure is made of compressive and tensile components held together in equilibrium [1,2]. Tensegrity structures were first used in art [3] and have then been applied in civil engineering [4] and robotics [5–7,15]. There are suitable to model muskuloskeleton structures where the bones are the compressive components and the muscles and tendons are the tensile elements [8]. A preliminary, planar bird neck robotic model is considered in this paper. This model is built upon stacking a series of Snelson's X-shape mechanisms [2]. Although simplified because it is planar, this model goes beyond the only available bird neck model in the literature that

© Springer Nature Switzerland AG 2019
B. Corves et al. (Eds.): EuCoMeS 2018, MMS 59, pp. 365–375, 2019.
https://doi.org/10.1007/978-3-319-98020-1_43

uses a simple planar articulated linkage [13]. Snelson's X-shape mechanisms have been studied by a number of researchers, either as a single mechanism [5,7,9] or assembled in series [10–12]. In this paper, the mechanism is actuated with two lateral tendons threaded through the spring attachment points like in [12]. This mechanism differs from to the ones analyzed in [5,7,9–11] in that its upper link is a rigid bar, a choice that aims at keeping to one the mobility of each mechanism. This mechanism is a tensegrity mechanism of class 2 (2 compressive elements linked together [4]) and the neck model resulting from stacking several such mechansims if of class 3. Finally, the mechanism is supposed to operate in a vertical plane and is thus subject to gravity, unlike in [12], where the mechanism was used in a snake-like manipulator moving on the ground. The goal of this study is to provide a preliminary design scheme of the elementary tensegrity mechanisms that are to be cascaded in a preliminary bird neck model prototype.

## 2   Description of the Neck Model and Design Strategy

We would like to build a prototype composed of a number of stacked elementary mechanisms or *segments* where each segment represents a vertebra of a bird neck. The number of vertebrae depends on the bird specie (from 10 in the parrot to 26 in the swan [14]). The design aims at determining the optimal link lengths and springs stiffness for a maximal wrench feasible workspace under prescribed maximal actuation force constraints. Moreover, the mechanism is constrained to stand in a prescribed rest configuration, so that, when all the mechanisms are stacked together, the resulting model takes the characteristic S-shape rest posture observed in all bird necks [14], see Fig. 1.

The design of the full neck model is conducted in statics only, which makes it possible to reduce the design problem to the sequential design of each segment, starting with the last one. Each segment must carry the sub-chain made of the mechanisms stacked overhead and the head (only the head for the last segment). Accordingly, the segment under design is subject to a vertical force $F_P$ acting on a point $P$, where $P$ is the center of gravity of the aforementioned sub-chain and $F_P$ is its weight. Each segment is described in its base reference frame, defined such that its origin coincides with hinge A and the x-axis is aligned with the base link 1 (Fig. 2). The coordinates $\hat{x}_P$, $\hat{y}_P$ of $P$ are defined in a frame attached to the upper link 4 as in Fig. 2. These coordinates are assumed known and their determination is out of the scope of this study as they rely on the determination of some critical configuration, which is not straightforward.

A single segment of the bird neck model consists of a symmetric four-bar mechanism with crossed links and two pre-tensioned springs, as shown in Fig. 2. The four links are rigid, homogeneous and linear bars of mass $m_i, i = 1...4$, jointed at A, B, C and D. The two crossed bars (resp. the upper and lower bar) are of equal length $L$ (resp. $b$). The two springs connect A and D, and B and C, respectively. Their free length is defined as $l_0 = L - b$, which corresponds to the minimal length of AD and BC reached when the mechanism is in its flat singular configurations. Since the mechanism is constrained to operate out of its singularities (we impose $-\pi < \alpha < \pi$), $l_0 = L - b$ ensures that the springs are always in tension.

The mechanism is actuated by tendons connected at D and C and threaded through the springs. The tendons are assumed infinitely stiff, hence input forces $F_1$ and $F_2$ are considered to act directly on D and C, respectively. Since the tendons cannot push, $F_1$ and $F_2$ are always positive and they are bounded by $F_{max}$, which depends on the actuators used.

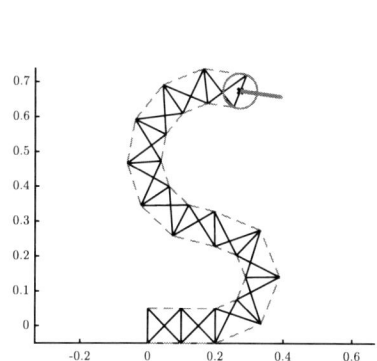

**Fig. 1.** Bird neck model at rest                **Fig. 2.** Model of a single segment.

The rigid bars will be 3D printed with ABS material of circular cross section with a fixed diameter. Thus, $m_2 = m_3 = \rho_m L$, $m_4 = m_1 = \rho_m b$, where $\rho_m$ is the mass per unit length. With a cross section diameter fixed to 0.01 m, $\rho_m = 0.0825$ kg/m. Two adjacent segments share the same bar, namely, the upper bar of the lower segment is the base bar of the upper segment.

The spring masses $m_{ri}, i = 1, 2$ cannot be neglected here as they turn out to be of the same order as the bar masses. In order to express $m_{ri}$ as a function of the stiffness $k_i$ and $L$, the two relations below are used:

$$k_i = \frac{Gd^5}{8D^3 l_0} \tag{1}$$

$$m_{ri} = \rho_s \frac{\pi^2 D d l_0}{4} \tag{2}$$

where $G$ and $\rho_s$ are the shear modulus and density, respectively; $D$ and $d$ are the spring and wire diameters, respectively and are linearly dependent: $D = \lambda d$. Combining (1) and (2), $m_{ri}$ can be expressed as $m_{ri} = \nu k_i l_0^2$, where $\nu$ is a parameter depending on the spring material and geometry. For a steel spring, choosing $\lambda = 8$ and knowing that $l_0 = L - b$, $m_{ri} = 0.008 k_i (L - b)^2$.

Without loss of generality, the length of the base and upper bars are fixed to $b = 0.1\,\mathrm{m}$ and the only length parameter to be optimized is $L$. For the optimization problem, finally, there are only three design parameters for each segment: $L$, $k_1$ and $k_2$.

## 3   Segment Modeling

The orientation of the mechanism upper bar $\alpha$ is chosen to specify the configuration of each segment. The orientation of movable links 2 and 3 are defined by $\phi$ and $\psi$, respectively (Fig. 2). The loop-closure constraint equation written in $A$ along $x$ and $y$:

$$b + L\cos(\psi) + b\cos(\alpha) - L\cos(\phi) = 0 \tag{3a}$$
$$L\sin(\psi) + b\sin(\alpha) - L\sin(\phi) = 0 \tag{3b}$$

allows one to write $\phi$ and $\psi$ as a function of $\alpha$:

$$\phi(\alpha) = 2\arctan\left[\frac{2bL\sin(\alpha) + S}{(2b^2 + 2bL)\left[\cos(\alpha) + 1\right]}\right] \tag{4a}$$

$$\psi(\alpha) = 2\arctan\left[\frac{-2bL\sin(\alpha) - S}{(2b^2 - 2bL)\left[\cos(\alpha) + 1\right]}\right] \tag{4b}$$

with $S = \sqrt{(-2bL\sin(\alpha))^2 + (-2bL\left[\cos(\alpha) + 1\right])^2 - (2b^2\left[\cos(\alpha) + 1\right])^2}$.

Using the cosine rule, the spring lengths can be expressed as follows:

$$l_1 = \sqrt{b^2 + L^2 + 2bL\cos(\psi)} \tag{5a}$$
$$l_2 = \sqrt{b^2 + L^2 - 2bL\cos(\phi)} \tag{5b}$$

The static model of the segment can be obtained with its potential energy $V$ and the potential function $E_{ex}$ associated to the external wrench, assumed to be conservative [4]. Here, the external wrench represents the two actuation forces $F_1, F_2$ and the weight of the overhead sub-chain $F_P$. The segment is in equilibrium when:

$$V' = E'_{ex} \tag{6}$$

where the $'$ means the partial derivative with respect to $\alpha$ : $\frac{\partial}{\partial\alpha}$. The potential energy $V$ is the sum of the potential energy $V_b$ associated with the bar masses and the potential energy $V_s$ associated with the springs stiffnesses and masses:

$$V_b = \tfrac{g}{2}\left[m_2 L\sin(\phi) + (m_3 + 2m_4)L\sin(\psi) + m_4 b\sin(\alpha)\right] \tag{7a}$$
$$V_s = \tfrac{1}{2}k_1(l_1 - l_0)^2 + \tfrac{1}{2}k_2(l_2 - l_0)^2 + \tfrac{g}{2}(m_{r1}l_1 + m_{r2}l_2)\cos(\tfrac{\alpha}{2}) \tag{7b}$$

Note that for writing simplification purposes, the base of the segment under study is supposed to be horizontal here. For any other base orientation, a component of the gravity vector along $x$ must be considered, with the only technical consequence of making (7a), (7b), (10c) and (10d) more lengthy.

The potential $E_{ex}$ is defined as:

$$E_{ex} = -F_1 l_1 - F_2 l_2 - F_p(L\sin(\psi) + \hat{x}_P \sin(\alpha) + \hat{y}_P \cos(\alpha)) \tag{8}$$

where $\hat{x}_P$ and $\hat{y}_P$ are the coordinates of $P$ in the frame attached to the upper link (Fig. 2).

Reporting Eqs. (8) and (7a), (7b) into (6) and using (5a), (5b) (4a) and (4b) yields the static equilibrium equation:

$$G_b(\alpha, L) + G_s(\alpha, L, k_1, k_2) = Z_1(\alpha, L)F_1 + Z_2(\alpha, L)F_2 + Z_p(\alpha, L)F_P \tag{9}$$

where:

$$Z_i(\alpha, L) = -l'_i \qquad \text{for } i = 1, 2 \tag{10a}$$

$$Z_P(\alpha, L) = -L\cos(\psi)\psi' - \hat{x}_P \cos(\alpha) + \hat{y}_P \sin(\alpha) \tag{10b}$$

$$G_b(\alpha, L) = \tfrac{g}{2}\left[m_2 L\cos(\phi)\phi' + (m_3 + 2m_4)L\cos(\psi)\psi' + m_4 b\cos(\alpha)\right] \tag{10c}$$

$$G_s(\alpha, L, k_1, k_2) = k_1\left[l'_1(l_1 - l_0) - \tfrac{g}{4}\nu l_0^2 l_1 \sin(\tfrac{\alpha}{2}) + \tfrac{g}{2}\nu l_0^2 l'_1 \cos(\tfrac{\alpha}{2})\right]$$
$$+ k_2\left[l'_2(l_2 - l_0) - \tfrac{g}{4}\nu l_0^2 l_2 \sin(\tfrac{\alpha}{2}) + \tfrac{g}{2}\nu l_0^2 l'_2 \cos(\tfrac{\alpha}{2})\right] \tag{10d}$$

The static equilibrium equations above depend only on $\alpha$ and on the design parameters $L$, $k_1$ and $k_2$. Indeed, $\psi$ and $\phi$, $l_1$ and $l_2$ can be expressed as a function of $\alpha$ using Eqs. (5a), (5b), (4a) and (4b) $l_0 = L - b$ and all mass parameters can be expressed as a function of $L$ (remember also that $b$ was fixed to $0.1\,\mathrm{m}$).

To avoid lengthy derivations in the next section, the above equation is rewritten in a more compact way below:

$$G_s(\alpha, L, k_1, k_2) = k_1 X_1(\alpha, L) + k_2 X_2(\alpha, L) \tag{11}$$

Two situations will be considered for the design process: the neck is at rest and the two actuation forces are active. The behavior of the mechanism in these two situations are studied in the next sections.

## 4    Behavior of the Mechanism at Rest

The behavior at rest is defined from (9) as the behavior without actuation: $F_1 = F_2 = 0\,\mathrm{N}$. Differentiating (9) with respect to $\alpha$ gives the mechanism stiffness $K_\alpha$ [5]:

$$K_\alpha(\alpha, L, k_1, k_2) = G'_b(\alpha, L) + G'_s(\alpha, L, k_1, k_2) - Z'_p(\alpha, L)F_P \tag{12}$$

The design should allow the mechanism to have a stable equilibrium in a prescribed rest configuration $\alpha_0$. Accordingly, its stiffness $K_\alpha$ should be strictly positive. For a better response to external perturbations, a minimal stiffness is even imposed: $K_\alpha(\alpha_0, L, k_1, k_2) > K_{min}$. The objective now is to determine the spring stiffnesses satisfying the minimal mechanism stiffness $K_{min}$ at the prescribed rest configuration $\alpha_0$, for a given dimension $L$. Combining (11) and (9) yields:

$$G_b + k_2 X_2 + k_1 X_1 = Z_p F_P \qquad (13)$$

where the dependency on $L$ and $\alpha_0$ is omitted to simplify the writting. Thus $k_2$ can be expressed as a function of $k_1, \alpha_0$ and $L$ :

$$k_2 = \frac{Z_p F_P - G_g - k_1 X_1}{X_2} \qquad (14)$$

Using (11), (12), the condition on the mechanism stiffness writes:

$$G_b' + k_2 X_2' + k_1 X_1' - Z_p' F_P > K_{min} \qquad (15)$$

Reporting (14) into (15) yields, after rearranging the terms:

$$k_1 \left[ X_1' - \frac{X_1}{X_2} X_2' \right] > K_{min} + Z_p' F_P - G_b' - \frac{Z_p F_P - G_g}{X_2} X_2' \qquad (16)$$

Since $\left[ X_1' - \frac{X_1}{X_2} X_2' \right] > 0$ the condition on $k_1$ can be written as $k_1 > k_{1min}$, with:

$$k_{1min} = \frac{(K_{min} + Z_p' F_P - G_b') X_2 - (Z_p F_P - G_b) X_2'}{X_2 X_1' - X_1 X_2'} \qquad (17)$$

Figure 3 shows the plots of $k_{1min}$ and $k_{2min}$ against the rest position $\alpha_0$ for several lengths $L$. Note that $k_{2min}$ was plotted using (14) with $k_1 = k_{1min}$ to confirm the symmetric behavior. $F_p$ was chosen here as $F_p = 5.9\,\mathrm{N}$, a weight that would represent the mass of about ten segments. The spring stiffness strongly depends on the link length $L$, which shows the direct influence of the geometric parameters on the mechanism stiffness $K_\alpha(\alpha, L, k_1, k_2)$. It is worth noting that if $\alpha_0$ is too large, namely if $\alpha_0 < -\pi/2$ or $\alpha_0 > \pi/2$, excessive spring stiffnesses are generally required for the mechanism minimal stiffness to be satisfied.

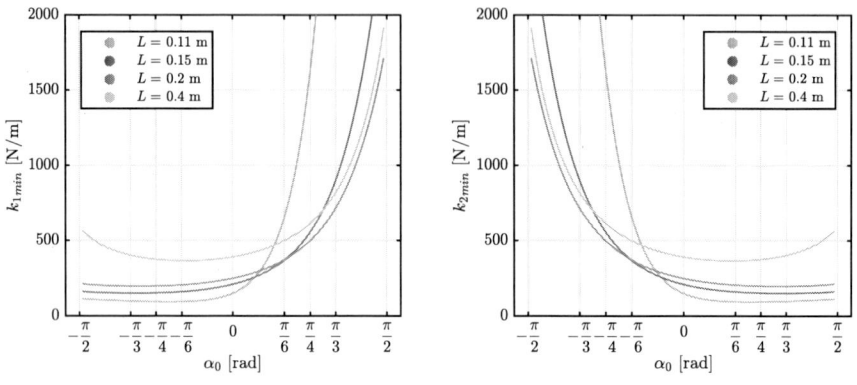

**Fig. 3.** Minimal spring stiffness $k_{1min}$ (left) and $k_{2min}$ (right) for different link lengths $L$, against the prescribed rest configuration $\alpha_0$ ($K_{min} = 1\,\mathrm{Nm/rad}$, $F_P = 5.9\,\mathrm{N}$ and $(\hat{x}_P, \hat{y}_P) = (0.05, 0.1)$).

## 5   Description of the Wrench Feasible Workspace and Influence of the Parameters

Using (9), and rearranging the terms, the static equilibrium condition is:

$$G_b(\alpha, L) + G_s(\alpha, L, k_1) - Z_p(\alpha, L)F_P = Z_1(\alpha, L)F_1 + Z_2(\alpha, L)F_2 \qquad (18)$$

The right-hand side of (18) is the actuation wrench. The wrench feasible workspace (WFW) must satisfy, by definition [9], the geometric constraints (3a), (3b) the static equilibrium (9) and the limits of the external forces $F_{max}$ and $F_{min}$ as introduced in Sect. 2. The coefficients $Z_1$ and $Z_2$ defined by (10a) can be shown to satisfy $Z_1(\alpha) > 0$ and $Z_2(\alpha) < 0$ for $-\pi < \alpha < \pi$.

Accordingly, it is possible to determine the limits of the actuation wrench as a function of the limits of the actuation forces as follows:

$$\begin{aligned}\Gamma_{max}(\alpha) &= Z_1(\alpha)F_{max} + Z_2(\alpha)F_{min}\\ \Gamma_{min}(\alpha) &= Z_1(\alpha)F_{min} + Z_2(\alpha)F_{max}\end{aligned} \qquad (19)$$

Figure 4 shows the bounds of the actuation wrenches for different link lengths as defined by (19). The area enclosed by the dashed curves indicates the feasible wrench area. Several instances of $G(\alpha) = G_b(\alpha, L) + G_s(\alpha, L, k_1) - Z_p(\alpha, L)F_P$ are present in the figures as well. The static equilibrium Eq. (9) can be satisfied for a range of $\alpha$ where $G$ is within the feasible wrench bounding curves:

$$\Gamma_{min}(\alpha) \leq G(\alpha) \leq \Gamma_{max}(\alpha) \qquad (20)$$

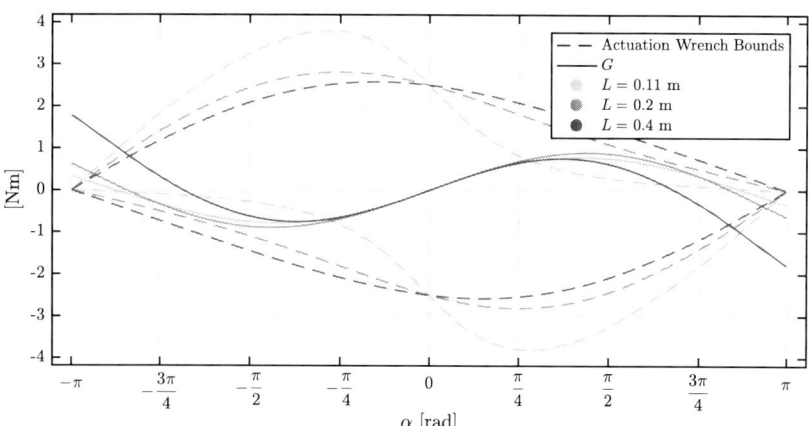

**Fig. 4.** Actuation wrench bounds and $G(\alpha)$ for varying link lengths $L$.

The limits of the WFW are determined by the intersections of $G(\alpha)$ with $\Gamma_{min}$ and $\Gamma_{max}$. If several intersections occur between $G(\alpha)$ and $\Gamma_{min}$ or $\Gamma_{max}$, the WFW is non-connected, meaning that it is not fully reachable in statics. Accordingly, the bounds of the WFW are taken as the first intersections that occur starting from the rest position $\alpha_0$. In Fig. 4, three instances of $G$ were plotted for different link lengths $L$ : for each length $L$, the minimal stiffnesses $k_{1min}$ and $k_{2min}$ were computed (here for $K_{min} = 1\,\mathrm{Nm/rad}$, $F_P = 5.9\,\mathrm{N}$, $(\hat{x_P}, \hat{y_P}) = (0.05, 0.1)$ and $\alpha_0 = 0$). $\Gamma_{min}$ and $\Gamma_{max}$ are influenced by $L$, $F_1$ and $F_2$ (here $F_{min} = 0\,\mathrm{N}$ and $F_{max} = 50\,\mathrm{N}$). It is apparent that the size of the WFW depends on $L$. Here the WFW is the smallest for $L = 0.11\,\mathrm{m}$, but is greater for $L = 0.2\,\mathrm{m}$ than for $L = 0.4\,\mathrm{m}$, which means that depending on the desired parameters ($\alpha_0$, $F_P$ and its application point $P(\hat{x}_P, \hat{y}_P)$), an optimal value $L$ can be found in order to maximize the WFW.

Figure 5 shows the influence of the different external parameters. Prescribing a non-zero rest position $\alpha_0$ shifts the WFW and reduces it; decreasing $F_P$ increases the size of the WFW. The bounds of the WFW are also influenced by the position of $P$ with respect to the upper bar.

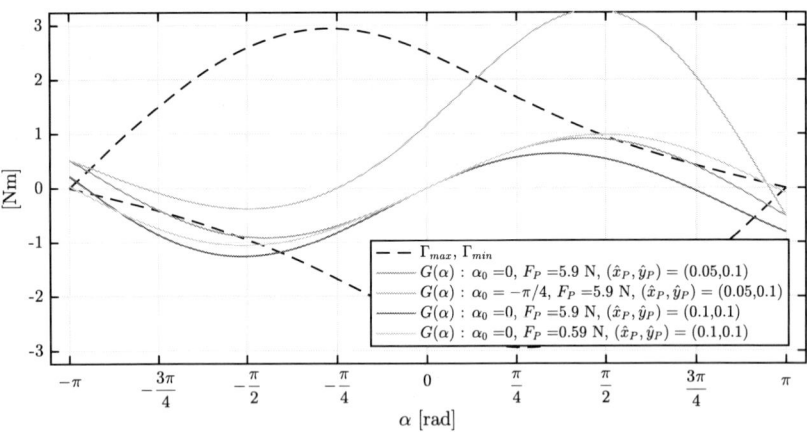

**Fig. 5.** Actuation wrench bounds and $G(\alpha)$ for varying desired parameters $\alpha_0$, $F_P$ and $(\hat{x}_P, \hat{y}_P)$. Blue and red (resp. blue and yellow, yellow and purple) curves $G(\alpha)$ show the effect of a change in $\alpha_0$ (resp. $\hat{x}_P$, $F_P$). (Color figure online)

## 6   Optimal Design of the Tensegrity Mechanism for Given Specifications

As shown in previous section, one can find a minimal stiffness for the springs depending on the link length $L$, given that the minimal segment stiffness $K_{min}$, the application point, the force $F_P$ and the rest position $\alpha_0$ are prescribed. Since the minimal and maximal wrenches applied by the cables also depend on $L$ ($F_{max}$ and $F_{min}$ are known), an optimal value of the link length $L$ can be found for the aforementioned prescribed parameters.

An algorithm has been written in order to compute the size of the WFW $W_s$ defined as : $W_s = |\alpha_{max} - \alpha_{min}|$. $\alpha_{max}$ (resp. $\alpha_{min}$) is obtained by computing the two first intersection of $G$ with $\Gamma_{max}$ and/or $\Gamma_{min}$ from the rest position $\alpha_0$. The optimal length $L$ is found when $W_s$ is maximal.

Figure 6 shows the size of the WFW and the optimal value of $L$ for different sets of prescribed parameters. When comparing the blue and the red curves, it is apparent that the optimal link length and the size of the WFW increase when $F_P$ decreases, which means that for higher external wrenches, the mechanism needs to be more compact and has a reduced WFW. For instance, the blue curve shows the optimal value for a rest configuration $\alpha_0 = 0$ of a segment that must carry around 10 segments above it. This value can be used to design a segment near the base of the neck. On the other hand, the red curve is computed for the same prescribed parameters but with a low external wrench, which can emphasize the case of a segment close to the bird head. When comparing the blue and the purple curves, plotted for $\alpha_0 = 0$ and $\alpha_0 = -\pi/4$, respectively, one can see that depending on the rest position, the optimal value of $L$ will be different. Those three curves were plotted for the minimal values of $k_1$ and $k_2$ found to satisfy (14) and (17). Nevertheless, higher stiffness values can be chosen, as illustrated by the yellow curve ($k_1 = 2k_{1min}$, $k_2 = k_1$ as $\alpha_0 = 0$). In this case, the size of the WFW is reduced but as the mechanism is stiffer, the optimal length value is increased. Accordingly, the minimal spring stiffness should be taken into account for the final design of the segment. It can also be observed in Fig. 6 that the size of the WFW does not change significantly when $L$ is increased from its optimal value. A drastic change is observed if $L$ is diminished for large workspaces but not for small workspaces.

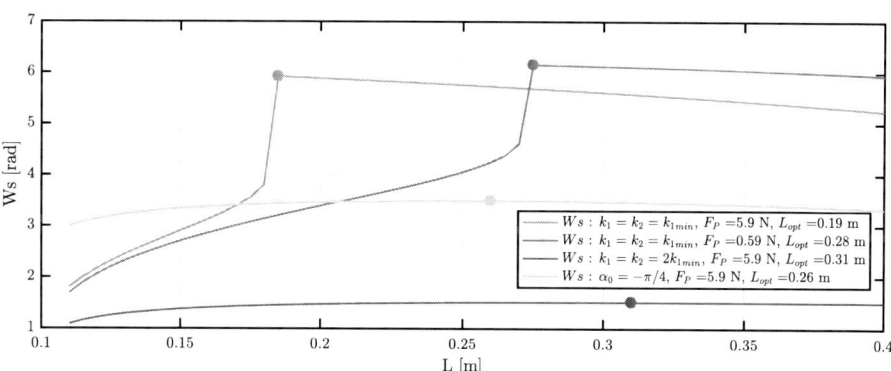

**Fig. 6.** Size of the WFW and optimal length $L$. For each case $K_{min} = 1\,\mathrm{Nm/rad}$ and $(\hat{x}_P, \hat{y}_P) = (0.05, 0.1)$. (Color figure online)

# 7   Conclusion

A methodology to design cascaded Snelson X-shape mechanisms with two lateral springs used in a bird neck model, has been proposed in this paper. The mechanisms are subject to their own weight and to an external wrench $F_P$. The link lengths and the spring stiffnesses of the mechanism are optimized as a function of a prescribed rest configuration with a given stiffness at rest and for a maximal wrench feasible workspace. The proposed methodology will be applied in an iterative scheme to design the complete S-shape bird neck model made of $n$ segments. The top segment (segment $n$) is first designed, where $F_P$ defines the head weight. Then segment $n - 1$ is designed and this time $F_P$ integrates the weight of segment $n$ in addition to the head weight, and so on until the base segment. In this work, constant limits on the actuation forces have been assumed. The choice of the actuation scheme of the complete mechanism is still an open issue and will be decided later in the light of the muscle organization of the bird neck, which is under investigation by our biologist partners.

**Acknowledgement.** This work was conducted with the support of the French National Research Agency (AVINECK Project ANR-16-CE33-0025).

# References

1. Motro, R.: Tensegrity systems: the state of the art. Int. J. Space Struct. **7**(2), 75–83 (1992)
2. Snelson, K.: Continuous tension, discontinuous compression structures, US Patent No. 3,169,611 (1965)
3. Fuller, R.B.: Tensile-integrity structures, United States Patent 3063521 (1962)
4. Skelton, R., de Oliveira, M.: Tensegrity Systems. Springer (2009)
5. Arsenault, M., Gosselin, C.M.: Kinematic, static and dynamic analysis of a planar 2-DOF tensegrity mechanism. Mech. Mach. Theory **41**(9), 1072–1089 (2006)
6. Crane, C., et al.: Kinematic analysis of a planar tensegrity mechanism with prestressed springs. In: Lenarcic, J., Wenger, P. (eds.) Advances in Robot Kinematics: Analysis and Design, pp. 419–427. Springer (2008)
7. Wenger, P., Chablat, D.: Kinetostatic analysis and solution classification of a planar tensegrity mechanism. In: Proceedings of 7th International Workshop on Computer Kinematics. Springer, pp. 422–431 (2017). ISBN 978-3-319-60867-9
8. Levin, S.: The tensegrity-truss as a model for spinal mechanics: biotensegrity. J. Mech. Med. Biol. **2**(3), 375–388 (2002)
9. Boehler, Q., et al.: Definition and computation of tensegrity mechanism workspace. ASME J. Mech. Robot. **7**(4), 044502 (2015)
10. Aldrich, J.B., Skelton, R.E.: Time-energy optimal control of hyper-actuated mechanical systems with geometric path constraints. In: 44th IEEE Conference on Decision and Control, pp. 8246–8253 (2005)
11. Chen, S., Arsenault, M.: Analytical computation of the actuator and Cartesian workspace boundaries for a planar 2-degree-of-freedom translational tensegrity mechanism. J. Mech. Robot. **4**, 011010 (2012)

12. Bakker, D.L., et al.: Design of an environmentally interactive continuum manipulator. In: Proceedings of 14th IFToMM World Congress in Mechanisms and Machine Science, Taipei, Taiwan (2015)
13. Zweers, G., Bout, R., Heidweiller, J.: Perception and Motor Control in Birds: An Eco- logical Approach. Springer (1994). ISBN 978-3-642-75869-0
14. von Boas, J.E.: Biologisch-anatomische Studien über den Hals der Vögel Andr. Fred. Host and Son, Köbenhavn (1929)
15. Boehler, Q., Vedrines, M., Abdelaziz, S., Poignet, P., Renaud, P.: Design and evaluation of a novel variable stiffness spherical joint with application to MR-compatible robot design. In: 2016 IEEE International Conference on Robotics and Automation (ICRA), pp. 661–667 (2016)

# Stationary Distance Between Spatial Conics

Paul Zsombor-Murray[(✉)]

Centre for Intelligent Machines/Department of Mechanical Engineering,
McGill University, Montreal, Canada
paul@cim.mcgill.ca

**Abstract.** Non-iterative computation of shortest and greatest distance between conics, other than circles, in arbitrary *spatial* disposition is apparently an unsolved problem. Although useful in detecting collision or interference between mechanical parts in motion the primary motivation for this endeavour is my obsession with trying to introduce elementary but little-known, at least to engineers, geometric methods to solve many types of problems in computational kinematics.

**Keywords:** Ellipses · 3D · Surface · Constraint · Equations
Gröbner · Basis

## 1 Introduction

The idea is to find points $P, Q$, one on each of two given conics in arbitrary pose, so that the length of the segment between them is minimum, maximum or some inflective value[1] This exercise has been carried out with respect to a pair of circles in [1,2]. As regards distances between quadric and other surfaces there is an article by Lopes *et al.* [3]. Now to address the case of a pair of conics, other than circles, one may begin by formulating an ideal setup without compromising generality as summarized in Fig. 1. Two conics $c_P$ and $c_Q$ are placed such that their plane normal vectors $\mathbf{e}, \mathbf{f}$ through their respective centre points $E, F$ are in horizontal planes spaced the same distance $s$ above and below the origin of a Cartesian frame. Coordinate directions $x, y, z$ are shown, as appropriate, in top, front and two first auxiliary projections. Similarly the axes given by $E, \mathbf{e}$ and $F, \mathbf{f}$ are displaced by the same angle $\sigma$ on either side of principal plane $y = 0$. The common normal between axes is on the coordinate $z$-axis. To ensure that the principal axes of the conics, ellipses in this case, assume general orientation one of each is specified as sloping at angles $\mu, \nu$ measured from the horizontal.

---

[1] In formulating this problem I did not set out to resolve a compelling issue in robotics rather to suggest a framework in which to discuss some questions that might arise in that context. *E.g.*, what symmetry does a pair of conics, arbitrarily disposed in space, posses? What's the biggest sphere that can be "pushed" through the "strait" so defined? What's the smallest spherical "bag" that may contain a pair of ellipses so disposed?

© Springer Nature Switzerland AG 2019
B. Corves et al. (Eds.): EuCoMeS 2018, MMS 59, pp. 376–383, 2019.
https://doi.org/10.1007/978-3-319-98020-1_44

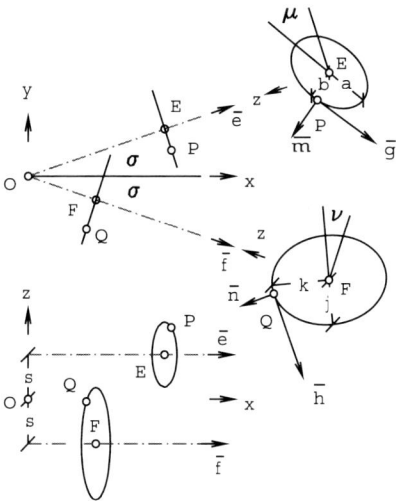

**Fig. 1.** Two ellipses in general but ideal pose

## 2    Placing Conics

Presentation shown in Fig. 1 was deemed preferable to having a conic in standard form. It is believed that symmetric disposition reduces numerical error. We begin by defining the conics, with given principal axis lengths $a, b$ and $j, k$, in standard form on the plane $x = 0$. Respective symmetric coefficient matrices $M_P, M_Q$ of two elliptical *cylinders* $k_P, k_Q$ with axes along the Cartesian $x$-axis are

$$M_P = \begin{bmatrix} -1 & 0 & 0 & 0 \\ 0 & 0 & 0 & 0 \\ 0 & 0 & A & 0 \\ 0 & 0 & 0 & B \end{bmatrix}, \quad M_Q = \begin{bmatrix} -1 & 0 & 0 & 0 \\ 0 & 0 & 0 & 0 \\ 0 & 0 & J & 0 \\ 0 & 0 & 0 & K \end{bmatrix}, \quad A = \frac{1}{a^2}, \ B = \frac{1}{b^2}, \ J = \frac{1}{j^2}, \ K = \frac{1}{k^2}.$$

The first task, then, is to rotate these about the $x$-axis by angles $\mu, \nu$ in a right hand sense, respectively, then similarly about the $z$-axis by $\pm\sigma$. Finally $M_P, M_Q$ are translated from the origin to $E(e_1, e_2, e_3), F(f_1, f_2, f_3)$ to assume their displaced form $[a_{ij}], [b_{ij}]$ with the following multiplications.

$$[a_{ij}] = [T_E][R_{+\sigma}][R_\mu][M_P][R_\mu]^\top [R_{+\sigma}]^\top [T_E]^\top$$
$$[b_{ij}] = [T_F][R_{-\sigma}][R_\nu][M_Q][R_\nu]^\top [R_{-\sigma}]^\top [T_F]^\top$$

where

$$[T_E] = \begin{bmatrix} 1 & -e_1 & -e_2 & -e_3 \\ 0 & 1 & 0 & 0 \\ 0 & 0 & 1 & 0 \\ 0 & 0 & 0 & 1 \end{bmatrix}, \quad [T_F] = \begin{bmatrix} 1 & -f_1 & -f_2 & -f_3 \\ 0 & 1 & 0 & 0 \\ 0 & 0 & 1 & 0 \\ 0 & 0 & 0 & 1 \end{bmatrix},$$

$$[R_\mu] = \begin{bmatrix} 1 & 0 & 0 & 0 \\ 0 & 1 & 0 & 0 \\ 0 & 0 & \cos\mu & -\sin\mu \\ 0 & 0 & \sin\mu & \cos\mu \end{bmatrix}, \quad [R_\nu] = \begin{bmatrix} 1 & 0 & 0 & 0 \\ 0 & 1 & 0 & 0 \\ 0 & 0 & \cos\nu & -\sin\nu \\ 0 & 0 & \sin\nu & \cos\nu \end{bmatrix},$$

$$[R_{+\sigma}] = \begin{bmatrix} 1 & 0 & 0 & 0 \\ 0 & \cos\sigma & -\sin\sigma & 0 \\ 0 & \sin\sigma & \cos\sigma & 0 \\ 0 & 0 & 0 & 1 \end{bmatrix}, \quad [R_{-\sigma}] = \begin{bmatrix} 1 & 0 & 0 & 0 \\ 0 & \cos\sigma & \sin\sigma & 0 \\ 0 & -\sin\sigma & \cos\sigma & 0 \\ 0 & 0 & 0 & 1 \end{bmatrix}$$

Without going into messy details, which will be saved for a numerically computed example, consider the resulting coefficient matrices $[a_{ij}], [b_{ij}]$ of the displaced conic cylinders.

$$[a_{ij}] = \begin{bmatrix} a_{00} & a_{01} & a_{02} & a_{03} \\ a_{01} & a_{11} & a_{12} & a_{13} \\ a_{02} & a_{12} & a_{22} & a_{23} \\ a_{03} & a_{13} & a_{23} & a_{33} \end{bmatrix}, \quad [b_{ij}] = \begin{bmatrix} b_{00} & b_{01} & b_{02} & b_{03} \\ b_{01} & b_{11} & b_{12} & a_{13} \\ b_{02} & b_{12} & b_{22} & b_{23} \\ b_{03} & b_{13} & b_{23} & b_{33} \end{bmatrix}$$

## 3   Two Point and Plane Pairs; Their Position and Normal Vectors

Quadric cylinders $k_P, k_Q$ are sectioned by axis-normal planes $e, f$ to yield conics $c_P, c_Q$. Position vectors of $P, Q$ are $\mathbf{p}, \mathbf{q}$ while normal vectors of $e, f$ are designated as $\mathbf{e}, \mathbf{f}$. Distances from $z$-axis measured along respective conic axes to $E, F$ are defined as $t, u$.

$$e\{E_0 : E_1 : E_2 : E_3\} \rightarrow \begin{bmatrix} E_0 \\ \mathbf{e} \end{bmatrix}$$

$$\mathbf{e} = \begin{bmatrix} \cos\sigma \\ \sin\sigma \\ 0 \end{bmatrix}, E\{1 : e_1 : e_2 : e_3\} = \{1 : t\cos\sigma : t\sin\sigma : s\}$$

Similarly

$$f\{F_0 : F_1 : F_2 : F_3\} \rightarrow \begin{bmatrix} F_0 \\ \mathbf{f} \end{bmatrix}$$

$$\mathbf{f} = \begin{bmatrix} \cos\sigma \\ -\sin\sigma \\ 0 \end{bmatrix}, \quad F\{1 : f_1 : f_2 : f_3\} = \{1 : u\cos\sigma : -u\sin\sigma : -s\}$$

So

$$E_0 + E_1 e_1 + E_2 e_2 + E_3 e_3 = 0 \rightarrow E_0 + t\cos^2\sigma + t\sin^2\sigma = 0 \rightarrow E_0 = -t$$
$$\text{similarly } F_0 = -u$$

while the six parameters to be determined are $p_i, q_i$.

$$P\{1 : p_1 : p_2 : p_3\}, \quad \mathbf{p} = [p_1 \ p_2 \ p_3]^\top, \quad Q\{1 : q_1 : q_2 : q_3\}, \quad \mathbf{q} = [q_1 \ q_2 \ q_3]^\top$$

## 4   Two Vector Pairs and Two Cross-Products

As described in [1], the shortest distance from any point in space to a circle is on the line that connects the point to a point on the circle curve and intersectes the line on the circle centre that is normal to the circle plane, *i.e.*, the circle axis. That is because such a connecting line is normal to the tangent to the circle at the point on the circle. Our problem is complicated somewhat in the case of general conics. There is no such unique axis. Nevertheless the direction of local tangent must be obtained in order to impose this necessary condition. Referring to Fig. 1 it is seen that the directions of the tangents on $P, Q$ are given by vectors $\mathbf{g}, \mathbf{h}$ where

$$\mathbf{g} = \mathbf{e} \times \mathbf{m}, \quad \mathbf{h} = \mathbf{f} \times \mathbf{n}$$

and $\mathbf{m}, \mathbf{n}$ are normal to planes $m, n$ to $k_P, k_Q$ at $P, Q$.

$$m = \begin{bmatrix} M_0 \\ \mathbf{m} \end{bmatrix}^\top = \begin{bmatrix} a_{00} + a_{01}p_1 + a_{02}p_2 + a_{03}p_3 \\ a_{01} + a_{11}p_1 + a_{12}p_2 + a_{13}p_3 \\ a_{02} + a_{12}p_1 + a_{22}p_2 + a_{23}p_3 \\ a_{03} + a_{13}p_1 + a_{23}p_2 + a_{33}p_3 \end{bmatrix}^\top = \begin{bmatrix} M_0 \\ M_1 \\ M_2 \\ M_3 \end{bmatrix}^\top, \quad \mathbf{m} = \begin{bmatrix} M_1 \\ M_2 \\ M_3 \end{bmatrix}$$

$$n = \begin{bmatrix} N_0 \\ \mathbf{n} \end{bmatrix}^\top = \begin{bmatrix} b_{00} + b_{01}q_1 + b_{02}q_2 + b_{03}q_3 \\ b_{01} + b_{11}q_1 + b_{12}q_2 + b_{13}q_3 \\ b_{02} + b_{12}q_1 + b_{22}q_2 + b_{23}q_3 \\ b_{03} + b_{13}q_1 + b_{23}q_2 + b_{33}q_3 \end{bmatrix}^\top = \begin{bmatrix} N_0 \\ N_1 \\ N_2 \\ N_3 \end{bmatrix}^\top, \quad \mathbf{n} = \begin{bmatrix} N_1 \\ N_2 \\ N_3 \end{bmatrix}$$

## 5   Six Constraint Equations

The complete set of constraint equations required to solve this problem, *viz.*, to find points pairs $P, Q$ that define segment lengths which correspond to stationary distances, can be expressed succinctly thus:-

- Point $P$ is on cylinder $k_P$.
- Point $Q$ is on cylinder $k_Q$.
- Point $P$ is on plane $e$.
- Point $Q$ is on plane $f$.
- Line segment between $P$ and $Q$ is normal to vector $\mathbf{g}$, the direction of the tangent at point $P$ on conic $c_P$.
- Line segment between $P$ and $Q$ is normal to vector $\mathbf{h}$, the direction of the tangent at point $Q$ on conic $c_Q$.

Here then is this sequence of the six equations in symbolic form expanded to the limits of practicality.

$$P \in k_P : \quad P[T_E][R_{+\sigma}][R_\mu][M_P][R_\mu]^\top [R_{+\sigma}]^\top [T_E]^\top P^\top = 0 \tag{1}$$

$$Q \in k_Q : \quad Q[T_F][R_{-\sigma}][R_\nu][M_Q][R_\nu]^\top [R_{-\sigma}]^\top [T_F]^\top Q^\top = 0 \tag{2}$$

$$P \in e: \quad -t + \cos\sigma p_1 + \sin\sigma p_2 = 0 \tag{3}$$

$$Q \in f: \quad -u + \cos\sigma q_1 - \sin\sigma q_2 = 0 \tag{4}$$

$$\mathbf{g} \perp (\mathbf{q} - \mathbf{p}) \equiv \mathbf{g} \cdot (\mathbf{q} - \mathbf{p}) = 0, \quad \mathbf{g} = \mathbf{e} \times \mathbf{m}:$$
$$\sin\sigma(a_{03} + a_{13}p_1 + a_{23}p_2 + a_{33}p_3)(q_1 - p_1)$$
$$- \cos\sigma(a_{03} + a_{13}p_1 + a_{23}p_2 + a_{33}p_3)(q_2 - p_2) \tag{5}$$
$$+ [\cos\sigma(a_{02} + a_{12}p_1 + a_{22}p_2 + a_{23}p_3)$$
$$- \sin\sigma(a_{01} + a_{11}p_1 + a_{12}p_2 + a_{13}p_3)](q_3 - p_3) = 0$$

$$\mathbf{h} \perp (\mathbf{q} - \mathbf{p}) \equiv \mathbf{h} \cdot (\mathbf{q} - \mathbf{p}) = 0, \quad \mathbf{h} = \mathbf{f} \times \mathbf{n}:$$
$$- \sin\sigma(b_{03} + b_{13}q_1 + b_{23}q_2 + b_{33}q_3)(q_1 - p_1)$$
$$- \cos\sigma(b_{03} + b_{13}q_1 + b_{23}q_2 + b_{33}q_3)(q_2 - p_2) \tag{6}$$
$$+ [\cos\sigma(b_{02} + b_{12}q_1 + b_{22}q_2 + b_{23}q_3)$$
$$+ \sin\sigma(b_{01} + b_{11}q_1 + b_{12}q_2 + b_{13}q_3)](q_3 - p_3) = 0$$

## 6   Numerical Example

Yes, a complete symbolic Gröbner basis solution is feasible. It was tried successfully but the sums of products in the coefficients were so voluminous so as to be quite useless; annoying even. Ten parameters to define this example are given as follows.

$$a = 2, \quad b = 1, \quad j = 3, \quad k = 2, \quad \mu = \pi/6, \quad \nu = \pi/3, \quad \sigma = \pi/4, \quad s = 2, \quad t = 7, \quad u = 3$$

Required trigonometric ratios are computed as

$$\cos\mu = \sqrt{3}/2, \ \sin\mu = 1/2, \ \cos\nu = 1/2$$
$$\sin\nu = \sqrt{3}/2, \ \cos\sigma = 1/\sqrt{2}, \ \sin\sigma = 1/\sqrt{2}$$

As an aid to any who may wish to algorithmically implement this procedure all matrices and constraint equations using the data given above are reproduced here regardless of how tedious this may seem to some readers.

$$M_P = \begin{bmatrix} -4 & 0 & 0 & 0 \\ 0 & 0 & 0 & 0 \\ 0 & 0 & 1 & 0 \\ 0 & 0 & 0 & 4 \end{bmatrix}, \quad M_Q = \begin{bmatrix} -36 & 0 & 0 & 0 \\ 0 & 0 & 0 & 0 \\ 0 & 0 & 4 & 0 \\ 0 & 0 & 0 & 9 \end{bmatrix}$$

$$T_E = \begin{bmatrix} \sqrt{2} & -7 & -7 & -2\sqrt{2} \\ 0 & \sqrt{2} & 0 & 0 \\ 0 & 0 & \sqrt{2} & 0 \\ 0 & 0 & 0 & \sqrt{2} \end{bmatrix}, \quad T_F = \begin{bmatrix} \sqrt{2} & -3 & 3 & 2\sqrt{2} \\ 0 & \sqrt{2} & 0 & 0 \\ 0 & 0 & \sqrt{2} & 0 \\ 0 & 0 & 0 & \sqrt{2} \end{bmatrix}$$

$$R_\mu = \begin{bmatrix} 2 & 0 & 0 & 0 \\ 0 & 2 & 0 & 0 \\ 0 & 0 & \sqrt{3} & -1 \\ 0 & 0 & 1 & \sqrt{3} \end{bmatrix}, \quad R_\nu = \begin{bmatrix} 2 & 0 & 0 & 0 \\ 0 & 2 & 0 & 0 \\ 0 & 0 & 1 & -\sqrt{3} \\ 0 & 0 & \sqrt{3} & 1 \end{bmatrix}$$

$$R_{+\sigma} = \begin{bmatrix} 2 & 0 & 0 & 0 \\ 0 & \sqrt{2} & -\sqrt{2} & 0 \\ 0 & \sqrt{2} & \sqrt{2} & 0 \\ 0 & 0 & 0 & 2 \end{bmatrix}, \quad R_{-\sigma} = \begin{bmatrix} 2 & 0 & 0 & 0 \\ 0 & \sqrt{2} & \sqrt{2} & 0 \\ 0 & -\sqrt{2} & \sqrt{2} & 0 \\ 0 & 0 & 0 & 2 \end{bmatrix}$$

$$[a_{ij}] = \begin{bmatrix} 72 & -6\sqrt{6} & 6\sqrt{6} & -52 \\ -6\sqrt{6} & 7 & -7 & 3\sqrt{6} \\ 6\sqrt{6} & -7 & 7 & -3\sqrt{6} \\ -52 & 3\sqrt{6} & -3\sqrt{6} & 26 \end{bmatrix}, \quad [b_{ij}] = \begin{bmatrix} -120 & -10\sqrt{6} & -10\sqrt{6} & 84 \\ -10\sqrt{6} & 31 & 31 & -5\sqrt{6} \\ -10\sqrt{6} & 31 & 31 & -5\sqrt{6} \\ 84 & -5\sqrt{6} & -5\sqrt{6} & 42 \end{bmatrix}$$

With satisfaction it is noted that $[a_{ij}], [b_{ij}]$ remain singular of rank 3 like $M_P, M_Q$, as one expects. Linear dependency is exhibited by the middle two rows of $[a_{ij}], [b_{ij}]$. Numerically, the constraint equations, Eq. 1 to Eq. 6, become

eq1 :  $72 - 12\sqrt{6}p_1 + 12\sqrt{6}p_2 - 104p_3 + 7p_1^2 - 14p_1p_2 + 6\sqrt{6}p_1p_3 + 7p_2^2$
$-6\sqrt{6}p_2p_3 + 26p_3^2 = 0$

eq2 :  $-120 - 20\sqrt{6}q_1 - 20\sqrt{6}q_2 + 168q_3 + 31q_1^2 + 62q_1q_2 - 10\sqrt{6}q_1q_3$
$+31q_2^2 - 10\sqrt{6}q_2q_3 + 42q_3^2 = 0$

eq3 :  $14 - \sqrt{2}p_1 - \sqrt{2}p_2 = 0$,  eq4 :  $6 - \sqrt{2}q_1 - \sqrt{2}q_2 = 0$

eq5 :  $(-52 + 3\sqrt{6}p_1 - 3\sqrt{6}p_2 + 26p_3)(q_1 - p_1 - q_2 + p_2) + 2(6\sqrt{6} - 7p_1 + 7p_2$
$-3\sqrt{6}p_3)(q_3 - p_3) = 0$

eq6 :  $(-84 + 5\sqrt{6}q_1 + 5\sqrt{6}q_2 - 234q_3)(q_1 - p_1 + q_2 - p_2) + 2(-10\sqrt{6} + 31q_1$
$+31q_2 - 5\sqrt{5}q_3)(q_3 - p_3) = 0$

The following Maple command

```
Groebner[Basis]([eq1,eq2,eq3,eq4,eq5,eq6],
plex(q3,q2,q1,p3,p2,p1));
```

returns a univariate of degree 16 in $p_1$. Four solutions of this particular problem are real. The remaining five basis polynomials are of degree 15 in $p_1$ and linear in but one of the other variables in the sequence $p_2, p_3, q_1, q_2, q_3$, the reverse order in the plex(...) argument specified. The real solutions are tabulated below.

| $p_1$ | $p_2$ | $p_3$ | $q_1$ | $q_2$ | $q_3$ |
|---|---|---|---|---|---|
| 3.7325 | 6.1670 | 3.0175 | 3.5659 | −0.6767 | 0.0281 |
| 3.7556 | 6.1439 | 3.0632 | 1.0330 | −3.2097 | −0.6949 |
| 6.1096 | 3.7899 | 0.8840 | 1.1079 | −3.1347 | −0.5478 |
| 6.1440 | 3.7555 | 0.9370 | 3.6145 | −0.6282 | −0.1125 |

## 7  Intersecting Surfaces, Line Segments and End Points

The first image, Fig. 2, shows the two cylinders of elliptical right section intersected by planes normal to their axes. This should help in interpreting Fig. 1 to show how the conics are placed as desired. Although the construction of the

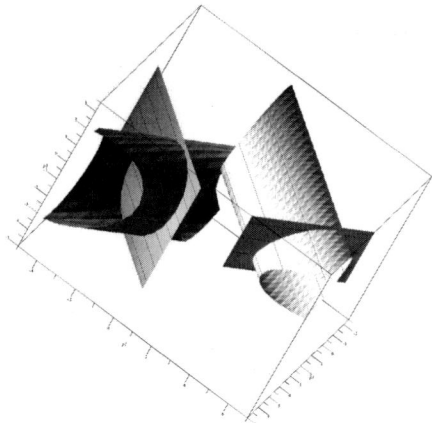

**Fig. 2.** Two conics on the intersection of cylinders and planes

latter figure is precise the numerical data is not identical to that used in the example above. The next image Fig. 3, generated by the computation, shows lines connecting points on curves of intersection between the elliptical cylinders. It maintains about the same view perspective as that of Fig. 2. Segments of all four lines and most of the eight end points are visible. More importantly one can see that the pale green line is the shortest shown while the red is longest. Blue and black segments are inflective. End points are on closer flank of the smaller, lighter shaded cylinder and on the far side of the larger, darker surface as regards the blue line. Close/far interchange relationship in the case of the black line. The red one is the longest segment connecting the conics. Figure 4 is

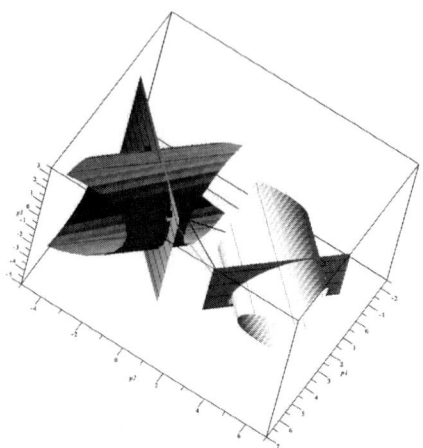

**Fig. 3.** Four line segments and eight endpoints

**Fig. 4.** View from underneath to show all points

a view from beneath both cylinders and is large enough to show all end points, pairs of which are close to each other, to advantage. Coordinate frame has been removed to make things a little clearer.

## 8    Conclusion

With this method an underlying geometry, not available if an iterative approach using parameterized space curves, is revealed. If a 16th degree univariate is acceptable for real time inverse kinematics of a 6R serial robot why not here too? here either.

## References

1. Zsombor-Murray, P.J., Pfurner, M.: Intrusion, proximity & stationary distance. In: Proceedings of CK2017, Poitiers France, 22–24 May 2017, pp. 475-482 (2017)
2. Zsombor-Murray, P.J., Hayes, M.J.D., Husty, M.L.: Extreme distance to a spatial circle. Trans. Can. Soc. Mech. Eng. **28**(2A), 221–235 (2004)
3. Lopes, D.S., Silva, M.T., Ambrósio, J.A., Paulo Flores, P.: A mathematical framework for rigid contact detection between quadric and superquadric surfaces. In: Multibody System Dynamics, pp. 255–280 (2010)
4. Zsombor-Murray, P.J.: Geometric thinking and the geometry of conics and quadrics. In: Proceedings of 19th ICGG 2018, Milano, 07 August 2018, 12 p., 25 May 2017, 4 p. (2018)

# Estimating the Probability of Failures of a 3R̲RR Manipulator Using a Metamodel

Hiparco Lins Vieira$^{(\boxtimes)}$ and Maíra M. da Silva

São Carlos School of Engineering, University of São Paulo, São Carlos, Brazil
hiparcolins@usp.br, mairams@sc.usp.br

**Abstract.** Parallel manipulators are known to provide greater dynamic performance and precision than serial manipulators. In contrast, the former have singularities within their workspace. Consequently, techniques are required to keep the manipulator away from the singularities. However, the location of singularities may be unavailable. In cases where the uncertainties of the system are taken into account, the exact location of the singularities is undetermined. This represents, in fact, a risk of failure for the manipulator, since it may accidentally assume a singular configuration. This paper tackles this problem. A method is presented to determine the configurations that provide a greater risk of failure. Additionally, it estimates the manipulator's probabilities of failure. Initially, these probabilities are determined for specific configurations of the manipulator's workspace. These values are used to design a metamodel in order to estimate the probabilities of failure for the entire workspace. The metamodel provides information that can be used to avoid configurations with considerable risks of failure. As a case study, a 3R̲RR manipulator is analyzed.

**Keywords:** Parallel manipulators · Uncertainties · Reliability
Metamodeling · Singularities

## 1 Introduction

Parallel Kinematic Machines (PKMs) have been increasingly applied for industrial purposes. The noteworthy reasons for their success are: increased accuracy [9]; improved dynamic performance [14]; and energy efficiency [8]. Unfortunately, these machines have a reduced workspace, which may compromise their applicability. To top that off, PKMs also have singularities in their workspace.

It is undeniable that singularities must be avoided during motion planning. Accordingly, some authors suggest treating them as obstacles [3,7], which are avoided through motion planning strategies. In many applications, however, the exact location of all singular regions is unavailable and their determination may be complex. Alternatively, an index to evaluate the closeness between the manipulator and singular regions can be used: the inverse of the condition number

© Springer Nature Switzerland AG 2019
B. Corves et al. (Eds.): EuCoMeS 2018, MMS 59, pp. 384–391, 2019.
https://doi.org/10.1007/978-3-319-98020-1_45

($iCN$) of the Jacobian matrix [2]. It measures the sensitivity of the output against errors or variations in system inputs [6]. A null value of $iCN$ represents a singularity, whereas a unit value stands for an isotropic configuration.

In this context, the workspace can be divided into three regions [17]: singularities, unsafe regions and safe regions. Singularities are always undesired and must be avoided. The unsafe regions are those near singularities, where the $iCN$ values are unbearable. Conversely, safe regions are those regions where values of $iCN$ is tolerable. A critical $iCN$, represented by $CiCN$, can be used to set the limits between safe and unsafe regions; its value is defined by the designer, based on the manipulator's performance requirements [17]. In real applications, however, the manipulator is influenced by uncertainties, which affects the effector's pose. Consequently, the effector may be misplaced in an unsafe or singular region, which is considered a failure.

Many researchers analyzed the impact of uncertainties in the performance and precision of parallel manipulators. In this scope, two techniques are generally explored: Monte Carlo [15,16] simulation and Interval Analysis [4,10,11]. Despite the importance of the aforementioned contributions, the determination of the probability of failures was not part of their studies. Recently, a method for uncertainty analysis was presented [17]. Using a Monte Carlo simulation, the technique computes the probability of failure for several configurations in the manipulator's workspace. Practical applications of this technique may be challenging, since the probabilities of failure of the entire workspace are unknown. In addition, the technique may be inappropriate for real-time applications because the Monte Carlo algorithm generally lacks efficiency. In this paper, the aforementioned technique is improved, using a metamodel [5]. The metamodel is constructed using the data provided by the Monte Carlo algorithm. As result, it estimates the probability of failure for any configurations in the manipulator's workspace, demanding a low computational cost. Maps of failure are drawn to compare both methods. A 3RRR parallel manipulator is used as case study, and only uncertainties in the manipulator's links are considered.

The paper is structured as follows: Sect. 2 describes the 3RRR kinematic model; Sect. 3 presents the methodology; Sect. 4 describes the results, and conclusions are drawn in Sect. 5.

## 2   Kinematic Model

This section describes the 3RRR kinematic model. The robot has three kinematic chains; each chain has an actuated joint (R), and two passive joints (RR), as shown in Fig. 1. The actuators are located in $A_i (i = 1, 2, 3)$, where $i$ symbolizes the number of the kinematic chain. The passive joints are situated in $B_i$, $C_i$,. The center of the workspace is represented by $O$. The orientations of $A_i B_i$ and $B_i C_i$ are expressed by $\theta_i$ and $\beta_i$, respectively. The effector's pose and the actuators inputs are represented, respectively, by $\mathbf{X} = [x, y, \alpha]^T$ and $\boldsymbol{\Theta} = [\theta_1, \theta_2, \theta_3]^T$. The orientation of $\overrightarrow{OA_i}$ is represented by $\gamma_i$.

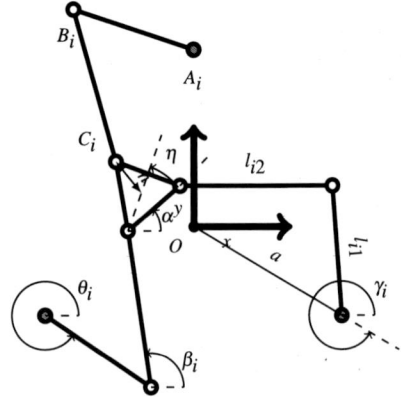

**Fig. 1.** Geometrical scheme of the 3R̲RR manipulator.

## 2.1   Inverse Kinematics

The inverse kinematics seeks to find the position of the actuators that take the effector to a given position. In order to obtain the kinematic equations of the manipulator, two variables $\psi_{xi}$ and $\psi_{yi}$ are initially defined as follows:

$$\begin{bmatrix} \psi_{xi} \\ \psi_{yi} \end{bmatrix} = \begin{bmatrix} x \\ y \end{bmatrix} + h \begin{bmatrix} \cos(\alpha + \eta) \\ \sin(\alpha + \eta) \end{bmatrix} - a \begin{bmatrix} \cos(\eta) \\ \sin(\eta) \end{bmatrix}. \tag{1}$$

From $B_iC_i$, it is possible to establish the following equation:

$$\left\| \begin{bmatrix} \psi_{xi} - l_{i1}\cos(\theta_i) \\ \psi_{yi} - l_{i1}\sin(\theta_i) \end{bmatrix} \right\| = l_{i2}, \tag{2}$$

which can be expanded to:

$$\psi_{xi}^2 + \psi_{yi}^2 + l_{i1}^2 - l_{i2}^2 - 2l_{i1}\psi_{xi}\cos(\theta_i) - 2l_{i1}\psi_{yi}\sin(\theta_i) = 0. \tag{3}$$

Three variables $e_{i1}$, $e_{i2}$ e $e_{i3}$ are defined, where $e_{i1} = \psi_{xi}^2 + \psi_{yi}^2 + l_{i1}^2 - l_{i2}^2$, $e_{i2} = -2l_{i1}\psi_{xi}$, and $e_{i3} = -2l_{i1}\psi_{yi}$. These values are substituted in Eq. 3. Next, the tangent half-angle substitution is applied. As a result, the expressions for $\theta_i$ and $\beta_i$ are obtained:

$$\theta_i = 2\arctan\left( \frac{-e_{i3} \pm \sqrt{e_{i2}^2 + e_{i3}^2 - e_{i1}^2}}{e_{i1} - e_{i2}} \right). \tag{4}$$

$$\beta_i = \arctan\left( \frac{\psi_{yi} - l_{i1}\sin(\theta_i)}{\psi_{xi} - l_{i1}\cos(\theta_i)} \right). \tag{5}$$

Deriving Eq. 3 in time leads to:

$$\dot{x}l_{i2}\cos(\beta_i) + \dot{y}l_{i2}\sin(\beta_i) + \dot{\alpha}l_{i2}h\sin(\beta_i - \eta - \alpha) - \dot{\theta}_i l_{i1}l_{i2}\sin(\beta_i - \theta_i) = 0. \tag{6}$$

Equation 3 can be restated as:

$$A\dot{X} = B\dot{\Theta}, \tag{7}$$

## 2.2   Normalization of the Jacobian Matrix

The Jacobian matrix $\mathbf{A}$ is not dimensionally homogeneous, for it relates different physical dimensions (translational and rotational). In this case, the $iCN$ values would not be consistent. To solve this matter, the Jacobian matrix has to be normalized, using a characteristic length $(L_c)$ [13]. For a 3R̲RR manipulator, the normalized matrix $\mathbf{A}$ is obtained by dividing its third column by $L_c$ [1], where $L_c = \sqrt{2}h$. Once the matrix is homogenized, its $iCN$ is obtained by the ratio between its minimum and maximum singular values.

# 3   Methodology

The methodology is divided into four steps: configuration sampling, Monte Carlo algorithm, determination of the probabilities of failure, and metamodeling.

The manipulator's $iCN$ index is configuration-dependent. Hence, it is unavoidable to specify which configurations $\mathbf{X}$ will be analyzed. To accomplish this, several configurations are sampled, using a three-dimensional grid $(x, y, \alpha)$. Firstly, $j$ values of $\alpha$ are defined. Then, for each value, the workspace is determined, and a regular sampling $(x, y)$ with resolution $r_s$ is performed. These samples are analyzed in a Monte Carlo simulation. The main purpose of the Monte Carlo algorithm is to provide information about how the robot's uncertainties impact on the $iCN$ values. Only uncertainties in the manipulator's links' lengths were considered. Moreover, the length of each link was assumed to have a Gaussian distribution.

Subsequently, for each sampled configuration, a Monte Carlo algorithm is performed, using $n_s$ random combinations of the link's lengths. In the end, for each configuration, a $iCN$ distribution is obtained. The distribution is used to calculate the probability of failure of each configuration.

The probability of failure indicates the probability of a configuration accidentally be in a singularity or unsafe region. Before computing these probabilities, the manipulator's failure modes must be specified. Failure modes represent the ways by which the manipulator may fail. In this context, two failure modes were considered:

1. Workspace failure - occurs when the manipulator was supposed to be in a safe region, but it accidentally lies in a singularity $(iCN = 0)$, or when the manipulator tries to achieve a pose that it can no longer assume $(\sigma_{max} = 0)$;
2. $iCN$ failure - occurs when the manipulator was supposed to lie in a safe region, but, instead, it lies in an unsafe region $(0 < iCN \leq CiCN)$.

As aforementioned, after the Monte Carlo algorithm, each sampled configuration will have an $iCN$ distribution. Then, a distribution fitting algorithm is employed to identify the probability density function that best fits the data. Due to the fine results, the continuous data was fit by a Gaussian distribution. Further details can be found in [17].

Then, the probability of failure $P_f$ of a given configuration $\mathbf{X}$ is computed by:

$$P_f(\mathbf{X}) = \frac{1}{n_s}\{N_{ws} + (1 - N_{ws})F(CiCN)\}, \tag{8}$$

where $N_{ws}$ is the number of workspace failures encountered, and $F(CiCN)$ is the Gaussian cumulative distribution function, representing the probability of $iCN(\mathbf{X}) \leq CiCN$. At the end of this step, every sampled configuration in the workspace has a probability of failure. The Monte Carlo algorithm, however, is computational inefficient, which hinders real-time applications. Thus, a meta-model is used to efficiently estimate the probabilities of failure for the entire workspace.

The metamodel was developed using an artificial neural network (ANN), with architecture Multi-layer Perceptron (MLP). The Levenberg-Marquardt training algorithm was adapted, in order to speed-up the training process [12]. Four inputs were defined: $\mathbf{X} = [x, y, \alpha]^t$ and $iCN(\mathbf{X})$. The ANN output is $P_f(\mathbf{X})$. The data obtained with the Monte Carlo algorithm was used in the neural network training and validation steps. When the training step is concluded. The ANN is used to estimate the probabilities of failure of any configuration in the workspace.

As aforementioned, the metamodel provides efficiency, but it decreases the estimation's accuracy. Consequently, there is a chance that the ANN will provide a bad estimation (an outlier). A solution for this problem is to explore the manipulator's rotational symmetry. The 3R̲RR manipulator has a three-fold rotational symmetry. If the manipulator is rotated, for instance, 120° or −120°, it will appear the same. Therefore, some configurations are equivalent. These configurations have the same $iCN$ values. In addition, if the uncertainties have the same distributions in each kinematic chain, the probabilities of failure of equivalent configurations are also the same. Thus, instead of estimating only the $P_f$ of a given configuration $\mathbf{X}$, the metamodel is used to estimate also the $P_f$ of its equivalent configurations. Consequently, the metamodel provides a set of three possible estimations. Thus, the median of this set is chosen as a credible estimation for $\mathbf{X}$.

## 4    Results

As stated, only uncertainties in the manipulator's links are considered. The links were assumed to vary according to a Gaussian distribution $\mathcal{N}(\mu; \sigma)$, where $\mu$ represents the mean and $\sigma$ the standard deviation. The parameters used in the simulation are described in Table 1. Only the Jacobian matrix $\mathbf{A}$ was analyzed. In addition, $n_s = 40,000$ and $CiCN = 0.1$. The selected ANN has four hidden layers, with 17 neurons in each. Each neuron has a logistic activation function.

Failure maps can be drawn using the probabilities of failure obtained in the simulations. These maps display the probability of failure of each sampled configuration. A map is drawn for each value of $\alpha$. The results using the metamodel are compared with those obtained from the Monte Carlo simulation (without using the metamodel). The Monte Carlo simulation provided the probabilities

**Table 1.** Parameters used in the simulations.

| Step | Parameter |
|------|-----------|
| Modelling | $l_{i1} = \mathcal{N}$ (0.191 m; 0.0006 m), $l_{i2} = \mathcal{N}$ (0.232 m; 0.0006 m) |
|  | $a = 0.2598$ m, $h = 0.0597$ m, $\gamma_1 = 0$ rad, $\gamma_2 = 2\pi/3$ rad, $\gamma_3 = 4\pi/3$ rad |
| Sampling | $r_s = 0.005$ m $j = 5$, $\alpha_1 = -\pi/6$ rad, $\alpha_2 = -\pi/12$ rad |
|  | $\alpha_3 = 0$ rad, $\alpha_4 = \pi/12$ rad, $\alpha_5 = \pi/6$ rad |

of failure for five different values of $\alpha$ (see Table 1). Some of these results are compared with those obtained from the ANN.

Figure 2 depicts two failure maps, obtained using the Monte Carlo technique without considering the metamodel. The regions in white have a high probability of failure ($P_f \cong 1$), while those in orange have low probability of failure ($P_f \cong 0$). An interesting aspect of the estimation lies in the limits between regions of high and low probability of failure. Apparently, these probabilities may change abruptly. However, in these same regions, it is possible to note some isolated configurations with medium probability of failure, represented in yellow. This suggests that the transition between regions with low and high probability of failure may not be depicted in these maps due to the grid resolution. Coincidentally, these discontinuities are also detected in the inner and outer workspace limits of all maps.

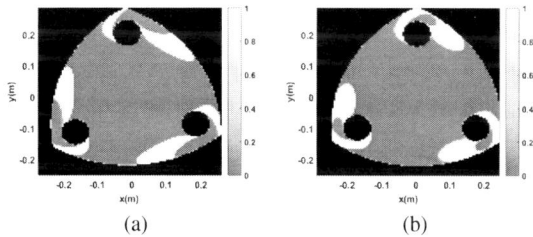

(a)                              (b)

**Fig. 2.** Failure maps obtained using the only Monte Carlo simulation, with resolution $r_s = 0.005$ m. (a) $\alpha_1 = \frac{-\pi}{12}$ rad, (b) $\alpha_2 = 0$.

Figure 3 displays the probabilities of failure estimated using the metamodel. It is possible to observe that the ANN estimations, in Fig. 3(a) and (c), reliably represent those acquired from the Monte Carlo algorithm. Another interesting information about these maps are the limits between the regions of high and low $P_f$. These limits, which were not clear (continuous) in Fig. 2, are clearly seen in Fig. 3. The same is observed in the workspace limits.

Figure 3(b) displays the failure map for $\alpha$: $-\pi/24$, obtained using the metamodel. The expectation is that this map shows the transition between both maps in Fig. 3(a) and (c). In addition to displaying the transition between maps, it

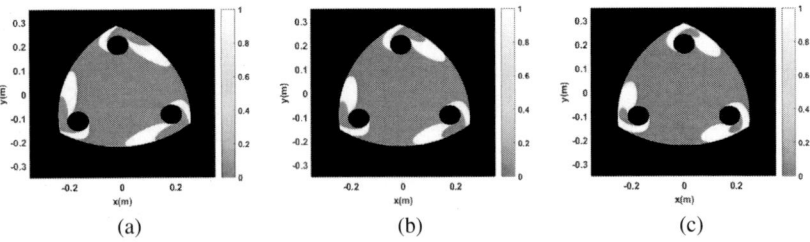

**Fig. 3.** Failure maps computed using the ANN estimation procedure, with resolution $r_s = 0.001$ m. (a) $\alpha_1 = \frac{-\pi}{12}$ rad, (b) $\alpha_6 = \frac{-\pi}{24}$ rad, (c) $\alpha_2 = 0$ rad.

satisfactorily exhibit the limits between high and low $P_f$ regions, even in the limits of the manipulator's workspace. The main advantage provided by the ANN metamodel is the reduction of the computational cost. While the Monte Carlo algorithm demands approximately 8.7 s to estimate the probability of failure of a single configuration, the metamodel needed almost 214 s to estimate the probability of failure of approximately 1,33 billion configurations; in average, the ANN estimates the probability of failure of a configuration in about 16 μs.

## 5   Conclusions

In this paper, the impact of the uncertainties of a 3RRR manipulator on the conditioning was analyzed. Uncertainties may cause failures, accidentally leading the manipulator to low conditioning regions. Probabilities of failure for some configurations in the workspace were computed, using a Monte Carlo algorithm. A metamodel was developed to estimate probabilities of failure for any configuration in the manipulator's workspace. When compared to the Monte Carlo estimation, the metamodel provided high accuracy with a considerable lower computational cost. In addition, the metamodel provided extra information, by showing the variation of the probabilities of failure in the workspace limits.

**Acknowledgements.** This research is supported by FAPESP 2014/01809-0. Moreover, H. L. Vieira is thankful for his CNPq grant.

## References

1. Alba-Gomez, O., Wenger, P., Pamanes, A.: Consistent kinetostatic indices for planar 3-DOF parallel manipulators, application to the optimal kinematic inversion. In: Volume 7: 29th Mechanisms and Robotics Conference, Parts A and B, vol. 2005, pp. 765–774. ASME (2005). https://doi.org/10.1115/DETC2005-84326
2. Cha, S.H., Lasky, T., Velinsky, S.: Determination of the kinematically redundant active prismatic joint variable ranges of a planar parallel mechanism for singularity-free trajectories. Mech. Mach. Theory **44**(5), 1032–1044 (2009). https://doi.org/10.1016/j.mechmachtheory.2008.05.010

3. Dash, A.K., Chen, I.M., Yeo, S.H., Yang, G.: Workspace generation and planning singularity-free path for parallel manipulators. Mech. Mach. Theory **40**(7), 776–805 (2005). https://doi.org/10.1016/j.mechmachtheory.2005.01.001

4. Gouttefarde, M., Daney, D., Merlet, J.P.: Interval-analysis-based determination of the wrench-feasible workspace of parallel cable-driven robots. IEEE Trans. Robot. **27**(1), 1–13 (2011). https://doi.org/10.1109/TRO.2010.2090064

5. Hendrickx, W., Dhaene, T.: Sequential design and rational metamodelling. In: 2005 Proceedings of the Winter Simulation Conference, pp. 9–pp (2005). https://doi.org/10.1109/WSC.2005.1574263

6. Kincaid, D., Cheney, W.: Numerical Analysis: Mathematics of Scientific Computing. Brooks/Cole Publishing Co., Pacific Grove (1991)

7. Lahouar, S., Zeghloul, S., Romdhane, L.: Singularity free path planning for parallel robots, pp. 235–242. Springer, Dordrecht (2008). https://doi.org/10.1007/978-1-4020-8600-7_25

8. Li, Y., Bone, G.M.: Are parallel manipulators more energy efficient? In: Proceedings 2001 IEEE International Symposium on Computational Intelligence in Robotics and Automation (Cat. No. 01EX515), pp. 41–46 (2001). https://doi.org/10.1109/CIRA.2001.1013170

9. Merlet, J.P.: Parallel Robots. Kluwer Academic Publishers, Norwell (2002)

10. Merlet, J.P.: Interval analysis and robotics. In: Robotics Research: The 13th International Symposium ISRR, pp. 147–156. Springer, Heidelberg (2011). https://doi.org/10.1007/978-3-642-14743-2_13

11. Merlet, J.P., Daney, D.: Dimensional synthesis of parallel robots with a guaranteed given accuracy over a specific workspace. In: Proceedings of the 2005 IEEE International Conference on Robotics and Automation, pp. 942–947 (2005). https://doi.org/10.1109/ROBOT.2005.1570238

12. Mohamad, N., Zaini, F., Johari, A., Yassin, I., Zabidi, A.: Comparison between Levenberg-Marquardt and scaled conjugate gradient training algorithms for breast cancer diagnosis using MLP. In: 2010 6th International Colloquium on Signal Processing its Applications, pp. 1–7 (2010). https://doi.org/10.1109/CSPA.2010.5545325

13. Mohammadi, H.R., Zsombor Murray, P.J.D., Angeles, J.: The isotropic design of two general classes of planar parallel manipulators. J. Robot. Syst. **12**(12), 795–805 (1995). https://doi.org/10.1002/rob.4620121204

14. Paccot, F., Andref, N., Martinet, P.: A review on the dynamic control of parallel kinematic machines: theory and experiments. Int. J. Robot. Res. **28**(3), 395–416 (2009). https://doi.org/10.1177/0278364908096236

15. Rugbani, A.: Modelling and analysis of the geometrical errors of a parallel manipulator micro-CMM. In: Precision Assembly Technologies and Systems: Proceedings of the 6th IFIP WG 5.5 International Precision Assembly Seminar, IPAS 2012, Chamonix, France, 12–15 February 2012, pp. 105–117. Springer, Heidelberg (2012). https://doi.org/10.1007/978-3-642-28163-1_14

16. Sovizi, J., Alamdari, A., Das, S., Krovi, V.: Random matrix based uncertainty model for complex robotic systems. In: 2014 IEEE International Conference on Robotics and Automation (ICRA), pp. 4049–4054 (2014). https://doi.org/10.1109/ICRA.2014.6907447

17. Vieira, H.L., de Carvalho Fontes, J.V., Beck, A.T., da Silva, M.M.: Robust critical inverse condition number for a 3RRR robot using failure maps, pp. 285–294. Springer (2018). https://doi.org/10.1007/978-3-319-67567-1_27

# Compliant Class 1 Tensegrity Structures
# for Gripper Applications

Susanne Sumi$^{(\boxtimes)}$, Valter Böhm, Philipp Schorr, Lena Zentner,
and Klaus Zimmermann

Ilmenau University of Technology, Ilmenau, Germany
{Susanne.Sumi,Valter.Boehm,Philipp.Schorr,Lena.Zentner,
Klaus.Zimmermann}@tu-ilmenau.de

**Abstract.** This paper describes concepts for finger-grippers based on compliant multistable class 1 tensegrity structures. Two of these concepts are selected and examined in detail. With theoretical investigations the member parameters and the resulting gripping forces are determined. There are done dynamical analyses of one of these grippers to obtain the behaviour with an actuation force. Moreover demonstrators of both grippers are built.

**Keywords:** Compliant tensegrity structure
Multiple states of self-equilibrium · Gripper application

## 1 Introduction

Mechanical compliant, prestressed structures are a recent discussed topic [1,3,5,10,11]. One special class of such structures are compliant tensegrity structures. These structures are consisting of compressed and tensioned members. The compressed members are indirectly connected through the compressed members. The name "tensegrity structures" was coined by Buckminster Fuller and Kenneth Snelson in the 1960s. It combines the words "tension" and "integrity". Tensegrity structures are filigree, lightweight, free standing, mechanical compliant and prestressed structures. They have a good strength-to-weight-ratio and external forces are distributed through the whole structure [6,12]. Due to these reasons compliant tensegrity structures are interesting for robotic applications.

Currently, there are several examples for the use of tensegrity structures in mobile robots [4,7,8]. Additionally, it is possible to use these structures in gripper applications. In [9] are grippers and manipulators based on tensegrity structures presented. These grippers are based on tensegrity structures with only one equilibrium configuration. It is possible, that a tensegrity structure has more than one (static stable) equilibrium configuration. Such tensegrity structures are called multistable. They seem to have advantages compared to monostable tensegrity structures, such as different possible geometrical shapes, smaller control efforts to achieve another shape/stable state. In [13–15] a two-finger-gripper

© Springer Nature Switzerland AG 2019
B. Corves et al. (Eds.): EuCoMeS 2018, MMS 59, pp. 392–399, 2019.
https://doi.org/10.1007/978-3-319-98020-1_46

based on a multistable tensegrity structure is presented. It is a class 2 tensegrity structure, meaning that some compressed members are connected directly, in contrast to class 1 tensegrity structures. The fingers of the gripper are connected directly with each other through compressed members.

In this article basic investigations on multistable class 1 tensegrity structures will be considered with respect to the future aim to realise grippers with an enhanced mechanical compliance.

Within this paper, the geometrical shape of a tensegrity structure is obtained with the form-finding algorithm from [2]. This algorithm uses a static Finte Element Method including geometric nonlinearity. The members are assumed to be massless 2D linear spring elements with constant axial stiffness. The characteristic equation $F(u) = K(u)u$ is solved with an incremental-iterative procedure. ($u$ vector of nodal displacements, $F(u)$ vector of nodal forces, $K(u)$ tangent stiffness matrix)

This work is composed as follows: After the introduction, in Sect. 2, concepts for the development of such grippers are introduced. Two of the concepts are selected and in Sect. 3 theoretical analyses with these structures are done. Afterwards in Sect. 4 demonstrators are presented. Finally some conclusions and further research directions are given.

## 2    Concepts

In this section four concepts (A, B, C and D) are presented. These concepts show, how planar multistable class 1 tensegrity structures may be used as finger-grippers. For that shall be used few compressed members to get structures which are simple. Two equilibrium configurations shall be used to represent the opened and the closed gripper. The four concepts are depicted in Fig. 1 and described in the following. It is a qualitative consideration without specific member parameters. For the existence of the considered equilibrium configuration, suitable member parameters have to be chosen (see Sect. 3.1).

**Variant A:** The simplest possible multistable tensegrity structure consists of two compressed members and four tensioned members. The member parameters of the compressed and the tensioned members are assumed to be equal, respectively. Adding gripper arms at node 2 and 3 turns this tensegrity structure into a gripper with two stable equilibrium configurations. The opening and closing of the gripper may be achieved by a rotation of member 2 of 180° (change between both equilibrium configurations).

**Variant B:** This gripper is based on variant A but in contrast to it the member parameters are not equal. The length of the compressed members are different (member 2 is shorter than member 1). The not rectangular angle between member 1 and 2 can be achieved with different parameters of the tensioned members. There are still two equilibrium configurations but the rotation angle of member 2 to change between these equilibrium configurations is less than 180°. The gripper arms are stiff connected to node 2 and 3. Member 2 rotates around its centre

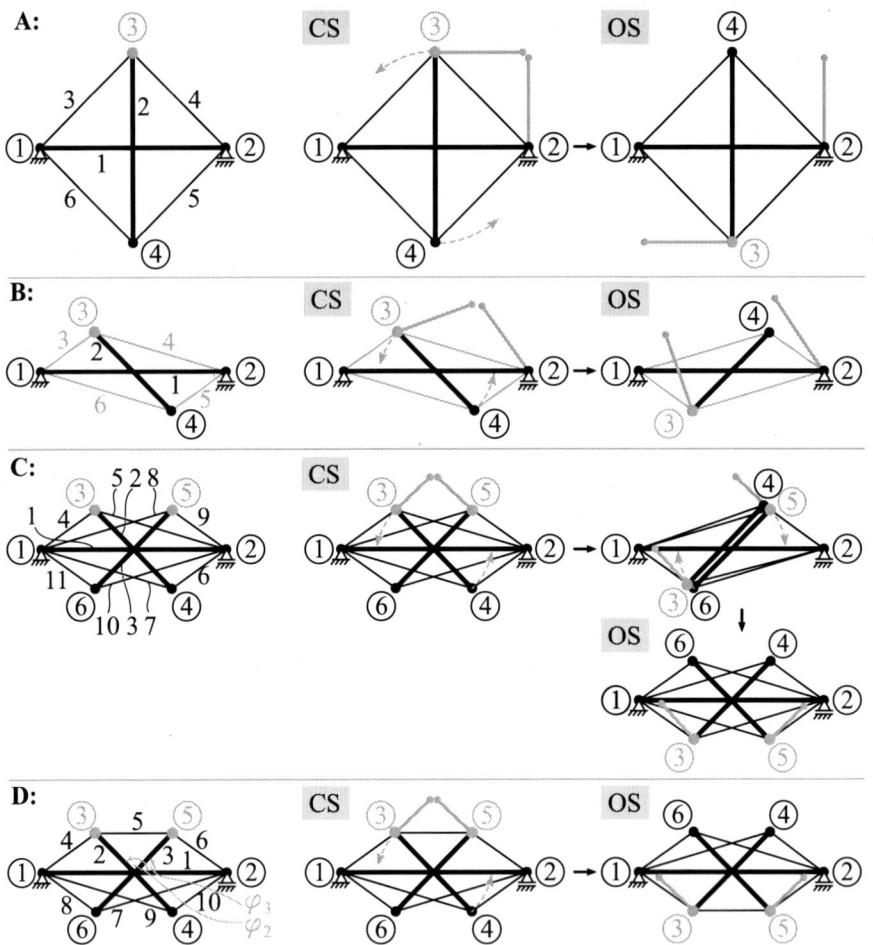

**Fig. 1.** Four concepts for the development of grippers based on tensegrity structures. Thick lines: compressed members, thin lines: tensioned members. CS - closed state, OS - opened state.

to open and close the gripper. A specific characteristic of this variant (and A) is that one gripper arm is fixed and only one gripper arm moves.

**Variant C:** Variant B is used twice in this variant. Gripper arms may be added at node 3 and node 5. Since member 2 and 3 are not direct connected the structure has four equilibrium configurations. Meaning that member 2 and 3 may move independently into the other equilibrium configuration. For that reason the gripper has four different states: closed, half opened to the left or right side and opened. In contrast to the variants A and B, no gripper arm is fixed. But there are two actuators needed to open and close the gripper.

**Variant D:** In comparison to variant C the tensioned members are connected in another way. Member 2 and 3 are connected with a tensioned member. For that reason these two members cannot move independently anymore. The tensegrity structure has two equilibrium configurations. The gripper arms may be connected as in variant C. It can be proved (see Sect. 3.3) that only one actuation at one member (2 or 3) is sufficient to move into the second equilibrium configuration. So a gripping of the object from two sides is possible and there is only one actuator needed.

Since in variant C as well as in variant D the gripper arms are decoupled from the base, these two variants will be considered in the following sections.

## 3    Theoretical Studies

In this section suitable parameters for the members are chosen, the gripping force is calculated and a dynamical analysis is done.

### 3.1    Member Parameters

The member parameters (free-lengths and stiffnesses) shall be determined so that the resulting tensegrity structure fulfils the listed properties. Since both variants are similar, the properties are the same.

- There are two equilibrium configurations.
- The tensegrity structure shall be axisymmetric to the centre of member 1.
- Member 2/3 may rotate 360° around its centre without touching node 1 or 2, to avoid collisions while opening and closing the gripper.
- The angle between member 1 and member 2/3 shall be between 110° and 160°.
- In the opened state the tips of the gripper arms shall be higher than member 6 and 11 (variant C) or member 8 and 10 (variant D), to avoid collisions between the gripper arms and the tensioned members.
- The gripper arms shall have their tips at the same point in the closed state.

The parameters of the tensioned members are set so that there are real springs available, see Table 1. Afterwards the lengths of the compressed members are varied systematically (member 1: $[100, 200]$ mm, member 2, 3: $[40, 100]$ mm, step-size 1 mm) to obtain a tensegrity structure with the listed properties. For every parameter set the form-finding algorithm from [2] is applied to get the corresponding equilibrium configurations. The obtained lengths of the members are: member 1: 140 mm, member 2 and 3: 60 mm for both variants. The gripper arms are chosen according to the last two points in the list: The gripper arms have an angle to member 2 and 3 of 80° in variant C and 70° in Variant D. The lengths of the gripper arms are set to 38.5 mm for variant C and 46.0 mm for Variant D. The opened and closed state of the gripper (which are equilibrium configurations) are depicted in Fig. 2.

Independent from the chosen member parameters the existence and stability of one equilibrium configuration is a consequence of the existence and stability of the other equilibrium configuration.

**Table 1.** Member parameters of variant C and D. $k$ - stiffness of the members, $l_0$ - free lengths of the members.

| Member | | 4 | 5 | 6 | 7 | 8 | 9 | 10 | 11 |
|---|---|---|---|---|---|---|---|---|---|
| Variant C | $k$ [N/mm] | 0.16 | 0.20 | 0.16 | 0.18 | 0.20 | 0.16 | 0.18 | 0.16 |
| | $l_0$ [mm] | 22.7 | 37.9 | 22.7 | 41.1 | 37.9 | 22.7 | 41.1 | 22.7 |
| Variant D | $k$ [N/mm] | 0.36 | 0.43 | 0.36 | 0.33 | 0.34 | 0.33 | 0.34 | - |
| | $l_0$ [mm] | 23.3 | 33.4 | 23.2 | 33.4 | 28.7 | 33.4 | 28.7 | - |

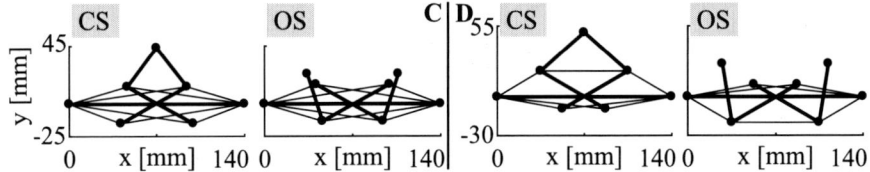

**Fig. 2.** Opened and closed state of the gripper (equilibrium configurations) variant C and D.

## 3.2   Gripping Forces

The gripping forces of the grippers have been calculated with the form finding algorithm from [2], too. The gripping force in dependency of the distance between the tips of the gripper arms is calculated and depicted in Fig. 3. If the distance of the tips reaches a critical value the tensegrity structure moves into the other equilibrium configuration, due to the snap-through-behaviour. So the maximal gripping forces and the maximal distances between the tips can be obtained from Fig. 3. The achievable gripping force of variant C is smaller (approx. a third) than of variant D and the possible distance between the tips is smaller, too.

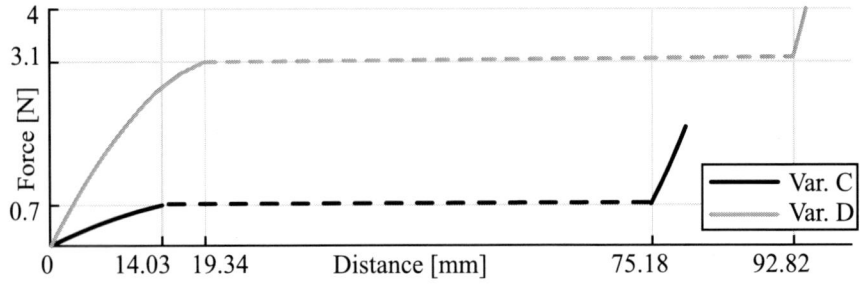

**Fig. 3.** Calculated gripping force of both variants. Snap-through happens at a distance of 14.03 mm (variant C), 19.34 mm (variant D).

There are several possibilities to enlarge the gripping force. The length of the gripper arm may be increased. So the gripper arms are prestressed if it is closed but the possible distance between the tips would decrease. Another possibility is to scale the stiffnesses of all members by the same factor. This increases or decreases the prestress of the gripper and the gripping forces. The achievable distances between the tips stay the same.

### 3.3    Dynamical Behaviour

To analyse the dynamical behaviour of the grippers the equations of motion have been derived. In the following a dynamical study with variant D is presented. It is investigated, if one external force at one node is enough to move into the other equilibrium configuration.

Member 1 is fixed on the plane with bearings. The compressed members are assumed to be stiff with a mass m1, m2, m3. The gripper arms are neglected. The generalised coordinates are set to be the $(x, y)$-coordinates and the angle $\varphi_i$ of the centre of member 2 and 3 regarding to the x-axis. The tensioned members are assumed to be linear springs. Parallel to each tensioned member a linear damper is modelled. Due to the large prestress of the structure the influence of the gravity is neglected. The friction between connected members is neglected, too. The equations of motion are derived with the Lagrange's equations of the second kind for non-conservative forces. They are given by:

$$\ddot{x}_i = \frac{1}{m_i}\left(-\frac{\partial U}{\partial x_i} - \frac{\partial D}{\partial \dot{x}_i} + Q_{x_i}\right) \quad \ddot{y}_i = \frac{1}{m_i}\left(-\frac{\partial U}{\partial y_i} - \frac{\partial D}{\partial \dot{y}_i} + Q_{y_i}\right)$$
$$\ddot{\varphi}_i = \frac{12}{m_i l_i^2}\left(-\frac{\partial U}{\partial \varphi_i} - \frac{\partial D}{\partial \dot{\varphi}_i} + Q_{\varphi_i}\right), \quad i = 2, 3, \tag{1}$$

where U is the potential energy and D is the dissipation function of the structure. $Q_{x_i}, Q_{y_i}, Q_{\varphi_i}$ are the generalised forces (which are no dissipative and conservative forces). It is an explicit, coupled, nonlinear differential equation system.

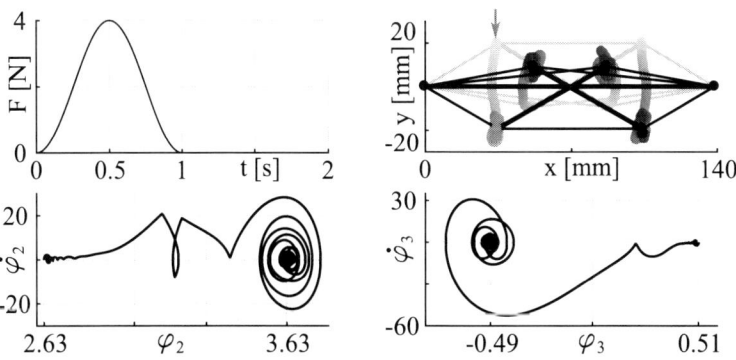

**Fig. 4.** Phase portraits and movement of the nodes during actuation.

An external force is applied at node 3 in negative y-direction. The actuation force is shown in the upper left diagram in Fig. 4. The movement of the structure is obtained with (1) and shown in the upper right diagram in Fig. 4. The phase portraits $(\varphi_2, \dot{\varphi}_2)$ and $(\varphi_3, \dot{\varphi}_3)$ are depicted in the lower diagrams in Fig. 4. It can be read off the upper right diagram (Fig. 4) that the tensegrity structure moves into the other equilibrium configuration and the angles in the phase portraits change to the corresponding stationary point. So these results show that the opening and closing of the gripper with only one actuation at only one member is possible.

## 4   Demonstrators

To verify the theoretical results demonstrators of the grippers have been built. The parameters from Sect. 3.1 are used. They are built with rods made of aluminium, tension springs made of steel and connecting pieces made of PLA printed in 3D. To obtain spatial stability all members have been doubled. Pictures of the demonstrators are presented in Fig. 5.

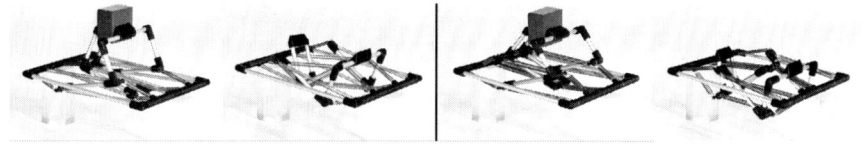

**Fig. 5.** Demonstrators of the tensegrity grippers.

They have the predicted equilibrium configurations and they are able to grip objects. With the help of the demonstrators it is proven, that the practical realisation of these finger-grippers is possible. Arrangements of the members have been found, which avoid collisions between each other while changing between the opened and closed state of the gripper.

## 5   Conclusions and Outlook

In this article four concepts for finger-grippers based on multistable class 1 tensegrity structures have been presented. Two of these concepts have been selected and examined in detail. With theoretical investigations the member parameters and the resulting gripping forces have been determined. With the equations of motion the behaviour of one variant with an actuation force has been investigated. It was demonstrated that one actuation force is enough to open and close the gripper. Further actuation principles for both variants could be evaluated in further investigations.

Moreover demonstrators of both grippers have been built. They have the predicted equilibrium configurations and a change between these equilibrium configurations is possible. Additional experiments with these grippers to verify theoretical results and the use in gripper applications have to be done. Afterwards actuators may be integrated into these demonstrators to open and close the gripper automatically. Further work will focus on the development of the considered grippers for specific applications.

**Acknowledgements.** This work is supported by the Deutsche Forschungsgemeinschaft (DFG project BO4114/2-1).

# References

1. Boehler, Q., et al.: Definition and computation of tensegrity mechanism workspace. J. Mech. Robot. **7**(4), 4 p. (2015). Paper No. JMR-14-1168
2. Böhm, V., et al.: Compliant multistable tensegrity structures. Mech. Mach. Theory **115**, 130–148 (2017)
3. Caluwaerts, K., et al.: Design and control of compliant tensegrity robots through simulation and hardware validation. J. R. Soc. Interface **11**, 13 p. (2014). Paper No. 20140520
4. Chen, L., et al.: Soft spherical tensegrity robot design using rod-centered actuation and control. ASME J. Mech. Robot. **9** (2017)
5. Henke, E.-F.M., et al.: Entirely soft dielectric elastomer robots. In: Electroactive Polymer Actuators and Devices (EAPAD) 2017. Proceedings of the SPIE, vol. 10163, p. 101631N, 10 May 2017
6. Juan, S.H., Mirats-Tur, J.M.: Tensegrity frameworks: static analysis review. Mech. Mach. Theory **43**(7), 859–881 (2008)
7. Kaufhold, T., et al.: Indoor locomotion experiments of a spherical mobile robot based on a tensegrity structure with curved compressed members. In: IEEE AIM, pp. 523–528 (2017)
8. Kim, K., et al.: Hopping and rolling locomotion with spherical tensegrity robots. In: IEEE/RSJ IROS, pp. 4369–4376 (2016)
9. Lessard, S., et al.: A bio-inspired tensegrity manipulator with multi-DOF, structurally compliant joints. In: Proceedings of the 2016 IEEE/RSJ International Conference on Intelligent Robots and Systems (IROS), pp. 5515–5520 (2016)
10. Rieffel, J.A., et al.: Morphological communication: exploiting coupled dynamics in a complex mechanical structure to achieve locomotion. J. R. Soc. Interface **7**, 613–621 (2010)
11. Santer, M., et al.: Compliant multistable structural elements. Int. J. Solids Struct. **45**, 6190–6204 (2008)
12. Skelton, R.E., et al.: An introduction to the mechanics of tensegrity structures. In: The Mechanical Systems Design Handbook: Modeling, Measurement, and Control. CRC Press (2001)
13. Sumi, S., et al.: A novel gripper based on a compliant multistable tensegrity mechanism. In: Proceedings of the MAMM 2016, Ilmenau, Germany, pp. 115–126 (2017)
14. Sumi, S., et al.: Compliant gripper based on a multistable tensegrity structure. In: New Trends in Mechanism and Machine Science: Theory and Industrial Applications, pp. 143–151 (2017)
15. Sumi, S., et al.: A multistable tensegrity structure with a gripper application. Mech. Mach. Theory **114**, 204–217 (2017)

# Design of 3-DOF Zero Coupling Degree Planar Parallel Manipulator Based on Coupling-Reducing and Its Kinematic Performance Improvement

Ke Xu[1], Jiaming Deng[1], Guanglei Wu[2], Ju Li[1], and Huiping Shen[1(✉)]

[1] School of Mechanical Engineering, Changzhou University,
Changzhou 213016, People's Republic of China
shp65@126.com
[2] School of Mechanical Engineering, Dalian University of Technology,
Dalian 116024, People's Republic of China
gwu@dlut.edu.cn

**Abstract.** This paper deals with the topological architectural optimization and results in the enhanced kinematic performance for a 3-dof planar traditional parallel manipulator (TPM). A 3-dof planar zero coupling degree manipulator (ZCDM) is proposed based on the method of reducing coupling degree, of which the forward kinematic problem is simplified, leading to partially decoupled input-output motion. Moreover, a comparative study between TPM and ZCDM is carried out with respect to the workspace and singularity, which shows the advantages of the ZCDM compared to TPM.

**Keywords:** Parallel mechanism · Coupling degree · Reducing coupling degree
Topological optimization · Kinematic performance

## 1 Introduction

The 3-RRR planar traditional parallel manipulator (TPM) suffers from two disadvantages due to its fully symmetrical structure. One problem is that its direct position problem is difficult to be solved analytically and the other one is that its input-output motion is not decoupled. In the literature, such a TPM have been extensively investigated on many aspects [1]. Oetomo et al. [2] set up three kinematic constraint equations and then got an eight-degree polynomial expression. Gosselin [3] conducted the parametric design optimization of the 3-RRR PM. Wu et al. [4] carried out comparison terms of statics and dynamics among the 4-RRR, 3-RRR and 2-RRR PMs. Taking prismatic pair as actuated one, the reachable workspace of 3-RRR PM was analyzed by Li et al. [5]. Gao et al. [6] systematically analyzed the relationship between branched chains' length and the workspace shape of 3-RRR PM.

However, accuracy analysis and design of this type of TPM are difficult and motion control is comparatively complex, wherein the reasons lies in that it is difficult to get its analytical expression of the direct kinematics and this TPM does not have input-output (*I-O*) motion decoupling.

© Springer Nature Switzerland AG 2019
B. Corves et al. (Eds.): EuCoMeS 2018, MMS 59, pp. 400–408, 2019.
https://doi.org/10.1007/978-3-319-98020-1_47

This paper presents a 3-dof planar zero coupling degree manipulator (ZCDM) based on the method of reducing coupling degree. Thanks to the previous characteristics, the analytical direct kinematics of this manipulator can be readily obtained and it has *I-O* motion decoupling property. Moreover, the workspace and singularity of the ZCDM are analyzed. This work shows that the overall performance of ZCDM is better than that of 3-RRR planar TPM.

## 2    3-RRR TPM and Its Topological Optimization Design

Typical 3-RRR planar TPM is shown in Fig. 1. The line and triangle labeled from 1 to 7 denoted the seven different links, respectively. The equilateral triangle moving platform 1 is connected with the base platform 0 through three RRR branch chains. The base coordinate system *o-xy* and the moving coordinate system *o′-x′y′* are established, with the origins located at the point o of the base and the center of the mobile platforms, respectively.

Lengths for each link of three branch chains are given as follows, respectively.

$$R_{11}R_{12} = l_1, R_{12}R_{13} = l_2, R_{21}R_{22} = l_7, R_{22}R_{23} = l_6, R_{31}R_{32} = l_5, R_{32}R_{33} = l_4.$$

The side length of the moving platform 1 is $l_3$, and its rotation angle $\gamma$ is described by spanning $x′$-axis from x-axis counterclockwise. The input angles of three actuated pairs $R_{11}$, $R_{21}$ and $R_{31}$ are denoted by $\theta_1$, $\theta_2$ and $\theta_3$, respectively.

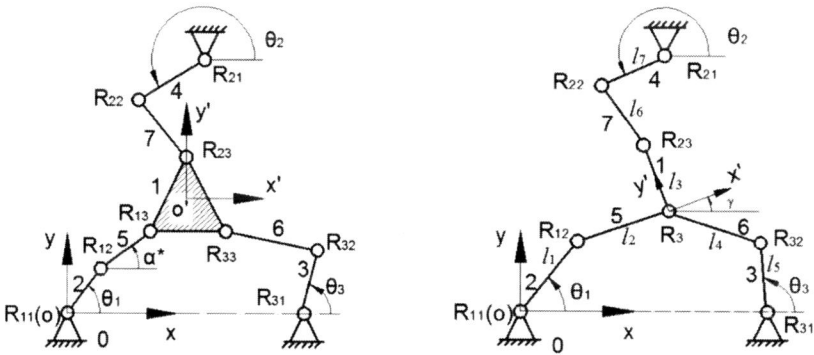

**Fig. 1.** Planar traditional PM (TPM)    **Fig. 2.** Zero coupling degree manipulator (ZCDM)

### 2.1    Coupling Degree (κ) of 3-RRR TPM

According to the principle of mechanisms composition based on the ordered single-open-chain (*SOC*) [7, 8], this 3-RRR TPM can be decomposed into following two *SOC*s. The restraint degree (Δ), an important topological characteristics, of each SOC is calculated as follows, respectively.

$$SOC_1\{-R_{11} - R_{12} - R_{13} - R_{33} - R_{32} - R_{31}-\}, \quad \Delta_1 = \sum_{i=1}^{6} f_i - I_1 - \xi_{L_1} = 6 - 2 - 3 = 1$$

$$SOC_2\{-R_{21} - R_{22} - R_{23}-\}, \quad \Delta_2 = \sum_{i=1}^{3} f_i - I_2 - \xi_{L_2} = 3 - 1 - 3 = -1$$

Here SOC unit is defined as serial connection of joints and links, which stands for real loop when kinematic analysis of a mechanism discussed.

The Coupling degree $k$ of the PM is calculated as [7, 8]

$$k = \frac{1}{2} \sum_{j=1}^{2} |\Delta_j| = \frac{1}{2}(|1| + |-1|) = 1$$

Where, $I_j$ is the number of inputs in the $j^{th}$ $SOC_j$, $f_i$ is the $DOF$ of the $i^{th}$ kinematic pairs, $\xi_{L_j}$ is the number of independent equations of $j^{th}$, and $\Delta_j$ is the constraint degree of $j^{th}$ $SOC_j$, respectively.

In the above equation, $k$ reveals the relationship between the unknown variables of independent loops inside of a PM. The larger the value of $k$ is, the more complex complexity of this mechanism is [7, 8]. Since the coupling degree of this TPM is $k = 1$, its numerical solutions of direct kinematics could be obtained by solving a polynomial equation with one unknowns. Then one-dimensional search method is utilized easily to obtain its numerical direct solutions. However the calculation procedure is complicated and time-consuming.

Moreover, since each output parameter $(x, y, \gamma)$ of moving platform 1 is related to all of three input angles $\theta_1$, $\theta_2$, and $\theta_3$, this manipulator does not have I-O motion decoupling characteristics, which is also undesirable for path planning and motion controlling.

## 2.2    Topological Optimum Design of the TPM

In order to overcome the two disadvantages stated above, we conduct an optimum design for topological structure of this manipulator. By using the method for reducing coupling degree of PM proposed by Shen [9], we combine two arbitrary joints on the moving platform 1, such as $R_{13}$ and $R_{33}$ in the Fig. 1, into one complex joint. This operation leads to an improved manipulator shown in the Fig. 2. Its topological analysis can be decomposed into following:

$$SOC_1\{-R_{11} - R_{12} - R_3 - R_{32} - R_{31}-\}, \quad \Delta_1 = \sum_{i=1}^{5} f_i - I_1 - \xi_{L_1} = 5 - 2 - 3 = 0$$

$$SOC_2\{-R_{21} - R_{22} - R_{23} - R_3-\}, \quad \Delta_2 = \sum_{i=1}^{3} f_i - I_2 - \xi_{L_2} = 4 - 1 - 3 = 0$$

Therefore, $k = \frac{1}{2} \sum\limits_{j=1}^{2} |\Delta_j| = \frac{1}{2}(|0| + |0|) = 0$

It means that the coupling degree of this changed PM is reduced form one to zero, thus, its analytical expression of direct kinematics can be readily obtained. We called the improved manipulator as 3-dof planar zero coupling degree manipulator (ZCDM). It is recently noted that the ZCDM happens to be the same as the architecture proposed in Ref. [10]. However ZCDM in this paper is obtained based on a systematic and theoretical method for reducing coupling degree proposed by the author [9] instead of intuition or occasional experience of designer, which is a contribution of this paper.

## 3   Kinematic Performance Improvement of ZCDM

### 3.1   Direct Kinematics

The problem of the direct kinematics of ZCDM can be described as: with three known input angles $\theta_1$, $\theta_2$, and $\theta_3$, it is required to solve the orientation angle $\gamma$ and position $(x, y)$ of revolute joint $R_3$ of the moving platform 1.

The base coordinate system $o\text{-}xy$ is shown in Fig. 2, which is the same as in Fig. 1.

The coordinates of $R_3$ are easily obtained by using the positions of $R_{12}$ and $R_{32}$ below.

$$\begin{cases} x_{R_3} = \frac{-E \pm \sqrt{E^2 - 4DF}}{2D} \\ y_{R_3} = \frac{C}{B} - \frac{A}{B} x_{R_3} \end{cases} \tag{1}$$

Where,

$A = 2(l_1 \cos\theta_1 - l_8 - l_5 \cos\theta_3)$, $B = 2(l_1 \sin\theta_1 - l_5 \sin\theta_3)$, $C = l_4^2 - l_2^2 + l_1^2 - l_8^2 - l_5^2 - 2l_5 l_8 \cos\theta_3$,
$D = A^2 + B^2$, $E = 2l_1 AB \sin\theta_1 - 2l_1 B^2 \cos\theta_1 - 2AC$, $F = C^2 + l_1^2 B^2 - 2l_1 BC \sin\theta_1 - l_2^2 B^2$.

The coordinate of joint $R_{23}$ can be easily obtained by using joints $R_{22}$ and $R_3$ below

$$\begin{cases} x_{R_{23}} = \frac{-e \pm \sqrt{e^2 - 4df}}{2d} \\ y_{R_{23}} = \frac{c}{b} - \frac{a}{b} x_{R_{23}} \end{cases} \tag{2}$$

Here,

$a = 2(x_{R_3} - l_9 - l_7 \cos\theta_2)$,     $b = 2(y_{R_3} - l_{10} - l_7 \sin\theta_2)$,
$c = l_6^2 - l_3^2 + x_{R_3}^2 + y_{R_3}^2 - l_9^2 - l_7^2 - l_{10}^2 - 2l_{10} l_7 \sin\theta_2 - 2l_9 l_7 \cos\theta_2$,
$d = a^2 + b^2$, $e = 2y_{R_3} ab - 2x_{R_3} b^2 - 2ac$, $f = c^2 + (x_{R_3}^2 + y_{R_3}^2 - l_3^2) b^2 - 2y_{R_3} bc$.

Sequentially, angle $\gamma$ is expressed as

$$\tan \gamma = (y_{R_{23}} - y_{R_3})/(x_{R_{23}} - x_{R_3}) \tag{3}$$

According to Eq. (1), the position of the moving platform 1, *i.e.*, $(x_{R3}, y_{R3})$, is determined by only two input angles $\theta_1$ and $\theta_3$. It is also known from Eq. (3) that rotation angle $\gamma$ is the function of the three input angles $\theta_1$, $\theta_2$, and $\theta_3$. Therefore this 3-dof planar ZCDM has partially decoupled *I-O* motion property.

Inverse kinematics analysis of the ZCDM is simpler, which is not presented for saving the space.

As shown in Fig. 2, the structural parameters in the unit of *mm* from Ref. [4] of the ZCDM are shown as follows.

$$l_1 = l_5 = l_7 = 400, \, l_8 = 600, \, l_2 = l_3 = l_4 = l_6 = 300, \, l_9 = 1054.1, \, l_{10} = 1045.4$$

Let the input angles $\theta_1$, $\theta_2$, and $\theta_3$ be 60°, 240° and 70°, respectively, substituting the known parameters into Eqs. (1)–(3) gives two sets direct solutions, namely, $x = 461.1$, $y = 494.1$, $\gamma = -104.8544°$ and $x = 461.1$, $y = 494.1$, $\gamma = -20.0847°$. It is easy to verify the validation of these direct kinematics solutions by using the inverse kinematics.

### 3.2   Workspace Analysis

**Reachable Workspace.** Lengths of this ZCDM are as follows (units: *mm*).

$$l_1 = l_5 = l_7 = 200, \, l_2 = l_4 = l_6 = 200, \, l_3 = 100\sqrt{3}, \, l_8 = 300, \, l_9 = 150, \, l_{10} = 150\sqrt{3}$$

Through programming computation on MATLAB, the reachable workspace area of 3-RRR TPM is $3.5868 \times 10^7$ mm$^2$, and its shape is shown in Fig. 3(a). The reachable workspace area of the ZCDM is $2.6095 \times 10^7$ mm$^2$, and its shape is shown in Fig. 3(b).

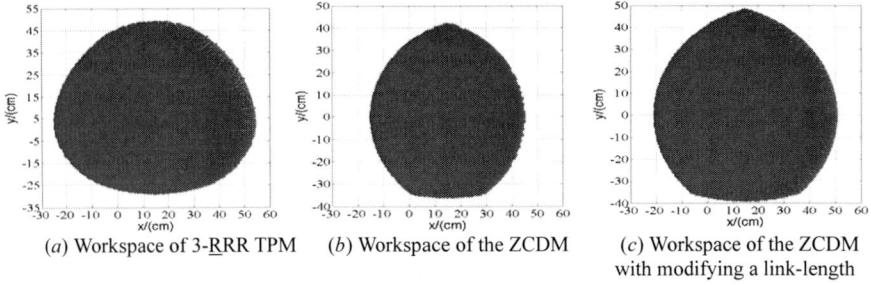

(a) Workspace of 3-RRR TPM     (b) Workspace of the ZCDM     (c) Workspace of the ZCDM with modifying a link-length

**Fig. 3.** Reachable workspace comparison between ZCDM and TPM

From the Fig. 3, it is found that the area of the ZCDM is 27.25% less than TPM. However, reachable workspace shape and size of the ZCDM can be changed and enlarged by means of increasing one or multiple link-length. For instance, when length of the links 2, 3, 4, 5, 6, and 7 are increased to one-sixth of the length of link 1, i.e., $l_3/6$, the area of the ZCDM will be $3.7948 \times 10^7$ mm², for which the workspace size is increased by 5.52% compared to the TPM. Moreover, the shape is symmetrical and forms a solid region, as shown in Fig. 3(c).

**Dexterous Workspace.** For the 3-<u>R</u>RR TPM shown in Fig. 1, the center of the moving platform 1 is taken as the base point $O'$. If the base point $O'$ in the range of dexterous workspace, the moving platform 1 can rotate completely around this base point. The dexterous workspaces of the ZCDM and 3-<u>R</u>RR TPM are calculated and shown in Fig. 4(a) and (b) respectively. We find that the dexterous workspace area of the ZCDM and TPM are $8.6567 \times 10^6$ mm², $6.3429 \times 10^6$ mm², respectively. Thus the former is 36.48% bigger than the later.

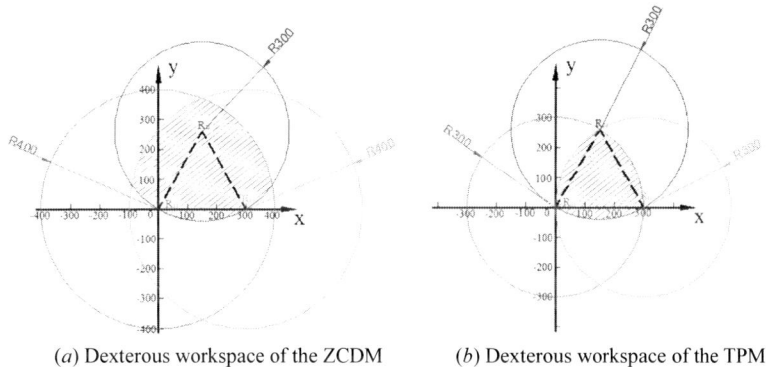

(a) Dexterous workspace of the ZCDM          (b) Dexterous workspace of the TPM

**Fig. 4.** Dexterous workspace comparison between the ZCDM and TPM

### 3.3   Singularity Analysis

**Calculate of $J_P$ and $J_q$ Matrix.** Let $V = [\dot{x} \quad \dot{y} \quad \dot{\gamma}]^T$ and $\omega = [\dot{\theta}_1 \quad \dot{\theta}_2 \quad \dot{\theta}_3]^T$ are the output velocity of the end-effect or and joint input angle velocity. Then we have

$$J_p V = J_q \omega \tag{4}$$

Where, $J_p = \begin{bmatrix} f_{11} & f_{12} & f_{13} \\ f_{21} & f_{22} & f_{23} \\ f_{31} & f_{32} & f_{33} \end{bmatrix}, J_q = \begin{bmatrix} u_{11} & & \\ & u_{22} & \\ & & u_{33} \end{bmatrix}$

$u_{11} = l_1(y_{R_{12}} - y_{R_3})\cos\theta_1 - l_1(x_{R_{12}} - x_{R_3})\sin\theta_1, u_{22} = l_7(y_{R_{22}} - y_{R_{23}})\cos\theta_2 - l_1(x_{R_{22}} - x_{R_{23}})\sin\theta_2,$

$u_{33} = l_5(y_{R_{32}} - y_{R_3})\cos\theta_3 - l_5(x_{R_{32}} - x_{R_3})\sin\theta_3;$

$f_{11} = x_{R_{12}} - x_{R_3}, f_{12} = y_{R_{12}} - y_{R_3}, f_{13} = 0;\ f_{21} = x_{R_{22}} - x_{R_{23}}, f_{22} = y_{R_{22}} - y_{R_{23}},$

$f_{23} = l_3(x_{R_{23}} - x_{R_{22}})\cos\gamma + l_3(y_{R_{23}} - y_{R_{22}})\sin\gamma;\ f_{31} = x_{R_{32}} - x_{R_3}, f_{32} = y_{R_{32}} - y_{R_3}, f_{33} = 0.$

The singular configuration of the PM could be classified into the following three types follows:

① When $\det(J_q) = 0$, input singularity occurs.
② When $\det(J_p) = 0$, output singularity occurs.
③ When $\det(J_q) = \det(J_p) = 0$, hybrid singularity occurs.

## Singularity Comparison Between the ZCDM and 3-RRR TPM

First, Input Singularity. When the input singularity happens, the movable platform 1 of this PM will lose its mobility along some directions. In this instantaneous moment, at least one motion chain reaches at the boundary of workspace, and we have $\det(J_q) = 0$.

The solution set $A$ of this equation is shown below

$$A = \{A_1 \cup A_2 \cup A_3\} \tag{5}$$

Here $A_1 = \{(y_{R_{12}} - y_{R_3})\cos\theta_1 - (x_{R_{12}} - x_{R_3})\sin\theta_1 = 0\}$, which means that three points $R_{11}$, $R_{12}$ and $R_3$ are collinear.

$A_2 = \{(y_{R_{22}} - y_{R_{23}})\cos\theta_2 - (x_{R_{22}} - x_{R_{23}})\sin\theta_2 = 0\}$, which means that three points $R_{23}$, $R_{22}$ and $R_{21}$ are collinear.

$A_3 = \{(y_{R_{32}} - y_{R_3})\cos\theta_3 - (x_{R_{32}} - x_{R_3})\sin\theta_3 = 0\}$, which means that three points $R_{31}$, $R_{32}$ and $R_3$ are collinear.

Second, Output Singularity. In this case, the movable platform 1 still has local motion when all actuated joints are locked. If a limited force is applied to the movable platform 1, the three input links need infinite actuated force to achieve force balance. Hence, we have $\det(J_P) = 0$, the solution of the set $B$ for this equation is shown below

$$B = \{B_1 \cup B_2\} \tag{6}$$

Here, $B_1 = \{(x_{R_{23}} - x_{R_{22}})\cos\gamma + (y_{R_{23}} - y_{R_{22}})\sin\gamma = 0\}$, which means that three points $R_{22}$, $R_{23}$ and $R_3$ are collinear. $B_2 = \{f_{12}f_{31} - f_{11}f_{32} = 0\}$, which means that three points $R_{12}$, $R_{32}$ and $R_3$ are collinear.

It is clear that from the discussion, the hybrid singularity analysis of the ZCDM is simpler than 3-RRR TPM.

The conclusion is obtained respectively by a comprehensive comparison of the input and output singularity of the two mechanisms, shown in Table 1.

Authors recently note that in addition to workspace analysis of ZCDM its joint force and transmission investigations were also been performed [10] and the ZCDM

**Table 1.** Performance comparison of the ZCDM and TPM

| Performance | ZCDM | TPM(3-RRR) |
|---|---|---|
| $k$ | 0 | 1 |
| Direct kinematics | Analytic | Numerical |
| *I-O* decoupling | Yes | No |
| Reachable workspace | Slightly smaller* | Bigger |
| Dexterous workspace | Big | Small |
| Singularity | Simple | Complex |

*Note: The size of reachable workspace of ZCDM could be improved or increased by magnified slightly the length of some links.

has a high rotation capability [11]. Further, an optimization using algebraic singularity equations gave more accurate and complete workspace analysis in Ref. [12].

From Table 1 and investigation results in Refs. [10–12], it is easy to found that the overall performance of the ZCDM is greatly improved and better that of 3-RRR TPM.

## 4   Conclusions

In this paper, two disadvantages of the traditional 3-RRR TPM could be overcome by a 3-dof planar ZCDM proposed, which is derived based on its topological structural optimization by using the method for reducing coupling degree of PM suggested by the author. The optimal result leads to the enhanced kinematic performance, (1) the analytical solutions for the direct kinematics of the ZCDM can be obtained because of $k = 0$. Its path planning and position control will be simpler due to the decoupled input-output motion, and (2) The overall kinematic performances of ZCDM are better than that of the 3-RRR TPM.

**Acknowledgments.** This research was possible thanks to support from the National Natural Science Foundation of China (Grants No. 51475050 and 51375062).

## References

1. Gosselin, C., Merlet, J.: Direct kinematics of planar parallel manipulators: special architectures and number of solutions. Mech. Mach. Theory **29**, 1083–1097 (1994)
2. Oetomo, D., Liaw, H.C., Alici, G., Shirinzadeh, B.: Direct kinematics and analytical solution to 3-RRR parallel planar mechanisms. In: International Conference on Control, pp. 1–6 (2006)
3. Gosselin, C.M., Angeles, J.: The optimal kinematic design of a planar three-degree-of-freedom parallel manipulator. J. Mech. Transm. Autom. Des. **110**(3), 35–41 (1988)
4. Wu, J., Wang, J.S., You, Z.: A comparison study on the dynamics of planar 3-DOF 4-DOF, 3-RRR and 2-RRR parallel manipulators. Robot. Comput.-Integr. Manuf. **27**(1), 150–156 (2011)

5. Li, D.H., Song, S.T.: Research of the reachable workspace of symmetrical planar 3-RRR parallel mechanism. J. Mech. Transm. (2015). ISSN 1004-2539, 09-0029-03
6. Gao, F., Liu, X.J., Chen, X.: The relationships between the shapes of the workspaces and the link lengths of 3-DOF symmetrical planer parallel manipulators. Mech. Mach. Theory **36**, 205–220 (2001)
7. Yang, T.L., Liu, A.X., Luo, Y.F.: Theory and Application of Robot Mechanism Topology. Science Press, Beijing (2012)
8. Yang, T., Liu, A.X., Shen, H.P., et al.: Topology Design of Robot Mechanisms. Springer, Singapore (2018)
9. Shen, H.P., Yang, L.J., Meng, Q.M., Yin, H.B.: Topological structure coupling-reducing of parallel mechanisms. In: The 14th IFToMM World Congress, Taipei, Taiwan, 25–30 October 2015 (2015)
10. Takeda, Y., Funabashi, H., Muramatsu, N.: Planar in-parallel actuated mechanism with three degrees of freedom with large practically dextrous working space (kinematic synthesis). Trans. JSME Ser. C **62**(599), 2920–2926 (1996)
11. Arakelian, V., Briot, S., Yatsun, S., Yatsun, A.: A new 3-DoF planar parallel manipulator with unlimited rotation capability. In: 13th World Congress in Mechanism and Machine Science, Guanajuato, México, 19–25 June 2011. IMD-123 (2011)
12. Chablat, D., Moroz, G., Arakelian, V., Briot, S., Wenger, P.: Solution regions in the parameter space of a 3-RRR decoupled robot for a prescribed workspace. In: Latest Advances in Robot Kinematics, pp. 357–364 (2012)

# Some Properties of the Irvine Cable Model and Their Use for the Kinematic Analysis of Cable-Driven Parallel Robots

Jean-Pierre Merlet$^{(\boxtimes)}$

HEPHAISTOS project, Université Côte d'Azur, Inria, Sophia-Antipolis, France
Jean-Pierre.Merlet@inria.fr

**Abstract.** Cable model has a strong influence on the complexity of the kinematic analysis of cable-driven parallel robots (CDPR). The most complete model relies on Irvine equation that takes into account both the elasticity and the deformation of the cable due to its own mass and has been shown to be very realistic. This model is complex, non algebraic and numerically ill-conditioned, thereby leading to difficulties when using it in a kinematic analysis involving several cables. We exhibit some properties of this model that may drastically improve the analysis computation time when used in kinematic studies.

**Keywords:** Cable-driven parallel robots · Cable model
Sagging cables

## 1  Introduction

In this paper we will consider the Irvine sagging cable model that has been proposed for elastic and deformable cable with mass [3] and that has been shown to be in very good agreement with experimental results [9]. This model assumes that the cable lies in a vertical plane, the *cable plane*, and is therefore a 2D model. A reference frame is defined in this plane with its origin at $A_i$, one of the extremity of the cable. The coordinates of the other cable extremity $B_i$ are $(x_b \geq 0, z_b)$ and we will assume that $B_i$ is below $A_i$ so that $z_b \leq 0$ (Assumption 1). Vertical and horizontal forces $F_z, F_x > 0$ are exerted on the cable at point $B_i$ (Fig. 1).

For a cable with length at rest $L_0$ the coordinates of $B$ are given by:

$$x_b = F_x \left( \frac{L_0}{EA_0} + \frac{sinh^{-1}(F_z) - sinh^{-1}(F_z - \frac{\mu g L_0}{F_x})}{\mu g} \right) \tag{1}$$

$$z_b = \frac{F_z}{EA_0} - \mu g L_0^2 / 2 + \frac{\sqrt{F_x^2 + F_z^2} - \sqrt{F_x^2 + (F_z - \mu g L_0)^2}}{\mu g} \tag{2}$$

where $E$ is the Young modulus of the cable material, $A_0$ the cable cross-section area and $\mu$ the cable linear density.

© Springer Nature Switzerland AG 2019
B. Corves et al. (Eds.): EuCoMeS 2018, MMS 59, pp. 409–416, 2019.
https://doi.org/10.1007/978-3-319-98020-1_48

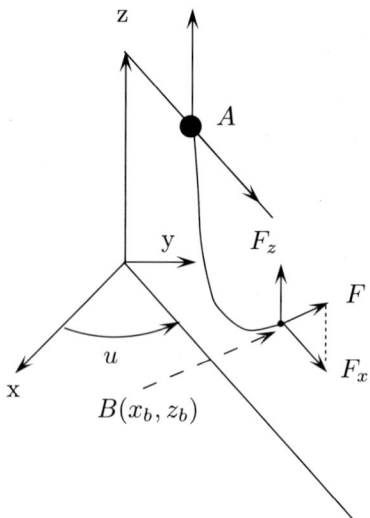

**Fig. 1.** Notation for a sagging cable

Studying the properties of the Irvine cable model is justified by its use in the kinematic analysis of cable-driven parallel robots (CDPR). The model will obviously play an essential role for the inverse and direct kinematics (IK and DK) which in turn plays a role in workspace and singularity analysis. This model also influences the static analysis whose purpose is to determine the tension in the cables [2]. Previous works have focused on the analysis of Eqs. (1, 2) whose unknowns are $x_b, z_b, L_0, F_x, F_z$, especially assuming that 3 of these 5 unknowns have a fixed value [7] in which case the solution is unique but has to be determined numerically (case 1). Other works have addressed the even more complex case of the IK of CDPR in which $n$ cables are attached to a rigid body in a known pose (hence the cable plane and the $x_b, z_b$ of each cable are known) with the purpose of determining $L_0$. Here we have a system of $2n$ Eqs. (1, 2) with $3n$ unknowns but the mechanical equilibrium of the platform imposes 6 additional equations. If $n = 6$ we end-up with a square system of equations [4,8,10] (case 2). For the IK solving authors have used optimization or have assumed that the solution is sufficiently close to the rigid leg case which is therefore used as initial guess for a solving based on the Newton scheme. However these methods cannot guarantee to find the solution in case 1 or all solutions in case 2 where there may be multiple solutions. The problem is even more complex for the DK of CDPR: in that case the kinematic constraints are always a square system that has usually multiple solutions (note: the DK assumes the measurement of $L_0$ while current systems provided the stressed length so that corrective steps should theoretically be applied). We have addressed these issues in previous publications using as solving method an interval analysis-based approach that is guaranteed to provide all solutions assuming that the unknowns are bounded [5,6]. However the efficiency of this approach is heavily dependent on a careful modeling and analysis of the equations at hand, transforming a problem that is almost intractable to one that may be solved in a few seconds.

Interval analysis is based on *interval evaluation* of a function $f$ in the unknowns $\{x_1, x_2, \ldots x_n\}$ that are supposed to be bounded i.e. for each $x_i$ we have $x_i \in [\underline{x_i}, \overline{x_i}]$ where $\underline{x_i}, \overline{x_i}$ are respectively the lower and upper bound for $x_i$. Such bounds define a *box* in the $n$-dimensional space of the unknowns. Being given such a box $\mathcal{B}$ the interval evaluation $\hat{f}$ of $f$ over $\mathcal{B}$ is an interval $[\underline{f}, \overline{f}]$ such that for any point $X$ in $\mathcal{B}$ we have $\underline{f} \leq f(X) \leq \overline{f}$. In other words $\underline{f}$ is either equal to or a minorant of the minimum $f_{min}$ of $f$ over $\mathcal{B}$ while $\overline{f}$ is equal to or a majorant of the maximum $f_{max}$ of $f$ over $\mathcal{B}$. The interval evaluation of $f$ is relatively easy to obtain if $f$ is expressed in terms of classical mathematical functions using the *natural evaluation* which basically consist in replacing the operators by interval equivalents. For example interval evaluation of the Irvine equations may be obtained by natural evaluation. However the efficiency of interval algorithms is drastically dependent upon the *tightness* of the interval evaluation: the closer $\underline{f}, \overline{f}$ are to $f_{min}, f_{max}$, the faster will be the algorithm. An interval evaluation will be denoted *tight* if $\hat{f} = [f_{min}, f_{max}]$. But the natural evaluation may lead to large under or overestimation of the minimum and maximum as soon as there are multiple occurrences of the unknowns in $f$ (it may be proven that if there is only a single occurrence of each unknown in $f$, then $\hat{f}$ is tight, up to round-off errors). The tightness will improve when the widths of the intervals for the unknowns decrease but an efficient way to improve the tightness of the evaluation is to consider the derivatives of $f$ and their own interval evaluation. Let $f_i$ be the derivative of $f$ with respect to $x_i$ and let $[\underline{f_i}, \overline{f_i}]$ be its interval evaluation over $\mathcal{B}$. If $\underline{f_i} > 0$ or $\overline{f_i} < 0$, then $f$ is monotonic with respect to $x_i$. Consequently $\hat{f}$ may be obtained as $[\text{Min}\hat{f}(\mathcal{B}_i), \text{Max}\hat{f}(\mathcal{B}_i)]$ where $\mathcal{B}_i$ are the boxes that are derived from $\mathcal{B}$ with $x_i$ set to $\underline{x_i}$ or $\overline{x_i}$. Note that this process has to be applied recursively. Indeed assume that there is a $j > 1$ such that $f$ is monotonic with respect to $x_j$ (implying that $\hat{f}$ will be obtained using $\mathcal{B}_j$), while for $i < j$ this was not the case. But for $i < j$ the monoticity has been evaluated using $\mathcal{B}$ and as we are now using the tighter $\mathcal{B}_j$ the monoticity test may give another result. Using this process we may tighten the interval evaluation of $f$ up to the point where $\hat{f} = [f_{min}, f_{max}]$ if $f$ is such that all $f_i, i \in [1, n]$ are positive or negative. We present in the next sections some interesting properties of the Irvine equations that can be used for analysis or solving purposes.

## 2    Properties of the Irvine Equations

A preliminary property will play an important role: we have assumed that $B$ has an altitude that is equal or lower to the one of $A$ with the direct consequence that $F_z \leq \mu g L_0/2$.

## 2.1    Derivatives of the Irvine Equations

The sign of the derivatives of the Irvine equations may be obtained with interval evaluation but it is interesting to determine beforehand if they may be inherently monotonic.

Under assumption 1 we may establish the sign of derivatives of Eqs. (1), (2) that will be presented without proof as they are trivial. We have

$$\frac{\partial z_b}{\partial L_0} < 0 \qquad \frac{\partial z_b}{\partial F_z} > 0 \qquad \frac{\partial z_b}{\partial F_x} > 0 \tag{3}$$

As all derivatives of $z_b$ have a constant sign, then its interval evaluation for interval values for $F_x, F_z, L_0$ will always be tight and can be computed efficiently using only floating point operators. This may have an impact on the IK solving in which $z_b$ has a fixed value: If $\hat{z}_b \cap z_b = \emptyset$, then (2) has no solution for the current $F_x, F_z, L_0$ box. We have also:

$$\frac{\partial x_b}{\partial L_0} > 0 \qquad \frac{\partial x_b}{\partial F_z} > 0 \qquad \frac{\partial x_b}{\partial F_x} > 0 \tag{4}$$

It may also be interesting to consider the distance $D = x_b^2 + z_b^2$ between $A$ and $B$. We have $\partial D / \partial L_0 > 0$ but no general monotonicity can be obtained with respect to $F_x, F_z$.

Let $F_x, F_z, L_0$ being bounded i.e. $F_x \in [\underline{F_x}, \overline{F_x}]$, $F_z \in [\underline{F_z}, \overline{F_z}]$, $L_0 \in [\underline{L_0}, \overline{L_0}]$. Let us assume that $z_b$ is fixed and consider the equation $f(L_0, F_z, F_x) - z_b = 0$. Using the implicit value theorem it may be shown that the solution of this equation satisfies

$$\frac{\partial F_x}{\partial L_0} > 0 \qquad \frac{\partial F_x}{\partial F_z} < 0$$

so that $F_x$ is restricted to lie in the interval $[\underline{F_x'}, \overline{F_x'}]$ where $\underline{F_x'}$ is the solution of (2) obtained for $L_0 = \overline{L_0}$, $F_z = \underline{F_z}$ and $\overline{F_x'}$ is the solution of (2) obtained for $L_0 = \underline{L_0}$, $F_z = \overline{F_z}$. The range for $F_x$ may therefore be calculated as $[\underline{F_x}, \overline{F_x}] \cap [\underline{F_x'}, \overline{F_x'}]$ and the equation has no solution if this intersection is empty. More generally if we consider (2) when 2 of the unknowns are fixed and denotes by $S$ its solution in the last unknown we get $[\underline{L_0'}, \overline{L_0'}] = [S(\overline{F_x}, \overline{F_z}), S(\underline{F_x}, \underline{F_z})]$ and $[\underline{F_z'}, \overline{F_z'}] = [(S(\underline{F_x}, \overline{L_0}), S(\overline{F_x}, \underline{L_0})]$.

## 2.2    New Forms for the Irvine Equation

A property of interval analysis is that two mathematically equivalent forms of $f$ may have different interval evaluations. For example $f^1 = x^2 + 2x + 1$ and $f^2 = (x+1)^2$ are equivalent but $\hat{f}^2$ will be tight with only one occurrence of $x$ while $\hat{f}^1$ will not if $\underline{x} < 0$. Therefore interval analysis algorithms are partly based on heuristics that compute various interval evaluations of the same $f$ expressed in different ways and returning $\hat{f}$ as their intersection. We present in this section various new relationships between the quantities appearing in the Irvine equations. They are usually expressed in implicit form $G(x_b, z_b, F_x, F_z, L_0) = 0$ and the interval evaluation of $G$ may allow one to discard boxes in an interval analysis algorithm but it may also happen that the analysis of $G$ provides bounds for one variable being given interval values for the other unknowns.

## Using the $z_b$ Equation

### $F_x$ as function of $z_b, F_z, L_0$. Let

$$a^2 = F_x^2 + F_z^2 \quad b^2 = F_x^2 + (F_z - \mu g L_0)^2 \quad a^2 - b^2 = \mu g L_0 (2F_z - \mu g L_0)$$

then $z_b$ may be written as

$$z_b = (\frac{(a-b)}{\mu g})(\frac{(a+b)}{2EA_0} + 1) = \frac{L_0(2F_z - \mu g L_0)}{2EA_0} + \frac{a-b}{\mu g} \tag{5}$$

Let us assume now that $z_b, F_x, L_0$ are given so that (2) has only $F_x$ as unknown. Our objective is to get an expression of this unknown. Let us define

$$a^2 = F_x^2 + F_z^2 \quad b^2 = F_x^2 + (F_z - \mu g L_0)^2 \quad U = \frac{L_0(F_z - \mu g L_0/2)}{EA_0} - z_b$$

so that Eq. (2) may be written as

$$U + \frac{(a-b)}{\mu g} = 0 \tag{6}$$

We have also

$$a^2 - b^2 = 2F_z \mu g L_0 - (\mu g L_0)^2 = V = (a+b)(a-b) = (a+b)(-U\mu g)$$

from which we get

$$b = -\frac{V}{U\mu g} - a$$

Reporting $b$ in (6) leads to

$$2a = -U\mu g - \frac{V}{U\mu g} = W \tag{7}$$

Note that $U, V$ are not function of $F_x$ so that $W$ is expressed only as a function of $F_z, L_0$. As $a^2 = (W/2)^2 = F_x^2 + F_z^2$ we get

$$F_x^2 = (W/2)^2 - F_z^2 \tag{8}$$

where the right-hand term is a function of $F_z, L_0$ only. This equation provides $F_x$ if $z_b, L_0, F_z$ are fixed. Let's assume now that $F_z$ has an interval value and consider $P = F_x^2 = (W/2)^2 - F_z^2$. Our problem is to determine the value of $F_z$ so that $P > 0$. The polynomial $P$ is of degree 4 in $F_z$ and factors out in 4 terms that are linear in $F_z$. The root of $P$ are

$$s_1 = \frac{\mu g L_0}{2} + \frac{\mu g A_0 E z_b}{2A_0 E + \mu g L_0} \quad s_2 = \frac{\mu g L_0}{2} + (z_b - L_0)\frac{A_0 E}{L_0}$$

and

$$s_3 = \frac{\mu g L_0}{2} + (z_b + L_0)\frac{A_0 E}{L_0} \quad s_4 = \frac{\mu g L_0}{2} - \frac{\mu g A_0 E z_b}{2A_0 E - \mu g L_0}$$

If we assume $2A_0 E > \mu g L_0$ then the roots in $F_z$ are ordered as $s_2, s_1(< \mu g L_0/2), s_4(> \mu g L_0/2), s_3$ and $P$ will be positive if $F_z \in [s_2, s_1]$. If $2A_0 E < \mu g L_0$ then the roots are ordered as $s_2, s_4(< \mu g L_0/2), s_1(< \mu g L_0/2), s_3$. Therefore there are 2 possible ranges for $F_z$ leading to a positive $P$: $[s_2, s_4]$, $[s_1, \mu g L_0/2]$.

$L_0$ **as function of** $z_b, F_z, F_x$. We are now interested in determining $L_0$ when $F_x, F_z, z_b$ are fixed. Let $U_1 = \sqrt{F_x^2 + F_z^2}$, $U_2 = \mu g F_z/EA_0$, $U_3 = (\mu g)^2/(2EA_0)$ and $U_4 = -\mu g z_b + U_1$. Equation (2) may be written as

$$U_2 L_0 - U_3 L_0^2 + U_4 = \sqrt{F_x^2 + (F_z - \mu g L_0)^2} \qquad (9)$$

Squaring the previous equation leads to

$$P_s = (U_2 L_0 - U_3 L_0^2 + U_4)^2 - (F_x^2 + (F_z - \mu g L_0)^2) = 0 \qquad (10)$$

As $U_1, U_2, U_3, U_4$ are not function of $L_0$ this equation is a fourth order polynomial in $L_0$. Using the Sturm sequences it is possible to show that $P_s$ has only 2 roots in the range $[0, \infty]$. But it may be seen that (10) leads to 2 possibilities for (9) namely

$$U_2 L_0 - U_3 L_0^2 + U_4 = \pm\sqrt{F_x^2 + (F_z - \mu g L_0)^2}$$

The negative version leads to $z_b > 0$ which is not valid under assumption 1 and therefore solving $P_s$ (whose roots may be obtained in analytical form) leads to a single solution for $L_0$.

$F_z$ **as function of** $z_b, F_x, L_0$. We consider determining $F_z$ for given $L_0, F_x, z_b$. Equation (8) is a 4th order polynomial $Q$ in $F_z$ with the constraint that $W > 0$. Using Budan-Fourier theorem [1] it is possible to show that $Q$ has 0 or 2 roots in the range $]-\infty, \mu g L_0/2]$ but only one this root will lead to a positive $W > 0$. The analysis of the sign of $W$ is complex but it may be shown that if $EA_0 \gg \mu g L_0$, then $F_z$ must belong to the range $[\mu g L_0/2 + EA_0 z_b/L_0, \mu g L_0/2 - \mu g z_b^2/(2L_0)]$.

**Using the** $x_b$ **and** $z_b$ **Equations**
$F_z$ **as function of** $x_b, z_b, F_x, L_0$. First we will consider the calculation of $F_z$ being given $F_x, L_0, x_b, z_b$. We define

$$u = \frac{F_z}{F_x} \quad v = \frac{F_z - \mu g L_0}{F_x}$$

so that Eq. (1) may be written as

$$(\frac{x_b}{F_x} - \frac{L_0}{EA_0})\mu g = sinh^{-1}(u) - sinh^{-1}(v) \qquad (11)$$

We define $H_1 = x_b/F_x - L_0/(EA_0)$ and considering that $sinh^{-1}(u) - sinh^{-1}(v) = sinh^{-1}(u\sqrt{1 + v^2} - v\sqrt{1 + u^2})$ and taking the hyperbolic sine of both terms of Eq. (11) we obtain:

$$H = sinh(H_1)\mu g = u\sqrt{1 + v^2} - v\sqrt{1 + u^2} \qquad (12)$$

We have already defined $a^2 = F_x^2 + F_z^2$, $b^2 = F_x^2 + (F_z - \mu g L_0)^2$ so that $a^2/F_x^2 = 1 + u^2$ and $b^2/F_x^2 = 1 + v^2$. Equation (12) may therefore be written as:

$$F_x H = ub - va \qquad (13)$$

Note that the left-hand term of this equation is not a function of $F_z$. We have already established in Sect. 2.2 the values of $a, b$ as function of $z_b, L_0, F_z$ while $u, v$ are functions of $F_x, L_0, F_z$. Hence the right-hand term of (13) is a function of $z_b, L_0, F_x, F_z$. This function is a third order polynomial $P_3$ in $F_z$. By using the Sturm sequence [1] and the constraint $a > 0$ it is possible to show that $P_3$ has a single real root in the range $[-\infty, \mu g L_0/2]$.

$z_b$ **as a function of** $z_b, F_x, F_z, L_0$.  Equation (2) provides a mean of calculating $z_b$ when $F_x, F_z, L_0$ are known but does not involve $x_b$. We provide here another form that involves $x_b$. Using the notation and result of Sect. 2.2 we get:

$$z_b = \frac{L_0}{F_z}\left(\sqrt{F_x^2 + F_z^2} - \frac{F_x^2}{\mu g L_0}sinh(\mu g(\frac{x_b}{F_x} - \frac{L_0}{EA_0}))\right) + \frac{L_0(F_x - \mu g L_0/2)}{EA_0} \qquad (14)$$

Note that we may also obtain a bound on the cable tension $\sqrt{F_x^2 + F_z^2}$ at $A$ as

$$\sqrt{F_x^2 + F_z^2} = \frac{F_x^2}{\mu g L_0}sinh(\mu g(\frac{x_b}{F_x} - \frac{L_0}{EA_0})) + F_z(\frac{z_b}{L_0} + \frac{\mu g L_0}{2EA_0} - \frac{F_z}{EA_0}) \qquad (15)$$

## 2.3   Using the Cable Tangents

Sensors may provide measurements of the cable tangents $v = (F_z - \mu g L_0)/F_x$ at $A$ and $u = F_z/F_x$ at $B$. Under the assumption that $u, v$ are known we get

$$F_x = \frac{\mu g L_0}{(u - v)} \qquad F_z = uF_x \qquad F_x^2 + F_z^2 = \left(\frac{\mu g L_0}{(u - v)}\right)^2(1 + u^2) \qquad (16)$$

A trivial transformation of (2) leads to:

$$\mu g L_0^2(u + v) + 2 A_0 E L_0(\sqrt{u^2 + 1} - \sqrt{v^2 + 1}) + 2 z_b EA_0(v - u) = 0 \qquad (17)$$

which is a quadratic polynomial in $L_0$ whose coefficients are functions of $u, v, x_b$. It is easy to show that this polynomial has a single positive root. Now Eq. (1) may be written as

$$F_x\left(\frac{L_0}{EA_0} + \frac{(sinh^{-1}(u) - sinh^{-1}(v))}{\mu g}\right) - x_b = 0 \qquad (18)$$

As we have $F_x = \mu g L_0/(u - v)$ this equation may be transformed in a second order polynomial in $L_0$ whose coefficients are functions of $u, v, x_b$. Here again it is easy to show that this polynomial has at most one positive root.

As $F_z = uF_x$ and $L_0 = (F_x(u - v))/(\mu g)$ Eqs. (11), (18) are polynomials in $F_x$ with coefficients that are function of $u, v$. The resultant of these equations in

$F_x$ establishes a polynomial relationship between $x_b$, $z_b$ which is a quadric, more precisely a *parabola* which is written as

$$(Ax_b + Cz_b)^2 + Dx_b + Fz_b = 0$$

with $R_1 = (sinh^{-1}(u) - sinh^{-1}(v))/(\mu g)$, $R_2 = \sqrt{1+u^2} - \sqrt{1+v^2}$, $A = \sqrt{\mu g}(u - v)(u + v)$, $C = -2\sqrt{\mu g}(u - v)$, $D = 2EA_0(u - v)(R_1\mu g(u + v) - 2R_2)R_2$, $F = -2\mu g EA_0(u - v)(R_1\mu g(u + v) - 2R_2)R_1$. Note that if $EA_0 \gg \mu g L_0$, then $A, C$ are small and $D, F$ very large so that the parabola is very close to a line.

## 3   Conclusion

We have presented in this paper various results regarding the Irvine equations that may be useful both for the analysis and solving of kinematic equations that rely on this cable model as they establish a more general view of the underlying structure of this model. We have already implemented some of these results in our CDPR IK and DK solver with a strong influence on the solving time as soon as we are interested in analyzing all aspects of the Irvine equations. Real-time computation is not really an issue whatever model is used as an initial guess for the solution will be known and guaranteed Newton scheme exists. Still open issues on the CDPR with sagging cables such as workspace and singularity analysis have to be investigated with this new approach to the Irvine equations.

## References

1. Ciarlet, P., Lions, J.L.: Handbook of Numerical Analysis: Solution of Equations in Rn (part 3), vol. 7. North-Holland, Amsterdam (2000)
2. Hui, L.: A giant sagging-cable-driven parallel robot of FAST telescope: its tension-feasible workspace of orientation and orientation planning. In: 14th IFToMM World Congress on the Theory of Machines and Mechanisms. Taipei, 27–30 October 2015
3. Irvine, H.M.: Cable Structures. MIT Press, Cambridge (1981)
4. Kozak, K., et al.: Static analysis of cable-driven manipulators with non-negligible cable mass. IEEE Trans. Robot. **22**(3), 425–433 (2006)
5. Merlet, J.P.: The kinematics of cable-driven parallel robots with sagging cables: preliminary results. In: IEEE International Conference on Robotics and Automation, Seattle, pp. 1593–1598, 26–30 May 2015
6. Merlet, J.P.: On the inverse kinematics of cable-driven parallel robots with up to 6 sagging cables. In: IEEE International Conference on Intelligent Robots and Systems (IROS), Hamburg, Germany, pp. 4536–4361, 28 September–2 October 2015
7. Papini, D.: On shape control of cables under vertical static loads. Master's thesis, Lund University, Lund (2010)
8. Riehl, N., et al.: Effects of non-negligible cable mass on the static behavior of large workspace cable-driven parallel mechanisms. In: IEEE International Conference on Robotics and Automation, Kobe, pp. 2193–2198, 14–16 May 2009
9. Riehl, N., et al.: On the determination of cable characteristics for large dimension cable-driven parallel mechanisms. In: IEEE International Conference on Robotics and Automation, Anchorage, pp. 4709–4714, 3–8 May 2010
10. Sridhar, D., Williams II, R.: Kinematics and statics including cable sag for large cables suspended robots. Global J. Res. Eng.: H Robot. Nano-Tec **17**(1), 1–18 (2017)

# Synthesis of Scalable Planar Scissor Linkages with Anti-parallelogram Loops

Şebnem Gür, Cevahir Karagöz, Gökhan Kiper$^{(\boxtimes)}$,
and Koray Korkmaz

İzmir Institute of Technology, İzmir, Turkey
{sebnemgur, cevahirkaragoz, gokhankiper,
koraykorkmaz}@iyte.edu.tr

**Abstract.** Scissor linkages are commonly used as mechanisms for scaling objects. They constitute a significant portion of deployable structures. Since 1960s many researchers sought to form novel structures using the scissor units. In 1990s Hoberman brought a new perspective to the field when he used the loops, not the units to form linkages. Starting from mid 2000s, other researchers joined into this new approach of design. One of the latest researches presented a design for scaling a circular forms with anti-parallelogram loops. This study shows that an anti-parallelogram loop assembly can also be used for scaling planar curves with variable curvature.

**Keywords:** Scalable scissor linkage · Anti-parallelogram loop
Loop assembly method

## 1 Introduction

Mechanisms that can transform between an open (deployed) and a closed (stowed) configuration are called deployable structures [8]. Scissor linkages are widely used for deployable structures. Scissor-like elements (SLEs) are ternary links with revolute joints. Two SLEs connected at the middle joint constitute a scissor unit. The first academic study on SLEs was conducted by Pinero in 1961 for his design of a deployable theater composed of pantographic elements [17]. Escrig defined the foldability conditions [2, 3]. The subject remains to be of interest today for many other researchers [13, 16, 22].

A special type of deployable structures is scaling mechanisms. As the name suggests these mechanisms maintain a specific form while they expand and shrink. Hoberman introduced angulated SLEs in 1990 for scaling mechanisms [10, 11]. Endpoints of an angulated SLE follows radial lines during deployment while preserving the angle in between provided that the SLE satisfies certain geometrical conditions. In 1997 You and Pellegrino [20] reported the necessary conditions for deployment with subtended angles and introduced Generalized Angulated Elements (GAEs) of Type I and Type II.

© Springer Nature Switzerland AG 2019
B. Corves et al. (Eds.): EuCoMeS 2018, MMS 59, pp. 417–424, 2019.
https://doi.org/10.1007/978-3-319-98020-1_49

Hoberman's invention also introduces a novel approach: assembling loops to synthesize structures. Unlike the commonly used method of assembling scissor units, Hoberman places rhombus loops on a closed curve and determines the link lengths of the angulated SLEs using the edges of the rhombi [9, 10]. Liao and Li [14], Kiper and Söylemez [12] also made use of rhombi for scaling planar graphs and polygons. Yar et al. [19] used kite and dart loops and Gür et al. [5] used anti-parallelogram loops to obtain planar linkages comprising SLEs.

## 2  Assembly of Loops

There are several ways to classify scissor units used in scissor linkages. In the most common classification scissor linkages are composed of three types of scissor units: transitional, polar and angulated units. Another classification by Zhang et al. [21] reveals three corresponding types: parallel, symmetric and isogonal units. Basically, these units are used for altering either the linear lengths, or the circular arc length or the curvature of a curve segment. Maden et al. [15] present the alternative assemblies of scissor units and provide analysis and design criteria. When the scissor linkages in the literature are examined, it is seen that rhombus, kite, dart, parallelogram, anti-parallelogram and general convex and concave shaped loops can be found. As an example, a translational linkage with rhombus loops and a polar linkage with kite loops are seen in Fig. 1.

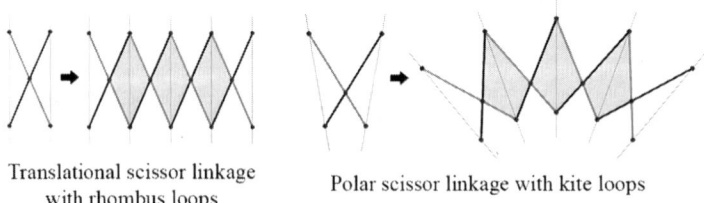

Translational scissor linkage
with rhombus loops

Polar scissor linkage with kite loops

**Fig. 1.** Loop geometries of two scissor structures produced with primary scissor elements

In order to form a scalable polygon, Hoberman [9] places rhombi on the edges of a polygon, which is usually a segmented approximation of a continuous curve (Fig. 2). In this study, anti-parallelogram loops are used to form scaling linkages.

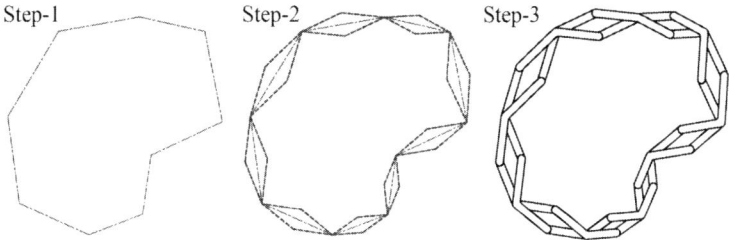

Step-1          Step-2          Step-3

**Fig. 2.** Loop assembly method as shown in MIT Class 6.S080 by Hoberman [9]

## 3    Anti-parallelogram Loop Assemblies

Anti-parallelogram, also known as crossed parallelogram or contraparallelogram, has two short and two crossing long edges of equal length. A linkage forming an anti-parallelogram loop can be folded flat in both directions of the motion. Mirror symmetry of the loop about a vertical axis is preserved during deployment. The symmetry axis passes through the crossing point of the long edges (Fig. 3).

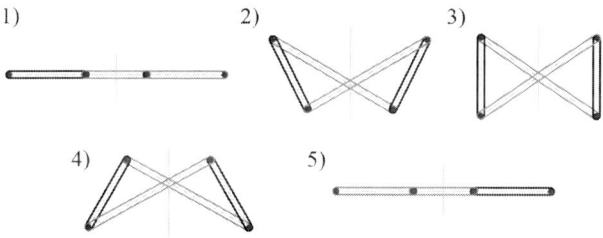

1)    2)    3)

4)    5)

**Fig. 3.** Motion of anti-parallelogram loop

The loops in an assembly are connected at the corners which symbolize the joint positions. There are many possible alternatives of connecting the loops depending on the choice of connection corner and also their rotation and configuration. The alternatives of anti-parallelogram loop arrays were studied by Gür in detail and a real-life application of one of the linkages resulting from the study was published in 2017 [4, 6]. In another former study of Gür et al. [5] an assembly with an alternating order of anti-parallelogram loops was used to form ring-like structures with radial deployment. This study presents the use of anti-parallelogram loops to form linkages for scaling planar curves with variable curvature using the same array type used in [5] for ring-like structures (Fig. 4). SolidWorks software is used to create a kinematic model and visualize the motion. It is observed that the curve form is preserved during the motion (Fig. 5).

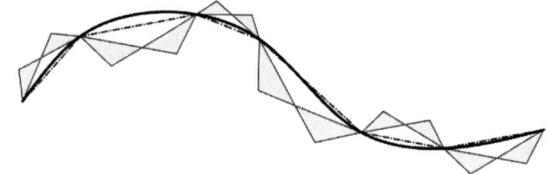

**Fig. 4.** Anti-parallelogram loop assembly on a polyline with variable curvature

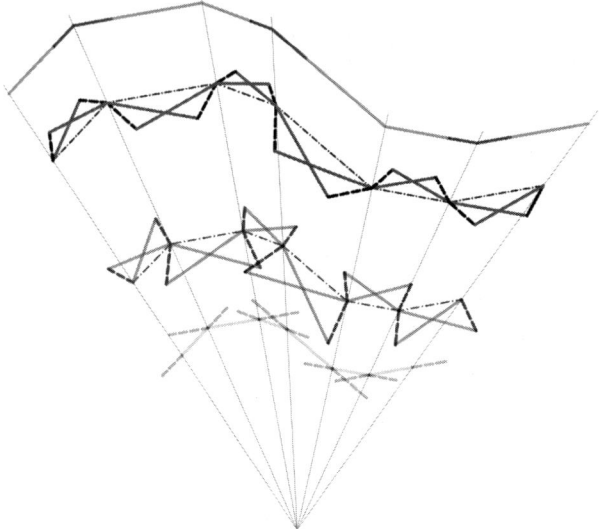

**Fig. 5.** Deployment of the linkage

When a planar curve is given, it can be discretized at some points to approximate it as a polyline. The discretization points can be selected to be equally spaced on the curve, or the spacing can be decided based on the change in curvature [7] or according to special design criteria of a specific problem. Since the main characteristics of a curve is how it is curved as one moves on the curve, the curvature must be somehow represented with the polyline approximation. This can be done considering the centers of the circles passing through every three consecutive points of the polyline or considering the angle between two consecutive line segments of the polyline (Fig. 6). The center of a circle passing through three points is the intersection of perpendicular bisectors of two pairs of points. Hoberman makes uses of these "normal lines" in [11] for his construction with rhombi. For most scissor units, the normal lines pass through joints, however this is not the general case. In [1, 21] unit lines are defined as the lines connecting the joint pairs on the two sides of a scissor unit, but these unit lines do not always coincide with normal lines and may not correspond to any geometrical characteristics of the approximated curve.

For an anti-parallelogram, the vertical symmetry axis, which does not pass through any joints but passes through the intersection of the crossing links, corresponds to the normal lines when the loops are assembled on a polyline.

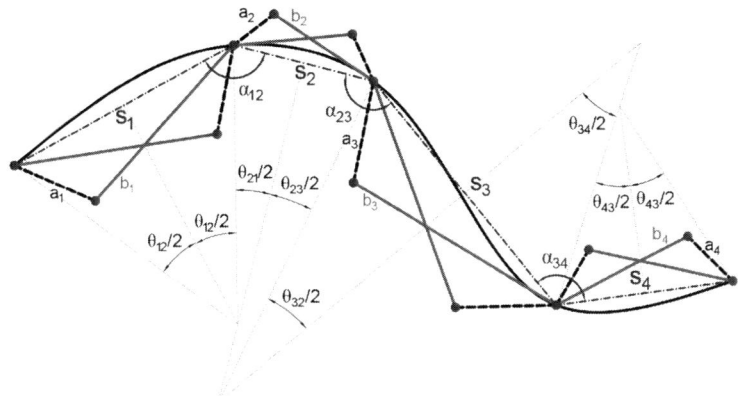

**Fig. 6.** Parameters of the linkage

Let the line segment lengths of a polyline be represented by $S_n$ for segment n. The ratio of $n^{th}$ segment length to the first one can be determined as $k_n$:

$$\frac{S_2}{S_1} = k_2, \quad \frac{S_3}{S_1} = k_3 \quad \cdots \quad \frac{S_n}{S_1} = k_n$$

The link lengths of the anti-parallelogram loops are determined using the ratio between the short ($a_1$) and long edge ($b_1$). The ratio R should be equal for all loops on every segment, hence resulting in GAEs of Type II, i.e. similar GAEs:

$$R = \frac{b_1}{a_1} = \frac{b_2}{a_2} = \cdots = \frac{b_n}{a_n}$$

At the fully deployed form, the sum of the link lengths is equal to segment lengths. Therefore, the kink angle $\alpha_{nn+1}$ of an angulated SLE meeting at the vertex of a polyline is simply the angle between the segments meeting at the vertex. The kink angle at a vertex is equal to the summation of halves of subtended angles of the neighboring segments. $\alpha_{nn+1} + (\theta_{nn+1} + \theta_{n+1n})/2 = 180°$ in Fig. 6. Since the kink angles of a pair of angulated SLEs meeting at a vertex are equal to each other, all anti-parallelogram loops deploy with the same ratio during the motion, hence resulting in a scaling linkage. A more detailed proof can be found in [5]. The remaining link lengths can be found as follows:

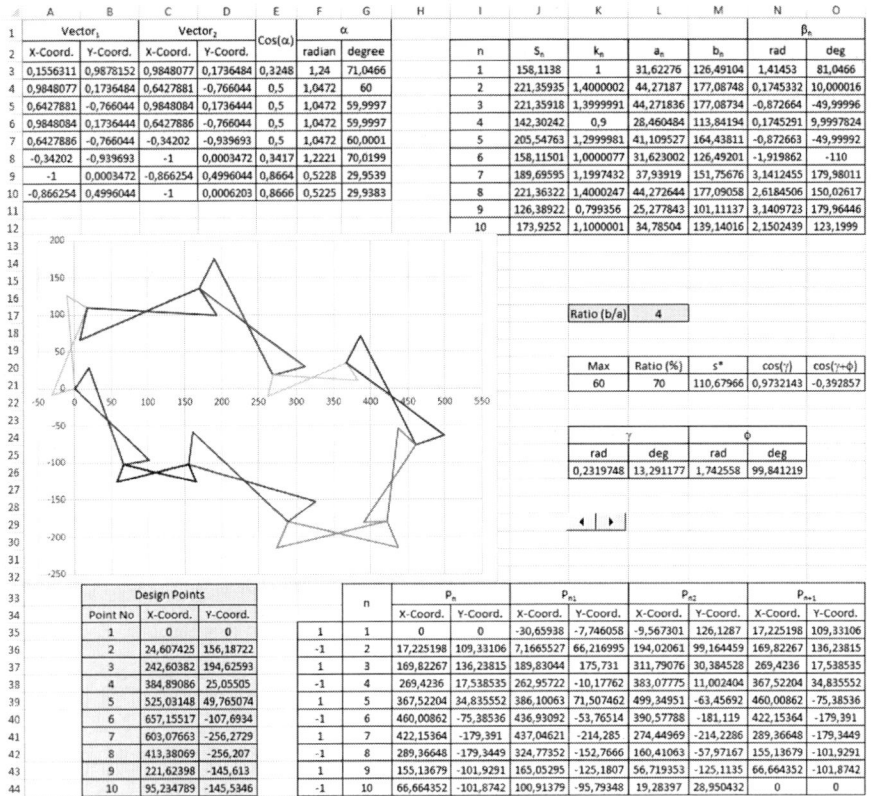

**Fig. 7.** Kinematic analysis in Excel

$$S_n = a_n + b_n = a_n(1 + R) \quad \Rightarrow \quad a_n = \frac{S_n}{1 + R} \quad , \quad b_n = S_n - a_n$$

Also the maximum deployed-to-compact form ratio can be found as:

$$\text{Compactness ratio: } 100 \times \frac{b_n - a_n}{b_n + a_n} = 100 \times \frac{R - 1}{R + 1}$$

The ratio R and compactness ratio can be utilized as design measures. Once the link lengths are decided, kinematic analysis of the resulting linkage can be performed. Derivation of the kinematic analysis formulations are straightforward (see for ex. [18]) and are not presented here for conciseness. The formulations are implemented in Microsoft Office Excel (Fig. 7). Cells highlighted in blue color are the inputs. The fully deployed and partially deployed forms of a polyline with a closed contour, i.e. a polygon, can be seen in Fig. 8. In Fig. 8, the link lengths ratio is R = 2.5, but any ratio can be selected by the designer.

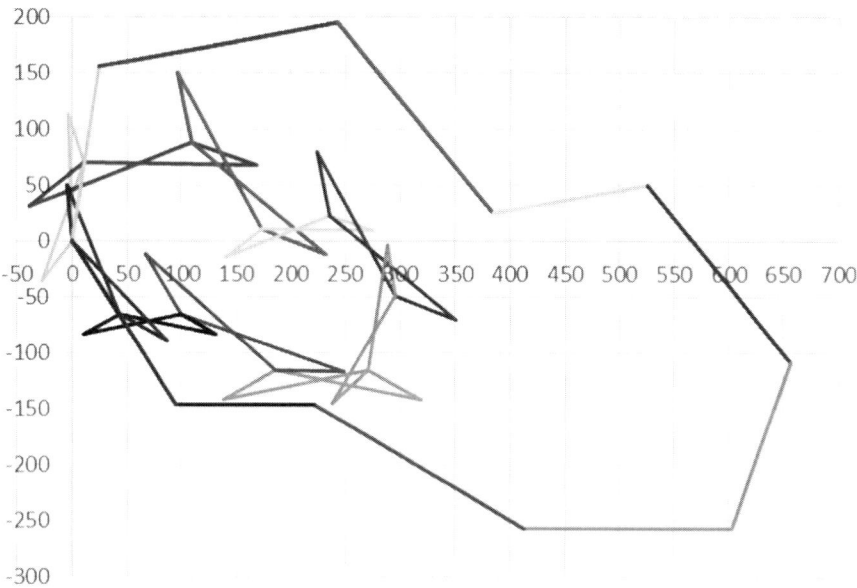

**Fig. 8.** Deployment of a linkage visualized in Excel graphics

# 4 Conclusions

Previous studies showed that a circular assembly of anti-parallelogram loops in a specific array, whether they are identical or similar, is capable of radial deployment. In this study an assembly of same array type is used for deployable structure that can scale a curve with variable curvature. SolidWorks and Excel softwares are used for modelling and calculation. Once the curve is approximated by a poly-line, there is only one free parameter, ratio R, in order to determine the link lengths of the linkage. It is demonstrated that such a linkage is capable of radial deployment with subtended angles. Along with many other scissor mechanisms, anti-parallelogram loop assemblies have potential applications in kinetic architecture.

**Acknowledgments.** This project has received funding from the European Union's Horizon 2020 research and innovation programme under the Marie Skłodowska-Curie grant agreement No 689983.

# References

1. De Temmerman, N.: Design and analysis of deployable bar structures for mobile architectural applications. Ph.D. thesis, Vrije Universiteit Brussel, Brussels, Belgium (2007)
2. Escrig, F.: Expandable space frame structures. In: Proceedings of the 3rd International Conference on Space Structures. Elsevier Applied Science Publishers, Guildford (1984)
3. Escrig, F.: Expandable space structures. Int. J. Space Struct. **1**(2), 79–91 (1985)
4. Gür, Ş.: Design of single degree-of-freedom planar linkages with antiparallelogram loops using loop assembly method. M.Sc. thesis, İzmir Institute of Technology (2017)
5. Gür, Ş., Korkmaz, K., Kiper, G.: Radially expandable ring-like structure with antiparallelogram loops. In: Proceedings of the International Symposium of Mechanism and Machine Science ISMMS-2017, Baku, Azerbaijan (2017)
6. Gür, Ş., Yar, M., Korkmaz, K.: A novel two degrees-of-freedom structural mechanism proposal for multi-functional transformable bridge. J. Fac. Eng. Archit. Gaz. **32**(4), 1379–1392 (2017)
7. Hamann, B., Chen, J.: Data point selection for piecewise linear curve approximation. Comput. Aided Geom. Des. **11**(3), 289–301 (1994)
8. Hernández Merchan, C.H.: Deployable Structures. M.Sc. thesis, Massachusetts Institute of Technology (1987)
9. Hoberman, C., Demaine, E., Rus, D.: MIT Class 6.S080 (Aus) - Mechanical Invention through Computation, Mechanism Basics (2013). MIT online courses http://courses.csail.mit.edu/6.S080/
10. Hoberman, C.: Reversibly Expandable Doubly-Curved Truss Structure, Patent no: US4942700A (1990)
11. Hoberman, C.: Radial Expansion/Retraction Truss Structures, Patent no: US5024031A (1991)
12. Kiper, G., Söylemez, E.: Irregular polygonal and polyhedral linkages comprising scissor and angulated elements. In: Proceedings of the 1st IFToMM Asian Conference on Mechanism and Machine Science (CD), Taipei (2010)
13. Langbecker, T.: Kinematic analysis of deployable scissor structures. Int. J. Space Struct. **14**(1), 1–15 (1999)
14. Liao, Q., Li, D.: Mechanisms for scaling planar graphs. Chin. J. Mech. Eng. **8**, 26 (2005)
15. Maden, F., Korkmaz, K., Akgun, Y.: A review of planar scissor structural mechanisms: geometric principles and design methods. Archit. Sci. Rev. **54**(3), 246–257 (2011)
16. Nagaraj, B.P., Pandiyan, R., Ghosal, A.: Kinematics of pantograph masts. Mech. Mach. Theory **44**(4), 822–834 (2009)
17. Piñero, E.P.: Project for a mobile theatre. Archit. Des. **12**(1), 154–155 (1961)
18. Söylemez, E.: Using computer spreadsheets in teaching mechanisms. In: Proceedings of EUCOMES 2008, pp. 45–53. Springer, Dordrecht (2008)
19. Yar, M., Korkmaz, K., Kiper, G., Maden, F., Akgün, Y., Aktaş, E.: A novel planar scissor structure transforming between concave and convex configurations. Int. J. Comput. Meth. Exp. Meas. **5**(4), 442–450 (2017)
20. You, Z., Pellegrino, S.: Foldable bar structures. Int. J. Solids Struct. **34**(15), 1825–1847 (1997)
21. Zhang, R., Wang, S.W., Chen, X.J., Ding, C., Jiang, L., Zhou, J., Liu, L.G.: Designing planar deployable objects via scissor structures. IEEE Trans. Vis. Comput. Graph. **22**(2), 1051–1062 (2016)
22. Zhao, J.S., Chu, F.L., Feng, Z.J.: The mechanism theory and application of deployable structures based on SLE. Mech. Mach. Theory **44**(2), 324–335 (2009)

# Author Index

© Springer Nature Switzerland AG 2019
B. Corves et al. (Eds.): EuCoMeS 2018, MMS 59, pp. 425–427, 2019.
https://doi.org/10.1007/978-3-319-98020-1